Lecture Notes in Computer Science   2260

Edited by G. Goos, J. Hartmanis, and J. van Leeuwen

Springer
Berlin
Heidelberg
New York
Barcelona
Hong Kong
London
Milan
Paris
Tokyo

Bahram Honary (Ed.)

Cryptography and Coding

8th IMA International Conference
Cirencester, UK, December 17-19, 2001
Proceedings

 Springer

Series Editors

Gerhard Goos, Karlsruhe University, Germany
Juris Hartmanis, Cornell University, NY, USA
Jan van Leeuwen, Utrecht University, The Netherlands

Volume Editor

Bahram Honary
Lancaster University
Faculty of Applied Sciences, Department of Communication Systems
Lancaster LA1 4YR, United Kingdom
E-mail: b.honary@lancaster.ac.uk

Cataloging-in-Publication Data applied for

Die Deutsche Bibliothek - CIP-Einheitsaufnahme

Cryptography and coding : ... IMA international conference ... ;
proceedings. - 5 (1995) [?]-. - Berlin ; Heidelberg ; New York ; Barcelona ;
Hong Kong ; London ; Milan ; Paris ; Singapore ; Tokyo : Springer, 1995 [?]-
    (Lecture notes in computer science ; ...)
    Erscheint zweijährl. - Früher verl. von Clarendon P., Oxford. - Wurde
    früher nicht angezeigt. - Bibliographische Deskription nach 8
    (2001)
    8. Cirencester, UK, December 17 - 19, 2001. - 2001
    (Lecture notes in computer science ; 2260)
    ISBN 3-540-43026-1

CR Subject Classification (1998): E.3-4, G.2.1, C.2, J.1

ISSN 0302-9743
ISBN 3-540-43026-1 Springer-Verlag Berlin Heidelberg New York

Springer-Verlag Berlin Heidelberg New York
a member of BertelsmannSpringer Science+Business Media GmbH

http://www.springer.de

© Springer-Verlag Berlin Heidelberg 2001
Printed in Germany

Typesetting: Camera-ready by author, data conversion by PTP-Berlin, Stefan Sossna
Printed on acid-free paper   SPIN 10846076   06/3142   5 4 3 2 1 0

Table of Contents

A Statistical Decoding Algorithm for General Linear Block Codes 1
  A. Al Jabri

On the Undetected Error Probability for Shortened Hamming Codes on
Channels with Memory 9
  Christoph Lange, Andreas Ahrens

The Complete Weight Enumerator for Codes over $\mathcal{M}_{n \times s}(\mathbb{F}_q)$ 20
  Irfan Siap

Further Improvement of Kumar-Rajagopalan-Sahai Coding Constructions
for Blacklisting Problem 27
  Maki Yoshida, Toru Fujiwara

A Simple Soft-Input/Soft-Output Decoder for Hamming Codes 38
  Simon Hirst, Bahram Honary

A Technique with an Information-Theoretic Basis for Protecting Secret
Data from Differential Power Attacks 44
  Manfred von Willich

Key Recovery Attacks on MACs Based on Properties of
Cryptographic APIs 63
  Karl Brincat, Chris J. Mitchell

The Exact Security of ECIES in the Generic Group Model 73
  N.P. Smart

A New Ultrafast Stream Cipher Design: COS Ciphers 85
  Eric Filiol, Caroline Fontaine

On Rabin-Type Signatures 99
  Marc Joye, Jean-Jacques Quisquater

Strong Adaptive Chosen-Ciphertext Attacks with Memory Dump
(or: The Importance of the Order of Decryption and Validation) 114
  *Seungjoo Kim, Jung Hee Cheon, Marc Joye, Seongan Lim,
  Masahiro Mambo, Dongho Won, Yuliang Zheng*

Majority-Logic-Decodable Cyclic Arithmetic-Modular AN-Codes in 1, 2,
and L Steps 128
  F. Javier Galán-Simón, Edgar Martínez-Moro, Juan G. Tena-Ayuso

Almost-Certainly Runlength-Limiting Codes 138
  David J.C. MacKay

Weight vs. Magnetization Enumerator for Gallager Codes 148
   Jort van Mourik, David Saad, Yoshiyuki Kabashima

Graph Configurations and Decoding Performance 158
   J.T. Paire, P. Coulton, P.G. Farrell

A Line Code Construction for the Adder Channel with Rates Higher than
Time-Sharing 166
   P. Benachour, P.G. Farrell, Bahram Honary

The Synthesis of TD-Sequences and Their Application to Multi-functional
Communication Systems 176
   Ahmed Al-Dabbagh, Michael Darnell

Improvement of the Delsarte Bound for τ-Designs in Finite Polynomial
Metric Spaces 191
   Svetla Nikova, Ventzislav Nikov

Statistical Properties of Digital Piecewise Linear Chaotic Maps and Their
Roles in Cryptography and Pseudo-Random Coding 205
   Shujun Li, Qi Li, Wenmin Li, Xuanqin Mou, Yuanlong Cai

The Wide Trail Design Strategy 222
   Joan Daemen, Vincent Rijmen

Undetachable Threshold Signatures 239
   Niklas Borselius, Chris J. Mitchell, Aaron Wilson

Improving Divide and Conquer Attacks against Cryptosystems by Better
Error Detection / Correction Strategies 245
   Werner Schindler, François Koeune, Jean-Jacques Quisquater

Key Recovery Scheme Interoperability – A Protocol for
Mechanism Negotiation 268
   Konstantinos Rantos, Chris J. Mitchell

Unconditionally Secure Key Agreement Protocol 277
   Cyril Prissette

An Efficient Stream Cipher Alpha1 for Mobile and Wireless Devices 294
   N. Komninos, Bahram Honary, Michael Darnell

Investigation of Linear Codes Possessing Some Extra Properties 301
   Viktoriya Masol

Statistical Physics of Low Density Parity Check Error Correcting Codes .. 307
   David Saad, Yoshiyuki Kabashima, Tatsuto Murayama, Renato Vicente

Generating Large Instances of the Gong-Harn Cryptosystem 317
   Kenneth J. Giuliani, Guang Gong

Lattice Attacks on RSA-Encrypted IP and TCP 329
  P.A. Crouch, J.H. Davenport

Spectrally Bounded Sequences, Codes, and States: Graph Constructions
and Entanglement 339
  Matthew G. Parker

Attacking the Affine Parts of SFLASH................................ 355
  Willi Geiselmann, Rainer Steinwandt, Thomas Beth

An Identity Based Encryption Scheme Based on Quadratic Residues 360
  Clifford Cocks

Another Way of Doing RSA Cryptography in Hardware................. 364
  Lejla Batina, Geeke Muurling

Distinguishing TEA from a Random Permutation: Reduced Round
Versions of TEA Do Not Have the SAC or Do Not Generate
Random Numbers.. 374
  *Julio César Hernández, José María Sierra, Arturo Ribagorda,
  Benjamín Ramos, J.C. Mex-Perera*

A New Search Pattern in Multiple Residue Method (MRM) and Its
Importance in the Cryptanalysis of the RSA.......................... 378
  *Seyed J. Tabatabaian, Sam Ikeshiro, Murat Gumussoy,
  Mungal S. Dhanda*

A New Undeniable Signature Scheme Using Smart Cards 387
  Lee Jongkook, Ryu Shiryong, Kim Jeungseop, Yoo Keeyoung

Non-binary Block Inseparable Errors Control Codes 395
  Alexandr Y. Lev, Yuliy A. Lev, Vyacheslav N. Okhrymenko

Cryptanalysis of Nonlinear Filter Generators with
$\{0,1\}$-Metric Viterbi Decoding..................................... 402
  Sabine Leveiller, Joseph Boutros, Philippe Guillot, Gilles Zémor

Author Index 415

A Statistical Decoding Algorithm for General Linear Block Codes

A. Al Jabri

EE Dept, College of Eng., P.O.Box 800
King Saud University, Riyadh 11421, Saudi Arabia
aljabri@ksu.edu.sa

Abstract. This paper introduces a new decoding algorithm for general linear block codes. The algorithm generates a direct estimate of the error locations based on exploiting the statistical information embedded in the classical syndrome decoding. The algorithm can be used to cryptanalyze many algebraic-code public-key crypto and identification systems. In particular results show that the McEliece public-key cryptosystem with its original parameters is not secure.

Keywords: Decoding, General Linear Block Codes, McEliece System, Statistical.

1 Introduction

The problem of decoding an arbitrary linear block codes is known to be **NP** hard. In practice, codes are usually designed with a certain algebraic structure that can be exploited to speed up the decoding process.

For general linear block codes with no obvious structure, one usually uses syndrome or probabilistic decoding algorithms [3, 2]. These are generally not efficient especially for large codes. Randomly generated codes, on the other hand, are typically good [5]. In fact, the minimum distance, d, of an (n, k) randomly generated code is related to the rate of the code by

$$1 - \frac{k}{n} \approx h(\frac{d}{n}),$$

where $h(x)$, $0 \le x \le 1$ is the binary entropy function defined by $x \log_2(1/x) + (1 - x) \log_2(1/(1 - x))$. For half rate codes, for example, the above relation can be approximated by $d \approx 011\, n$ or up to $0.55\, n$ errors can be corrected by these codes. for such codes or codes with no obvious structure, one usually decode using using syndrome or probabilistic decoding algorithms [2,3]. This later class includes the widely used information set decoding algorithm. Generally, these algorithms are not computationally and/or storage efficient especially for large codes.

An (n, k) linear block code of length n and k information bits is usually characterized by its generator matrix, G, or equivalently by its parity-check matrix, H. These matrices are related by the following relationship

B. Honary (Ed.): Cryptography and Coding 2001, LNCS 2260, pp. 1–8, 2001.
© Springer-Verlag Berlin Heidelberg 2001

$$GH^T = 0,$$

where T denotes the transpose operation. Here we assume binary block codes and that the code can correct up to t errors. To encode, the information vector $\mathbf{u}$ is multiplied by $\mathbf{G}$ to obtain the codeword vector $\mathbf{c}$. That is,

$$\mathbf{c} = \mathbf{uG}.$$

The received vector $\mathbf{y}$ is the codeword $\mathbf{c}$ plus a random binary error vector $\mathbf{e}$. Or,

$$\mathbf{y} = \mathbf{c} \oplus \mathbf{e}.$$

For $\mathbf{e}$ to be correctable, its Hamming weight must be less than or equal to t.

In classical syndrome decoding, a table of all possible correctable error vectors $\mathbf{e}$ and the corresponding syndromes $\mathbf{s} = \mathbf{eH}^T$ is first constructed. To decode, the received vector $\mathbf{y}$ is first multiplied by $\mathbf{H}^T$ to obtain $\mathbf{s}$ and hence $\mathbf{e}$ from the lookup table. This follows, since

$$\mathbf{y}H^T = (\mathbf{c} \oplus \mathbf{e})H^T = \mathbf{e}H^T = \mathbf{s}$$

and the fact that $\mathbf{c}H^T = 0$.

2   The Statistical Decoding Concept

Let $\mathcal{C}$ and $\mathcal{H}$ be the sets of all codewords generated by the code $\mathbf{G}$ and its dual H, respectively. For any $\mathbf{c} \in \mathcal{C}$ and $\mathbf{h} \in \mathcal{H}$ the following holds

$$\mathbf{ch}^T = 0.$$

If $\mathbf{yh}^T = 1$, then this particular $\mathbf{h}$ is said to have provided odd error detection for $\mathbf{y}$. Similarly, if $\mathbf{yh}^T = 0$, then $\mathbf{h}$ is said to have provided even error detection for y. This later case also includes the no error case.

Now consider the case when the odd detection process is restricted to those elements of $\mathcal{H}$ with large weights. Let this subset be denoted by $\mathcal{H}_w$. Note that Small weights can be used instead as well. In general, the vector $\mathbf{h}$, in the process $\mathbf{yh}^T$, acts as a mask or a filter on $\mathbf{y}$. If $\mathbf{yh}^T = 1$ or equivalently $\mathbf{eh}^t = 1$, then it is very likely that the $\mathbf{h}$ vector components at the error positions will be ones.

If all the $\mathbf{h} \in \mathcal{H}_w$ *vectors, satisfying the condition* $\mathbf{yh}^T$ *for the given* $\mathbf{y}$ *, are added (in* $\mathbf{R}^n$*), then positions with error will generally have higher frequencies than those with no errors.* Asymptotic expression for these frequencies will be given later. Let the resulting vector be $\mathbf{v}$. That is,

$$\mathbf{v} = \sum_{\mathbf{h} \in \mathcal{H}_w} (\mathbf{yh}^T)\mathbf{h}.$$

The operation $\mathbf{eh}^T = 1$ splits $\mathcal{H}_w$ into two subsets. Both provide information about the error positions. One subset, however, contains the error positions information within the highest m ($\geq t$) values of $\mathbf{v}$, while the other has the information in the least m values for some m, $t \leq m \leq n - k$. The selection of the highest or the lowest depends respectively on whether t is odd or even. The m is proposed here to count for the ambiguity in locating the t error positions. In fact, due to the statistical nature of the computation process of $\mathbf{v}$, the errors will not be confined to the t maximum values but will generally be among a larger number of positions; m in this case. This value (threshold) m should be set to guarantee that all error patterns are correctable.

Since $weight(\mathbf{e})$ is not known apriori, the decoder has to assume two scenarios. The first is for odd values of $weight(\mathbf{e})$ and the errors are within the positions corresponding to the highest m values of $\mathbf{v}$. The second is to assume even $weight(\mathbf{e})$ and that the errors are within the lowest m values of $\mathbf{v}$. Once the m error candidate positions are determined, one can then choose a subvector $\mathbf{y}_k$ of k bits from the remaining error free positions of $\mathbf{y}$ and the corresponding submatrix $\mathbf{G}_k$ from $\mathbf{G}$ and calculate $\mathbf{y}_k \mathbf{G}_k^{-1}$, if exists, or try another selection. Let $\mathbf{u}_1$ and $\mathbf{u}_2$ be the corresponding solutions to the two selections. That is,

$$\mathbf{u}_i = \mathbf{y}_{ki} \, \mathbf{G}_{ki}^{-1} \quad i = 1, 2.$$

The decoder can find the correct solution by checking the weight of $\mathbf{u}_1 \mathbf{G} \oplus \mathbf{y}$ and $\mathbf{u}_2 \mathbf{G} \oplus \mathbf{y}$ then selecting the $\mathbf{u}$ that yields a weight less than or equal to t.

The above is summarized in the following algorithm.

The Statistical Decoding Algorithm:

Input:

$\mathcal{H}_w{}^1$, $\mathbf{y}$.

Output u $'$; an estimate of the the information vector $\mathbf{u}$.

1. Calculate the error-locating vectors $\mathbf{v}$.

$$\mathbf{v} = \sum_{h \in \mathcal{H}_w} (\mathbf{y}\mathbf{h}^T)\mathbf{h}.$$

2. Calculate $\mathbf{u}_1'$ and $\mathbf{u}_2'$.

$$\mathbf{u}_i' = \mathbf{y}_{ki} \, \mathbf{G}_{ki}^{-1}, \quad i = 1, 2.$$

3. Check

$$weight(\mathbf{u}_i' \mathbf{G} \oplus \mathbf{y}) \quad i = 1, 2,$$

and choose $\mathbf{u}_i'$ that yields weight $\leq t$ and set the result to $\mathbf{u}$.

---

[1] The set $\mathcal{H}_w$ has to be generated and stored in advance. There are many efficient algorithms for such generation. (See for example [2,7]).

3 Applications

In what follows we use equivalent codes to some well known codes to test the performance of the proposed algorithm. Without loss of generality, we assume, in all cases, that **G** is in the format $[I : P]$ or $\mathbf{H} = [P^T : I]$.

a. $(7, 4, 3)$ Hamming Code:
This is a trivial case and is presented to clarify the idea.
Consider a $(7, 4)$ Hamming code. The Parity check matrix of the code is

$$P^T = \begin{bmatrix} 0 & 1 & 1 & 1 \\ 1 & 1 & 1 & 0 \\ 1 & 0 & 1 & 1 \end{bmatrix}$$

Here the weight distribution is $w_4 = 7$ and $\mathcal{H}_4$ is

$$\mathcal{H}_4 = \{(1011001), (0101011), (0010111), (1110010), (1001110), (0111100), (1100101)\}.$$

In this case $m = t = 1$. The error location in this case corresponds to position in **v** with the highest value. For example a single error at position 2 will result in

$$\mathbf{v} = (2422222).$$

b. (23,12,3) Golay Code:

This code can correct up to 3 errors. For this code

$$P = \begin{bmatrix}
1 & 1 & 0 & 0 & 0 & 1 & 1 & 1 & 0 & 1 & 0 \\
0 & 1 & 1 & 0 & 0 & 0 & 1 & 1 & 1 & 0 & 1 \\
1 & 1 & 1 & 1 & 0 & 1 & 1 & 0 & 1 & 0 & 0 \\
0 & 1 & 1 & 1 & 1 & 0 & 1 & 1 & 0 & 1 & 0 \\
0 & 0 & 1 & 1 & 1 & 1 & 0 & 1 & 1 & 0 & 1 \\
1 & 1 & 0 & 1 & 1 & 0 & 0 & 1 & 1 & 0 & 0 \\
0 & 1 & 1 & 0 & 1 & 1 & 0 & 0 & 1 & 1 & 0 \\
0 & 0 & 1 & 1 & 0 & 1 & 1 & 0 & 0 & 1 & 1 \\
1 & 1 & 0 & 1 & 1 & 1 & 0 & 0 & 0 & 1 & 1 \\
1 & 0 & 1 & 0 & 1 & 0 & 0 & 1 & 0 & 1 & 1 \\
1 & 0 & 0 & 1 & 0 & 0 & 1 & 1 & 1 & 1 & 1 \\
1 & 0 & 0 & 0 & 1 & 1 & 1 & 0 & 1 & 0 & 1
\end{bmatrix}$$

The dual code has 253 codewords of weight 16. These vectors are sufficient to locate all the error patterns of weight 3 or less. The following table shows the values of the **v** components in the error and the error free positions for different error weights.

The number of possible error patterns is $\binom{23}{1} + \binom{23}{2} + \binom{23}{3} = 2047$. For syndrome decoding, one needs to store all these vectors and their syndromes.

Table 1. Frequency in **v**

t	Error positions	Correct positions
1	176	120
2	56	80
3	96	88

Comparing this with the 253 vectors needed in the statistical decoding approach indicates the efficiency of the proposed algorithm. Note that for this case $m = t$. This, however, is not generally the case for all codes as shown in the following example.

c. $(31, 11, 5)$ **BCH Codes:**

Consider a code equivalent to the (31,11,5) BCH code obtained by permuting the columns of the generator matrix. The **P** matrix is given below.

$$P = \begin{bmatrix}
1 & 0 & 1 & 1 & 0 & 0 & 0 & 1 & 0 & 0 & 1 & 1 & 0 & 1 & 1 & 0 & 1 & 0 & 1 & 0 \\
0 & 1 & 0 & 1 & 1 & 0 & 0 & 0 & 1 & 0 & 0 & 1 & 1 & 0 & 1 & 1 & 0 & 1 & 0 & 1 \\
1 & 0 & 0 & 1 & 1 & 1 & 0 & 1 & 0 & 1 & 1 & 1 & 1 & 0 & 1 & 1 & 0 & 0 & 0 & 0 \\
0 & 1 & 0 & 0 & 1 & 1 & 1 & 0 & 1 & 0 & 1 & 1 & 1 & 1 & 0 & 1 & 1 & 0 & 0 & 0 \\
0 & 0 & 1 & 0 & 0 & 1 & 1 & 1 & 0 & 1 & 0 & 1 & 1 & 1 & 1 & 0 & 1 & 1 & 0 & 0 \\
0 & 0 & 0 & 1 & 0 & 0 & 1 & 1 & 1 & 0 & 1 & 0 & 1 & 1 & 1 & 1 & 0 & 1 & 1 & 0 \\
0 & 0 & 0 & 0 & 1 & 0 & 0 & 1 & 1 & 1 & 0 & 1 & 0 & 1 & 1 & 1 & 1 & 0 & 1 & 1 \\
1 & 0 & 1 & 1 & 0 & 1 & 0 & 1 & 1 & 1 & 0 & 1 & 1 & 1 & 0 & 1 & 0 & 1 & 1 & 1 \\
1 & 1 & 1 & 0 & 1 & 0 & 1 & 1 & 1 & 1 & 0 & 1 & 1 & 0 & 0 & 0 & 0 & 0 & 0 & 1 \\
1 & 1 & 0 & 0 & 0 & 1 & 0 & 0 & 1 & 1 & 0 & 1 & 1 & 0 & 1 & 0 & 1 & 0 & 1 & 0 \\
0 & 1 & 1 & 0 & 0 & 0 & 1 & 0 & 0 & 1 & 1 & 0 & 1 & 1 & 0 & 1 & 0 & 1 & 0 & 1
\end{bmatrix}$$

It can be shown that the dual code has 186 codewords of weight 26 and that these vectors are sufficient to correct all error patterns of weight 5 or less. For this example, m is found to be 10 and is obtained by a computer search. The number of error patterns needed in the syndrome table construction is 206367. Exhaustive search of the codewords, on the other hand, will require the test of 2^{11} codewords. In any case, comparing this with 186 vectors needed in the proposed algorithm indicates the high efficiency of the proposed algorithm.

Next we consider a more serious case.

d. $(32, 16, 2)$ **Random code:**

This is a randomly generated $(32, 16, 5)$ code. It can be shown (exhaustive search) that the minimum distance for this code is 5. That is, the code can correct 2 random errors. The P matrix is

$$P = \begin{bmatrix}
0 & 1 & 1 & 1 & 1 & 1 & 1 & 0 & 1 & 0 & 1 & 1 & 0 & 0 & 1 & 0 \\
0 & 0 & 1 & 1 & 0 & 1 & 0 & 0 & 1 & 0 & 0 & 1 & 1 & 0 & 0 & 0 \\
0 & 0 & 0 & 1 & 0 & 0 & 0 & 0 & 0 & 1 & 0 & 1 & 1 & 1 & 1 & 1 \\
0 & 1 & 1 & 0 & 0 & 1 & 0 & 1 & 1 & 0 & 0 & 0 & 1 & 1 & 1 & 0 \\
1 & 1 & 0 & 1 & 1 & 0 & 0 & 1 & 1 & 0 & 0 & 1 & 1 & 1 & 1 & 1 \\
1 & 0 & 0 & 1 & 0 & 1 & 0 & 0 & 0 & 1 & 1 & 1 & 1 & 0 & 0 & 1 \\
1 & 0 & 1 & 1 & 0 & 0 & 1 & 0 & 0 & 0 & 0 & 1 & 1 & 0 & 0 & 1 \\
1 & 1 & 1 & 1 & 0 & 0 & 1 & 1 & 0 & 1 & 1 & 1 & 0 & 1 & 1 & 0 \\
1 & 0 & 1 & 1 & 1 & 1 & 0 & 0 & 0 & 1 & 1 & 1 & 1 & 1 & 1 & 1 \\
0 & 1 & 1 & 1 & 1 & 1 & 1 & 1 & 0 & 1 & 1 & 1 & 0 & 1 & 1 \\
0 & 0 & 1 & 0 & 1 & 1 & 0 & 0 & 1 & 0 & 1 & 1 & 1 & 1 & 1 \\
1 & 1 & 1 & 0 & 0 & 1 & 1 & 0 & 0 & 0 & 1 & 0 & 0 & 1 & 0 & 1 \\
1 & 1 & 1 & 1 & 1 & 0 & 1 & 1 & 1 & 1 & 1 & 0 & 0 & 1 & 0 & 1 \\
1 & 1 & 1 & 1 & 1 & 1 & 0 & 0 & 1 & 1 & 1 & 1 & 0 & 0 & 0 & 1 \\
0 & 1 & 1 & 0 & 1 & 1 & 0 & 1 & 1 & 1 & 1 & 0 & 1 & 0 & 0 & 1 \\
0 & 1 & 1 & 1 & 1 & 1 & 1 & 1 & 1 & 0 & 1 & 0 & 1 & 0 & 1
\end{bmatrix}$$

A random search is performed for $\mathcal{H}_w$ and it is found that the following 28 vectors are sufficient for decoding all errors of weight ≤ 2. Here m is taken to be 14.

Suppose a single error has occurred in position 10. In this case $\mathbf{v}$ will be

14 16 18 14 15 16 13 15 17 **21** 15 15 15 16 17 15

15 14 17 15 17 17 15 13 15 17 17 15 17 17 16 15

Note that position 10 has the highest value (21 in this case) among the values of $\mathbf{v}$. Now suppose that two errors have occurred in position 10 and 20. In this case $\mathbf{v}$ will be

9 9 7 11 9 10 10 12 10 **6** 9 8 8 10 9 10

8 10 8 **6** 10 8 9 9 9 8 9 9 9 9 9 11

Note again that position 10 and position 20 have the smallest values (6 in this case) among the values of $\mathbf{v}$. Similar calculations can be done for other error patterns. One can find the different thresholds to detect and locate different kind of correctable error patterns. From this a single threshold can be found for all error patterns.

ASYMPTOTIC BEHAVIOR OF THE ALGORITHM

Here we test the performance of the proposed algorithm on large codes usually used in designing cryptosystem. The decoding problem of general linear block codes is known to be **NP** hard. This fact is used as a basis for designing many public-key and identification systems [4, 7]. A typical code size for these applications is $(1024, 512)$.

d. McEliece Public-key Cryptosystem: (Goppa (1024,512,51))

In this system, the public-key $\mathbf{G}$ is a $k \times n$ matrix composed of three private matrices: a $k \times k$ scrambling matrix $\mathbf{S}$, a $k \times n$ Goppa code generator matrix $\mathbf{G}'$ and a $n \times n$ permutation matrix $\mathbf{P}$ such that $\mathbf{G} = \mathbf{SG}'\mathbf{P}$. To encrypt, the information vector is first multiplied by $\mathbf{G}$ then an error vector $\mathbf{e}$ of weight t or less is added. That is, the ciphertext vector $\mathbf{y}$ is given by

$$\mathbf{y} = \mathbf{uG} \oplus \mathbf{e}.$$

Knowledge of the private keys enables one to easily decode $\mathbf{y}$ while this is not the case with $\mathbf{G}$ alone. Codes of large sizes are proposed for this system. These codes, generally, behave like randomly generated codes and one can reasonably assume that the binomial distribution is a good approximation for the weights [5].

Using this approximation, it can be shown that the probability, p, in erroneous positions of $\mathbf{v}$ is given by

$$p = \frac{\sum_{m \ odd} \binom{n-t}{w-m}\binom{t-1}{m-1}}{\sum_{m \ odd} \binom{t}{m}\binom{n-t}{w-m}}$$

and the probability, q, of error free positions ion $\mathbf{v}$ is given by

$$q = \frac{\sum_{m \ odd} \binom{n-t-1}{w-m-1}\binom{t}{m}}{\sum_{m \ odd} \binom{t}{m}\binom{n-t}{w-m}}.$$

We have interest in estimating the number of $\mathbf{h}$ vectors, N, required for there to be a 0.95 probability that the relative frequency estimate for the probability of an error event would be within ϵ of P. For large codes, it is noticed that the difference between p and q is very small. This puts some restriction on the value of ϵ. One can use the central limit theorem to bound this number. Let $f_e(N)$ be the relative frequency of the error event in some position. Since $f_e(n)$ has mean p and variance $p(p-1)$, then it can be shown [6] that for a 0.95 probability that

$$N = 625 \times 10^{-6}p(1-p)\epsilon^{-2}.$$

Using this, the number of $\mathbf{h}$ vectors required to identify the erroneous places in McEliece system is 2^{38}. Our experimentation shows that for a single 700 MHz PENTIUM II processor, one needs a round 2^9 hrs per decryption. This can be significantly reduced using parallel computations. One needs to note that a large number of vectors has to be stored. This, however, is within the reach of today's technology.

4 Conclusion

In this paper, a new algorithm for decoding general linear block codes is proposed. The algorithm is of a different flavor from classical decoding algorithms and is much more efficient. Preliminary results of its performance have been presented. This can be used as a solution to decode many good linear codes that can be constructed with no obvious algebraic structure. This also suggests a possible class of attacks on crypto and identification systems based on the difficulty of the decoding problem of general linear block codes.

Acknowledgment. The author would like to thank Prof. Paddy Farrel for the discussion regarding the subject of the paper.

References

1. A. Al Jabri,"A new Class Of Attacks On McEliece Public-Key and Related Cryptosystems," The 2001 Canadian Workshop On Information Theory, Vancouver, British Columbia, June 3-6, 2001.
2. A. Canteaut and F. Chabaud,"A New Algorithm for Finding Minimum Weight Words in a Linear Code: Application to McEliece's Cryptosystem and to Narrow-Sense BCH Codes of Length 611", IEEE Trans. Inform. Theory, vol. IT-44(1), pp. 367-378, 1998.
3. P.J. Lee and E.F. Brickell,"An Observation on the Security of McEliece's Public-Key Cryptosystem", in Lecture Notes in Computer Science 330, Advances in Cryptology: Proc. Eurocrypt'88, C.G. Gunther, Ed., Davos, Swizerland, May 25-27, 1988, pp. 275-280, Berlin: Springer-Verlag, 1988.
4. R. J. McEliece, "A Public-Key Cryptosystem Based on Algebraic Coding Theory, DSN Progress Report 42-44, pp. 114-116, Jet Propulsion Laboratory, CA, Jan-Feb 1978.
5. F.J. McWilliams and N.J. Sloane,"The theory of error correcting codes", North Publishing Co. 3rd ed.,North Mathematical Library , Vol. 16, Netherlands 1983.
6. A. Papoulis, "Probability, Random Variables, and Stochastic Processes," McGraw-Hill, New York, 1965.
7. J. Stern "A method for finding codewords of small weight," in Coding Theory and Applications, G. Cohen and J. Wolfmann , Eds, New York, Springer-Verlag, 1989, pp. 106-113.

On the Undetected Error Probability for Shortened Hamming Codes on Channels with Memory

Christoph Lange and Andreas Ahrens*

Rostock University
Department of Electrical Engineering and Information Technology
Institute of Communications and Information Electronics
Richard-Wagner-Str. 31, 18119 Rostock, FRG
christoph.lange@ntie.e-technik.uni-rostock.de
andreas.ahrens@ntie.e-technik.uni-rostock.de

Abstract. In this contribution, the probability of undetected errors on channels with memory is determined. Different error detection strategies (e.g. shortened hamming codes), which have been adopted as international standards, were analyzed. A setup is presented, which is based on a classification of error patterns in blocks by error-length and error-weight. This approach is used to determine the probability of an error pattern being a valid codeword and consequently not detectable. For these investigations a digital channel model, whose characteristics were found by analyzing real shortwave channel connections, was used. As a result of the investigation it is shown, that generator polynomials of the same code rate on channels with memory are the more efficiently the more equidistantly the exponents in the generator polynomial are distributed.

1 Introduction

Hamming codes or shortened Hamming codes are widely used for error detection in data communication systems. For example, a distance-4 cyclic Hamming code with 16 bits for error detection has been adopted by the CCITT (Recommendation X.25) for use in packet-switched data networks [6], [5]. The code is generated either by the polynomial

$$G_1(z) = (z + 1)(z^{15} + z^{14} + z^{13} + z^{12} + z^4 + z^3 + z^2 + z + 1)$$
$$= z^{16} + z^{12} + z^5 + 1 \quad \text{(CRC-CCITT, Code-1)} \tag{1}$$

or by the polynomial

$$G_2(z) = (z + 1)(z^{15} + z^{14} + 1)$$
$$= z^{16} + z^{14} + z + 1 \quad \text{(Code-2)} \tag{2}$$

---

* This contribution was presented in parts at the Nordic HF Conference, Fårö/Sweden, August 2001.

B. Honary (Ed.): Cryptography and Coding 2001, LNCS 2260, pp. 9–19, 2001.

where $z^{15} + z^{14} + z^{13} + z^{12} + z^4 + z^3 + z^2 + z + 1$ and $z^{15} + z^{14} + 1$ are primitive polynomials of degree 15. The natural length of this code is $n = 2^{15} - 1 = 32767$. In practice, the length of a data packet is no more than a few thousand bits, which is much shorter than the natural length of the code. Consequently, a shortened version of the code is used. Often, the length of a data packet varies from a few hundred bits to a few thousand bits [5].

An undetected error occurs, if a transmitted codeword is distorted by the channel and appears as a different valid codeword on the receiver side [2]. The probability of an undetected error is a commonly used indicator for the reliability of a given code-channel combination [12]. To be able to analyze the performance of such error detection strategies, time-consuming simulations are necessary. Often, either the weight distribution of the code or its dual code is used to determine the reliability of a given code-channel combination [6], [7], [11], [13] or numerical results can be found in the literature [12]. Asymptotic bounds are an insufficient measure since they do not take the real channel conditions into account [13]. By analyzing real shortwave channel connections, it was found in [8] and [9], that a classification of error patterns by the error-length and the error-weight leads to better results than to work only with the error-weight distribution. The reason for this is, that not all error patterns appear equally distributed on the real HF channel. For the determination of the performance of different error detection strategies, the error patterns in blocks of the length n are classified in this paper by the error-length l and the error-weight g. In Fig. 1 an error pattern of the length $l = 6$ with the weight $g = 3$ in a block of length $n = 9$ is shown.

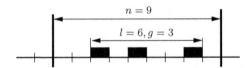

Fig. 1. Error pattern in a block of the length n

With the digital channel model derived in [3], we are able to generate typical shortwave error patterns. This model is based on the assumption

$$p_B(n) = p_e \cdot n^\alpha \quad \text{for} \quad n \le n_{\max} , \tag{3}$$

where the block-error rate $p_B(n)$ is described as a function of the block-length n, the bit error rate p_e (BER) and the error concentration value $(1 - \alpha)$ within the parameters of $0 \le (1 - \alpha) \le 0.5$ [3], [9]. The value of $n_{\max}$ indicates the maximum block length, to which the model assumption can be maintained with $p_B(n = n_{\max}) = 1$. This digital channel model takes the interdependence of errors into account, as it can be expected on many HF channels.

2 Basics

The error detection coding fails, if a transmitted codeword is distorted by the channel and appears as a different valid codeword at the receiver side. In this case, the error pattern in the block is a valid codeword. The undetected error probability $p_{BR}(n)$, as a commonly used indicator for the reliability of a given code-channel combination, can be defined as

$$p_{BR}(n) = R \cdot p_B(n) \ , \tag{4}$$

and describes the ratio of undetected faulty blocks to transmitted blocks. With a coding strategy the reduction factor R, as the ratio of undetected faulty blocks to faulty blocks, should be kept as little as possible. The encoding sets, which number $v_{l,g}$ of error patterns of the length l with the weight g are valid codewords and therefore not detectable. In a block of the length n

$$t_{l,g} = (n - l + 1) \cdot \binom{l-2}{g-2} \tag{5}$$

patterns with the criteria l and g can be found. If the two outer erroneous bits terminate the error pattern, there exist $\binom{l-2}{g-2}$ possible error patterns. They may appear in total $(n-l+1)$ times in a block of length n (Fig. 2). Altogether $2^n - 1$ other error patterns are possible.

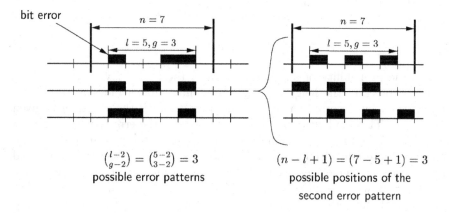

$$\binom{l-2}{g-2} = \binom{5-2}{3-2} = 3$$

possible error patterns

$$(n - l + 1) = (7 - 5 + 1) = 3$$

possible positions of the second error pattern

Fig. 2. Possible error patterns in a block of the length $n = 7$ with the length $l = 5$ and weight $g = 3$

Example 1. Block length, error weight and error length

As an example, a block of length $n = 7$ with an error pattern of length $l = 5$ and weight $g = 3$ is considered. In such a case, there may appear

$$t_{l,g} = t_{5,3} = 3 \cdot \binom{3}{1} = 9$$

possible error patterns. This is illustrated in Fig. 2.

The rate of patterns with the criterion l and g is then given by

$$a_{l,g} = \frac{t_{l,g}}{2^n - 1} \ . \tag{6}$$

From the $\binom{l-2}{g-2}$ possible patterns are now $v_{l,g}$ not detectable, since they are valid codewords. The generation of the valid codewords and the determination of their length-weight distribution will be explained in section 3. The probability $y_{l,g}$ that an error pattern of the length l with the weight g is not detectable is given by

$$y_{l,g} = \frac{v_{l,g}}{\binom{l-2}{g-2}} \ . \tag{7}$$

Considering error patterns of the same probability, the rate

$$a_{l,g} \cdot y_{l,g} \tag{8}$$

is not detectable. In the real channel, $a_{l,g}$ is now substituted by

$$x_{l,g} = \frac{P(l,g,n)}{p_B(n)} \ , \tag{9}$$

which is the probability of an error-pattern of the length l with its weight g in a faulty block of the length n. The probability, that a block of the length n is distorted by an error pattern of the length l with the weight g is given by $P(l,g,n)$. This modification is necessary, because the error patterns are not equally distributed on real channels. In Fig. 3 and 4 the probabilities $x_{l,g}$ on channels without and with memory are shown, respectively. With the model used here, the classical memoryless channel is specified by an error concentration value of $(1 - \alpha) = 0$, whereas an error concentration value bigger than zero describes a channel with memory [3]. The reduction factor R by coding can be defined as

$$R = \sum_{l=k+1}^{n} \sum_{g=2}^{l} x_{l,g} \cdot y_{l,g} \ , \tag{10}$$

since with the assumption of cyclic (n, m) coding (m bits of information and $(n - m) = k$ parity bits) patterns up to the length k are detectable [9], [1]. The value $x_{l,g}$ in (10) considers the channel characteristic and the value $y_{l,g}$ the code characteristic.

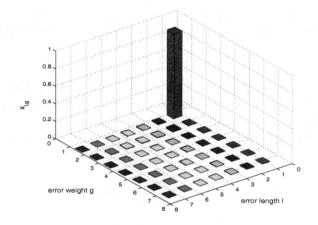

Fig. 3. Probability $x_{l,g}$ with $p_e = 10^{-2}, (1 - \alpha) = 0.0, n = 7,$

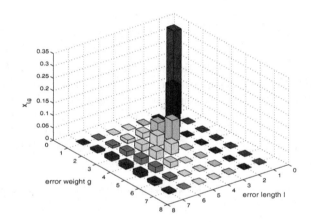

Fig. 4. Probability $x_{l,g}$ with $p_e = 10^{-2}, (1 - \alpha) = 0.5, n = 7,$

3 Determination of the Code Characteristics

For the determination of the undetected error probability, the valid codewords must be classified by their lengths and their weights. It is assumed, that the pattern in Fig. 1 represents a valid codeword. It has the length $l = 6$ and weight $g = 3$. If the generator polynomial $G(z)$ is known, the valid codewords $C(z)$ can be constructed by multiplying it by the information polynomial $N(z)$:

$$C(z) = G(z) \cdot N(z) \ . \tag{11}$$

For the generation of the codewords with the criteria l and g, only the information polynomials $N_c(z) \in N(z)$ have to be taken into account, which are not divisible by z, since cyclic shifting does not affect the codeword length-weight distribution. Thus, the information polynomials $N(z)$ are considered, for which

$$N(z) \bmod z \neq 0 \tag{12}$$

holds. This is illustrated in Fig. 5 for a (7,4) Hamming code (generator polynomial $G(z) = z^3 + z + 1$) with the exemplary considered information polynomials $N_1(z) = 1$ and $N_2(z) = z$. In both cases the codewords have the same characteristic. The resulting codeword length-weight distribution for a $(7,4)$ Hamming code is indicated in Tab. 1. The total number of codewords is obtained by a

Table 1. Elements $v_{l,g}$ of the $(7,4)$ Hamming code

g \ l	4	5	6	7
3	1	-	1	1
4	-	1	1	2
7	-	-	-	1

multiplication of $v_{l,g}$ by $(n - l + 1)$, since a pattern with the criteria l and g can altogether occur $(n - l + 1)$ times within a codeword with cyclic shifting. This approach leads to the well-known weight spectrum of codewords [10].

The probability $y_{l,g}$ of not detecting an error pattern with the criteria l and g, is obtained after (7). An error is not detectable, if it corresponds to a valid codeword. Due to the error detecting properties of cyclic codes, the position of the patterns with the criteria l and g within the block is of no interest.

Example 2. Determination of $y_{l,g}$ for $l = 4$, $g = 3$ for the $(7,4)$ Hamming Code

$$\binom{l-2}{g-2} = \binom{2}{1} = 2 \quad \text{and} \quad v_{l,g} = v_{4,3} = 1 \quad \text{(see Tab. 1)}$$

$$\Rightarrow y_{l,g} = y_{4,3} = \frac{1}{2}$$

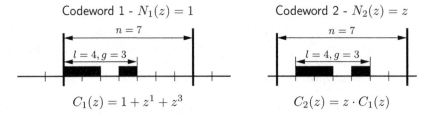

Codeword 1 - $N_1(z) = 1$ Codeword 2 - $N_2(z) = z$

$C_1(z) = 1 + z^1 + z^3$ $C_2(z) = z \cdot C_1(z)$

Fig. 5. Characteristics of the valid codewords for a $(7,4)$ Hamming-Code

Only one of the two possible error patterns with the criteria $l = 4$ and $g = 3$ is a valid codeword. Therefore the probability of not detecting an error pattern with $l = 4$ and $g = 3$ amounts to 0.5. The probability of the appearance of such an error pattern is ruled by the channel characteristics.

For the investigated polynomials (Code-1 and Code-2) the probabilities $y_{l,g}$ are depicted in Fig. 6 and 7. The importance of the generator codeword ($l = 17, g = 4$) for the determination of the reduction factor and the undetected error probability becomes obvious.

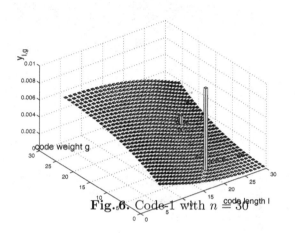

Fig. 6. Code-1 with $n = 30$

4   Results

In this section, the reliability of different code-channel combinations is analyzed. For the consideration of the channel characteristics the error patterns in faulty blocks must be classified by their error-length and error-weight. The searched probabilities $x_{l,g}$ can be determined by simulations, using the digital channel model suggested in [3]. The digital channel model is available at [4].

Results of the analysis of both generator-polynomials Code-1 and Code-2 are shown in Fig. 8 and 9. For the investigations, an error concentration value of $(1 - \alpha) = 0.3$ was chosen, as it was found by analyzing real short wave channel connections in Central Europe [3], [5], [9]. An error concentration value of $(1 - \alpha) = 0$ describes the classical memoryless channel. The superiority of the Code-1 on channels with memory is illustrated. This analysis led to the result, that codes of the same length are efficient on channels with memory, if the exponents in the generator polynomial are equidistantly distributed. With the Code-2, this is not as fulfilled as when the Code-1 is considered. For this reason,

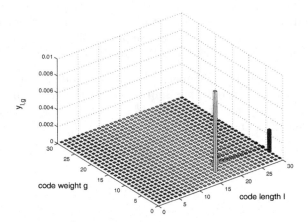

Fig. 7. Code-2 with $n = 30$

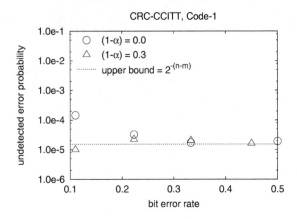

Fig. 8. Code-1 with $n = 30$

the superiority of the Code-1 becomes obvious on channels with memory over the Code-2. The Code-1 generally leads to a lower undetected error probability on channels with memory than the Code-2. The analysis of the code characteristic has shown, that the structure of the generator polynomial has a dominating influence on the determination of the undetected error probability. For the analysis of these properties, the error-gap density function $v(b) = P(X = b)$ on channels with memory was used, which describes the probability that after an error the gap to the next error is b intervals long [3]. This point makes obvious why the Code-2 works for $(1 - \alpha) = 0.3$ worse than the Code-1. The probability, that after an error in the distance $b = 0$ an error reappears, is increased with increasing error concentration value (Fig. 10). But exactly such a

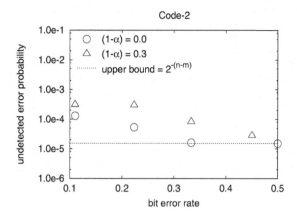

Fig. 9. Code-2 with $n = 30$

combination (parts z^1 and $z^0 = 1$ in the generator polynomial of Code-2) leads to a valid codeword at the Code-2. A behaviour exactly reverse can be found at the Code-1. With an increased error concentration value $(1-\alpha)$, the value of $v(0)$ increases, too, but this result does not lead to a valid codeword. Thus, a generator polynomial with equidistantly distributed exponents results in codes with good undetected error probability characteristics on channels with memory, since the generator polynomial dominates the undetected error probability. Looking at Fig. 9 it becomes obvious, too, why the Code-2 works on channels with memory worse than on memoryless channels. Based on the aspect, that with increasing error concentration value, the value of $v(0)$ increases and that such a constellation (two or more adjacent exponents in the generator polynomial are not zero) leads to a valid codeword, it can be explained, why we have a loss in the efficiency. In Code-1 such a combination was missed and approximately the same performance was achieved.

5   Conclusion

By the classification of the error pattern suggested in this work by the error-length and weight we are able to achieve an optimal combination of a given code-channel characteristic. The proposed setup takes the channel characteristics of the shortwave channel at the determination of the undetected error probability into account in an optimal way. It is shown, that codes of the same rate and length are efficient, if the exponents in the generator polynomial are equidistantly distributed. For the data transmission over shortwave channels the CRC-CCITT code (Code-1) shows the best results.

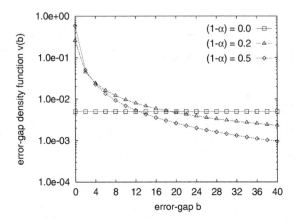

Fig. 10. Error-gap density function for short error-gaps b

Acknowledgement. We thank Prof. R. Kohlschmidt and Prof. R. Rockmann of Rostock University for their support of our work and Dr. C. Wilhelm for valuable comments.

References

1. Ahrens, A., Wilhelm, C.: Bestimmung des Reduktionsfaktors bei Einsatz zyklischer Codes auf Kanälen mit gebündelt auftretenden Übertragungsfehlern. 10. Symposium Maritime Elektronik, Arbeitskreis Maritime Mess- und Informationselektronik, June 2001, Universität Rostock, Germany, (141–144)
2. Ahrens, A., Lange, C.: On the Probability of Undetected Error in Protocol Structures. Nordic Shortwave Conference (Nordic HF 01), August 2001, Fårö, Sweden, (3.1.1–3.1.8)
3. Ahrens, A.: A New Digital Channel Model Suitable for the Simulation and Evaluation of Channel Error Effects. Colloquium on Speech Coding Algorithms for Radio Channels, IEE Electronics and Communications, April 2000, London, UK, Reference Number 2000/030
4. http://www-nt.e-technik.uni-rostock.de/nt/english/fg_nt.html
5. Ahrens, A., Greiner, G.: A Radio Protocol for TCP/IP Application. Colloquium on Frequency Selection and Management Techniques for HF Communication, IEE Electronics and Communications, March 1999, London, UK, Reference Number 1999/017, 11/1-11/6
6. Fujiwara, T., Kasami, T.; Kitai, A., Lin, S.: On the Undetected Error Probability for Shortened Hamming Codes. IEEE Transactions on Communications **33** (1985) 570–574
7. Leung, C.: Evaluation of the Undetected Error Probability of Single Parity-Check Product Codes. IEEE Transactions on Communications **31** (1983) 250–253
8. Wilhelm, C.: Über den Zusammenhang zwischen Kanal und Codierung bei der Datenfernübertragung. Nachrichtentechnik-Elektronik **17** (1967) 386–392
9. Wilhelm, C.: Datenübertragung. Militärverlag, Berlin (1976)

10. Wozencraft, J. M., Jacobs, I. M.: Principles of Communication Engineering. Waveland Press, Prospect Heigths (Illinios) (1990)
11. Wolf, J., Michelson, A., Levesque, A.: On the Probability of Undetected Error for Linear Block Codes. IEEE Transactions on Communications **30** (1982) 317–324
12. Wong, B., Leung, C.: On Computing Undetected Error Probabilities on the Gilbert Channel. IEEE Transactions on Communications **43** (1995) 2657–2661
13. Witzke, B., Leung, C.: A Comparison of some Error Detecting CRC Code Standards. IEEE Transactions on Communications **33** (1985) 996–998

The Complete Weight Enumerator for Codes over $\mathcal{M}_{n \times s}(\mathbb{F}_q)$

Irfan Siap

Sakarya University, Sakarya, Turkey.
isiap@sakarya.edu.tr,
WWW home page: http://www.sakarya.edu.tr/~ isiap

Abstract. The MacWilliams identity for codes over $\mathcal{M}_{n \times s}(\mathbb{F}_q)$ endowed with a non-Hamming metric is proved in [1]. We introduce a complete weight enumerator for these codes and prove a MacWilliams identity with respect to this new metric for the complete weight enumerator.

1 Introduction

Let $\mathbb{F}_q = \{0 = \alpha_0, \alpha_1, \ldots, \alpha_{q-1}\}$ denote the finite field with q elements. Let $\mathcal{M}_{n \times s}(\mathbb{F}_q)$ denote the set of $n \times s$ matrices over $\mathbb{F}_q$. A new non-Hamming ρ metric on linear spaces over finite fields has been recently introduced in [3]. Let $\omega = (p_0, p_1, \ldots, p_{s-1}) \in \mathcal{M}_{1 \times s}(\mathbb{F}_q)$. Then,

$$\rho(\omega) = \begin{cases} \max\{i | p_i \neq 0\} + 1, & p_i \neq 0, \\ 0, & \omega = 0. \end{cases} \tag{1}$$

Let $\Omega = (\omega_1, \omega_2, \ldots, \omega_n)^T \in \mathcal{M}_{n \times s}(\mathbb{F}_q)$ and define $\rho(\Omega) = \sum_{i=0}^{n} \rho(\omega_i)$. ρ is a metric over $\mathcal{M}_{n \times s}(\mathbb{F}_q)$.

Definition 1. *A linear subspace C of $\mathcal{M}_{n \times s}(\mathbb{F}_q)$ is called a linear code.*

Let $C \subset \mathcal{M}_{n \times s}(\mathbb{F}_q)$ be a linear code. $w_r(C) = |\{\Omega \in C | \rho(\Omega) = r\}|$, $0 \leq r \leq ns$ is called the *ρ weight spectrum* of the code, and the *ρ weight enumerator* is defined as follows:

$$W(C|z) = \sum_{r=0}^{ns} w_r(C) z^r = \sum_{\Omega \in C} z^{\rho(\Omega)}. \tag{2}$$

Let $\omega_1 = (p_0, p_1, \ldots, p_{s-1}), \omega_2 = (q_0, q_1, \ldots, q_{s-1}) \in \mathcal{M}_{1 \times s}(\mathbb{F}_q)$. The inner product of ω_1 and ω_2 is defined by

$$\langle \omega_1, \omega_2 \rangle = \sum_{i=0}^{s-1} p_i q_{s-1-i} \tag{3}$$

B. Honary (Ed.): Cryptography and Coding 2001, LNCS 2260, pp. 20–26, 2001.

and this is extended to inner product of $\Omega_1 = (\omega_1, \ldots, \omega_n)^T, \Omega_2 = (\mu_1, \ldots, \mu_n)^T \in \mathcal{M}_{n \times s}(\mathbb{F}_q)$ as

$$\langle \Omega_1, \Omega_2 \rangle = \sum_{i=1}^{n} \langle \omega_i, \mu_i \rangle. \tag{4}$$

The dual code of C is defined by

$$C^{\perp} = \{\Omega_2 \in \mathcal{M}_{n \times s}(\mathbb{F}_q) | \langle \Omega_2, \Omega_1 \rangle = 0 \text{ for all } \Omega_1 \in C\}, \tag{5}$$

and $C^{\perp}$ is also a linear code of length n.

For $s = 1$ and arbitrary n the ρ metric coincides with Hamming metric and the MacWilliams identity is given in [2]. For $n = 1$ and arbitrary s the MacWilliams identity is given in [4]. Now we consider the following example given in [1],

Example 1:

$$C_1 = \{\begin{pmatrix} 0\,0 \\ 0\,0 \end{pmatrix}, \begin{pmatrix} 1\,0 \\ 1\,0 \end{pmatrix}\}, \quad C_2 = \{\begin{pmatrix} 0\,0 \\ 0\,0 \end{pmatrix}, \begin{pmatrix} 0\,0 \\ 0\,1 \end{pmatrix}\}. \tag{6}$$

The ρ weight enumerator of the above codes is $1 + z^2$.

The dual codes of C_1 and C_2 are

$$C_1^{\perp} = \{\begin{pmatrix} 0\,0 \\ 0\,0 \end{pmatrix}, \begin{pmatrix} 0\,1 \\ 0\,1 \end{pmatrix}, \begin{pmatrix} 1\,1 \\ 1\,1 \end{pmatrix}, \begin{pmatrix} 1\,1 \\ 0\,1 \end{pmatrix},$$
$$\begin{pmatrix} 0\,1 \\ 1\,1 \end{pmatrix}, \begin{pmatrix} 1\,0 \\ 0\,0 \end{pmatrix}, \begin{pmatrix} 0\,0 \\ 1\,0 \end{pmatrix}, \begin{pmatrix} 1\,0 \\ 1\,0 \end{pmatrix}\}$$

and

$$C_2^{\perp} = \{\begin{pmatrix} 0\,0 \\ 0\,0 \end{pmatrix}, \begin{pmatrix} 1\,0 \\ 0\,0 \end{pmatrix}, \begin{pmatrix} 0\,1 \\ 0\,0 \end{pmatrix}, \begin{pmatrix} 0\,0 \\ 0\,1 \end{pmatrix},$$
$$\begin{pmatrix} 1\,1 \\ 0\,0 \end{pmatrix}, \begin{pmatrix} 1\,0 \\ 0\,1 \end{pmatrix}, \begin{pmatrix} 0\,1 \\ 0\,1 \end{pmatrix}, \begin{pmatrix} 1\,1 \\ 0\,1 \end{pmatrix}\}.$$

The ρ weight enumerators of $C_1^{\perp}$ and $C_2^{\perp}$ are

$$W(C_1^{\perp} \mid z) = 1 + 4z^4 + 2z + z^2, \quad W(C_2^{\perp} \mid z) = 1 + 2z^4 + z^3 + 3z^2 + z. \tag{7}$$

As seen above, although the ρ weight enumerators of the codes C_1 and C_2 are the same, the ρ weight enumerators of the duals are different. To overcome this problem, in [1], the orbits of a linear group which preserves the metric ρ are considered. It is shown that The MacWilliams identity over such orbits holds [4]. In the next section, we propose a complete weight enumerator which overcomes this problem and carries more information for the code.

2   The complete weight enumerator

Let us consider the matrices $A = \begin{pmatrix} 0 & 1 \\ 0 & 0 \end{pmatrix}$, and $B = \begin{pmatrix} 1 & 0 \\ 1 & 0 \end{pmatrix}$. $\rho(A) = \rho(B) = 2$. However, the rows of the matrices A and B have different structures and it is natural to expect the problem occured in the above example. The weight ρ strongly depends in the order of the elements of the rows. It is possible to overcome the problem that occured in the above example by defining a weight enumerator that preserves the order of the entries of the matrices and carries more information about the code. Before we give the definition of the complete weight enumerator we make the following identification:

$$\varphi_1 : \mathcal{M}_{1 \times s}(\mathbb{F}_q) \to \mathbb{F}_q[x]/(x^s)$$
$$P = (p_0, p_1, \dots, p_{s-1}) \to p_0 + p_1 x + \cdots + p_{s-1} x^{s-1}.$$

Let $\mathbf{P} = (P_1, \dots, P_n)$ where $P_i = (p_{i0}, p_{i1}, \dots, p_{i,s-1})$ for $1 \leq i \leq n$. We extend φ_1 to

$$\varphi : \mathcal{M}_{n \times s}(\mathbb{F}_q) \to \mathcal{M}_{n \times 1}(\mathbb{F}_q[x]/(x^s))$$
$$\mathbf{P} \to (p_{00} + p_{01} x + \cdots + p_{0,s-1} x^{s-1}, \dots, p_{n0} + p_{01} x + \cdots + p_{n,s-1} x^{s-1})^T.$$

The maps defined above are vector isomorphisms over $\mathbb{F}_q$. The ρ weight of a polynomial $p(x) \in \mathbb{F}_q[x]/(x^s)$ is simply $\deg(p(x)) + 1$, i.e.

$$\rho(p(x)) = \deg(p(x)) + 1. \tag{8}$$

Let $p(x) = p_0 + \cdots + p_{s-1} x^{s-1} \in \mathbb{F}_q[x]/(x^s)$. The lth $(0 \leq l \leq s-1)$ coefficient of $p(x)$ is defined by

$$c_l(p(x)) = p_l. \tag{9}$$

Let $P(x) = (P_1(x), \dots, P_n(x))^T$, and $Q(x) = (Q_1(x), \dots, Q_n(x))^T \in \mathcal{M}_{n \times 1}(\mathbb{F}_q[x]/(x^s))$ where $P_i(x) = p_{i0} + p_{i1} x + \cdots + p_{i,s-1} x^{s-1}$, and $Q_i(x) = q_{i0} + q_{i1} x + \cdots + q_{i,s-1} x^{s-1}$. The inner product of $P(x)$ and $Q(x)$ defined above becomes:

$$\langle P(x), Q(x) \rangle = \sum_{i=0}^{n} c_{s-1}(P_i(x) Q_i(x)). \tag{10}$$

The *Hamming weight* of an element $a \in F_q$ is defined by

$$w(a) = \begin{cases} 0, & \text{if } a = 0 \\ 1, & \text{otherwise.} \end{cases} \tag{11}$$

Let $C \subset \mathcal{M}_{n \times s}(\mathbb{F}_q)$ be a linear code with size m. For simplification purposes, let $C = \{A^{(0)}, A^{(1)}, \ldots, A^{(m)}\}$. Also, let

$$A^{(i)} = \begin{pmatrix} a_{10}^{(i)} & a_{11}^{(i)} & \cdots & a_{1,s-1}^{(i)} \\ a_{20}^{(i)} & a_{21}^{(i)} & \cdots & a_{2,s-1}^{(i)} \\ & & \vdots & \\ a_{n0}^{(i)} & a_{n1}^{(i)} & \cdots & a_{n,s-1}^{(i)} \end{pmatrix}, \qquad 0 \le i \le m.$$

Let

$$Y_{ns} = (y_{10}, \ldots, y_{1,s-1}, \ldots, y_{n0}, \ldots, y_{n,s-1}).$$

We define the *complete ρ weight enumerator* of a code C by

$$W_C(Y_{ns}) = \sum_{i=0}^{m} y_{10}^{w(a_{10}^{(i)})} \cdots y_{1,s-1}^{w(a_{1,s-1}^{(i)})} \cdots y_{n0}^{w(a_{n0}^{(i)})} \cdots y_{n,s-1}^{w(a_{n,s-1}^{(i)})}. \qquad (12)$$

Note that the complete ρ weight enumerator is a polynomial of ns variables. Further, it is possible to obtain the ρ weight enumerator by specializing the complete ρ weight enumerator.

Example 2: The ρ complete weight enumerators of the codes C_1, C_2, $C_1^{\perp}$ and $C_2^{\perp}$ (Example 1) are

$$\begin{aligned} W_{C_1}(Y_{22}) =& 1 + y_{10}y_{20}, \\ W_{C_2}(Y_{22}) =& 1 + y_{21}, \\ W_{C_1^{\perp}}(Y_{22}) =& 1 + y_{11}y_{21} + y_{10}y_{11}y_{20}y_{21} + y_{10}y_{11}y_{21} + y_{11}y_{20}y_{21} \\ & + y_{10} + y_{20} + y_{10}y_{20}, \\ W_{C_2^{\perp}}(Y_{22}) =& 1 + y_{10} + y_{11} + y_{21} + y_{10}y_{11} + y_{10}y_{20} + y_{11}y_{21} + y_{10}y_{11}y_{21}. \end{aligned}$$

By letting, $y_{10}^{i_0}y_{11}^{i_1} = z^{2i_1+(1-i_1)i_0}$ and $y_{20}^{i_0}y_{21}^{i_1} = z^{2i_1+(1-i_1)i_0}$, we obtain the ρ weight enumerators (6).

The following lemmas are going to play an important role in the proof of the main theorem.

Lemma 1. *[2] Let χ be a nontrivial additive character of $\mathbb{F}_q$. Let β be a fixed element of $\mathbb{F}_q$. Then,*

$$\sum_{i=0}^{q-1} \chi(\beta\alpha_i) = \begin{cases} q, & \beta = 0, \\ 0, & \beta \ne 0. \end{cases} \qquad (13)$$

Lemma 2. *Let χ be a nontrivial additive character of $\mathbb{F}_q$. Then,*

$$\sum_{P(x) \in C} \chi(\langle P(x), Q(x) \rangle) = \begin{cases} 0, & Q(x) \notin C^{\perp}, \\ |C|, & Q(x) \in C^{\perp}. \end{cases}$$

Proof. If $Q(x) \in C^{\perp}$, then it is clear. If $Q(x) \notin C^{\perp}$, then there exists $P(x) \in C$ such that $\langle P(x), Q(x) \rangle \neq 0$. Let $\langle P(x), Q(x) \rangle = \gamma$. Then, the map

$$\varphi_{Q(x)} : C \to \mathbb{F}_q$$

$$P(x) \to \langle P(x), Q(x) \rangle = \sum_{i=0}^{n} c_{s-1}(P_i(x)Q_i(x))$$

is $\mathbb{F}_q$-linear and onto. Thus, $C/Ker(\varphi_{Q(x)}) \cong \mathbb{F}_q$. Hence,

$$\sum_{P(x) \in C} \chi\left(\langle P(x), Q(x) \rangle\right) = \frac{|C|}{q} \sum_{\alpha \in \mathbb{F}_q} \chi(\alpha) = 0, \quad \text{Lemma 1. } \square$$

Lemma 3. *Let χ be a nontrivial additive character of $\mathbb{F}_q$ and i, j be fixed. Let $p(x) = p_{i0} + p_{i1}x + \cdots + p_{i,s-1}x^{s-1} \in \mathbb{F}_q[x]/(x^s)$.*

$$\sum_{\alpha \in \mathbb{F}_q} \chi(\langle p(x), \alpha x^j \rangle) y_{ij}^{w(\alpha)} = (1 + (q-1)y_{ij})^{1-w(p_{i,s-1-j})} (1 - y_{ij})^{w(p_{i,s-1-j})}.$$

Proof.

$$\sum_{\alpha \in \mathbb{F}_q} \chi(\langle p(x), \alpha x^j \rangle) y_{ij}^{w(\alpha)} = \sum_{\alpha \in \mathbb{F}_q} \chi(c_{s-1}(p(x)(\alpha x^j))) y_{ij}^{w(\alpha)}$$

$$\sum_{\alpha \in \mathbb{F}_q} \chi(p_{i,s-1-j}\alpha) y_{ij}^{w(\alpha)} = 1 + \sum_{k=1}^{q-1} \chi(p_{i,s-1-j}\alpha_k) y_{ij}^{w(\alpha_k)}$$

$$= \begin{cases} 1 + (q-1)y_{ij}, & p_{i,s-1-j} = 0, \\ 1 - y_{ij}, & p_{i,s-1-j} \neq 0. \end{cases} \quad \text{(Lemma 1.)} \square$$

Lemma 4. *Let $f : \mathcal{M}_{n \times 1}(\mathbb{F}_q[x]/(x^s)) \to \mathbb{C}[y_{10}, \ldots, y_{n,s-1}]$ and χ be a nontrivial additive character of $\mathbb{F}_q$. Then,*

$$\sum_{Q(x) \in C^{\perp}} f(Q(x)) = \frac{1}{|C|} \sum_{P(x) \in C} \hat{f}(P(x))$$

where $\hat{f}(P(x)) = \sum_{Q(x) \in \mathcal{M}_{n \times 1}(\mathbb{F}_q[x]/(x^s))} \chi\left(\langle P(x), Q(x) \rangle\right) f(Q(x))$, $P(x) = (P_1(x), \ldots, P_n(x))^T$ and $Q(x) = (Q_1(x), \ldots, Q_n(x))^T$.

Proof. Let $P_i = p_{i0} + \cdots + p_{i,s-1}x^{s-1}$ and $Q_i(x) = q_{i0} + \cdots + q_{i,s-1}x^{s-1}$ for $1 \le i \le n$.

$$\sum_{P(x) \in C} \hat{f}(P(x)) = \sum_{P(x) \in C} \sum_{Q(x) \in \mathcal{M}_{n \times 1}(\mathbb{F}_q[x]/(x^s))} \chi\left(\langle P(x), Q(x) \rangle\right) f(Q(x))$$

$$= \sum_{P(x) \in C} \sum_{Q(x) \in C^{\perp}} \chi\left(\langle P(x), Q(x) \rangle\right) f(Q(x))$$

$$+ \sum_{P(x) \in C} \sum_{Q(x) \notin C^{\perp}} \chi\left(\langle P(x), Q(x) \rangle\right) f(Q(x))$$

$$= |C| \sum_{Q(x) \in C^{\perp}} f(Q(x)). \quad \text{(by Lemma 2.)} \square$$

Theorem 1.

$$\sum_{Q(x) \in C^{\perp}} y_{10}^{w(q_{10})} \cdots y_{1,s-1}^{w(q_{1,s-1})} \cdots y_{n0}^{w(q_{n0})} \cdots y_{n,s-1}^{w(q_{n,s-1})}$$

$$= \frac{1}{|C|} \left(\prod_{i=1}^{n} \prod_{j=0}^{s-1} (1 + (q-1)y_{ij}) \right) \sum_{P(x) \in C} \prod_{k=1}^{n} \prod_{l=0}^{s-1} \left(\frac{1 - y_{kl}}{1 + (q-1)y_{kl}} \right)^{w(p_{k,s-1-l})}.$$

Proof. We take

$$f((Q_1(x), \ldots, Q_n(x))) = y_{10}^{w(q_{10})} \cdots y_{1,s-1}^{w(q_{1,s-1})} \cdots y_{n1}^{w(q_{n0})} \cdots y_{n,s-1}^{w(q_{n,s-1})}.$$

in Lemma 4. Then,

$$\hat{f}(P(x)) = \sum_{Q(x) \in \mathcal{M}_{n \times 1}(\mathbb{F}_q[x]/(x^s))} \chi\left(\langle P(x), Q(x) \rangle\right) y_{11}^{w(q_{10})} \cdots y_{1s}^{w(q_{1,s-1})} \cdots y_{n0}^{w(q_{n0})} \cdots y_{ns}^{w(q_{n,s-1})}$$

$$= \sum_{Q(x) \in \mathcal{M}_{n \times 1}(\mathbb{F}_q[x]/(x^s))} \prod_{i=1}^{n} \chi\left(\langle P_i(x), Q_i(x) \rangle\right) y_{11}^{w(q_{10})} \cdots y_{1s}^{w(q_{1,s-1})} \cdots y_{n0}^{w(q_{n0})} \cdots y_{ns}^{w(q_{n,s-1})}$$

$$= \sum_{q_{10} \in \mathbb{F}_q} \chi(\langle P_1(x), q_{10} \rangle) y_{10}^{w(q_{10})} \cdots \sum_{q_{1,s-1} \in \mathbb{F}_q} \chi(\langle P_1(x), q_{1,s-1}x^{s-1} \rangle) y_{1,s-1}^{w(q_{1,s-1})}$$

$$\cdot \sum_{q_{20} \in \mathbb{F}_q} \chi(\langle P_2(x), q_{20} \rangle) y_{20}^{w(q_{20})} \cdots \sum_{q_{2,s-1} \in \mathbb{F}_q} \chi(\langle P_2(x), q_{2,s-1}x^{s-1} \rangle) y_{2,s-1}^{w(q_{2,s-1})}$$

$$\vdots$$

$$\cdot \sum_{q_{n0} \in \mathbb{F}_q} \chi(\langle P_n(x), q_{n0} \rangle) y_{n0}^{w(q_{n0})} \cdots \sum_{q_{n,s-1} \in \mathbb{F}_q} \chi(\langle P_n(x), q_{n,s-1}x^{s-1} \rangle) y_{n,s-1}^{w(q_{n,s-1})}$$

Applying Lemma 3,

$$\hat{f}(P(x)) = \prod_{l=0}^{s-1}(1 + (q-1)y_{1l})^{1-w(p_{1,s-1-l})}(1-y_{1l})^{w(p_{1,s-1-l})}.$$

$$\vdots$$

$$= \prod_{l=0}^{s-1}(1 + (q-1)y_{nl})^{1-w(p_{n,s-1-l})}(1-y_{nl})^{w(p_{n,s-1-l})}$$

$$= \left(\prod_{i=1}^{n}\prod_{j=0}^{s-1}(1 + (q-1)y_{ij})\right)\prod_{k=1}^{n}\prod_{l=0}^{s-1}\left(\frac{1-y_{kl}}{1+(q-1)y_{kl}}\right)^{w(p_{k,s-1-l})}.\square$$

References

1. MacWilliams-type Theorems for a Non-Hamming Metric, Steven T. Dougherty and Maxim M. Skriganov, preprint.
2. F.J. MacWilliams and N.J.A Sloane, *The Theory Of Error Correcting Codes*, North-Holland Pub. Co., 1977.
3. M. Yu Rosenbloom and M. A. Tsfasman, *Codes for the m-metric*, Problems of Information Transmission, Vol. 33. No. 1, 45-52, 1997.
4. M.M. Skriganov, *Coding theory and uniform distributions*, St. Petersburg Math. J. Vol 143, No. 2, 2002.

Further Improvement of Kumar-Rajagopalan-Sahai Coding Constructions for Blacklisting Problem

Maki Yoshida and Toru Fujiwara

Department of Informatics and Mathematical Science, Osaka University,
1-3 Machikaneyama, Toyonaka, Osaka, 560-8531 Japan
{maki-yos, fujiwara}@ics.es.osaka-u.ac.jp

Abstract. Solutions based on error-correcting codes for the blacklisting problem of a broadcast distribution system have been proposed by Kumar, Rajagopalan and Sahai. We have optimized their schemes by choosing the parameters properly. In this paper, we propose the further improvement for their schemes. On the transmission, the improved schemes with the parameters chosen for the optimized original schemes are more efficient than the optimized original schemes. The average amount of transmission is 70 percents of that for the optimized original scheme in one of the typical cases, while the amount of the storage is the same.

1 Introduction

Consider the distribution of digital contents over a broadcast channel where the contents should be available only to subscribers among the users of the broadcast channel, e.g., pay-TV. A data supplier gives each subscriber a decoder containing a secret decryption key and broadcasts the contents in encrypted form. For each distribution, every decoder decrypts the broadcasted contents in encrypted form by using its secret decryption key. We will call a system for such distribution a *broadcast distribution system* (BDS).

As pointed out in [3], there is a desired property for a BDS: For every distribution, the data supplier can prevent subscribers from decrypting the broadcasted contents in encrypted form without renewing any secret decryption key where no coalition of prevented subscribers can recover the contents from the broadcast. A BDS which meets the desired property is suitable for the environment where for every distribution the data supplier identifies those whom the contents are to be distributed, e.g., pay-TV with various programs. The problem to construct a BDS which meets the property and allows prevention of a limited number of subscribers is called the *blacklisting problem* [1, 6–11].

In the general model of a BDS, the broadcasted contents in encrypted form consists of an enabling part and a cipher part. The cipher part is the symmetric encryption of the contents under a session key. For each distribution, a session key is chosen randomly. The enabling part contains the information to obtain

B. Honary (Ed.): Cryptography and Coding 2001, LNCS 2260, pp. 27–37, 2001.

the session key by using a secret decryption key. Under the assumption that the used symmetric encryption algorithm is secure, the security of distributing the contents is reduced to that of distributing a session key. Then, we consider the problem to distribute a session key securely.

In this paper, we focus on the unconditionally secure solutions for the black-listing problem. In [1, 8], the efficient solutions based on threshold secret sharing are proposed. In [6], Kumar et al. proposes the solutions based on an error correcting code and a cover-free family. The solutions in [6], called the KRS schemes in this paper, are unconditionally secure. Although the solutions in [1, 8] are asymptotically more efficient than the KRS schemes, it is important to improve the efficiency of the KRS schemes. A reason given in [4] is that the KRS schemes are more appropriate than the solutions in [1, 8] for a long-lived broadcast encryption. In fact, the long-lived broadcast encryption schemes based on the KRS schemes are proposed in [4].

In [10], we have analyzed the KRS schemes detailedly and presented a method to choose the values of parameters such that the amount of transmission is minimized (the results in [10] include that in [9]). In this paper, the KRS schemes with such choice of parameters are called optimized KRS schemes. The performance of the optimized KRS schemes is shown in [10]. From the results in [10], the KRS schemes are suitable and practical for the sizes of a blacklisting problem such that the relatively small amount of information, say just a single key, is sent, and the proportion of the maximum number of the excluded subscribers to that of all subscribers is small, say not more than 1%.

Our goal is to improve the KRS schemes furthermore. The improvement means to decrease the amount of the transmission, while that of the storage is the same. We define the enabling part more properly by using the property of the cover-free family. By improving the KRS schemes, there are cases where the amount of transmission becomes only the several times as much as that in [1, 8].

In Sect. 2, we show the definition of a blacklisting problem. The difference between the improved KRS schemes and the previous KRS schemes considered in [4, 6, 7, 9, 10] is the definition of the enabling part. Then, in Sect. 3, we describe the original KRS schemes in [6]. In Sect. 4, we present the improvement and the detailed analysis of the improved KRS schemes. Finally, in Sect. 5, we show the numerical results of the improvement.

2   Blacklisting Problem

In this paper, we consider an (N, K)-blacklisting problem with $0 < K < N$. In the model, there is the data supplier and the group of subscribers. N denotes the number of the subscribers. The subscribers are assumed to be numbered from 1 to N. K denotes the maximum number of subscribers whom the data supplier can prevent from recovering the session key contained in the enabling part. The prevented subscribers are called *excluded subscribers*. All subscribers except for excluded subscribers are called *unexcluded subscribers*.

An (N, K)-blacklisting problem is a problem to construct a BDS which satisfies the following three requirements:

(R1) Without knowing the broadcasted enabling part, no subset of subscribers has any information about the session key, even given all the secret decryption keys of the subscribers;

(R2) An unexcluded subscriber can uniquely determine the session key from the broadcasted enabling part and his secret decryption key;

(R3) After receiving the broadcasted enabling part, no subset of excluded subscribers has any information about the session key.

The performance of a solution for an (N, K)-blacklisting problem is evaluated by the following three complexity measures:

(CM1) The size of storage required by the data supplier;

(CM2) The size of storage required by a subscriber;

(CM3) The size of a broadcasted enabling part, called the necessary increase of bandwidth.

3   Kumar-Rajagopalan-Sahai Coding Constructions

The KRS schemes proposed in [6] have the following common framework based on an error correcting code: A session key is encoded into one or more codewords; For the set of excluded subscribers, some code symbols are removed and the remaining code symbols are encrypted in such a way that every unexcluded subscriber can obtain the enough code symbols to recover the codewords by using his decryption key and any excluded subscriber cannot obtain any code symbol. In this framework, a cover-free family takes important role. The schemes differ in the construction of the cover-free family, and three constructions of the cover-free family are proposed in [6]. In the following, we show the detailed description of the common framework of the KRS schemes and the role of the cover-free family.

The KRS scheme for an (N, K) blacklisting problem has five auxiliary parameters n, p, α, k and q. The scheme uses the (N, K, n, p, α)-cover-free family, the (n, k) maximum distance separable (MDS) code over $GF(q)$ with $k = \lfloor \alpha p \rfloor$, and some unconditionally secure cryptosystem.

First, we show the definition of the (N, K, n, p, α)-cover-free family. Consider a family of sets

$$\mathcal{S} = \{S_1, S_2, \ldots, S_N\},$$

where S_i with $1 \le i \le N$ is a subset with size p of the universe U with $|U| = n$. Then, $\mathcal{S}$ is an (N, K, n, p, α)-cover-free family if for any $S', S'_1, S'_2, \ldots S'_K \in \mathcal{S}$,

$$\left| S' \backslash \bigcup_{i=1}^{K} S'_i \right| \ge \alpha |S'|.$$

For simplicity, we use the (n, k) MDS code over $GF(q)$, where q is a power of 2. This assumption, q is a power of 2, is natural, since we consider the case where the session key is represented as a binary string.

In [10], it is shown that the choice of parameters is optimum, i.e., minimizes the amount of transmission, only if the number of generated codewords in the KRS scheme is only one. In this paper, we also choose the parameters such that the generated codeword is only one codeword. Then, in the KRS schemes, the number of encryption keys of the used unconditionally secure cryptosystem is n. Let

$$U = \{k_1, k_2, \ldots, k_n\},$$

be the set of the n encryption keys, which is the universe in the cover-free family. For a message M, its ciphertext, which is encrypted with the encryption key k_h, is denoted $E_{k_h}(M)$. In the following, we assume that $E_{k_h}(v_h)$ is defined as $k_h + v_h$ and k_h is the random element of $GF(q)$.

The KRS scheme consists of two phases, an initial set-up phase and a distributing phase. At the initial set-up phase, the data supplier constructs an (N, K, n, p, α)-cover-free family $\mathcal{S} \triangleq \{S_1, S_2, \ldots, S_N\}$. The i-th user has S_i as his secret decryption key, called a secret key set. The data supplier knows the secret key set of every subscriber.

Next, consider the distributing phase. A set of the excluded subscribers are represented by $X \subset \{1, 2, \ldots, N\}$. A session key is represented as a k-tuple over $GF(q)$, denoted $\boldsymbol{u}$, and the enabling part for the session key is constructed as follows.

(Step 1) $\boldsymbol{u}$ is encoded by using the (n, k) MDS code over $GF(q)$ into the codeword $\boldsymbol{v} \triangleq (v_1, v_2, \ldots, v_n)$.

(Step 2) For the set X of excluded subscribers, the set $S(X)$ of used encryption keys is the set of all encryption keys except for the keys held by excluded subscribers in X, that is,

$$S(X) \triangleq \left\{ k_h \,\middle|\, 1 \leq h \leq n, k_h \notin \bigcup_{j \in X} S_j \right\}.$$

(Step 3) The set EP of ciphertexts is the enabling part where

$$EP \triangleq \{E_{k_h}(v_h) | 1 \leq h \leq n, k_h \in S(X)\}.$$

For EP, every excluded subscriber can decrypt no ciphertext in EP while an unexcluded subscriber i can decrypt the ciphertexts $\{E_{k_h}(v_h) | k_h \in S_i \cap S(X)\}$ and retrieve the codeword, i.e. the session key, from the code symbols $\{v_h | k_h \in S_i \cap S(X)\}$.

Security of the KRS scheme

(R1) is satisfied since a session key is chosen randomly and independent of the encryption keys. The property of the cover-free family implies that (R2) is satisfied, since for every X with $0 \leq |X| \leq K$ and each unexcluded subscriber $i \notin X$, the number of code symbols that the unexcluded subscriber i can obtain, i.e., $|\{E_{k_h}(v_h) | k_h \in S_i \cap S(X)\}|$, is not less than $k = \lfloor \alpha p \rfloor$. For every X with

$0 \leq |X| \leq K$ and each excluded subscriber $i \in X$, there is no code symbol that the excluded subscriber i can obtain, thus and (R3) is satisfied.

Complexity of the KRS scheme

(CM1') The size of storage required by the data supplier is that to store n encryption keys. The size of n encryption keys is $n \cdot \log q$.

(CM2') The secret decryption key is the set of p encryption keys. Then, the size of storage required by a subscriber is $p \cdot \log q$.

(CM3') For a set X of excluded subscribers, the necessary increase of bandwidth is not only the size of $|S(X)|$ ciphertexts. To inform which code symbols are removed, we add the 1-bit flag for each symbol. The total length of the flags is n-bit. Therefore, for the set X of excluded subscribers, the necessary increase of bandwidth is $|S(X)| \cdot \log q + n$.

4   Proposed Improvement and Its Analysis

In this section, we propose the improvement to decrease the necessary increase of bandwidth while the amount of the storage is the same. For a set X of excluded subscribers, the size of the enabling part is proportional to the size of $S(X)$. We give the more proper definition of $S(X)$ to decrease its size as the improvement of the KRS schemes.

4.1   Proposed Improvement

Our idea is simple. To form $S(X)$, we exclude not only all encryption keys of the excluded subscribers but also some keys of unexcluded subscribers from $S(X)$. The definition of $S(X)$ does not relate to whether the requirement (R1) is satisfied or not. The requirements (R2) and (R3) are satisfied if $S(X)$ is defined as follows: No encryption key held by excluded subscribers in X is in $S(X)$, and $\lfloor \alpha p \rfloor$ encryption keys of each unexcluded subscriber is in $S(X)$.

In the original KRS schemes, if $|X| < K$, the property of the (N, K, n, p, α)-cover-free family implies that many subscribers may have more than $\lfloor \alpha p \rfloor$ encryption keys in $S(X)$ to withstand the exclusion of one or more subscribers. We exclude such redundant encryption keys from $S(X)$.

For the given set X of excluded subscribers, we present the precise construction of $S(X)$. (Step 2) in the KRS schemes described in Sect. 3 is replaced by the following steps (Step 2-1)–(Step 2-3).

(Step 2-1) A subset of unexcluded subscribers are chosen randomly from $\{1, 2, \dots, N\} \setminus X$, denoted Y. In order for the improved KRS scheme using any cover-free family to be secure, the size of Y should be at most $K - |X|$.

(Step 2-2) For each $i \in Y$, the $\lfloor \alpha p \rfloor$ encryption keys are chosen from $S_i \setminus \bigcup_{j \in X} S_j$ as shown in Fig. 1. The set of keys chosen from S_i is denoted T_i. Some greedy approaches are applicable to decrease the number of chosen keys.

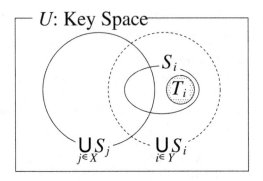

Fig. 1. The subset T_i of S_i.

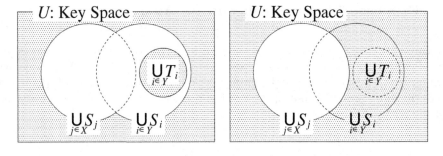

(a) The improved KRS scheme   (b) The original KRS scheme

Fig. 2. The set $S(X)$ of chosen encryption keys.

(Step 2-3) $S(X)$ is defined by

$$S(X) \triangleq \left\{ k_h \left| 1 \leq h \leq n, k_h \notin \bigcup_{j \in X \cup Y} S_j \right. \right\} \bigcup \left\{ k_h \left| 1 \leq h \leq n, k_h \in \bigcup_{i \in Y} T_i \right. \right\},$$

as shown in Fig. 2(a).

Further excluded encryption keys of unexcluded subscribers are the subset of $(\bigcup_{i \in Y} S_i \setminus \bigcup_{j \in X} S_j)$ (see Fig. 2). The size of $S(X)$ is close to that of $S(Z)$ with $|Z| = K$, i.e. the minimum size. In the following, $|Y|$ is defined as $K - |X|$ to decrease the necessary increase of bandwidth as much as possible.

4.2   Analysis

In this section, for the improved KRS scheme, we analyze the security and the complexity.

We can show that the requirements (R1) and (R3) are satisfied by the same way as that for the original KRS scheme. The requirement (R2) is satisfied if the followings (R2a) and (R2b) are satisfied:
(R2a) For each unexcluded subscriber $i \in Y$, $|S_i \cap S(X)| \geq \lfloor \alpha p \rfloor$;
(R2b) For each unexcluded subscriber $i \notin Y$, $|S_i \cap S(X)| \geq \lfloor \alpha p \rfloor$.
From (Step 2-2) in Sect. 4.1, (R2a) is satisfied, since

$$S_i \cap S(X) = S_i \cap \left\{ k_h \,\middle|\, 1 \leq h \leq n, k_h \in \bigcup_{i \in Y} T_i \right\}$$

$$\supseteq S_i \cap T_i = T_i,$$

and $|S_i \cap S(X)| \geq |T_i| = \lfloor \alpha p \rfloor$. From the remark in (Step 2-1), $|Y|$ is not more than $K - |X|$. Then (R2b) is also satisfied, since the excluded encryption keys $U \setminus S(X)$ are chosen from the K or less secret key sets and the property of the (N, K, n, p, α)-cover-free family.

The complexities of the storages (CM1') and (CM2') are the same as those in the original KRS schemes. The complexity (CM3'), the necessary increase of bandwidth, is given by $|S(X)| \cdot \log_2 q + n$. Though $|S(X)|$ depends on the set X of excluded subscribers and the used cover-free family, the following lemma gives a simple upper bound on $|S(X)|$ which is derived from the definition of $S(X)$ in Sect. 4.1.

Lemma 1. *For X with $0 \leq |X| \leq K$ and Y with $X \cap Y = \emptyset$ and $|Y| = K - |X|$, $|S(X)| \leq |S(X \cup Y)| + (K - |X|) \cdot \lfloor \alpha p \rfloor$.*

We will derive the average of $|S(X \cup Y)|$ when the family S is chosen from a given Ω, to obtain an upper bound on the average of $|S(X)|$. Since $|X \cup Y| = K$, it is enough to derive the average of $|S(Z)|$ with $|Z| = K$. Let $\sigma(Z, S)$ denote the size of $S(Z)$ when the cover-free family S is used. We will give the average of $\sigma(Z, S)$ over the family Ω of the cover-free families under the assumption that each family in Ω is equiprobable. The average is denoted $\bar{\sigma}(Z, \Omega)$.

In [6], three constructions of cover-free family are proposed. One construction is based on algebraic geometry code, and it is easy to see that the generated cover-free family is impractical [9, 10].

First, we consider the improved KRS scheme using the randomized construction proposed in [6]. In the randomized construction, for given N and K, the (N, K, n, p, α)-cover-free family $S = \{S_1, S_2, \dots, S_N\}$ is constructed as follows:
(1) Choose a positive integer p and its multiple n with $n \geq 2p$. Generate the set U of the n encryption keys.
(2) Partition the set U into p disjoint subsets $U_1, U_2, \dots, U_p$. Each subset contains $\frac{n}{p}$ keys.
(3) S_i with $1 \leq i \leq N$ is obtained by choosing one key randomly from each subset U_h with $1 \leq h \leq p$.
This is a probabilistic construction. To obtain the cover-free family with enough high probability, the parameters should be chosen properly.

Let Ω_R be the set of all families constructed by the randomized construction. The following lemma shows $\bar{\sigma}(Z, \Omega_R)$. The average is the approximation since Ω_R includes the families which are not cover-free families. However, if the success probability is enough close to 1, then the great majority of the families are the cover-free families, and the accuracy of this approximation is sufficient for evaluating the complexity.

Lemma 2. *Let Ω_R be the set of all families constructed by the randomized construction. For a set Z of K excluded subscribers,*

$$\bar{\sigma}(Z, \Omega_R) = n \cdot \left(1 - \frac{p}{n}\right)^K. \tag{1}$$

(Proof) Since $U_1, U_2, \ldots, U_p$ are disjoint subsets of U, $|S(Z)|$ is equal to $|S(Z) \cap U_1| + |S(Z) \cap U_2| + \cdots + |S(Z) \cap U_p|$.

From Eq. (8.10) in [12], the average sum of sizes $|S(Z) \cap U_h|$ with $1 \le h \le p$ is equal to the sum of these averages. For Z, the each average size $|S(Z) \cap U_h|$ with $1 \le h \le p$ is equal to another average size $|S(Z) \cap U_{h'}|$ with $1 \le h' \le p$ and $h' \ne h$. Then, $\bar{\sigma}(Z, \Omega_R)$ is derived from the average size of $S(Z) \cap U_1$.

The set $S(Z) \cap U_1$ is constructed by removing all the encryption keys in the K secret key sets from U_1. For any encryption key, the probability that the key is not in some secret key set is equal to $1 - \frac{p}{n}$. The reason is that for each of K secret key sets, one encryption key in U_1 is chosen randomly from n/p encryption keys. For U_1, using the principle of inclusion and exclusion, the average number of encryption keys which are not in any of K secret key sets is equal to $\frac{n}{p}(1 - \frac{p}{n})^K$. This average is equal to the average size of $S(Z) \cap U_1$. Then, $\bar{\sigma}(Z, \Omega_R) = p \cdot \frac{n}{p}(1 - \frac{p}{n})^K = n(1 - \frac{p}{n})^K$. $\square$

From Lemmas 1 and 2, the upper bound of the average necessary increase of bandwidth is derived.

Theorem 1. *For a set X of K or less excluded subscribers, the average necessary increase of bandwidth is upper bounded by*

$$\left(n \cdot \left(1 - \frac{p}{n}\right)^K + (K - |X|) \cdot \lfloor \alpha p \rfloor\right) \cdot \log q + n, \tag{2}$$

under the assumption that any family constructed by the randomized construction is used in the improved KRS scheme.

Secondly, we consider the improved KRS scheme using the polynomial construction of the cover-free family proposed in [6]. In the polynomial construction, n is chosen as $n = p^2$, and N distinct polynomials are generated to decide which p out of p^2 encryption keys are assigned to each secret key set. Let $\Omega_{P,i}$ be the set of all cover-free families constructed by the polynomial construction where the degree of all generated polynomials is limited to i or less. The following lemma shows $\bar{\sigma}(Z, \Omega_{P,1})$.

Lemma 3. *For a set Z of K excluded subscribers,*

$$\bar{\sigma}(Z, \Omega_{P,1}) = \begin{cases} p^2 - Kp + \sum_{i=2}^{K}(-1)^i \binom{K}{i} p \prod_{j=1}^{i-1} \frac{p-j}{p^2-j}, & \text{for } K \geq 2, \\ p^2 - p, & \text{otherwise.} \end{cases} \tag{3}$$

It should be noted that $\bar{\sigma}(Z, \Omega_R)$ given by (1) is a good approximation of $\bar{\sigma}(Z, \Omega_{P,1})$ when $\binom{K}{i}p^{2-i}$ approaches 0 rapidly as i increases. For $\Omega_{P,2}$, a similar formula can be derived. For many practical cases, the optimum choices of parameters imply that the cover-free family in $\Omega_{P,i}$ with $i = 1, 2$ is used. In the next section, we show the numerical results when the both improved and original KRS schemes with the polynomial construction.

5   Numerical Results

In this section, we show the numerical results for the proposed improvement. As discussed in Sect. 4.2, compared with the original KRS scheme, the complexities (CM1') and (CM2') are the same, and (CM3') is not worse than that of the original KRS scheme. We focus on the results for (CM3'). The average amount of the necessary increase of bandwidth is decreased significantly when the number of excluded subscribers is often much smaller than K. If the number of excluded subscribers is always K, then the average amount is the same. These two cases are extremes. We show the results for more practical cases such that, for every θ with $0 \leq \theta \leq K$, the probability with $|X| = \theta$ is the same, i.e., is equal to $\frac{1}{K+1}$.

Generally, the size of the session key, denoted ℓ, is defined by what the session key is used for. Then, it is assumed that the value of ℓ with $\ell > 0$ is also given as those of N and K. We consider the case that the size of the session key is relatively small, as in [4, 9, 10].

In Fig. 3, we show the average necessary increases of bandwidth for $N = 10^4$, K with $50 \leq K \leq 260$ and $\ell = 128$. We compare the improved KRS scheme with the optimized original KRS scheme in [10] and the most efficient schemes based on threshold secret sharing with threshold $K + 1$ in [1,8], called the TSS scheme here. The reason that we compare the KRS scheme with the TSS scheme is that the performance of the KRS scheme can be close to that of the most efficient schemes even if the latter is asymptotically more efficient than the former. In [1], it is shown that the asymptotic complexities of transmission for the KRS scheme and the TSS scheme are $O(K^2)$ and $O(K)$, respectively.

The five parameters n, p, α, k and q in the improved KRS scheme and the original KRS scheme are determined based on the method in [10] which minimizes the worst necessary increase of bandwidth in the original KRS scheme. The parameters in the TSS scheme are chosen properly for N, K and ℓ. The precise necessary increase of bandwidth is $K \cdot (\ell + \lceil \log N \rceil)$ bits.

From the comparison, we see that there are the practical sizes, say $(N, K, \ell) = (10^4, 100, 128)$, such that the average necessary increase of bandwidth in the improved KRS scheme is 70 percents of that for the optimized original KRS scheme.

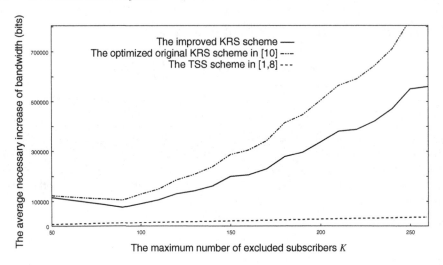

Fig. 3. The average necessary increase of bandwidth for the fixed $N = 10,000$ and $\ell = 128$.

This means that the average amount of the necessary increase of bandwidth becomes to be 6.2 times as much as that in the TSS scheme, while for the optimized original KRS scheme, it is 8.9 times.

6   Conclusion

The KRS schemes in [6] are improved. The amount of transmission is decreased significantly when the number of excluded subscribers is often much smaller than K. The amount of transmission in the worst case is the same as that in the original scheme. For some typical sizes of a blacklisting problem, the average amount of transmission is at most several times as much as the most efficient solutions for a blacklisting problem even if the KRS schemes are asymptotically less efficient than the most efficient schemes. From the results, we see that there are sizes that the KRS schemes are enough efficient to be of practical use.

References

1. J. Anzai, N. Matsuzaki and T. Matsumoto, "Quick Group Key Distribution Scheme with Entity Revocation," Advances in Cryptology—ASIACRYPT '99 (LNCS 1716), pp.333–347, Springer-Verlag, 1999.
2. C. Blundo, L. F. Mattos, and D. R. Stinson, "Generalized Beimel-Chor Schemes for Broadcast Encryption and Interactive Key Distribution," Theoretical Computer Science, vol.200, nos 1–2, pp.313–334, 1998.
3. A. Fiat and M. Naor, "Broadcast Encryption," Advances in Cryptology—CRYPTO '93 (LNCS 773), pp.480–491, Springer-Verlag, 1994.

4. J.A. Garay, J. Staddon and A. Wool, "Long-Lived Broadcast Encryption," Advances in Cryptology—CRYPTO 2000 (LNCS 1880), pp.333–352, Springer-Verlag, 2000.
5. E. Gafni, J. Staddon and Y.L. Yin, "Efficient methods for integrating traceability and broadcast encryption," Advances in Cryptology—CRYPTO '99 (LNCS 1666), pp.372–387, Springer-Verlag, 1999.
6. R. Kumar, S. Rajagopalan and A. Sahai, "Coding Constructions for Blacklisting Problems without Computational Assumptions," Advances in Cryptology—CRYPTO '99 (LNCS 1666), pp.609–623, Springer-Verlag, 1999.
7. K. Kurosawa, T. Yoshida and Y. Desmedt, "Inherently Large Traceability of Broadcast Encryption Scheme," Proc. of 2000 IEEE International Symposium on Information Theory, p.464, Sorrento, Italy, 2000.
8. M. Naor and B. Pinkas, "Efficient Trace and Revoke Schemes," Financial Cryptography 2000, Anguilla, British West Indies, February 2000.
9. M. Yoshida, T. Horikawa and T. Fujiwara, "An Improvement of Coding Constructions for Blacklisting Problems," Proc. of the 2000 International Symposium on Information Theory and Its Applications, pp.493–496, Honolulu, Hawaii, U.S.A., November 2000.
10. M. Yoshida and T. Fujiwara, "Analysis and Optimization of Kumar-Rajagopalan-Sahai Coding Constructions for Blacklisting Problem," IEICE Transactions of Fundamentals, vol.E84-A, no.9, pp.2338–2345, September 2001.
11. M. Yoshida and T. Fujiwara, "An Efficient Traitor Tracing Scheme for Broadcast Encryption," Proc. of 2000 IEEE International Symposium on Information Theory, p.463, Sorrento, Italy, 2000.
12. R.L. Graham, D.E. Knuth and O. Patashnik, Concrete Mathematics, Addison-Wesley, 1989.

A Simple Soft-Input/Soft-Output Decoder for Hamming Codes

Simon Hirst and Bahram Honary

Department of Communication Systems, Lancaster University, Lancaster, LA1 4YR, UK.
{s.a.hirst, b.honary}@lancs.ac.uk

Abstract. In an earlier paper, bit flipping methods for decoding low-density parity-check (LDPC) codes on the binary symmetric channel were adapted for generalised LDPC (GLDPC) codes with Hamming sub-codes. We now employ the analysis of this weighted bit flipping method to develop a simple soft-input/soft-output decoder for Hamming codes based on a conventional hard decision decoder. Simulation results are presented for decoding of both product codes and GLDPCs on the AWGN channel using the proposed decoder. At higher rates the indications are that good performance is possible at very low decoding complexity.

1 Introduction

In recent years, iterative coding schemes have proved successful in approaching channel capacity limits with low decoded bit error rates (BERs) and practical decoding complexity [1][5][6]. We include GLDPCs in this statement: these are defined by random sparse graphs with constraint nodes that relate to error-correcting codes [5][8]. Simple bit flipping (BF) decoding methods for LDPCs have been principally investigated on the binary symmetric channel (BSC) [4][6][7] (but have also been extended to the AWGN channel [4]).

We have successfully generalised BF methods for LDPCs to a weighted bit flipping (WBF) method for GLDPCs with Hamming sub-codes [2][3]. Applying WBF to codes of long length, the cut-off rate R_0 of the BSC may be approached at relatively low decoding cost. In this paper, we extend the iterative WBF method to the AWGN channel.

2 Motivation

We first review the WBF approach for Hamming-based schemes [2][3]. For the product code (PC) and GLDPC constructions employed, all symbols will belong to two sub-codes as shown in Fig. 1. Two interlocking (7,4,3) Hamming component codewords are shown with only one symbol in common.

B. Honary (Ed.): Cryptography and Coding 2001, LNCS 2260, pp. 38-43, 2001.

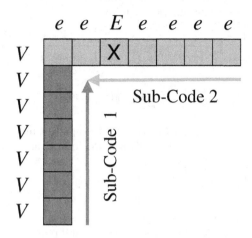

Fig. 1. Example Votes from Sub-Code Decoders.

Component hard decision decoders (HDDs) are used to generate symbol votes as illustrated. The decoders need only allow for two instances,

- all-zero syndromes ⇒ vote V sent to all symbols (see sub-code 1 in Fig. 1).
- non-zero syndromes ⇒ vote E sent to error position and vote e sent to all other positions (see sub-code 2 in Fig. 1 where '**X**' indicates error position).

Votes V, e and E are given numerical values such that the vote pairs of symbols, either VV, eV, ee, EV, Ee or EE, may be ranked in terms of reliability by the sum of their constituent votes. At each iteration the WBF technique proceeds by: i). applying the HDDs ii). ranking symbols by their vote pairs and iii). flipping all bits of lowest reliability before a new iteration begins. For extended Hamming codes an additional vote D is also required for all symbols of codewords where double errors are detected.

Experimental optimisation and Bayesian analysis both point to weights of $V = +1$, $e = 0$ and $E = -1$ as producing lowest BERs where we use larger weights to represent higher reliability [3]. Analysis in terms of an APP decoding strategy yields identical optimum weights [2]. It is this approach that will be extended.

3 HDD-SISO Decoder

The proposed soft-input/soft-output (SISO) decoder is based around the single application of a HDD and hence we will refer to it as a HDD-SISO decoder. The decoder is derived here for the simple case of a (7,4,3) Hamming code using the code's full weight distribution. For larger codes the analysis may be extended using only the first term in the distribution.

In the first stage of the decoder, for each symbol position j there is an 'averaging' process over the other $(n-1)$ LLR inputs i.e. all $L(u_i)$ where $i \in [1,n]$ and $i \neq j$. An average probability of bit error in the other symbol positions, p_j, may be approximated in terms of a least reliable input, $L^{min}(u_j)$, as follows,

$$p_j = \frac{1}{n-1}\sum_{i=1,i\neq j}^{n}\frac{1}{1+\exp(|L(u_i)|)} \cong \frac{1}{n-1}\cdot\frac{1}{1+\exp(L^{min}(u_j))} \tag{1}$$

where,

$$L^{min}(u_j)\triangleq \min_{i\in[1,n],i\neq j}\{|L(u_i)|\}.$$

Subsequently, the magnitude of an average soft input for the other $(n-1)$ positions may be found by representing the above BER in the form of an LLR,

$$L(p_j)\triangleq \log\frac{1-p_j}{p_j} \cong L^{min}(u_j)+\log(n-1). \tag{2}$$

In the second stage of the decoder, we employ analysis of an optimal SISO component decoder on the BSC with transition probability p to derive extrinsic LLRs [2]. In this case, the extrinsic LLR per symbol, $L^e(u_j)$, may be determined using the voting system of Section 2 and the following expressions,

$$\frac{L^e(u_j)}{S(u_j)} = A(p)\cdot\begin{cases}+1 & \text{for symbol vote } V\\ 0 & \text{for symbol vote } e\\ -1 & \text{for symbol vote } E\end{cases} \tag{3}$$

where,

$$A(p)=\log\frac{(1-p)^6+4p^3(1-p)^3+3p^4(1-p)^2}{3p^2(1-p)^4+4p^3(1-p)^3+p^6} \cong 2\cdot\log\frac{1-p}{p}-\log 3$$

and,

$$S(u_i)=\begin{cases}+1 & \text{if received bit } r_i = 0\\ -1 & \text{if received bit } r_i = 1\end{cases}.$$

In the HDD-SISO decoder, we combine the two stages as follows. Firstly, the signs of the input LLRs are used as the received bits in the above equation i.e. we replace $S(u_j)$ in (3) with $\text{sgn}(L(u_j))$. Secondly, the average BER in the inputs not including position j replaces the BSC transition probability i.e. we replace $\log\{(1-p)/p\}$ in (3) with $\log\{(1-p_j)/p_j\}$. After making these substitutions the proposed decoder may be described in terms of the following equations,

$$\frac{L^e(u_j)}{S(u_j)} = A(p_j)\cdot\begin{cases}+1 & \text{for symbol vote } V\\ 0 & \text{for symbol vote } e\\ -1 & \text{for symbol vote } E\end{cases} \tag{4}$$

where,

$$A(p_j)\cong 2\cdot L^{min}(u_j)+\log\frac{(n-1)^2}{3}$$

and,

$$S(u_j) = \mathrm{sgn}(L(u_j)) \ .$$

Hence, the decoder may be implemented as follows. Firstly, the two minimum absolute LLRs are found at the decoder's inputs. Next, the HDD is applied to the hard decision-limited inputs and the vote labels V, e and E assigned in the same manner as for WBF. Finally, equation (4) is employed to map the minimum LLR inputs to extrinsic values. For the symbol position with smallest LLR input, the second smallest input is employed in $L^{min}(u_j)$ whereas for all other positions the smallest input is employed in $L^{min}(u_j)$. We note that the relationship between $L^{min}(u_j)$ and $L^e(u_j)$ is linear.

4   Simulation Results

The decoder was employed in simulations of a (64,57,4) extended Hamming product code on the AWGN channel. An iterative APP decoding strategy rather than a serial turbo decoding strategy was used. In this case, row and column extrinsic are determined in parallel and combined to form a single a priori input to the next iteration [2]. Extrinsic LLRs are attenuated by both scaling and clipping such that decoded BERs may be optimised. The results of Fig. 2 show that for iteration 16 performance is close to iteration 2 of turbo decoding employing a more complex Chase SISO decoder. (E_b/N_0 is the ratio of energy per information bit-to-noise spectral density and P_b is the decoded BER). A 'fairer' comparison is with the Chase decoder when employed in an identical APP strategy and in this case performance is very close. Gains over the purely hard decision WBF are around 1.3 dB.

The same decoder has been successfully applied to (64,57,4) extended Hamming-based GLDPCs with rate $R = 50/64$ and various block lengths. In Fig. 3 we can see that for each GLDPC, 32 iterations of the proposed method shows a loss of $\cong 0.85$ to 1.00 dB over 16 iterations of turbo decoding. Average gains of the APP strategy using the HDD-SISO decoder over hard decision WBF are $\cong 1.35$ dB. For the long block length GLDPCs, we highlight the steepness of the BER curves and that good performance is possible within approximately 1.80 dB of the appropriate Shannon limit. For lower rate coding schemes based on (15,11,3) Hamming sub-codes the decoder is found to be much less successful.

5   Conclusions

Using analysis of weighted bit flipping decoding of Hamming-based codes on the BSC, we have developed a simple SISO decoder for Hamming codes. The operation of the HDD-SISO decoder can be described in simple terms: the corrective action of the hard decision decoder, as represented by the extrinsic LLRs, is simply weighted to mirror the average reliability of the decoder's inputs using $L^{min}(u_j)$. For PCs and GLDPCs based on high rate Hamming sub-codes, the decoder achieves good performance within $\cong 2.0$ dB of the Shannon limit. Importantly, the gradients of the

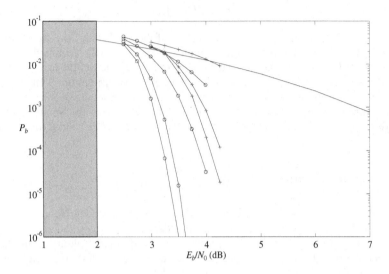

Fig. 2. Iterative Decoding of (64,57,4) Extended Hamming Product Code. Shows Turbo/Chase, Iterations 1, 2, 4 and 6 (*lines with* 'o's) and APP/HDD-SISO, Iterations 1, 8 and 16 (*lines with* '+'s). Also uncoded (*solid line*) and Shannon limit for rate $R = 0.793$ (*shaded area*)

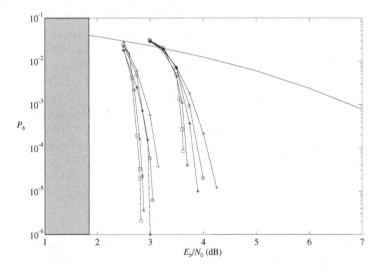

Fig. 3. Iterative Decoding of (64,57,4) Extended Hamming-based GLDPCs. Shows Turbo/Chase, Iteration 16 (*lines to left*) and APP/HDD-SISO, Iteration 32 (*lines to right*). Block lengths are 4096 ('+'s), 8192 ('o's), 16384 ('*'s), 32768 ('×'s), 65536 ('•'s) bits. Also uncoded (*solid line*) and Shannon limit for rate R = 0.781 (*shaded area*)

BER plots are found to be steep. The indications are that very low BERs may be achieved with computational cost per iteration little more than that of using a conventional HDD. To properly assess the usefulness of the new decoder, ongoing research is concentrated on performance bounds.

Acknowledgements. The authors would like to thank Tandberg Television UK for their support.

References

1. Berrou, C., and Glavieux, A.,"Near optimum error correcting coding and decoding: turbo-codes", *IEEE Trans. Commun.*, vol. 44, no. 10, pp. 1261-1271, Oct. 1996.
2. Hirst, S., and Honary, B., "Approaching cut-off rate on the binary symmetric channel using weighted bit flipping", Proc. of *6th Int. Symp. Commun. Theory & Applic.* (ISCTA '01), pp. 375-380, Ambleside, UK, July 2001.
3. Hirst, S., and Honary, B., "Decoding of generalised low-density parity-check codes using weighted bit flip voting", To appear in *IEE Proc. Commun..*
4. Kou, Y., Lin, S., and Fossorier, M. P. C., "Low density parity check codes based on finite geometries: a rediscovery and more", Submitted to *IEEE Trans. Inform. Theory..*
5. Lentmaier, M., and Zigangirov, K. Sh., "On generalised low-density parity-check codes based on Hamming component codes", *IEEE Commun. Letters*, vol. 3, no. 8, pp. 248-250, Aug. 1999.
6. MacKay, D. J. C., "Good error-correcting codes based on very sparse matrices", *IEEE Trans. Inform. Theory*, vol. 45, no. 2, pp. 399-431, Mar. 1999.
7. Sipser, M., and Spielman, D. A., "Expander codes", *IEEE Trans. Inform. Theory*, vol. 42, no. 6, pp. 1710-1722, Nov. 1996.
8. Tanner, R. M., "A recursive approach to low complexity codes", *IEEE Trans. Inform. Theory*, vol. IT-27, no. 5, pp. 533-547, Sept. 1981.

A Technique[1] with an Information-Theoretic Basis for Protecting Secret Data from Differential Power Attacks

Manfred von Willich

Box 411439, Craighall, Johannesburg 2024, South Africa
`manfred@cypherix.co.za`

Abstract. The classic "black-box" view of cryptographic devices such as smart cards has been invalidated by the advent of the technique of Differential Power Analysis (DPA) for observing intermediate variables during normal operation through side-channel observations. An information-theoretic approach leads to optimal DPA attacks and can provide an upper bound on the rate of information leakage, and thus provides a sound basis for evaluating countermeasures. This paper presents a novel technique of random affine mappings as a DPA countermeasure. The technique increases the number of intermediate variables that must be observed before gleaning any secret information and randomly varies these variables on every run. This is done without duplication of the processing of variables, allowing very efficient DPA resistant cipher implementations where the ciphers are designed to minimise overheads. A real-world system has been developed within the tight computational constraints of a smart card to exhibit first-order DPA-resistance for all key processing.

1   Introduction

Differential Power Analysis (DPA) is a major new threat to the secrecy of data processed in secure tokens such as smart cards. A thorough understanding of the attack and how to defend against it needs a formal approach, for which an information-theoretic approach provides a solid basis.

This paper illustrates this approach and presents a solution useful in this context primarily for cryptographic applications and certain types of processing, including some common operations including data storage.

1.1   What Is DPA?

The classic "black box" cryptographic security assumption on the processing of secret data in purpose-built device, such as a tamper-resistant smart card, has been invalidated with the advent of Differential Power Analysis (DPA). The technique of DPA exploits the fact that electronic hardware leaks information via side-channels as depicted in Fig. 1, from which secret data may be extracted with little effort. Some

---

[1]   Patent pending

B. Honary (Ed.): Cryptography and Coding 2001, LNCS 2260, pp. 44-62, 2001.
© Springer-Verlag Berlin Heidelberg 2001

other attacks exist that challenge the black box assumption, such as Differential Fault Analysis, but none are so effective or difficult to defend against as DPA.

In the example of analysing secret data being processed by a smart card, DPA generally seeks to extract the tiny data-dependent differences in observable waveforms (such as the current drawn by a device, emitted and conducted radiation), and to relate these to the internal values being processed.

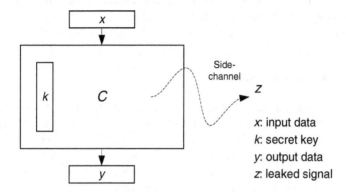

Fig. 1. A cryptographic operation with side-channel information

Cryptographic algorithms lend themselves to such analysis because their intermediate variables are usually not correlated to anything (they tend to display "ideal" randomness) except a correct hypothesis on the values in the calculation. It follows that valid hypotheses on internal values may be verified by finding such a correlation. This rather general approach (due to Kocher [2]) identifies a correlation with intermediate variables predicted by hypotheses on small numbers of bits of the cryptographic key. An attacker who has the specification of a cryptographic algorithm typically has a good chance of selecting an intermediate variable that has a physical analogue in an implementation. The attacker can locate a point in a sequence of waveforms of the current consumed by a smart card that exhibits a correlation with an intermediate variable deduced from a correct hypothesis on the key and certain known data (usually ciphertext). Often a few hundred known ciphertexts and associated side-channel observations are sufficient to determine a cryptographic key.

This approach is frequently illustrated in the context of the well-known DES algorithm. In the first round of processing, where the input data to the cipher is known (or the last round, where the output is known) for a sequence of differing data inputs but using the same key, each output bit of each lookup table (s-box) may be used for this purpose. Each bit depends on the changing input (output) and only six bits of the key. For each of the 64 possible combinations of these six bits (these being the hypotheses from which we wish to determine the correct one), the lookup table output bit may be determined. Using the bit value calculated from the data and the hypothesised key bits, we can classify a set of say a thousand observed current waveforms into two sets, and find the difference of the average of each of the resulting two sets of waveforms. We will then have 64 "differential" waveforms, which we can search for peaks, which result from the correlation mentioned and correspond to the point in time when the lookup table output bit of interest is being processed. Generally, one of the 64 differential waveforms will have significantly larger peaks than the others have. Repeating this for one output bit of each lookup table allows determination of the 48 bits of the key used during the first (last) round. Using the known values of these bits, the remaining 8 bits of the key may be found by extending the analysis to the next round. Many authors suggest that these bits may be found using a brute force attack, but this requires a known plaintext/ciphertext pair.

1.2   Information Leakage Rate

Although Kocher's style of attack is impressive in its effectiveness, it is not the most sensitive DPA attack possible, and hence is not suited to determining the effectiveness of a countermeasure. In particular, an approach of optimal detection of a signal in noise will extract the maximum information theoretically available from the observation, under the assumption that the attacker has full knowledge of the device characteristics excluding the secret information of interest.

An information-theoretic approach gives a sound basis for determining the maximum amount of information that can leak about a datum from a circuit using a realistic model of the way in which observable waveforms are determined by internal activities and noise. Suresh et al [4] suggests such an approach, which can be rigorously formalised. Without needing to put accurate figures to the leakage process, we can determine how certain strategies affect the upper bound on the information leaked about a cryptographic key or other data. Strategies that reduce this bound to acceptable levels (e.g. to a defined fraction of a secret key during the life of the key) can then be found.

Both the information-theoretic model and DPA analyses on standard hardware performed by us and others show that it is disconcertingly easy to extract data from a typical smart card using simple equipment and without operating the smart card in a fashion that it could detect as abnormal.

If we need to observe several intermediate variables to assemble any information about the original data and each of these variables has maximal entropy, the amount of observation needed to reduce the uncertainty in the original data increases with the power of the number of variables that must be observed. Thus if we need 100 noisy observations to obtain enough information on a physical intermediate variable, we would need 1 000 000 similar observations if we are forced to observe a minimum of three distinct intermediate variables to obtain the same information.

1.3   The Technique

The technique presented here comprises applying changing random affine mappings to the data being operated upon, thus making internal variables unpredictable ("mapped" or "obscured"). The operators themselves generally remain unchanged, but the content of lookup tables must be replaced to operate on the mapped data. The concept of random affine mappings is in itself not new. The well-known idea of a random affine constant (e.g. $x' = x + b$) for data hiding as defined in relation to an operator is extended here to include an invertible linear constant (e.g. $x' = ax + b$). This extension applies to essentially any operator to which the concept of "linearity" can be extended. Thus, finite field addition and multiplication both are clear candidates. The linearity in relation to a multiplicative operator generally results in an expression of the form $x' = bx^a$, and in relation to the exponentiation operator is as for multiplication for the base and appears to be restricted to multiplication for the exponent. (For RSA exponentiation, it may be more effective to express the exponent as an arbitrary sum of two shares due to the need to avoid use of the applicable modulus, $\varphi(n)$). The mapping is chosen such that the desired result can still be extracted knowing the mappings that apply to the input data. The exclusive-or operator appears to allow more entropy in the selection of the mapping for a given size of operand than any other operator does – 70.2 bits of random freedom when mapping an octet (mostly due to a random linear matrix A in the expression $x' = Ax + c$). In contrast, modulo-256 addition allows only 15 bits of freedom (the constant a being restricted to odd values). There is merit in having large entropy of mapping selection when many variables are to be hidden using the same or a closely related mapping (as may be useful, e.g. to avoid re-mapping of data).

Upon each use (at times chosen by design), the random mappings on the data are replaced, removing any fixed target data to observe, but in a way that retains the relations in the mapped data that allow reconstruction of the underlying data. Mappings are applied and removed at the input and output only – e.g. to incoming ciphertext to be deciphered and to outgoing ciphertext to be transmitted, and internal plaintext that need not remain secret but must be processed. Cryptographic keys are at no point represented in their original form in the field (even internally and when downloaded), and only a difference mapping is applied to the key data.

This technique substantially reduces the effectiveness of DPA. With care in the choice of the computational primitives, the added computational and storage costs can be remarkably low. The technique has been demonstrated to be effective, practical and economic.

As is to be expected, the technique is not effective for hardware that is shielded so poorly that intermediate variables may be determined from a single waveform.

Data most readily protected using this technique are cryptographic keys and data for storage. One significant advantage of the approach is that there is no system-level impact (such is the introduction of message sequence numbering).

2   Discussion of Information-Theoretic Analysis

This section illustrates the information-theoretic approach that can be used to evaluate the effectiveness of a defence against any DPA attack based on a mathematical model of signal leakage.

2.1   Defining Information Leakage

We wish to define an accurate measure of information leakage for purposes of determining the vulnerability of any device to a DPA attack. Measurements made on real devices can allow quantification of the amount of information potentially leaked by those devices, and in particular, the gains provided by a countermeasure can be determined.

For the purposes of this topic, the information leaked about a key by an observation will be defined in the information-theoretic sense. We define it as the reduction in entropy of the key when given the observation. The entropy of a variable is a measure of how much is not yet known about it – it is the amount of information required to fully specify it from what is already known. This definition ignores any complexity barriers that the attacker may face, and it is appropriate when trying to establish an upper limit to the information obtainable from side-channels from the device.

The entropy H of a key (in bits) may be expressed as follows. The expression uses the probability P_i that the attacker would assign to the ith possible value of the key with all the knowledge at his disposal including his prior knowledge and all observations. This is a standard and well-established definition of entropy:

$$H = -\sum_i P_i \log_2 P_i \tag{1}$$

2.2   A Formal Model of Leakage

A formal model of information leakage requires an accurate description of the relationship of the internal digital state of the device and the possible observed waveforms. For each of the possible internal states (by which is meant a distinct possible sequence of all internal variables), there will be an associated ensemble of waveforms due to added noise. An ensemble is essentially a set with an associated probability or probability density for each element of the set. When the waveform is in the form of a sequence of n samples, The ensemble may be mathematically described as joint probability density function in an n-dimensional state space, with each sample determining one of the n coordinates. There is one such joint probability density function associated with each possible internal state of the device.

A waveform observed by the attacker determines a set of coordinates in the state space, from which a probability density value may be determined for each possible internal state. The probabilities of each of the possible internal states given the observation are the same, normalised to sum to unity. These probabilities will then be modified by subsequent observed waveforms (by multiplying with the density values and scaling), producing an overall probability of each possible internal state. Summing the probabilities of all the internal states associated with a particular value of a key gives the probability of that value of the key. An attacker need only observe as many waveforms as is necessary to make the probability that the key takes on one value approach certainty.

In evaluating the vulnerability of a device to DPA, we must determine the average reduction in the entropy in such an attack. For this, we must find the ensemble average (i.e. weighted by the probability of each possible set of waveforms) of the information leaked (the reduction in entropy) over all the possible observed waveforms.

It must be stressed here that since this is a general measure of information leakage, this model assesses the vulnerability of a device to the *optimum* DPA attack (i.e. the *order* of the attack is immaterial).

Since these states are distinguished by a large amount of data unknown to the attacker (such as key values and internally generated random data) and each joint probability function has enormous dimensionality, this description as it stands would appear to be intractable. Yet it serves as a rigorous starting point. From this the necessary understanding may be derived, and realistic simplifying assumptions may be added to make the analysis more tractable.

2.3   A Realistic Simplified Model to Aid a DPA Attack

For the purposes of analysis, we can make a number of realistic simplifying assumptions. Foremost amongst these is the assumption that the noise is wide-band (spectrally white), has a normal distribution and is additive. Even when the noise does not have a flat spectrum (or equivalently that the noise contribution at different points in a waveform are not correlated), this may generally be arranged by means of a suitable filter (the noise probability density function at each point will remain a normal distribution). A further simplifying assumption is that the noise is stationary (of constant amplitude), and this too can be arranged by pre-processing if necessary.

Under these assumptions, standard matched filtering may be used to correctly determine the probability of each possible internal state. In this method, the expected (ensemble average) waveform for each internal state is multiplied with the observed waveform and integrated. Finding the mean square error is equivalent. The resultant set of scalar values (one value for each internal state) can be used to update the probability associated with all the internal states.

This approach remains intractable when considering all possible key values when the number of bits in the key is in a normal range (e.g. 56 bits for DES). However, this approach does suggest a viable attack on small portions of the key as they become incorporated into the processing. In the interests of simplifying an attack, the waveforms may be processed in small portions, at each step refining the probability estimates of initially only a few bits of the key. Only the most probable values of the bits in question need to be considered, since the option of backtracking exists when it appears that incorrect choices have been made. This approach is very similar to a standard signal decoding technique.

It may be noted at this point that unknown random data may influence the processing sequence of the processor, and is part of the internal state. These data would have to be determined as well as the key bits. Yet this fits neatly into the process of progressively refining the estimation of the internal state of the device. Where there is some ambiguity in this data, it will be reflected in several internal states having moderate probability. This type of data however typically has the most visible effect on a waveform, and it is difficult to disguise it in one waveform.

At the cost of sensitivity of the technique, the contribution of all digital processes not considered may be included in the effective noise. In a real attack, a device may be characterised while it is processing known data with known keys, which may have been determined by other means. This will provide a good estimate of the waveforms to be used in the matched filtering. This reduction in sensitivity is often tolerable, while the simplification is well worthwhile.

A point being made here is that the information-theoretic model suggests a DPA attack that is viable and has near-optimal sensitivity). The information-theoretic bound on the information leakage is therefore appropriate for evaluating DPA countermeasures.

2.4   Determining the Information Leakage

The attacker can accurately determine, in principle and in practice, the waveforms generated by the possible digital processes and the probability density function associated with the waveform samples. Unrelated digital processes in the device add considerably to the variation in the observed signal, but as this contribution is (at least in principle) predictable, it cannot be relied upon to add to the unpredictability of the observation.

It is assumed that the unpredictable component of the waveform (the additive noise) is large relative to the portion of the signal that allows differentiation between distinct waveforms. If this were not the case, the data being processed would be deducible from a single waveform. The applicable criterion here is the energy in the differential signal divided by the noise power spectral density rather than simple amplitudes.

Prior to any observations, we assume that the entropy of the key equals the number of bits in the key. The entropy of the key prior to an observation is calculated from the *a-priori* probabilities of each possible value of the key. The probabilities, as modified by the observation, are the *a-postiori* probabilities of the key for the observation. We treat the *a-postiori* probabilities for one observation as the *a-priori* probabilities of the next observation. Provided the observations are not correlated other than through the digital process being modelled, the *a-postiori* probabilities are the *a-priori* probabilities multiplied by the corresponding probability densities of the observation after normalisation.

The information leakage due to a set of observations (or equivalently the reduction in the entropy) is the ensemble-average (weighted by the probability density of the waveforms given the *a-priori* probabilities of the values of the key) over all waveforms of the reduction in entropy. The magnitude of this leakage can be used to rate the effectiveness of a strategy as a DPA countermeasure.

2.5   Information Leakage: An Example

Two simple information leakage scenarios of DPA leakage are presented in Table 1.

Table 1. An example of reducing information leakage

A: DPA on one bit	B: DPA on data hiding of one bit
Secret data $x = x_1 = x_2$ Noise amplitude $\sigma_{noise} = 1$ Signal amplitudes $c_1 = c_2 = 1$ A-*priori* probabilities $P(x=0) = P(x=1) = 0.5$. Entropy $H(x) = \log_2 0.5 = 1$ bit. Observed samples y_1 and y_2.	Secret data $x = x_1 \oplus x_2$ Noise amplitude $\sigma_{noise} = 1$ Signal amplitudes $c_1 = c_2 = 1$ A-*priori* probabilities $P(x=0) = P(x=1) = 0.5$ Entropy $H(x) = \log_2 0.5 = 1$ bit. Observed samples y_1 and y_2.
The explicit form of the distributions is: $$p(y_1, y_2 \mid x = 0)$$ $$= \frac{1}{\sqrt{2\pi\sigma^2}} e^{-\frac{(y_1-c_1)^2+(y_2-c_2)^2}{2\sigma^2}}$$ $$p(y_1, y_2 \mid x = 1)$$ $$= \frac{1}{\sqrt{2\pi\sigma^2}} e^{-\frac{(y_1+c_1)^2+(y_2+c_2)^2}{2\sigma^2}}$$	The explicit form of the distributions is: $$p(y_1, y_2 \mid x_1 = 0, x_2 = 0)$$ $$= \frac{1}{2\pi\sigma^2} e^{-\frac{(y_1-c_1)^2+(y_2-c_2)^2}{2\sigma^2}}$$ $$p(y_1, y_2 \mid x_1 = 0, x_2 = 1)$$ $$= \frac{1}{2\pi\sigma^2} e^{-\frac{(y_1-c_1)^2+(y_2+c_2)^2}{2\sigma^2}}$$ $$p(y_1, y_2 \mid x_1 = 1, x_2 = 0)$$ $$= \frac{1}{2\pi\sigma^2} e^{-\frac{(y_1+c_1)^2+(y_2-c_2)^2}{2\sigma^2}}$$ $$p(y_1, y_2 \mid x_1 = 1, x_2 = 1)$$ $$= \frac{1}{2\pi\sigma^2} e^{-\frac{(y_1+c_1)^2+(y_2+c_2)^2}{2\sigma^2}}$$
Average contribution by (y_1, y_2) to leakage $I(x;y_1,y_2)p(y_1,y_2)$	*Average* contribution by (y_1, y_2) to leakage $I(x;y_1,y_2)p(y_1,y_2)$
$H(x) = 1$ bit $R(x) = 0.75$ bits	$H(x) = 1$ bit $R(x) = 0.25$ bits

With reference to these scenarios, equation $I(x; y_1, y_2)$ expresses the reduction in the entropy of x obtained by observing y_1 and y_2.

$$I(x; y_1, y_2) = H(x) - H(x \mid y_1, y_2) \qquad (2)$$

Let $R(x)$ denote the mean reduction in the entropy of x. In order to determine $R(x)$, the above function must be multiplied by[2] $p(y_1, y_2)$ and the result must be integrated over all (y_1, y_2) – obviously using the definite integral, as shown here.

$$R(x) = \iint p(y_1, y_2) I(x; y_1, y_2) dy_1 dy_2 \qquad (3)$$

In attempting to combat DPA attacks, the objective would clearly be to *minimise* $R(x)$. In the case of the two scenarios, it is evident that Scenario **B** (which uses data hiding) leaks relatively little information (0.25 bits) after two observations. In contrast, Scenario **A** leaks 0.75 bits of information, with the same number of samples (we could have used one sample in Scenario **A**, which would have led to a value closer to 0.5 bits). It can be shown that for low leakage, the leakage rate is proportional to SNR^k for a k-share exclusive-or mechanism, where SNR is the ratio of the signal power to noise power in the samples of the example. As the SNR is reduced, the information leakage rate of Scenario **A** reduces roughly linearly, whereas that of Scenario **B** reduces roughly quadratically and is hence substantially less at low SNR.

3   Simple Defences

A number of defenses have been proposed to enhance DPA-resistance. These include measures such as making the device draw a constant current, using shielding, randomly varying the processing time of any particular intermediate variable, adding random noise to the observable waveforms, and reducing amplitude the signal generated by the hardware by careful design.

These defences can have no greater effect that to reduce the effective signal-to-noise ratio (SNR, defined in relation to power or energy) of the observed waveforms. Even combined together in a device such as a smart card, they do no more than make the attacker's task more difficult (each multiplying the number of waveforms the attacker needs by a small number), and should not be considered as a final solution. Computational defences are needed for effective defense against DPA, as suggested in [5].

Such defences (those that effectively reduce the signal-to-noise ratio) have value when combined with computational techniques. The computational techniques effectively put the signal-to-noise ratio to a small power, e.g. squaring it, along with making the attack more complex (e.g. by making any first-order attack such as

---

[2] Since the joint observations (y_1, y_2) occur with varying degrees of probability, a joint probability is required.

Kocher's general attack non-viable). A first-order DPA attack has this distinguishing characteristic – any number of waveforms may be averaged in each of the groups into which they are divided, and its effectiveness climbs in relation to the effective improvement in signal-to-noise ratio produced by this averaging.

4   The Proposed Technique

The technique proposed here provides a practical and effective modification of cryptographic processes, based on data encryption through frequent varying of the mapping of all data being processed, onto an obscured form for computation and storage. Examples of such data are cryptographic keys, stored and communicated data.

Both the mapped (obscured) data and the chosen mapping must be partially known before any information about the secret data becomes accessible.

Secret data, such as cryptographic keys, are never needed in the non-obscured form, and should be randomly re-mapped on each use to avoid data repetition that would facilitate a DPA-attack. An example of when this technique will have high value is in smart cards, where DPA can provide an unauthorised party with a cryptographic key in use within minutes, entirely though analysis of the leaked signals.

The method includes the following steps:

- Design of algorithms, particularly ciphers, for maximum benefit from this technique;
- Modifying the algorithm implementation to operate on mapped data;
- Initial mapping of data (especially cryptographic keys) for storage;
- Frequently changing the data mapping from the prior data mapping by use of a secondary (or delta) mapping;
- Mapping incoming data for input to the modified algorithm implementation; and
- Mapping of data output from the modified algorithm for further use.

4.1   Cipher Design

Care must be exercised in choice of a cipher algorithm. Suitable cipher design can result in the next step (cipher modification) adding very little processing overhead. Choosing the set of operations that are used in the cipher is important for minimising complexity and maximising data hiding. Understanding of the following aspects of the technique is essential during the design.

4.2   Cipher Modification

A different mapping may be used for every operation throughout the cipher, thus necessitating a change of the mapping on every data path. Alternatively, the mapping may be left unchanged between two operations. The latter is typically not possible when the two operations are unrelated, but when possible is useful in keeping the degree of complexity low.

Every operation is substituted with one that performs the equivalent operation with all values being mapped, as illustrated in Fig. 2. The output mapping (f_c) is determined by the input mappings $(f_a$ and $f_b)$ and any changes to the core operation. For example, adding a random value to each of the inputs of an addition is reversed on the output by subtracting the sum of the random values from the output.

The non-obscured values $(a, b$ and $c)$ still occur in Fig. 2, and at this point are only obscured during each individual operation, but are removed in the next step. The operation performed on the mapped values will normally be the same operation as before (e.g. addition), but may be different when the operator does not have the necessary properties (e.g. an arbitrary lookup table).

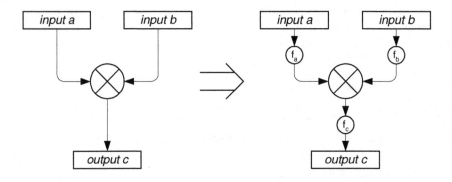

Fig. 2. Replacement of a two-input operation with a data-hiding equivalent

The next step is to combine consecutive mappings from adjacent operations into a single mapping that does not, even as an intermediate calculation value, derive the non-obscured data. Occurrence of the non-obscured data would provide a primary target for a DPA attack. Where the consecutive mappings are closely related the composite mapping may be somewhat simpler or even become the identity operation (and hence can be omitted, as is preferable). This is illustrated in Fig. 3.

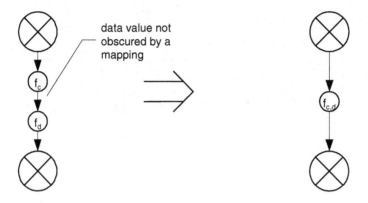

Fig. 3. Combining of consecutive mappings

If necessary, the combined mapping $(f_{c,d})$ is implemented using a look-up table. If cascaded lookup tables occur in this process, they may be combined into one lookup table. After this step, aside from the input data, key-data and output data, the data in all computations are obscured by the mappings. These external mappings are treated separately in subsequent steps.

Special care must be taken with lookup tables. Where there may be a non-linear interaction (in the analogue sense) between the input and output, certain "parasitic" intermediate variables may be produced. An example is in a microprocessor that has a multiplexed bus such as the 8051 derivatives, where the low-order eight address bits on the single bus are replaced by the looked-up value. A transition on a bit of the bus at this point results in a pulse of charging current that is the exclusive-or combination of the two values. For this reason, the input and output of the lookup table should at minimum be masked with unrelated affine constants. A second point to note is that a lookup table is used many times before it is re-mapped, there may be a second-order DPA attack based on the structure of the table itself leaking information about the mappings used. This will not normally be significant compared to the information leaked about the mappings in the input and output mappings, but where these mappings are cascaded to provide protection against a higher-order DPA attack, this leakage must be considered.

With careful choice of cipher design and restrictions on the chosen mappings, the resulting complexity need not be much greater than that of the original cipher. Additionally, computation relating to the mapping used in each computation may be kept to minimum. The resulting mathematically equivalent cipher is shown in Fig. 4.

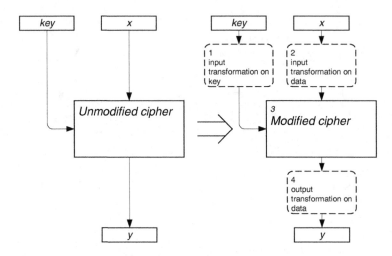

Fig. 4. Replacement of a cipher by its modified equivalent

4.3   Initial Storage of Keys

In Fig. 4, the key, input data and output data are still shown in non-hidden form, and may still be target of a DPA attack when they are read and written. The cryptographic key must be stored in an obscured form where the choice of mapping includes randomness. Additionally, the information encoding the choice of mapping must be stored. This initial step is only needed once with initial or master keys downloaded (typically in a protected environment), and never for keys downloaded using encrypted messages.[3] This may be expressed as storing the key k in its secret form $k_0 = f_0(k)$, as well as information identifying the choice of mapping, f_0. This mapping will most commonly be chosen in relation to the operators used in the cipher in which the key is used to avoid unnecessary re-mapping. Refer to Fig. 5.

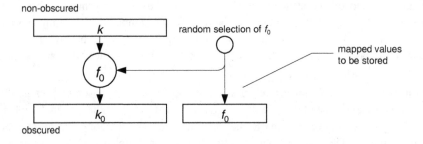

Fig. 5. Initial mapping of the key for storage

---

[3] Refer to the Section 4.6: *Cipher output data mapping.*

4.4   Per-Use Key Mapping

Even stored in obscured form as described in Section 4.3, repeated retrieval would allow both the secret data and the mapping information to be reconstructed through DPA techniques. Consequently, prior to using a cryptographic key, the mapping must be replaced with a fresh, randomly chosen mapping subject to the constraints imposed by the cipher – preferably every time the key is to be used. It is important that the original value of the key should not be computed during this process, even as a temporary variable. This leads to a derivation of values in the form $k_i = g_i(k_{i-1})$ and $f_i = g_i \circ f_{i-1}$. The latter is equivalent to saying $f_i(q) = g_i(f_{i-1}(q))$ for any q. The values k_i and f_i will replace the stored values k_{i-1} and f_{i-1}. They remain related by the identity $k_i = f_i(k)$. This process is illustrated in Fig. 6.

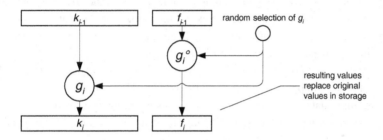

Fig. 6. Iterative mapping of a key

4.5   Cipher Input Data Mapping

The input data (x in Fig. 4) is first mapped using the mapping chosen for those inputs. This is analogous to the initial mapping of the key (as in Section 4.3), but may occur with all data to be processed, such as received ciphertext to be decrypted, or plaintext to be encrypted for transmission. Where sensitive information (e.g. keys) is to be encrypted, it must already be in secret form and a mapping substitution should be performed where appropriate (as in Section 4.4).

4.6   Cipher Output Data Mapping

The output may be mapped to its non-obscured value where it is not critical to hide its value (e.g., where ciphertext to be transmitted has been generated). Where this data must remain hidden (e.g. secret data, especially received cryptographic keys being downloaded, etc.), it should be stored without mapping it back to the non-obscured form, along with the mapping information. This mapping information must correspond to the required form for use with a key. Thus, the initial mapping of the key mentioned in Section 4.3 does not occur explicitly with received and decrypted keys. This makes the process of downloading keys resistant to DPA.

4.7   Higher-Order DPA Defences

Where the mapping data is used only externally to the cipher (as can be the case with careful design), multiple independent mappings may be applied in cascade (one after the other). This diminishes the usefulness of observations of the random mapping data used. Thus, an attack exploiting the mapping process may easily be defended against to ant desired degree.

This does not ensure that the aggregate mapping data cannot be obtained from the ciphering operation. Such information may be obtained from the structure of mapped lookup tables, for example. Cascading lookup tables does not alleviate this problem, as only the input and the output of the cascade are needed to determine the structure of the cascade and hence the cascaded mapping.

Care must be taken to avoid low-order information leakage, for example where multiple vectors are mapped with the same mappings. This is most significant where key data and input data share the same mapping – something best avoided. A simple example is finding a correlation between mapped input data values and mapped key data processed in the same data paths in a second-order attack.

5   A Practical Design

A proprietary real-world cryptographic cipher was designed within tight computational constraints to exhibit first-order DPA-resistance using the insights and techniques presented here. There was no computational overhead due to the affine mapping in the cipher itself, while mapping of input and output data added a small percentage to the overall processing time. A non-prohibitive fixed preparation overhead was also needed for random number generation and pre-calculation. Despite the restrictive design criteria, the resultant cipher has survived classic linear and differential cryptanalysis with flying colours.

As the most promising operator, an 8-bit exclusive-or was combined with 8-bit lookup tables for a practical cipher design. Input keys are already randomly pre-mapped on a per-octet basis for storage, whereas incoming ciphertext and plaintext are mapped prior to processing in the cipher proper. Lookup tables must be recalculated according to the applicable mappings applied. Every octet x is replaced in the real computation by its mapped equivalent $x' = Ax + b$. The 8-by-8 matrix A is a randomly chosen non-singular matrix used throughout the decryption or encryption of one ciphertext, and b is one of many random 8-bit constants chosen independently where practical for mapping different data being processed in the cipher. The matrix *A* is attractive in that it introduces 60.2 bits of randomness into the mapping, which allows each *A* to be used in more than one mapping.

Economy of design was crucial, and the design was thus constrained so that only one lookup table was to be used and the random mapping data was not referenced at all during the execution of the cipher. The result was a cryptographically strong cipher that executed at the same speed whether random mapping data was to be used or not, excluding the pre-and post-calculation relating to the random mappings.

The matrix multiplication of the calculation $x' = Ax + b$ may be implemented as a matrix multiplication or as a lookup table. Similarly, determination of the inverse matrix may be done by matrix inversion or by generating an inverting lookup table. As the amount of data to be mapped is small, we found both matrix multiplication and inversion to be more economic in our implementation.

A DPA attack was mounted against the design, with full knowledge of the implementation and exactly when every datum is being processed. The attack is based on averaged waveforms, and is hence a true first-order DPA attack. This attack is successful against the implementation when the mapping randomness is disabled with fewer than 10 observed waveforms used to produce each averaged waveform. The same attack is unsuccessful when the mapping randomness is enabled with 10^4 observed waveforms used to produce each averaged waveform.

6   An Example: Making an XOR-Based Cipher DPA-Resistant

In this example, a simplistic cipher is used constructed entirely from modulo-2 addition of octets (vectors of eight bits each) and a single lookup table that produces an 8-bit output value for each 8-bit input value. A single mapping is used to obscure every data octet in this example, of the form $d_{n,i} = A_i d_n + b_i$.

d_n is a typical octet of the data set $d = (d_0, d_1, \ldots)$, which may be the key, input and output of the cipher; A_i is a randomly selected non-singular 8-by-8 matrix of bits; and b_i is a randomly selected octet. The subscript *n* indicates a selected octet of the data set.

In Fig. 7, these operations have been combined to illustrate the example. A typical cryptographic cipher (encryption or decryption) would use many more operations and the data sizes of *k*, *x* and *y* would generally each be at least 64 bits. Each arrow represents the flow of one octet. The diagram shows equivalent operations with mapping of the data, but does not show the incremental mapping of the key (described in Section 4.4).

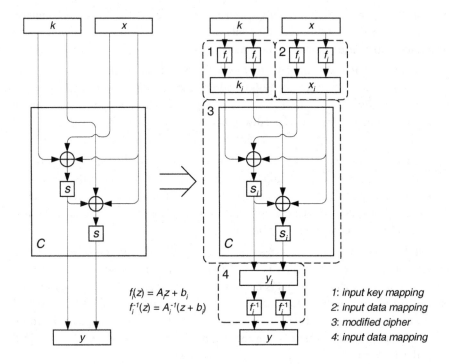

Fig. 7. Simplistic cipher illustrating the mapping process (excluding per-use key mapping)

The initially mapped key $k_{n,0} = A_0 k_n + b_0$ and the mapping $f_0 = (A_0, b_0)$ are stored. Prior to any use of the key, a fresh mapping is performed by choosing new G_i and h_i. Following this, $k_{n,i-1}$ is replaced by $k_{n,i} = G_i k_{n,i-1} + h_i$, A_{i-1} by $A_i = G_i A_{i-1}$ and b_{i-1} by $b_i = G_i b_{i-1} + h_i$. This has been omitted from Fig. 7 for simplicity.

Every lookup-table, s, is replaced by its equivalent s_i for operation on mapped values, defined by $s_i(z) = A_i s(A_i^{-1}(z + b_i)) + b_i$. (It is not necessary to generate the inverse of the matrix when generating s_i.) The input data octets, x_n, are then mapped using the same mapping, $x_{n,i} = A_i x_n + b_i$. The input is ciphered using the original cipher, except for the substituted lookup table. Aside from the per-key mapping, the substituted lookup table, the initial mapping and final mapping, there is no change to the computation involved in the cipher.

Finally, where the output, y, is to remain secret, such as with the value of a key, y_i, A_i and b_i are used instead of y. If it is to be mapped into its non-obscured state, this may be expressed as $y_n = A_i^{-1} y_{n,i} + b_i$.

A pivotal observation to be made is that due to the large number $(2^{70.2})$ of possible mappings, the same mapping can be used for hiding more than one octet of data effectively. This allows the modified cipher to remain simple. A simpler mapping may not hide multiple bytes adequately against DPA.

Since the mapping (A_i, b_i) and the mapped data d_i are changed on every use, the processed data (including the key) is not correlated with the original (non-obscured) data. Only a moderately complex function of several bits of data and the mapping is correlated to the original data. Each bit of the original data can be expressed as a function of the 17 bits being processed, making attacks complex. The matrix A does not in itself increase the order of a DPA-attack – it merely makes it more complex and less sensitive. This may be seen from the fact that there is a correlation between the hamming weight before and after multiplication by A, and that the zero vector in particular remains unchanged. On its own, it does not rule out a first-order DPA attack.

This example, except applied to the design of a cryptographically strong cipher, may be used effectively in smart cards available today, including those that use 8-bit processors and modest quantities of storage space.

7   Conclusions

This is a practical defence against DPA known to the author that needs neither a system-level design change (such as key-use sequence numbering) nor a doubling of processing cost (such as a share-based mechanism), although it is best introduced in association with cipher redesign. A practical smart card implementation of a strong algorithm for a real system has withstood determined and sensitive searches for first-order DPA vulnerabilities.

The technique may be applied to existing ciphers (such as DES and IDEA). DES, due to the need to use mapping data extensively throughout the modified cipher, results in something similar to the "duplication method" of Louis Goublin *et al.* IDEA, due to the way in which it combines incompatible operators, needs insertion of several pre-calculated re-mapping tables, but would be practical and necessary in combating emitted-RF DPA attacks on PC-based ciphering. The structure of the AES winner, Rijndael, has a pleasing structure for application of this technique.

References

1. Paul C. Kocher: Timing Attacks on Implementations of Diffie-Hellman, RSA, DSS, and Other Systems, *Advances in Cryptology: Proceedings of CRYPTO '96*, Springer-Verlag, August 1996, 104-113.
2. Paul Kocher, Joshua Jaffe and Benjamin Jun: Differential Power Analysis, *Advances in Cryptology: Proceedings of CRYPTO '99*, Springer-Verlag, August 1999, 388-397.
3. Paul Kocher, Joshua Jaffe and Benjamin Jun: Introduction to Differential Power Analysis and Related Attacks, http://www.cryptography.com/dpa/technical/index.html, 1998.

4. Suresh Chari, Charanjit S. Jutla, Josyula R. Rao and Pankaj Rohatgi: Towards Sound Approaches to Counteract Power-Analysis Attacks, *Advances in Cryptology: Proceedings of CRYPTO '99*, Springer-Verlag, August 1999, 398-412.
5. Louis Goublin and Jacques Patarin: "DES and Differential Power Analysis – The Duplication Method", *Cryptographic Hardware and Embedded Systems International Workshop*, August 1999, 158-172.

Key Recovery Attacks on MACs Based on Properties of Cryptographic APIs

Karl Brincat[*1] and Chris J. Mitchell[2]

[1] Visa International EU, PO Box 253, London W8 5TE, UK
brincatk@visa.com
[2] Information Security Group, Royal Holloway, University of London,
Egham, Surrey TW20 0EX, UK
C.Mitchell@rhul.ac.uk

Abstract. This paper is concerned with the design of cryptographic APIs (Application Program Interfaces), and in particular with the part of such APIs concerned with computing Message Authentication Codes (MACs). In some cases it is necessary for the cryptographic API to offer the means to 'part-compute' a MAC, i.e. perform the MAC calculation for a portion of a data string. In such cases it is necessary for the API to input and output 'chaining variables'. As we show in this paper, such chaining variables need very careful handling lest they increase the possibility of MAC key compromise. In particular, chaining variables should always be output in encrypted form; moreover the encryption should operate so that re-occurrence of the same chaining variable will not be evident from the ciphertext.

Keywords: Message Authentication Code, cryptographic API, cryptanalysis

1 Introduction

MACs, i.e. *Message Authentication Codes*, are a widely used method for protecting the integrity and guaranteeing the origin of transmitted messages and stored files. To use a MAC scheme it is necessary for the sender and recipient of a message (or the creator and verifier of a stored file) to share a secret key K, chosen from some (large) keyspace. The data string to be protected, D say, is input to a MAC function f, along with the secret key K, and the output is the MAC. We write

$$\mathrm{MAC} = f_K(D).$$

The MAC is then sent or stored with the message, i.e. the string which is transmitted or stored is $D||f_K(D)$ where $x||y$ denotes the concatenation of data items x and y.

---

[*] The views expressed in this paper are personal to the author and not necessarily those of Visa International

B. Honary (Ed.): Cryptography and Coding 2001, LNCS 2260, pp. 63–72, 2001.
© Springer-Verlag Berlin Heidelberg 2001

In this paper we are concerned with a particular class of cryptographic APIs, namely those providing access to the functions of a cryptographic module. Such modules will typically provide a variety of cryptographic operations in conjunction with secure key storage, all within a physically secure sub-system. We moreover assume that the API is designed so that use may be made of the cryptographic functions without access being given to internally stored keys. We are concerned here with attacks which may, in certain circumstances, enable a user with temporary access to the cryptographic API to use its functions to discover an internally stored key.

2   APIs, MACs, and Chaining Variables

Most cryptographic modules have relatively limited amounts of internal memory. They also typically require the data they process to be passed as parameters in an API procedure call, since they typically will not have the means to directly access memory in their host system. As a result they can normally only compute a MAC on a data string of a certain maximum length. Thus provisions are often made in the cryptographic API for the module to compute a 'part MAC' on a portion of a data string. Computation of a MAC on the complete data string will then require several calls to the module.

Given that it is always desirable to minimise the stored state within the module, it is therefore necessary to provide the means to input/output 'partial MAC computation state information' to/from the module. This is normally achieved by providing for the input/output of 'Chaining variables' within the MAC processing procedure calls of the API. That is, it should be possible to input and output the value of H_i for a certain i, as defined in Section 3 below.

Based on this idea, we can identify four different types of MAC computation call to a cryptographic module:

- *Type A: ('all')* where the entire data string to be MACed is passed in a single call and a MAC is output (no chaining variables are input or output),
- *Type B: ('beginning')* where the first part of a data string is passed to the module (and a chaining variable is output but not input),
- *Type M: ('middle')* where the central part of a data string is passed to the module (and chaining variables are input and output), and
- *Type E: ('end')* where the last part of a data string is passed to the module (and a chaining variable is input but not output, and a MAC is output).

3   On the Computation of MACs

MACs are most commonly computed using a block cipher in a scheme known as a CBC-MAC (for *Cipher Block Chaining* MAC). There are a number of variants on the basic CBC-MAC idea, although the following general model (see [1,2]) covers most of these variants.

The computation of a MAC on a data string D, assumed to be a string of bits, using a block cipher with block length n, is performed with the following steps.

1. *Padding and splitting.* The data string D is subjected to a padding process, involving the addition of bits to D, the output of which (the *padded string*) is a bit string of length an integer multiple of n (say qn). The padded string is divided (or 'split') into a series of n-bit blocks, $D_1, D_2, \ldots, D_q$.
2. *Initial transformation.* An Initial transformation I, possibly key-controlled, is applied to D_1 to yield the first *chaining variable* H_1, i.e. $H_1 = I(D_1)$.
3. *Iteration.* Chaining variables are computed as $H_i = e_K(D_i \oplus H_{i-1})$ for i successively equal to $2, 3, \ldots, q$, where K is a block cipher key, and where $\oplus$ denotes bit-wise exclusive-or of n-bit blocks.
4. *Output transformation.* The n-bit *Output block* $G = g(H_q)$, where g is the output transformation (which may optionally be key-controlled).
5. *Truncation.* The MAC of m bits is set equal to the leftmost m bits of G.

The relevant international standard, namely ISO/IEC 9797-1, [1], contains six different CBC-MAC variants. Four are based on combinations of two Initial and three Output transformations. The various Initial and Output transformations have been introduced to avoid a series of attacks possible against the 'original' CBC-MAC (using Initial transformation 1 and Output transformation 1).

- Initial transformation 1 is: $I(D_1) = e_K(D_1)$ where K is the key used in the Iteration step, i.e. it is the same as the Iteration step.
- Initial transformation 2 is: $I(D_1) = e_{K''}(e_K(D_1))$ where K is the key used in the Iteration step, and $K'' \neq K$.
- Output transformation 1 is: $g(H_q) = H_q$, i.e. the identity transformation.
- Output transformation 2 is: $g(H_q) = e_{K'}(H_q)$, where $K' \neq K$.
- Output transformation 3 is: $g(H_q) = e_K(d_{K'}((H_q))$, where $K' \neq K$.

These options are combined in the ways described in Table 1 to yield four of the six different CBC-MAC schemes defined in ISO/IEC 9797-1, [1].

Table 1. CBC-MAC schemes defined in ISO/IEC 9797-1

Algorithm number	Input transformation	Output transformation	Notes
1	1	1	The 'original' CBC-MAC scheme, [3].
2	1	2	K' may be derived from K.
3	1	3	CBC-MAC-Y, [4]. The values of K and K' shall be chosen independently.
4	2	2	K'' shall be derived from K' in such a way that $K' \neq K''$.

Finally note that three Padding Methods are defined in [1]. Padding Method 1 simply involves adding between 0 and $n - 1$ zeros, as necessary, to the end of the data string. Padding Method 2 involves the addition of a single 1 bit at the end of the string followed by between 0 and $n-1$ zeros. Padding Method 3 involves prefixing the string with an n-bit block encoding the bit length of the string, with the end of the string padded as in Method 1. When using a MAC algorithm it is necessary to choose one of the padding methods and the degree of truncation.

4   Attacks on CBC-MACs

There are two main types of attack on MAC schemes. In a *MAC forgery* attack, an unauthorised party is able to obtain a valid MAC on a message which has not been produced by the holders of the secret key. Typically the attacker will use a number of valid MACs and corresponding messages to obtain the forgery. A *key recovery* attack enables the attacker to obtain the secret MAC key. The attacker will typically need a number of MACs to perform such an attack, and may require considerable amounts of off-line computation. Note that a successful key recovery attack enables the construction of arbitrary numbers of MAC forgeries. In this paper we are exclusively concerned with key recovery attacks.

All attacks require certain resources (e.g. one or more MACs for known data strings). Clearly the less resources that are required for an attack, the more effective it is. As a result we introduce a simple way of quantifying an attack's effectiveness.

Following the approach used in [1], we do this by means of a four-tuple which specifies the size of the resources needed to the attacker. For each attack we specify the tuple $[a, b, c, d]$ where a is the number of off-line block cipher encipherments (or decipherments), b is the number of known data string/MAC pairs, c is the number of chosen data string/MAC pairs, and d is the number of on-line MAC verifications. The reason to distinguish between c and d is that, in some environments, it may be easier for the attacker to obtain MAC verifications (i.e. to submit a data string/MAC pair and receive an answer indicating whether or not the MAC is valid) than to obtain the genuine MAC value for a chosen message.

5   Key Recovery Attacks

In order to understand why the large number of CBC-MAC variants exist, we need to discuss some elementary key recovery attacks.

MAC algorithm 1 can be attacked given knowledge of one known message/MAC pair (assuming that $m > k$). The attacker simply recomputes the MAC on the message with every possible key, until the key is found giving the correct MAC. This attack has complexity $[2^k, 1, 0, 0]$, which is feasible if k is sufficiently small. E.g., if the block cipher is DES then $k = 56$, and it has been shown, [5], that a machine can be built for a few hundred thousand dollars which will search through all possible keys in a few days.

MAC algorithm 2 is subject to a similar attack. However, MAC algorithm 3 (CBC-MAC-Y) is not subject to this attack, which is one reason that it has been adopted for standardisation. In fact the best known key recovery attack on this MAC algorithm has complexity $[2^{k+1}, 2^{n/2}, 0, 0]$, as described in [6]. An alternative key recovery attack, requiring only one known MAC/data string pair, but a larger number of verifications, is presented in [7]; this attack has complexity $[2^k, 1, 0, 2^k]$.

Whilst there are many situations where MAC Algorithm 3 is adequate, in some cases the above-referenced key recovery attacks pose a threat to the secrecy of the MAC key. One example of a scenario where this might be true is where users of a service (e.g. banking or mobile telephony) are issued with tamper-resistant devices containing unique MAC keys. It might be possible for an individual responsible for shipping these devices to consumers to interrogate devices for a considerable period of time before passing them to their intended user. This interrogation might involve generating and/or verifying large numbers of MACs. If such an interrogation could be used to obtain the secret key, then the security of the system might be seriously compromised. This is precisely the case for GSM implementations using the COMP128 algorithm, where, as described in recent postings on the web, [8], access to a SIM (Subscriber Identity Module) by a retailer could enable the authentication key to be discovered prior to a SIM being issued to a customer.

This motivates the development of MAC algorithm 4 (first described in [7]), which was designed to offer improved security relative to MAC algorithm 3 at the same computational cost. Unfortunately, as shown in [9], MAC algorithm 4 only offers a significant gain in security with respect to key recovery attacks if Padding Method 3 is used.

6 Chaining Variable Protection

6.1 The Need for Encryption

Suppose the chaining variable H_i is passed to and from the module in unencrypted form. Suppose also that an attacker has temporary access to the API of the cryptographic module, and wishes to recover the MAC key.

We first consider the case of MAC algorithm 3, where we suppose that an independent pair of keys (K, K') is used. A single use of the 'Type B' call to the MAC computation function will return a chaining variable which depends only on the first key K (the chaining variable is essentially a MAC as generated by MAC algorithm 1). Knowledge of this one chaining variable, and of the data string used to yield it, will be sufficient to recover K in an attack of complexity $[2^k, 1, 0, 0]$. Given an additional MAC/data string pair, the other key K' can be recovered with a similar 'brute force' search, and hence we have recovered the entire key at a total complexity of $[2^{k+1}, 2, 0, 0]$. That is, MAC algorithm 3 is now no more secure than MAC algorithms 1 and 2.

The same attack applies to MAC algorithm 4, and thus, for APIs supporting MAC algorithms 3 and/or 4, encryption of the chaining variable is essential.

6.2   Simple Encryption Is Not Enough

We next see that, for MAC algorithm 4, encryption is not always sufficient. As we noted above, MAC algorithm 4 is best used in conjunction with Padding Method 3. However, although use of this padding method prevents the attacks described in [9], it does not prevent attacks made possible by the availability of encrypted chaining variables. This is based on the assumption that if the same chaining variable is output twice, then the encrypted version of that chaining variable will be the same on the two occasions. This will enable us to find a 'collision', which can be used as part of a key recovery attack. We next describe such an attack.

The attacker first chooses two n-bit blocks: D_1 and D_1'. These will represent the first blocks of padded messages, and hence they will encode the bit length of the messages (since Padding Method 3 is being used). Typically one might choose D_1 to be an encoding of $4n$ and D_1' to be an encoding of $3n$, which will mean that D_1 will be the first block of a 5-block padded message and D_1' will be the first block of a 4-block padded message. For the purposes of the discussion here we suppose that D_1 encodes a bit-length resulting in a $(q+1)$-block padded message, and D_1' encodes a bit-length resulting in a q-block padded message $(q \geq 4)$.

The attacker now acquires $2^{n/2}$ 'Type B' MAC computation calls to the cryptographic module for a set of $2^{n/2}$ two-block 'part messages' of the form D_1, X, where X varies at random. The attacker also acquires a further $2^{n/2}$ 'Type B' calls for the $2^{n/2}$ two-block 'part messages' of the form D_1', Y, where Y varies at random. By routine probabilistic arguments (called the 'birthday attack', see [10]), there is a good chance that two of these 'part messages' will yield the same chaining variable. We are assuming that this will be apparent from the encrypted versions of the chaining variables.

The attacker now has a pair of n-bit block-pairs: (D_1, D_2) and (D_1', D_2') say, which should be thought of as the first pair of blocks of longer padded messages, with the property that the 'partial MACs' for these two pairs are equal, i.e. so that if

$$
\begin{aligned}
H_1 &= e_{K''}(e_K(D_1)), \\
H_2 &= e_K(D_2 \oplus H_1), \\
H_1' &= e_{K''}(e_K(D_1')), \quad \text{and} \\
H_2' &= e_K(D_2' \oplus H_1'),
\end{aligned}
$$

then $H_2 = H_2'$.

The remainder of the attack is a modified version of Attack 1 from [9]. The attacker next acquires $2^{n/2}$ 'Type A' calls to the cryptographic module for a set of $2^{n/2}$ $(q+1)$-block padded messages of the form

$$
D_1, D_2, X_1, X_2, \ldots, X_{q-1},
$$

where $X_1, X_2, \ldots, X_{q-1}$ can be arbitrary. As previously, there is a good chance that two of these messages will yield the same MAC. Suppose the two padded strings are $D_1, D_2, E_3, E_4, \ldots, E_{q+1}$ and $D_1, D_2, E'_3, E'_4, \ldots, E'_{q+1}$, and suppose that the common MAC is M.

Now submit the following two padded strings for MACing, namely:

$$D'_1, D'_2, E_3, E_4, \ldots, E_q$$

and

$$D'_1, D'_2, E'_3, E'_4, \ldots, E'_q.$$

If we suppose that the MACs obtained are M' and M'' respectively, then we know immediately that

$$d_{K'}(M') \oplus E_{q+1} = d_K(d_{K'}(M)) = d_{K'}(M'') \oplus E'_{q+1}.$$

Now run through all possibilities L for the unknown key K', and set $x(L) = d_L(M')$ and $y(L) = d_L(M'')$. For the correct guess $L = K'$ we will have $x(L) = d_{K'}(M')$ and $y(L) = d_{K'}(M'')$, and hence $E_{q+1} \oplus x(L) = E'_{q+1} \oplus y(L)$. This will hold for $L = K'$ and probably not for any other value of L, given that $k < n$ (if $k \geq n$ then either a second 'collision' or a larger brute force search will probably be required).

Having recovered K', we do an exhaustive search for K using the relation $d_{K'}(M') \oplus E_{q+1} = d_K(d_{K'}(M))$ (which requires 2^k block cipher encryptions). Finally we can recover K'' by exhaustive search on any known text/MAC pair, e.g. from the set of $2^{n/2}$, which again will require 2^k block cipher encryptions.

It follows that the above attack has complexity $[2^{k+2}, 0, 3.2^{n/2}, 0]$, which is not significantly greater than the complexity of the best known key recovery attacks against MAC algorithm 3.

Thus it is important that the encryption is performed in such a way that if the same chaining variable is output twice then this is not evident from the ciphertext. This can be achieved in a variety of ways. One possibility is to encrypt the chaining variable using a randomly generated session key, and to encrypt this session key with a long term key held internally to the cryptographic module. The encrypted session key can be output along with the encrypted chaining variable.

6.3   Message Length Protection

If Padding Method 3 is used, then it is clearly necessary to inform the cryptographic module of the total message length when the first 'Type B' call is made, so that the Padding Block can be constructed and used to perform the first part of the MAC calculation. We next show that, along with the chaining variable, it is important for the on-going MAC calculation to 'keep track' of this message length, and of the amount of data so far processed.

Suppose this is not the case. Then a simpler variant of the attack described in Section 6.2 above becomes possible. This operates as follows.

The attacker arranges for the part MAC to be computed on the padded two-block 'part message' (D_1, D_2) using a 'Type B' MAC computation call to obtain the resulting chaining variable H_2, which may or may not be an encrypted version of the chaining variable resulting from the MAC operation. The attacker then assembles approximately $2^{n/2}$ distinct message strings $X_1, X_2, \ldots, X_{q-1}$ ($q \geq 3$) and arranges for their MACs to be computed using as many 'Type M' and 'Type E' calls as necessary, and with H_2 as the initial chaining variable input to the first 'Type M' call. The attacker effectively finds the MACs for the message strings $D_1, D_2, X_1, \ldots, X_{q-1}$. There is a good chance that two of these messages will yield the same MAC. Suppose that the two padded strings are $D_1, D_2, E_3, \ldots, E_{q+1}$ and $D_1, D_2, E'_3, \ldots, E'_{q+1}$. Denote the common MAC by M.

Since we are assuming that the cryptographic module does not keep track of the message length, it is not important whether the contents of block D_1 match the true length of the rest of the message. The MAC values obtained may not be true MACs of the (padded) input strings using Padding Method 3. Our objective is to recover the secret keys and not to present verifiable MAC forgeries. The fact that the calculated MAC values are not true MACs is unimportant and bears no consequence on the key recovery attack to be described.

The attacker now submits as many 'Type M' and 'Type E' calls as necessary to obtain the 'MAC' value for the two strings $D_1, D_2, E_3, \ldots, E_q$ and $D_1, D_2, E'_3, \ldots, E'_q$ using the partial MAC H_2 for the input pair (D_1, D_2) as chaining variable input to the first 'Type M' call in each case. Denote the final 'MAC' outputs by M' and M'' respectively. As before, we know immediately that

$$d_{K'}(M') \oplus E_{q+1} = d_K(d_{K'}(M)) = d_{K'}(M'') \oplus E'_{q+1}.$$

The attack now proceeds as in the previous case to recover K', K and finally K''. The complexity of this version of the attack is $[2^{k+2}, 0, 2^{n/2}, 0]$, which is less than the complexity of the attack described previously.

The main conclusion we can draw from this attack is that, when Padding Method 3 is used, a further variable should be input and output along with the chaining variable. This variable should indicate the number of data bits remaining to be processed. In addition, integrity protection should be deployed over the entire set of input/output variables, to prevent modifications being made.

7   Other Issues

The attacks described above are only examples of possible weaknesses of APIs when 'partial' MAC calculations are performed. We now mention one other area where problems may arise.

In some applications it is desirable to have a cryptographic module which can exist in two operational modes. In one mode MAC calculations are possible, but in the other mode only MAC verifications are possible. Indeed, by implementing such a scheme, it is possible to gain some of the desirable properties of a digital signature, simply by using a MAC.

To implement such a scheme will require two different sets of API procedure calls, one for MAC computation and one for MAC verification. This is simple enough, except when we consider the issue of verifying a MAC on a data string which is too long to be handled in one call to the module. In such a case it will be necessary to implement 'Types B, M and E' procedure calls for MAC verification. However, it should be clear that the 'Type B' and 'Type M' calls will be indistinguishable from the corresponding calls for a MAC computation. This poses significant risks to the separation of MAC computation and verification, and will need to be analysed carefully in the design of any module API.

8   Summary and Conclusions

There exist Cryptographic APIs and Cryptographic Modules following the model discussed in this paper. Sometimes Cryptographic APIs do not pass a chaining variable value but rather a pointer to the memory location of the chaining variable (eg [11]). It is not clear whether this memory is accessible or not by a potential attacker – if this memory is part of the cryptographic module, then presumably it is not accessible; otherwise, the attacks described here are possible. It has not always been possible to obtain the necessary detail about the format of the chaining variables produced by the investigated cryptographic modules and APIs (eg [12,11,13,14]). This is understandable as this information may be regarded as proprietary and sensitive by the manufacturers. The recent publication of [1] means that MAC algorithm 4, as described there, has probably not yet been implemented by many manufacturers, and thus the comments and attacks presented here should be considered as illustrations of what could happen, rather than actual attacks on any APIs and cryptographic modules in use today. However, the general comments on block-cipher based MACs may be relevant even to products in use today, especially if the chaining variables are not encrypted or their integrity is not protected.

When chaining variables have to be imported and exported from a cryptographic module, it is important that they be output in encrypted form. Moreover, the encryption should operate in such a way that if the same chaining variable is ever output twice, then this will not be evident from the ciphertext.

Moreover, when Padding Method 3 is used, a further variable should be input and output along with the chaining variable. This variable should indicate the number of data bits remaining to be processed. In addition, integrity protection should be deployed over the entire set of input/output variables, to prevent modifications being made.

References

1. International Organization for Standardization Genève, Switzerland: ISO/IEC 9797–1, Information technology — Security techniques — Message Authentication Codes (MACs) — Part 1: Mechanisms using a block cipher. (1999)

2. Preneel, B., van Oorschot, P.: On the security of iterated Message Authentication Codes. IEEE Transactions on Information Theory **45** (1999) 188–199
3. American Bankers Association Washington, DC: ANSI X9.9–1986 (revised), Financial institution message authentication (wholesale). (1986)
4. American Bankers Association Washington, DC: ANSI X9.19, Financial institution retail message authentication. (1986)
5. Electronic Frontier Foundation: Cracking DES: Secrets of encryption research, wiretap politics & chip design. O'Reilly (1998)
6. Preneel, B., van Oorschot, P.: A key recovery attack on the ANSI X9.19 retail MAC. Electronics Letters **32** (1996) 1568–1569
7. Knudsen, L., Preneel, B.: MacDES: MAC algorithm based on DES. Electronics Letters **34** (1998) 871–873
8. Wagner, D.: GSM cloning.
   `http://www.isaac.cs.berkeley.edu/isaac/gsm-faq.html` (1999)
9. Coppersmith, D., Mitchell, C.: Attacks on MacDES MAC algorithm. Electronics Letters **35** (1999) 1626–1627
10. Menezes, A., van Oorschot, P., Vanstone, S.: Handbook of Applied Cryptography. CRC Press, Boca Raton (1997)
11. IBM: (IBM PCI Cryptographic Coprocessor)
    `http://www-3.ibm.com/security/cryptocards/html/overcca.shtml.`
12. Baltimore: (KeyTools Overview) `http://www.baltimore.com/keytools/.`
13. Microsoft: (CryptoAPI Tools Reference)
    `http://msdn.microsoft.com/library/psdk/crypto/cryptotools_0b11.htm.`
14. RSA Laboratories: PKCS#11 Cryptographic Token Interface Standard. (1997) Version 2.01, `http://www.rsasecurity.com/rsalabs/pkcs/pkcs-11.`

The Exact Security of ECIES in the Generic Group Model

N.P. Smart

Department of Computer Science,
University of Bristol,
Merchant Venturers Building,
Woodland Road,
Bristol, BS8 1UB,
United Kingdom.
nigel@cs.bris.ac.uk

Abstract. In this paper we analyse the ECIES encryption algorithm in the generic group model of computation. This allows us to remove the non-standard interactive intractability assumption of the proof of security given in the literature. This is done at the expense of requiring the generic group model of computation.

1   Introduction

The area of 'provable security' has in recent years become increasingly important in cryptography. This has been driven by two forces. Firstly the increased international standardisation effort has led cryptographers interested in deployed systems to increasingly depend on systems and protocols for which the security is based on real scientific proof rather than ad hoc arguments. Secondly, the tools for provable security have advanced so that schemes as efficient as those currently employed can now be fully analysed.

The most important encryption scheme to come out of this effort has been RSA-OAEP, as used in almost all standards. The scheme RSA-OAEP replaced an earlier RSA encryption scheme, called PKCS v1.2, which was shown to be weak by Bleichenbacher [6]. The RSA-OAEP scheme was a great advance. Although it cannot be proved secure in the standard model of computation, if one is prepared to accept proofs in the random oracle model then one can give a proof of security against active adversarys, see [5], [12] and [19].

There have been a number of attempts to provide "good" encryption schemes based on discrete logarithms. The original El Gamal encryption algorithm [11] can be shown to be secure against passive adversaries, but it is totally insecure against active adversaries. A number of schemes have been given which are secure under the standard model of computation, e.g. the Cramer-Shoup scheme [9]. However, this later scheme is less efficient than currently deployed schemes.

In [3] Abdalla, Bellare and Rogaway present an encryption scheme, called DHIES, which is secure in the standard model of computation. The DHIES

B. Honary (Ed.): Cryptography and Coding 2001, LNCS 2260, pp. 73–84, 2001.

scheme is particularly important since it is as efficient as El Gamal. The combined effect of a security proof and its efficiency has led DHIES to be standardised in a number of standards including ANSI X9.63 [1] and SECG [2]. DHIES is particularly suited to elliptic curve groups, in which case the scheme is slightly simplified and is called ECIES.

Although the proof of security of ECIES is in the standard model of computation, it relies on a non-standard intractability assumption. Namely, an interactive hash based version of the Decision Diffie-Hellman problem. This has led some authors to criticise the security proof. An earlier version [4] of [3] had a claim of a proof of security of ECIES based on the concept of plain-text awareness, and hence also in the random oracle model, this proof was later withdrawn by the authors, see [17].

In this paper we adapt the proof of security of ECIES so that it does not depend on any non-standard intractability assumptions. This is done at the expense of working in the generic group model, a model which is often used to reason about elliptic curve systems, see for example [14], [7] and [8]. The generic group model is like the random oracle model except that instead of an idealised hash function being modelled we model an idealised finite abelian group of prime order. In particular we shall prove

Theorem 1. *In the generic group model*

$$\mathrm{InSec}^{\mathrm{CCA2}}(\mathrm{ECIES}; t, v, w, \mu, m) \leq \mathrm{InSec}(\mathrm{SYM}; t_1, 0, m, m')$$
$$+ 2\,v\,\mathrm{InSec}(\mathrm{MAC}; t_3, v-1) + \frac{2(v+w)^2}{q}.$$

For the exact definition of each of these quantities we refer to the next sections.

The author would like to thank Alf Menezes for suggesting this problem to him and Dan Brown for useful conversations during which the work on this paper was carried out.

2   The ECIES Scheme

Let G denote a group of prime order q with generator g. We require a symmetric encryption scheme $\mathrm{SYM} = (E_k, D_k)$, a MAC function MAC_k, and a key derivation function V. The key space of SYM will be denoted K_1 and the key space of the MAC will be denoted K_2. The key derivation function V will map group elements to the key space of both the encryption and MAC functions. The key size of both SYM and MAC will be around $n/2$ where $n = \log_2(q)$, hence the output of V should be of size approximately n.

The properties required of the key derivation function V will be quite mild when working in the generic group model, namely the output of V could simply return the bit representation of the input group element. When we turn to real world implementations we see a real difference between the real world and the generic group model. We shall return to this in a latter section.

The scheme ECIES is defined as a triple of randomised algorithms, {**keygen, enc, dec**}.

keygen	**enc**(pk, m)	**dec**(sk, e)
1. $v \leftarrow \{1, \ldots, q\}$.	1. $x \leftarrow \{1, \ldots, q\}$.	1. Parse e as $u\|r\|c$
2. $\text{pk} \leftarrow g^v$.	2. $u \leftarrow g^x$, $t \leftarrow \text{pk}^x$.	2. $t \leftarrow u^{\text{sk}}$
3. $\text{sk} \leftarrow v$	3. $(k_1, k_2) \leftarrow V(t)$	3. $(k_1, k_2) \leftarrow V(t)$
4. Return (pk, sk).	4. $c \leftarrow E_{k_1}(m)$	4. If $r \neq \text{MAC}_{k_2}(c)$
	5. $r \leftarrow \text{MAC}_{k_2}(c)$	$\quad$ Return **Invalid**
	6. Return $e \leftarrow u\|r\|c$.	5. $m \leftarrow D_{k_1}(c)$.
		6. Return m.

The use of an arbitrary symmetric key encryption function allows one to encrypt arbitrary long messages. The use of the MAC function is to protect against chosen ciphertext attacks.

3   Definitions

It is generally accepted that the "correct" notion of security for public key encryption functions is one of indistinguishability under adaptive chosen ciphertext attack, see [13], [15], [16] and [10]. This is defined by a game which the adversary A plays, made up of two phases. In the first phase, denoted **find**, the adversary outputs two messages m_1 and m_2. Then, hidden from the adversaries view a bit b is chosen. The message m_b is encrypted, to give e_b, and the encryption is given to the adversary. In the second **guess** stage the adversary tries to guess the bit b. The adversary is declared successful if the guessed bit is b.

In an adaptive chosen ciphertext attack the adversary has access to a decryption oracle which he can use to query any ciphertext of this choosing, except for the target ciphertext e_b. Since we are in the public key setting the adversary clearly has access to a encryption oracle.

More formally we define the advantage of the adversary A as

$$\text{Adv}_A^{\text{CCA2}}(\text{ECIES}) = 2 \Pr \left[\begin{array}{l} (\text{pk}, \text{sk}) \leftarrow \text{ECIES.keygen}; \\ (m_0, m_1, s) \leftarrow A^{\text{ECIES.dec}}(\textbf{find}, \text{pk}); \\ b \leftarrow \{0, 1\}; e \leftarrow \text{ECIES.enc}(\text{pk}, m_b) : \\ A^{\text{ECIES.dec}}(\textbf{guess}, \text{pk}, s, e) = b \end{array} \right] - 1.$$

Note, the advantage is considered to lie between 0 and 1. A value of zero indicating an adversary who does no better than guess the bit b, whilst a value of one indicating that the adversary always guesses correct. The security of ECIES is then defined to be

$$\text{InSec}^{\text{CCA2}}(\text{ECIES}; t, v, w, \mu, m) = \max_A \left\{ \text{Adv}_A^{\text{CCA2}}(\text{ECIES}) \right\},$$

where the maximum is over all adversaries A running in time t, making at most v queries to the decryption oracle of ECIES and w queries to the generic group oracle. All the decryption queries total at most μ bits, with size of the challenge messages m_0 and m_1 being at most m bits. An oracle query to the generic group oracle will consist of a call to either the group operation oracle, the equality test oracle, the inversion oracle or the oracle to generate a new random group element. Each of these oracle queries will be assumed to have the same cost.

One can define security of the symmetric encryption scheme $SYM = (E_k, D_k)$ in a similar way, except now one gives the adversary access to an encryption rather than a decryption oracle, for the target key K. Thus we are measure the security of SYM to withstand an adaptive chosen plaintext attack. Formally we define

$$\text{Adv}_A(SYM) = 2\Pr\left[\begin{array}{l} K \leftarrow SYM.\text{keygen}; (m_0, m_1, s) \leftarrow A^{E_K}(\mathbf{find}); \\ b \leftarrow \{0,1\}; c \leftarrow E_K(m_b) : A^{E_K}(\mathbf{guess}, s, c) = b \end{array}\right] - 1.$$

The security of SYM is then defined to be

$$\text{InSec}(SYM; t, \mu, m, m') = \max_A \{\text{Adv}_A(SYM)\},$$

where the maximum is over all adversaries running in time at most t, asking queries of E_K totalling at most μ bits in length. The output of the find stage is at most m bits, whilst m' is an upper bound on a ciphertext corresponding to a plaintext of m bits in length.

Just as in [3] we define security of the MAC function in terms of an adversary who is given an oracle for the MAC function with respect to some hidden key K. The adversary is deemed successful if it can output a (message, MAC) pair which has not yet been asked of the MAC oracle. Formally

$$\text{Succ}_A(MAC) = \Pr\left[\begin{array}{l} K \leftarrow MAC.\text{keygen}; (x, \tau) \leftarrow A^{MAC(K, \cdot)} : \\ (x, \tau) \text{ is unasked and } MAC(K, x) = \tau \end{array}\right].$$

The security of MAC is defined to be

$$\text{InSec}(MAC; t, v) = \max_A \{\text{Succ}_A(MAC)\},$$

where the maximum is over all adversaries A running in time at most t and making at most v oracle queries to the MAC oracle.

In all the above definitions the time t refers to the maximal number of steps that the algorithm requires in some fixed model of computation, and where each oracle query shall count as a single step.

We end this section by discussing the standard Decision Diffie-Hellman problem. This is the problem: Given g^x, g^y and g^z determine whether $z \equiv xy$ (mod q). Formally we have

$$\text{Adv}_A^{DDH}(G) = \Pr[u, v \leftarrow \{1, \ldots, q\} : A(g^u, g^v, g^{uv}) = 1]$$
$$- \Pr[u, v \leftarrow \{1, \ldots, q\}; h \leftarrow G : A(g^u, g^v, h) = 1].$$

Note, by results of Shoup [18, Theorem 4], we have

Theorem 2. *In the generic group model*

$$\text{Adv}_A^{DDH}(G) = w^2/q,$$

where w is the number of queries A makes to the generic group oracle.

4   Proof

The proof technique is exactly the same as the proof technique in [3]. The only difference is that we replace the interactive calls to the hash Diffie-Hellman oracle by calls to a generic group operation. This allows us to remove the need for the non standard intractability assumption, at the expense of assuming the generic group model.

We present the full proof, rather than simply referring to [3], since we wish to point out exactly how the proof depends on our assumption of a generic group model. In particular we wish to keep track of the number of oracle queries to the generic group oracle.

Let A denote an adversary attacking ECIES in the sense above. Assume it has running time at most t, makes v queries to its decryption oracle and w queries to its generic group oracle. At the end of its **find** stage we assume it outputs a string of length at most m. Let m' denote an upper bound on the length of an encryption under SYM of a plaintext of length at most m. We wish to use A to attack the security of the decision Diffie-Hellman problem in our generic group, the underlying symmetric encryption system and the underlying MAC. To do this we will need to give A a simulation of its decryption oracle. We shall first describe the basic simulator, which we shall then modify below, the simulator will maintain an internal list of group pairs L, which "represent" elements and their discrete logarithms. These are not "correct" discrete logarithms but the adversary will not be able to tell the difference.

Decryption Simulator(e)

1. Parse e as $u\|r\|c$.
2. If $(u, t') \in L$ for some t', set $t \leftarrow t'$, else $t \leftarrow G$.
3. $L \leftarrow L \cup \{(u, t)\}$.
4. $(k_1, k_2) \leftarrow V(t)$
5. If $r \neq \text{MAC}_{k_2}(c)$ return **Invalid**
6. $m \leftarrow D_{k_1}(c)$.
7. Return m.

Notice that algorithm A, when we supply it with the above decryption simulator will now make at most $v + w$ calls to the generic group oracle.

We now describe an algorithm B which will use A to break the encryption scheme. Recall from the previous definition that B has access to an oracle for encryption, but not decryption and that it runs in two stages.

Algorithm B$^{\mathcal{O}}$(find)

1. pk $\leftarrow G$
2. $(m_0, m_1, s) \leftarrow A(\textbf{find}, \text{pk})$.
3. Output $(m_0, m_1, (m_0, m_1, s, \text{pk}))$.

Algorithm B$^{\mathcal{O}}$(guess,c',s')

1. Parse s' as (m_0, m_1, s, pk).
2. $u \leftarrow G$.
3. $k_2 \leftarrow K_2$.
4. $r \leftarrow \text{MAC}_{k_2}(c')$.
5. $e \leftarrow u\|r\|c'$.
6. $b \leftarrow A(\textbf{guess}, \text{pk}, s, e)$.
7. Return b.

Clearly B's running time is

$$t_1 = O(t + 2\text{TIME}_G + \text{TIME}_{MAC.gen}(m')).$$

When applying algorithm B we use the decryption simulator given above, with one minor modification which we shall describe in a moment. Hence, B requires v queries to the decryption simulator and a total of $2+v+w$ queries to the group oracle. The time for these queries is already accounted for in the time t needed to run algorithm A. Note, that although algorithm B has access, by definition, to an encryption oracle $\mathcal{O}$ it does not actually use this oracle.

Due to the definition of security of SYM we need to modify the decryption simulator in one crucial way. This is because B is not allowed access to the decryption algorithm D_k for the key k "corresponding" to the group element u in the target cipher text y. We call a **Type Q** query, of the decryption simulator, one for which the input is of the form

$$u\|r'\|c'$$

for some r' and c' and for which the simulator does not output **Invalid**. As we are in the generic group model, we see that it is impossible for B to construct a non **Type Q** query which will result in the access of the decryption algorithm D_k for the key k, hence it is only the **Type Q** queries which we need to avoid. Hence, we modify the simulator as follows. If a **Type Q** query is made then Algorithm B will terminate by returning $b \leftarrow \{0,1\}$.

We shall now consider an algorithm C which will use algorithm A to attempt to solve the Decision Diffie-Hellman problem in G. It will take as input g^a, g^b and g^c, and output one if it believes $c = ab$, and zero otherwise.

Algorithm C(g^a,g^b,g^c)

1. $u \leftarrow g^a$.
2. $\text{pk} \leftarrow g^b$.
3. $(k_1, k_2) \leftarrow V(g^c)$.
4. $(m_0, m_1, s) \leftarrow A(\textbf{find}, \text{pk})$.
5. $b \leftarrow \{0,1\}$.
6. $c \leftarrow E_{k_1}(m_b)$.
7. $r \leftarrow \text{MAC}_{k_2}(c)$.
8. $e \leftarrow u\|r\|c$.
9. $b' \leftarrow A(\textbf{guess}, \text{pk}, s, e)$.
10. If $b' = b$ Return one, else Return zero.

Algorithm C's running time is bounded by

$$t_2 = O(t + \text{TIME}_{SYM.enc}(m) + \text{TIME}_{MAC.gen)}(m'))$$

and it makes v queries to algorithm A's decryption oracle, which means that using the above simulator it requires $v + w$ queries to the generic group oracle.

When algorithm C is given a valid Diffie-Hellman triple as input it runs algorithm A just as one would run A to mount an attack against ECIES. Hence,

$$\Pr\left[a, b, \leftarrow \{1, \ldots, q\}; c \leftarrow a \cdot b \ : \ C(g^a, g^b, g^c) = 1\right]$$
$$= \frac{1 + \mathrm{Adv}_A^{\mathrm{CCA2}}(\mathrm{ECIES})}{2}. \qquad (1)$$

Now suppose C is not given a valid Diffie-Hellman triple. When A does not make a **Type Q** query, C runs A in the same way that B runs A. Hence,

$$\Pr\left[a, b, c \leftarrow \{1, \ldots, q\} : C(g^a, g^b, g^c) = 1 \wedge \neg\textbf{Type Q}\right] = \frac{1 + \mathrm{Adv}_B(\mathrm{SYM})}{2}.$$

Now since B makes zero encryption queries and runs in time at most t_1 we obtain

$$\Pr\left[a, b, c \leftarrow \{1, \ldots, q\} \ : \ C(g^a, g^b, g^c) = 1 \wedge \neg\textbf{Type Q}\right]$$
$$\leq \frac{1 + \mathrm{InSec}(\mathrm{SYM}; t_1, 0, m, m')}{2}. \qquad (2)$$

When A makes a **Type Q** query of its decryption oracle we see that C runs A just as the following algorithm D runs A to break the MAC function. We first present algorithm D and then present the modification to the decryption simulator which is needed. Note that algorithm D is given access to an oracle $\mathcal{O}$ for the keyed hash function, for the key "corresponding" to the group element u in the target ciphertext.

Algorithm $D^{\mathcal{O}}$

1. pk, $u \leftarrow G$.
2. $k_1 \leftarrow K_1$.
3. $j \leftarrow \{1, \ldots, v\}$.
4. $(m_0, m_1, s) \leftarrow A(\textbf{find}, \mathrm{pk})$.
5. $b' \leftarrow \{0, 1\}$.
6. $c \leftarrow E_{k_1}(m_{b'})$.
7. $r \leftarrow \mathcal{O}(c)$.
8. $e \leftarrow u \| r \| c$.
9. Run $A(\textbf{guess}, \mathrm{pk}, s, e)$.
10. Return W.

The modified decryption simulator runs as follows. Each call to the decryption simulator is given an index $i \in \{1, \ldots, v\}$.

Decryption Simulator(e_i)

1. Parse e_i as $u' \| r' \| c'$.
2. If $(u', t') \in L$ for some t', set $t \leftarrow t'$, else $t \leftarrow G$.
3. $L \leftarrow L \cup \{(u', t)\}$.
4. $(k'_1, k'_2) \leftarrow V(t)$
5. If $i \neq j$ and $u' = u$ then

6.   If $\mathcal{O}(c') = r'$ then
7.   Return $D_{k_1}(c')$
8.   Else Return **Invalid**.
9. Else if $i \neq j$ and $u' \neq u$ then
10.   If $\text{MAC}_{k_2'}(c') = r'$ then
11.   Return $D_{k_1'}(c')$
12.   Else Return **Invalid**.
13. Else
14. $W \leftarrow (c', r')$.
15.   Return $D_{k_1}(c')$.

Suppose algorithm A makes a **Type Q** query e' to its decryption oracle. Let e_i denote the number of one such query, with $e_i = u' \| r' \| c'$. Now if j in Algorithm D takes this value of i then D succeeds in breaking MAC since (c', r') will be a valid pair. However, the equality $j = i$ can happen with probability at most $1/v$, we obtain

$$\Pr\left[\text{Algorithm } A \text{ makes a query of } \textbf{Type Q}\right] \leq v \, \text{Succ}_D(\text{MAC}).$$

Algorithm D makes at most $v - 1$ queries to its MAC oracle $\mathcal{O}$ and runs in time

$$t_3 = O(t + 2\text{TIME}_G + \text{TIME}_{\text{MAC}.gen}(m') + \text{TIME}_{\text{SYM}.enc}(m)),$$

since the time to cope with the decryption oracle queries is included in t. We therefore have

$$\Pr\left[a, b, c \leftarrow \{1, \ldots, q\} \; : \; C(g^a, g^b, g^c) = 1 \wedge \textbf{Type Q}\right]$$
$$\leq v \, \text{InSec}(\text{MAC}; t_3, v - 1). \tag{3}$$

Finally combining inequalities (1), (2) and (3) we obtain

$$\text{Adv}_C^{DDH}(G) \geq \frac{\text{Adv}_A^{\text{CCA2}}(\text{ECIES})}{2} - \frac{\text{InSec}(\text{SYM}; t_1, 0, m, m')}{2}$$
$$- v \, \text{InSec}(\text{MAC}; t_3, v - 1).$$

Recall that C makes $v + w$ queries to the generic group oracle and so by Shoup's result we obtain

$$\frac{(v + w)^2}{q} \geq \frac{\text{Adv}_A^{\text{CCA2}}(\text{ECIES})}{2} - \frac{\text{InSec}(\text{SYM}; t_1, 0, m, m')}{2}$$
$$- v \, \text{InSec}(\text{MAC}; t_3, v - 1).$$

In other words

$$\text{Adv}_A^{\text{CCA2}}(\text{ECIES}) \leq \text{InSec}(\text{SYM}; t_1, 0, m, m') + 2 \, v \, \text{InSec}(\text{MAC}; t_3, v - 1)$$
$$+ \frac{2(v + w)^2}{q}.$$

Finally since A was an arbitrary adversary subject to the constraint that it run in t steps, made v calls to its decryption oracle all of which total at most μ bits, w calls to its generic group oracle and the size of the output from its **find** stage was at most m we conclude

$$\mathrm{Adv}_A^{\mathrm{CCA2}}(\mathrm{ECIES}) = \mathrm{InSec}^{\mathrm{CCA2}}(\mathrm{ECIES}; t, v, w, \mu, m).$$

The main theorem now follows.

5   Practical Considerations

The original DHAES paper suggested using a key derivation function of the form $V(u, t)$. Otherwise this can lead to trivial malleability of the ciphertext in the case when a group of non-prime order is used. To see this, assume we have a ciphertext of the form $c = u\|r\|c$ corresponding to the plaintext m, where u is of prime order q but the underlying group is of order $2q$. If the private key x is even then one trivially obtains the following valid ciphertext, corresponding to the plaintext m,

$$u \cdot h \| r \| c$$

where h is an element of order two. To avoid this trivial malleability, which is not known to cause a problem in practice, we need to either use the key derivation function $V(u, t)$ or check that for each received ciphertext u is an element of order q.

In our analysis we have assumed a generic group of prime order. When we instantiate our generic group to the type of elliptic curves used in practical systems we can avoid this problem in a number of ways:

Large Prime Characteristic : In this case the elliptic curve group order is usually chosen to have prime order. Hence, simply checking that the point u lies on the curve will avoid the above malleability issue.

Characteristic Two : In this case the elliptic curve group order is usually chosen to have order twice a prime, i.e. $\#E(\mathbb{F}_{2^m}) = 2q$ and the value of m is prime. Assume the elliptic curve is given in the form

$$Y^2 + XY = X^3 + X^2 + b.$$

The following Lemma lets us easily check that u has prime order q in this case.

Lemma 1. *Assume m is odd, $\#E(\mathbb{F}_{2^m}) = 2q$ with q prime and $P = (x, y) \in E(\mathbb{F}_{2^m})$. The point P has order q if and only if*

$$\mathrm{Tr}_{\mathbb{F}_2}(x) = 1 \quad and \quad \mathrm{Tr}_{\mathbb{F}_2}(b/x^2) = 0.$$

Proof. If the point P has order q then one can find a point $Q = (x_1, y_1)$ such that

$$P = [2]Q.$$

We then have that
$$\lambda = x_1 + \frac{y_1}{x_1} \in \mathbb{F}_{2^m}$$
where λ is a root of the equation
$$\lambda^2 + \lambda + 1 + x = 0.$$

Now the existence of $\lambda \in \mathbb{F}_{2^m}$ is equivalent to $\mathrm{Tr}_{\mathbb{F}_2}(x) = 1$.

We are left with showing that given $\lambda \in \mathbb{F}_{2^m}$ the existence of $x_1, y_1 \in \mathbb{F}_{2^m}$. We obtain two simultaneous equations, in x_1 and y_1, given by
$$y_1 = x_1^2 + \lambda x_1$$
$$0 = y_1^2 + x_1 y_1 + x_1^3 + x_1^2 + b$$

Eliminating y_1 we obtain the equation, with $\tau = x_1^2$,
$$0 = \tau^2 + (\lambda^2 + \lambda + 1)\tau + b$$
$$= \tau^2 + x\tau + b$$

which will have a solution if and only if $\mathrm{Tr}_{\mathbb{F}_2}(b/x^2) = 0$.

We now turn to a discussion of the function V. In the generic group model there was little restriction required on such a function, and we could assume that it simply output the bit representation of the underlying group element. For any particular group, i.e. elliptic curves, one needs to be more careful. For example one should not just take the x-coordinate as the output of the key derivation function, one should also take a contribution from the y coordinate. This is to stop the obvious collision $V(P) = V(-P)$ when only the x-coordinate is used as input to V.

But, even if we used the bit representation of the compression of the point P, our proof still does not apply when the generic group was replaced by an elliptic curve group. This is because at one crucial point we assumed that it was hard for algorithm B to solve the following problem: Given public $u \in G$ and private $x \in \mathbb{Z}$ find a $u' \in G$ such that
$$V|_{\mathrm{SYM}}(u^x) = V|_{\mathrm{SYM}}(u'^x),$$
where $V|_{\mathrm{SYM}}$ is the function V where the output is projected down to the key space of the symmetric encryption function only. It is clear that this problem is easy when $V|_{\mathrm{SYM}}(P)$ is defined to be the first $n/2$ bits of the point compression of the elliptic curve point P, namely the first $n/2$ bits of the x-coordinate of P. Since then
$$V|_{\mathrm{SYM}}([x]P) = V|_{\mathrm{SYM}}([x](-P)).$$

Hence, our technique of using the generic group to remove the interactive Decision Diffie-Hellman assumption from the proof in [3] has essentially hidden the problem of a non-trivial interaction between the key derivation function V and the group. In [3] this is also hidden within the interactive Decision Diffie-Hellman assumption. Hence, when instantiating the protocol to a real life situation some care needs to be taken. For example V should be instantiated using a hash function such as SHA-1.

Perhaps the most interesting conclusion from this discussion is that, whilst ECIES is secure in the generic group model with a trivial definition for V, when we instantiate the group with an elliptic curve group, the protocol (with the same trivial definition of V) becomes insecure. Hence, this is possible evidence that elliptic curve groups should not be modelled by generic groups.

References

1. ANSI. ANSI X9.63-2001. Key agreement and key transport using elliptic curve cryptography. *ANSI Standards Committee X9*, Working Draft, 2001.
2. SECG. SEC 1: Elliptic Curve Cryptography, Version 1.0. *Standards for Efficient Cryptography Group*, 2000.
3. M. Abdalla, M. Bellare and P. Rogaway. DHAES: An encryption scheme based on the Diffie-Hellman problem. Submission to *P1363a : Standard specifications for Public-Key-Cryptography: Additional techniques*, 2000.
4. M. Bellare and P. Rogaway. Minimizing the use of random oracles in authenticated encryption schemes. In *Information and Communications Security*, Springer-Verlag LNCS 1334, 1–16.
5. M. Bellare and P. Rogaway. Optimal asymmetric encryption. In *Advances in Cryptology - EUROCRYPT '94*, Springer-Verlag LNCS 950, 92–111, 1995.
6. D. Bleichenbacher. Chosen ciphertext attacks against protocols based on the RSA encryption standard PKCS#1. In *Advances in Cryptology - CRYPTO '98*, Springer-Verlag LNCS 1462, 1–12, 1998.
7. D.R.L. Brown. Concrete lower bounds on the security of ECDSA in the Generic Group Model. Preprint, 2001.
8. D.R.L. Brown and D.B. Johnson. Formal security proofs for a signature scheme with partial message recovery. In *Topics in Cryptology : CT-RSA 2001*, Springer-Verlag LNCS 2020, 126–142, 2001.
9. R. Cramer and V. Shoup. A practical public key cryptosystem provably secure against adaptive chosen ciphertext attack. In *Advances in Cryptology - CRYPTO '98*, Springer-Verlag LNCS 1462, 13–25, 1998.
10. D. Dolev, C. Dwork and M. Naor. Non-malleable cryptography. In *23rd Annual ACM Symposium on Theory of Computing*, 542–552, 1991.
11. T. ElGamal. A public key cryptosystem and a signature scheme based on discrete logarithms. In *Advances in Cryptology - CRYPTO '94*, Springer-Verlag LNCS 196, 10–18, 1985.
12. E. Fujisaki, T. Okamoto, D. Pointcheval and J. Stern. RSA–OAEP is Secure Under the RSA Assumption. In *Advances in Cryptology - CRYPTO 2001*, Springer-Verlag LNCS 2139, 259–273, 2001.
13. S. Goldwasser and S. Micali. Probabilistic encryption. *Journal of Computer and System Sciences*, **28**, 270–299, 1984.
14. M. Jakobsson and C.P. Schnorr. Security of signed ElGamal encryption. In *Advances in Cryptology - ASIACRYPT 2000*, Springer-Verlag LNCS 1976, 73–89, 2000.

15. M. Noar and M. Yung. Public key cryptosystems provably secure against chosen ciphertext attacks. In *22nd Annual ACM Symposium on Theory of Computation*, 426–437, 1990.

16. C. Rackoff and D. Simon. Noninteractive zero-knowledge proof of knowledge and chosen ciphertext attack. In *Advances in Cryptology - CRYPTO '91*, Springer-Verlag LNCS 576, 434–444, 1991.

17. P. Rogaway. *Review of SEC 1*. Letter to SECG, 1999. Available from **http://www.secg.org/**.

18. V. Shoup. Lower bounds for discrete logarithms and related problems. In *Advances in Cryptology - EUROCRYPT '97*, Springer-Verlag 1233, 256–266, 1997.

19. V. Shoup. OAEP Reconsidered. In *Advances in Cryptology - CRYPTO 2001*, Springer-Verlag LNCS 2139, 238–258, 2001.

A New Ultrafast Stream Cipher Design: COS Ciphers

Eric Filiol[1,2] and Caroline Fontaine[3]

[1] ESAT/DAESR,
B.P. 18, 35998 Rennes, FRANCE
[2] INRIA, projet CODES, Domaine de Voluceau
78153 Le Chesnay Cédex, FRANCE
`Eric.Filiol@inria.fr`
[3] USTL, LIFL
59655 Villeneuve d'Ascq Cédex, FRANCE
`Caroline.Fontaine@lifl.fr`

Abstract. This paper presents a new stream cipher family whose output bits are produced by blocks. We particularly focus on the member of this family producing 128-bit blocks with a 256-bit key. The design is based on a new technique called *crossing over* which allows to vectorize stream ciphering by using nonlinear shift registers. These algorithms offer a very high cryptographic security and much higher speed encryption than any existing stream ciphers or block ciphers, particularly the AES candidates. A 1000 euros rewarded cryptanalysis challenge is proposed.

Keywords: stream cipher, nonlinear feedback shift register, vectorized cipher, high speed encryption, Boolean functions, block cipher.

1 Introduction

The transformation by composition of cryptographic primitives allows to build some other primitives. One famous example is that of block ciphers which can simulate stream ciphers through output feedback mode [4,18].

The main advantage of this approach is to obtain provable security. Attack on such simulated stream ciphers (*e.g.*) would, indeed, be equivalent to cryptanalyze the underlying block cipher. Such an approach is described in [3] for the ANSI message authentication code, built from block cipher (however there are few exceptions: for example, CBC mode revealed vulnerabilities independently from the quality of the underlying block cipher).

However building stream ciphers from block ciphers or hash functions represents a major drawback: block ciphers are usually much slower than real, true stream ciphers. Moreover they are not as suitable as stream ciphers, for VLSI implementations. The explosion of today's need for secure communications asks for very high encryption speed that block ciphers achieve with some difficulty. By now, the fastest block cipher software implementation runs at nearly 260

B. Honary (Ed.): Cryptography and Coding 2001, LNCS 2260, pp. 85–98, 2001.

Mbits/s (RC6 [2]). Even the few known software-optimized stream ciphers, such as SEAL [22], offer relatively limited speed encryption.

On the other hand, security of most frequent and common variants of stream ciphers (based on linear feedback shift registers and suitable Boolean functions) becomes little by little challenged and questionned. Recent new powerful attacks [5,6,8,12] proved that effective, real-life cryptanalysis become more and more reachable. In terms of security, recent developments of cryptology (NIST call for AES [19] is the most obvious) tends to be in favour of block ciphers.

One approach which has not been very often considered so far, is to consider and take the best of both worlds: to simulate block cipher by vectorizing stream ciphers. It nevertheless would combine the advantage of both sides without their respective drawbacks. First known attempts have been presented in [13,15] with keyed hash functions built from stream ciphers and in [1] with block ciphers built from SHA-1 and SEAL.

In this paper, we present a completely new approach in cipher design. By using only the known cryptographically strongest Boolean functions and Non Linear Feedback Shift Registers (NLFSR), we build a family of vectorized binary additive stream ciphers (namely where output bits are produced as a block of L bits in a row) called *COS ciphers*, with arbitrary output block size. It is based on a new construction called *crossing over* allowing to consider the internal state at instant t as constituent of the ciphering output blocks or vectors. These latter are bitwise xored to blocks of plaintext to produce blocks of ciphertext.

For each system of the COS family two modes are possible. The first one (mode I) considers any kind of plaintext. The second one (mode II) is restricted to plaintext without redundancy (*e.g.* compressed). This latter has been chosen for the Internet Film Independent Cinema (IFIC) project of the European PRIAMM call [21].

Given fast software implementation of Boolean functions and shift registers, and given that VLSI implementations require fewer gates than for block ciphers, our scheme reveals itself far better than previous ones in practical applications requiring large block sizes and high encryption speed. With 128-bit block size, experiments have reached very high speed encryption: about 330 Mbits/s with 128-bit block size and nearly 1Gbits/s in a 512-bit block size setting (for mode II).

In terms of security, we primarily rely on the cryptographic strength of the constituent primitives. The crossing over mechanism allows then to combine them in a very secure way. The key size can be 128, 192 and 256 bits long. These aspects insure a high level of security.

This paper is organized as follows. We first present in Section 2 the COS cipher family itself, and particularly the 128-bit block size version. Section 3 will discuss security aspects. Section 4 presents the performance of our implementations (encryption speed and code size). Section 5 finally presents the rewarded cryptanalysis challenge we propose.

More materials and data on this cipher (C implementation, test vectors, ...) can be found in [7].

2   The COS Ciphers

2.1   General Description

The COS cipher design exclusively centers on $n + 1$ NonLinear Feedback Shift Registers (NLFSR): $M, L_1, L_2, \ldots, L_n$.

Consider the $2L$-bit block size version. The M register is devoted to the key setup, that is to say to the generation of the $L_1, L_2, \ldots, L_n$ register initial states. M is $4L$ bits long. The L_i registers as to them generate the output ciphering blocks through the crossing over mechanism. They each are $2L$ bits long and are irregularly clocked.

The feedback Boolean functions of these registers L_i have been chosen according to their cryptographic properties. They are optimally *balanced, highly nonlinear* and have high *correlation-immunity order*. Moreover, their Algebraic Normal Form (ANF), that is to say their representation as multivariate polynomials, has to present some particular structure to meet important security requirements defined in [10]. We have considered the strongest known functions meeting all these criteria. After numerous simulations, a careful choice has been done among functions proposed in [9,16,17,24]. Optimal functions have been taken: an eleven-variables function for the M register feedback and nine-variables functions for the L_i register feedback.

The central point of COS design lies in the crossing over technique inspired by the same mechanism as in chromosome genetic differentiation (hence the name COS standing for Crossing Over System). Consider n registers $L_1, L_2, \ldots, L_n$ of length $2L$. If $MSB(L_i)$ and $LSB(L_i)$ denote respectively the L most significant bits and the L least significant bits of L_i, then one output block generation step can be summarized as follows:

1. Clock L_i at least L times.
2. Generate the $2(n-1)$ output L-bit blocks B_k $(k = 1, \ldots, 2(n-1))$ as follows, for $j = 1, \ldots, n$ such that $j \neq i$

$$MSB(B_k) = MSB(L_i) \oplus LSB(L_j)$$
$$LSB(B_k) = LSB(L_i) \oplus MSB(L_j)$$

3. $i = i + 1 \bmod n$

In mode II all the L-bit blocks are used whereas in mode I, only two randomly chosen L-bit blocks will be used.

It is obvious that the security of such a mechanism rely on the nonlinearity of the feedback. This aspect is discussed in Section 3.2. In terms of performance, we see that n $2L$-bit registers produce in each step, $2(n - 1)$ L-bit blocks with only (at least) L register clockings. Performance evaluation will be exposed in Section 4.

From a general point of view we then will speak of a $(n, 2L)$ COS cipher to describe one particular member of the COS family. Appendix B presents the general diagram of the $(2, 2L)$ COS ciphers. We now present more deeply the $(2, 128)$ cipher. Details of implementation and specifications can be found in [7].

2.2   Key Setup of the $(2, 128)$ COS Cipher

Whatever may be the value of n and L, any $(n, 2L)$ COS cipher works with an internal key K of 256 bits and a message key MK of 32 bits. So the first step in key setup expands shorter keys of either 128 or 192 bits to 256 bits.

Let us represent the M register (256 bits) by eight 32-bit words $M[0], M[1], \dots, M[7]$. In the same way, we will represent the key by $K[0], \dots, K[s]$ where $s = 3, 5, 7$ according to the user's key size, and $L_1[0], \dots, L_1[3], L_2[0], \dots, L_2[3]$ as the L_1 and L_2 registers.

We first fill up the M register with the user's key:

$$M[i] = K[i] \qquad i = 0, \dots, s$$

The expansion step then takes the user's key to produce lacking bits to fill up remaining part of the M register $M[s + 1], \dots, M[7]$. A look-up table T will give 8-bit blocks from 8-bit blocks of the user's key. Due to lack of space, this table will be found in [7]. To be more precise, if K_i denotes the 8-bit block of the user's key starting at bit i, we have:

- $s = 3$ (user's key has 128 bits)

$$M[4] = 2^{24} * T[K_{248}] + 2^{16} * T[K_{240}] + 2^8 * T[K_{232}] + T[K_{224}]$$
$$M[5] = 2^{24} * T[K_{216}] + 2^{16} * T[K_{208}] + 2^8 * T[K_{200}] + T[K_{192}]$$
$$M[6] = 2^{24} * T[K_{184}] + 2^{16} * T[K_{176}] + 2^8 * T[K_{168}] + T[K_{160}]$$
$$M[7] = 2^{24} * T[K_{152}] + 2^{16} * T[K_{144}] + 2^8 * T[K_{136}] + T[K_{128}]$$

- $s = 5$ (user's key has 192 bits)

$$M[6] = 2^{24} * T[K_{248} \oplus K_{216} \oplus K_{184}] + 2^{16} * T[K_{240} \oplus K_{208} \oplus K_{176}]$$
$$+ 2^8 * T[K_{232} \oplus K_{200} \oplus K_{168}] + T[K_{224} \oplus K_{192} \oplus K_{160}]$$
$$M[7] = 2^{24} * T[K_{152} \oplus K_{120} \oplus K_{80}] + 2^{16} * T[K_{144} \oplus K_{112} \oplus K_{80}]$$
$$+ 2^8 * T[K_{136} \oplus K_{104} \oplus K_{72}] + T[K_{128} \oplus K_{96} \oplus K_{64}]$$

The 32-bit message key MK is then combined with initial state of register M:

$$M[0] = M[0] \oplus MK$$

This key MK is to be changed for every different message. Since the intrinsic security of the cipher does not rely on MK, it can be transmitted with the message. Its role is to prevent different messages to be sent with the same internal 256-bits base key.

The M register is then clocked 256 times. The eleven-variables feedback function $f11$ takes bits $2, 31, 57, 87, 115, 150, 163, 171, 201, 227$ and 255 of M as input bits and outputs one feedback bit (see Figure 1). Function $f11$ having a too big representation cannot be given here due to lack of space. It will be found in [7]. Its properties will be exposed in Section 3.1.

After these 256 clockings, M register internal state provides initialization of L_1 register in this way:

$$L_1[i] = M[i + 4] \qquad i = 0, \dots, 3$$

Fig. 1. Clocking of the M register

M is then clocked 128 times and L_2 is initialized as follows:

$$L_2[i] = M[i] \qquad i = 0, \dots, 3$$

2.3   Encryption - Decryption of the $(2, 128)$ COS Cipher

The output blocks of the cipher are xored with either plaintext blocks (encipherment) or ciphertext blocks (decipherment) we just need to describe how these 128-bit blocks are generated. Each step S_i outputs one 128-bit block and decomposes as follows:

1. Compute clocking value d.
   a) Compute $clk = 2 * lsb(L_2) + lsb(L_1)$.
   b) $d = C[clk]$ where $C[0, \dots, 3] = \{64, 65, 66, 64\}$.
2. Clock L_1 if i even or L_2 if i odd, d times.
3. Produce a 128-bit block B_i in this way (see Figure 2):

$$B_i = (L_1[1] \oplus L_2[3]) + 2^{32} * (L_1[0] \oplus L_2[2])$$
$$+2^{64} * (L_1[3] \oplus L_2[1]) + 2^{96} * (L_1[2] \oplus L_2[0])$$

4. (Mode I only) Compute block indices j and k and keep only blocks j and k. If $lsb_2(L)$ and $msb_2(L)$ denotes the two most, respectively least significant bits of L, then block indices i and j (mode I only) are computed as follows:

$$j = lsb_2(L_1) \oplus lsb_2(L_2) \qquad k = msb_2(L_1) \oplus msb_2(L_2)$$

To prevent equality between j and k that is to say repeated output blocks, we fix $k = j \oplus 1$ when $j = k$.

Feedback Boolean functions are in 9 variables ($f9a$ and $f9b$) and both use bits $2, 5, 8, 15, 26, 38, 44, 47$ and 57 of L_1 and L_2. Interested readers will find the complete functions in [7].

3   Security Evaluation

The security of COS cipher lies primarily on the feedback Boolean functions and on the crossing over mechanism. The key size is large enough to resist exhaustive key search. In such kind of design with NLFSR, correlation attacks [5,6,8,12,20,

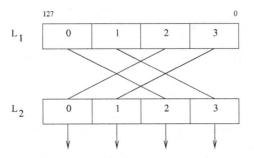

Fig. 2. Output block generation

23], linear syndrome attacks [25,26], ... are no longer working and even have no longer significance. Indeed they can be used only when register feedback is linear.

Thus concerning security, only two aspects have to be considered and controlled: randomness and presence of cycles. Both have great importance to prevent predictability on the ciphering output blocks.

3.1   The Boolean Functions

They primarily are important to yield the best possible randomness. Three Boolean functions have been chosen for the registers feedback: *f11* for the M register and *f9a*, *f9b* for the L_1 and L_2 registers. Functions *f11* is common to all members of the COS family. Additionnal functions, for $n > 2$ have been chosen for the other members [7].

The Boolean functions presenting the best cryptographic properties trade-offs (balancedness, correlation-immunity, nonlinearity, degree) have been presented in [9,16,17,24]. Different simulations have been conducted to choose the most suitable functions. Table 1 summarizes their main characteristics for the $(2, 128)$ COS cipher.

Table 1. Characteristics of feedback functions of the $(2, 128)$ COS Cipher

	Balanced	Correlation Immunity Order	Non Linearity	Degree
f11	yes	CI(2)	960	7
f9a	yes	CI(2)	224	6
f9b	yes	CI(2)	224	6

The strong trade-off between correlation-immunity and nonlinearity suppresses any exploitable correlation first, between L_1 and L_2 initializations and

K, and secondly, between output blocks and K. Moreover, experiments tend to show that nonlinearity (and to a certain extent correlation immunity order) has an important impact on the randomness properties. Particularly, we observe that the best statistical results were obtained for functions having an as high as possible algebraic degree. Thus, since the Boolean functions we have chosen satisfy the best trade-offs among all these criteria, they provide results as good as possible concerning randomness (see Section 3.3).

The Algebraic Normal Form structure has some importance too, if we consider the following proposition:

Proposition 1 *[10, page 115] A Feedback Shift Register (FSR) with feedback function $f(x_{j-1}, x_{j-2}, \dots, x_{j-L})$ is non singular (or equivalently every of its output sequence is periodic) if and only if f is of the form:*

$$f(x_{j-1}, x_{j-2}, \dots, x_{j-L}) = x_{j-L} \oplus g(x_{j-1}, x_{j-2}, \dots, x_{j-L+1})$$

for some Boolean function g.

This very important property prevents degeneration in the output sequence of the FSR (see [10, chap. VI and VII]) for details). All the feedback functions we chose meet this requirement.

3.2   The Crossing over Mechanism Security

Its aim is to use L_i internal states to directly build up output block and thus to obtain very high encryption speed. Moreover this mechanism directly takes part in strengthening the overall scheme security.

First, probability to have cycles in the output sequence is greatly reduced. Forecasting presence of cycles of length strictly less than 2^{L-1} where L is the register length, still remains a general open problem for NLFSR [11]. Only statistical testing (when tractable) can be envisaged up to now to give insight on possible cycles. Chosen Boolean functions of COS cipher family meet the few existing results in this area, that are exposed in [10].

Suppose however that registers L_1 and L_2 present cycles of length respectively l_1 and l_2 for some initializations. The crossing over mechanism will then obviously produce a cycle of length $lcm(l_1, l_2)$ which is confidently supposed to be extremely large. In the general case of a $(n, 2L)$ cipher, cycles we be of length $lcm(l_1, \dots, l_n)$.

One other important role of the crossing over mechanism is to suppress any exploitable correlation between output blocks B and internal states of the L_1 and L_2 registers at any time. Since register contents are cross-xored together and since respective bit probability of each register internal bit is exactly $\frac{1}{2}$, it is impossible to guess any internal bit of both L_1 and L_2, by knowing bits of B. In other words, if L_j^i denotes the i-th bit of register L_j and B^k the k-th bit of B then:

$$P[L_1^i = a | B^k = b] = P[L_2^l = a \oplus b | B^k = b] = \frac{1}{2}$$

where $B^k = L_1^i \oplus L_2^l$.

Ciphertext only attack (particularly for mode II) could supposedly suggest possible interesting results. Indeed one could try to xor several ciphertext blocks together to eliminate the output blocks and try to recover at least the underlying plaintext. In fact, this gives only access to a modulo 2 sum of at least four plaintext blocks and then is of no use.

More precisely, suppose that $B_t, B_{t+1}, B_{t+2}, B_{t+3}$ are combined with plaintext blocks $M_t, M_{t+1}, M_{t+2}, M_{t+3}$ producing ciphertexts blocks

$$C_t = B_t \oplus M_t, C_{t+1} = B_{t+1} \oplus M_{t+1}, C_{t+2} = B_{t+2} \oplus M_{t+2}, C_{t+3} = B_{t+3} \oplus M_{t+3}.$$

We know that B_{t+j} can be written as (if $L_i^{(t+j)}$ denotes internal states of register L_i at time $t + j$):

$$B_{t+j} = (L_1^{(t+j)}[1] \oplus L_2^{(t+j)}[3]) + 2^{32} * (L_1^{(t+j)}[0] \oplus L_2^{(t+j)}[2])$$
$$+2^{64} * (L_1^{(t+j)}[3] \oplus L_2^{(t+j)}[1]) + 2^{96} * (L_1^{(t+j)}[2] \oplus L_2^{(t+j)}[0]) \quad (1)$$

where $j = 0, \ldots, 3$. Since crossing over gives (in first approach let us forget irregularly clocking of the registers) for register L_i:

$$L_i^{(t+1)}[2] = L_i^{(t)}[0] \qquad L_i^{(t+1)}[3] = L_i^{(t)}[1] \quad (2)$$

at any instant time t, with Equations 1 and 2 we obviously have:

$$C_t \oplus C_{t+1} \oplus C_{t+2} \oplus C_{t+3} = M_t \oplus M_{t+1} \oplus M_{t+2} \oplus M_{t+3} \quad (3)$$

Thus xoring together four ciphertext blocks eliminates the underlying output blocks, yielding modulo 2 sum of four plaintext blocks. This sum cannot in any way be used to retrieve any of the four plaintext blocks. In other words, Equation 3 shows that to recover one plaintext block, the cryptanalyst must guess or know the three other plaintext blocks, that is to say he has to know 75% of the plaintext. When possible this situation makes encryption meaningless, above all for plaintext whose redundancy has been suppressed.

By recalling that, in addition, registers are irregularly clocked, even such a sum is impossible to obtain. To resume, the crossing over mechanism greatly increases the security of the COS ciphers.

3.3   Statistical Tests

We have used the statistical tests defined by D.E. Knuth [14] to check randomness of some COS Ciphers (mainly $(2, 128)$ and $(3, 128)$).

These tests are: frequency test, serial test, gap test, poker test, coupon collector's test, permutation test, run test, serial correlation test.

We did not notice any significant bias. Randomness results are rather very good. Output bits behave as coin-tossing experiment, independently from their neighbours.

4   Performance Analysis (Mode II)

A $(n, 2L)$ version of COS cipher will produce a $2(n-1)L$-bit block at each iteration (mode II) with only L register shifts. Hence COS ciphers are extremely fast ciphers. Performance analysis is presented in Table 2. It has been conducted

Table 2. Approximative encryption speed of some COS ciphers

	Mean Encryption speed
$(2, 128)$	120 Mbits/sec.
$(3, 128)$	253 Mbits/sec.
$(4, 128)$	330 Mbits/sec.
$(3, 256)$	500 Mbits/sec.
$(2, 512)$	510 Mbits/sec.
$(3, 512)$	978 Mbits/sec.

on a Pentium III with 500 MHz CPU, under Linux. Compiler was gcc-egcs-2.91.66 (egcs-1.1.2 release). Approximative mean encrypion speed (in Mbits/sec.) is given. Performance analysis has been conducted on an relatively optimized C implementation. Better results are bound to be obtained either with truly optimized C implementation or assembler implementation. Code size after complete compilation is less than 20 Kbytes.

5   Conclusion and Challenge

COS ciphers are new, ultrafast, (vectorized) stream ciphers built up entirely with non linear feedback shift registers. They offer a very high speed encryption and a very high cryptographic strength. They particularly exhibit extremely good randomness properties.

However, since public evaluation only can confirm the very high quality of this cipher, we propose a 1000 euros cryptanalysis challenge with no time limit. The first who breaks the (2, 128) COS cipher wins the prize. Challenge rules can be found in [7]. We hope that it will promote leading research in NLFSR theory.

These ciphers are placed under the GNU General Public Licence.

References

1. R. Anderson, E. Biham, Two Practical and Provable Secure Block Ciphers: BEAR and LION, In *Fast Software Encryption 96*, number 1039 in Lecture Notes in Computer Science, pp 113–120, Springer Verlag, 1997.
2. K. Aoki, H. Lipmaa, Fast Implementation of AES Candidates, In *Third AES Conference*, April 13-14th, 2000, New York.

3. M. Bellare, J. Killian, P. Rogaway, The security of Cipher Block Chaining, In *Advances in Cryptology - CRYPTO'94*, number 839 in Lecture Notes in Computer Science, pp 341–358, Springer Verlag.
4. C.M. Campbell, Design and Specification of Cryptographic Capabilities, *IEEE Computer Society Magazine*, Vol. 16, Nr. 6, pp 15–19, November 1979.
5. A. Canteaut, M. Trabbia, Improved Fast Correlation Attacks using Parity-check Equations of weight 4 and 5. In *Advances in Cryptology - EUROCRYPT'2000*, number 1807 in Lecture Notes in Computer Science, pp 573–588, Springer Verlag 2000.
6. V. Chepyzhov, T. Johansson, B. Smeets. A Simple Algorithm for Fast Correlation Attacks on Stream Ciphers. In *Fast Software Encryption 2000*, Lecture Notes in Computer Science 1978, Springer Verlag, 2001.
7. http://www-rocq.inria.fr/codes/Eric.Filiol/English/COS/COS.html
8. E. Filiol, Decimation Attack of Stream Ciphers, In *Progress in Cryptology - INDOCRYPT'2000*, number 1977 in Lecture Notes in Computer Science, pp 31–42, Springer Verlag, 2000.
9. E. Filiol, C. Fontaine, Highly Nonlinear Balanced Boolean Functions with a Good Correlation-Immunity, In *Advances in Cryptology - EUROCRYPT'98*, number 1403 in Lecture Notes in Computer Science, pp 475–488, Springer Verlag, 1998.
10. S.W. Golomb, Shift Register Sequences, Agean Park Press, 1982.
11. S.W. Golomb, On the Cryptanalysis of Nonlinear Sequences, In *7th IMA Conference on Cryptography and Coding*, number 1746 in Lecture Notes in Computer Science, pp 236–242 Springer Verlag, 1999.
12. T. Johansson, F. Jönsson, Improved Fast Correlation Attack on stream Ciphers via Convolutional Codes, In *Advances in Cryptology - EUROCRYPT'99*, number 1592 in Lecture Notes in Computer Science, pp 347–362, Springer Verlag, 1999.
13. H. Krawczyk, LFSR-based Hashing and Authentication, In *Advances in Cryptology - CRYPTO'94*, number 839 in Lecture Notes in Computer Science, pp 129–139 Springer Verlag.
14. D.E. Knuth, The Art of Computer Programming, , Addison-Wesley, 1984.
15. X.J. Lai, R.A. Rueppel, J. Woolven, A fast Cryptographic Checksum Algorithm based on Stream Ciphers, In *Advances in Cryptology - AUSCRYPT'92*, number 718 in Lecture Notes in Computer Science, pp 338–348 Springer Verlag.
16. S. Maitra, P. Sarkar, Constructions of Non Linear Boolean Functions with important Cryptographic Properties, In *Advances in Cryptology - EUROCRYPT'2000*, number 1807 in Lecture Notes in Computer Science, pp 485–506, Springer Verlag.
17. S. Maitra, P. Sarkar, New Directions in Design of Resilient Boolean Functions, In *Advances in Cryptology - CRYPTO'00*, number 1880 in Lecture Notes in Computer Science, pp 515–532 Springer Verlag.
18. National Bureau of Standards, NBS FIPS PUB 81, DES Modes of Operation, U.S. Department of Commerce, Dec 1980.
19. http://www.nist.gov/aes/
20. M. Mihaljevic, J.D. Golic, A Fast Iterative Algorithm for a Shift-Register Initial State Reconstruction given the Noisy Output Sequence, In *Advances in Cryptology - AUSCRYPT'90*, number 453 in Lecture Notes in Computer Science, pp 165–175, Springer Verlag, 1990.
21. http://www.industrie.gouv.fr/pratique/aide/appel/sp_priam.htm
22. P. Rogaway, D. Coppersmith, A Software-optimized Encryption Algorithm. In In *Fast Software Encryption 1993*, Lecture Notes in Computer Science 809, pp 56–63, Springer Verlag, 1994.

23. T. Siegenthaler, Decrypting a Class of Stream Ciphers using Ciphertext Only, *IEEE Transactions on Computers*, C-34, 1, pp 81–84, 1985.
24. Y. Tarannikov, On resilient Boolean Functions with Maximal Possible Nonlinearity, http://eprint.iacr.org/2000/005.ps.
25. K. Zheng, M. Huang, On the Linear Syndrome Method in Cryptanalysis, In *Advances in Cryptology - CRYPTO'88*, number 405 in Lecture Notes in Computer Science, pp 469–478, Springer Verlag, 1990.
26. K. Zeng, C.H. Yang, T.R. Rao, An Improved Linear Syndrome Algorithm in Cryptanalysis with Applications, In *Advances in Cryptology - CRYPTO'90*, number 537 in Lecture Notes in Computer Science, pp 34–47, Springer Verlag, 1991.

A   C Implementation

We give a readable but not optimized C code of the $(2, 128)$ COS cipher. Functions f_M, f_{L_1} and f_{L_2} and look-up table T (due to lack of space) will be found in [7].

```c
void setkey(unsigned long int * L1, unsigned long int * L2)
 {
  unsigned long int i,a,M[8],feed;

  /* M register common part initialization */
  M[0] = K1; M[1] = K2; M[2] = K3; M[3] = K4;

  /* M register user's key dependent part initialization */
  if(KEYSIZE == 256) {M[4] = K5; M[5] = K6; M[6] = K7; M[7] = K8;}
  if(KEYSIZE == 192)
    {
    M[4] = K5; M[5] = K6; a = K1 ^ K2 ^ K3;
    M[6]  = T[(a & 0xFF)] | (T[((a >> 8) & 0xFF)] << 8);
    M[6] |= (T[((a >> 16) & 0xFF)] << 16) | (T[a >> 24] << 24);
    a = K4 ^ K5 ^ K6;
    M[7]  = T[(a & 0xFF)] | (T[((a >> 8) & 0xFF)] << 8);
    M[7] |= (T[((a >> 16) & 0xFF)] << 16) | (T[a >> 24] << 24);
    }
  if(KEYSIZE == 128)
    {
    M[4]  = T[(K1 & 0xFF)] | (T[((K1 >> 8) & 0xFF)] << 8);
    M[4] |= (T[((K1 >> 16) & 0xFF)] << 16) | (T[K1 >> 24] << 24);
    M[5]  = T[(K2 & 0xFF)] | (T[((K2 >> 8) & 0xFF)] << 8);
    M[5] |= (T[((K2 >> 16) & 0xFF)] << 16) | (T[K2 >> 24] << 24);
    M[6]  = T[(K3 & 0xFF)] | (T[((K3 >> 8) & 0xFF)] << 8);
    M[6] |= (T[((K3 >> 16) & 0xFF)] << 16) | (T[K3 >> 24] << 24);
    M[7]  = T[(K4 & 0xFF)] | (T[((K4 >> 8) & 0xFF)] << 8);
    M[7] |= (T[((K4 >> 16) & 0xFF)] << 16) | (T[K4 >> 24] << 24);
    }
    M[0] ^= MK;            /* Message key introduction */

  /* Shift M register 256 times */
```

```
for(i = 0;i < 256;i++)
 {
  feed  = ((M[0] & 0x80000000L) >> 21);
  feed |= ((M[0] & 0x8L) << 6);
  feed |= ((M[1] & 0x200L) >> 1);
  feed |= ((M[2] & 0x800L) >> 4);
  feed |= ((M[2] & 0x8L) << 3);
  feed |= ((M[3] & 0x400000L) >> 17);
  feed |= ((M[4] & 0x80000L) >> 15);
  feed |= ((M[5] & 0x800000L) >> 20);
  feed |= ((M[6] & 0x2000000L) >> 23);
  feed |= ((M[7] & 0x80000000L) >> 30);
  feed |= ((M[7] & 0x4L) >> 2);
  M[7] = (M[7] >> 1) | ((M[6] & 1) << 31);
  M[6] = (M[6] >> 1) | ((M[5] & 1) << 31);
  M[5] = (M[5] >> 1) | ((M[4] &1) << 31);
  M[4] = (M[4] >> 1) | ((M[3] &1) << 31);
  M[3] = (M[3] >> 1) | ((M[2] &1) << 31);
  M[2] = (M[2] >> 1) | ((M[1] &1) << 31);
  M[1] = (M[1] >> 1) | ((M[0] &1) << 31);
  M[0] = (M[0] >> 1) | (fM[feed] << 31);
 }
*L1++ = M[4]; *L1++ = M[5]; *L1++ = M[6]; *L1   = M[7];

/* Clock M register 128 times */
for(i = 0;i < 128;i++)
 {
  feed  = ((M[0] & 0x80000000L) >> 21);
  feed |= ((M[0] & 0x8L) << 6);
  feed |= ((M[1] & 0x200L) >> 1);
  feed |= ((M[2] & 0x800L) >> 4);
  feed |= ((M[2] & 0x8L) << 3);
  feed |= ((M[3] & 0x400000L) >> 17);
  feed |= ((M[4] & 0x80000L) >> 15);
  feed |= ((M[5] & 0x800000L) >> 20);
  feed |= ((M[6] & 0x2000000L) >> 23);
  feed |= ((M[7] & 0x80000000L) >> 30);
  feed |= ((M[7] & 0x4L) >> 2);
  M[7] = (M[7] >> 1) | ((M[6] &1) << 31);
  M[6] = (M[6] >> 1) | ((M[5] &1) << 31);
  M[5] = (M[5] >> 1) | ((M[4] &1) << 31);
  M[4] = (M[4] >> 1) | ((M[3] &1) << 31);
  M[3] = (M[3] >> 1) | ((M[2] &1) << 31);
  M[2] = (M[2] >> 1) | ((M[1] &1) << 31);
  M[1] = (M[1] >> 1) | ((M[0] &1) << 31);
  M[0] = (M[0] >> 1) | (fM[feed] << 31);
 }
*L2++ = M[0]; *L2++ = M[1]; *L2++ = M[2]; *L2   = M[3];
 return;
}
```

```
void cos(unsigned long int * L1, unsigned long int * L2,
         unsigned long int * block, unsigned long int flag)
{
 unsigned long int av[4] = {64,65,66,64},clk,feed,i;

 clk = (L1[3] & 1) | ((L2[3] & 1) << 1);
 if(flag) {
   for(i = 0L;i < av[clk];i++)
     {
      feed  = ((L1[3] & 0x4) >> 2);
      feed |= ((L1[3] & 0x20L) >> 4);
      feed |= ((L1[3] & 0x100L) >> 6);
      feed |= ((L1[3] & 0x8000L) >> 12);
      feed |= ((L1[3] & 0x4000000L) >> 22);
      feed |= ((L1[2] & 0x40L) >> 1);
      feed |= ((L1[2] & 0x1000L) >> 6);
      feed |= ((L1[2] & 0x8000L) >> 8);
      feed |= ((L1[2] & 0x2000000L) >> 17);
      L1[3] = (L1[3] >> 1) | ((L1[2] & 1) << 31);
      L1[2] = (L1[2] >> 1) | ((L1[1] & 1) << 31);
      L1[1] = (L1[1] >> 1) | ((L1[0] & 1) << 31);
      L1[0] = (L1[0] >> 1) | (fL1[feed] << 31);
     }}
 else {
   for(i = 0L;i < av[clk];i++)
     {
      feed  = ((L2[3] & 0x4) >> 2);
      feed |= ((L2[3] & 0x20L) >> 4);
      feed |= ((L2[3] & 0x100L) >> 6);
      feed |= ((L2[3] & 0x8000L) >> 12);
      feed |= ((L2[3] & 0x4000000L) >> 22);
      feed |= ((L2[2] & 0x40L) >> 1);
      feed |= ((L2[2] & 0x1000L) >> 6);
      feed |= ((L2[2] & 0x8000L) >> 8);
      feed |= ((L2[2] & 0x2000000L) >> 17);
      L2[3] = (L2[3] >> 1) | ((L2[2] & 1) << 31);
      L2[2] = (L2[2] >> 1) | ((L2[1] & 1) << 31);
      L2[1] = (L2[1] >> 1) | ((L2[0] & 1) << 31);
      L2[0] = (L2[0] >> 1) | (fL2[feed] << 31);
     }}
 *block++ = (L2[0] ^ L1[2]); *block++ = (L2[1] ^ L1[3]);
 *block++ = (L2[2] ^ L1[0]); *block   = (L2[3] ^ L1[1]);
 return;
}
```

B   General Description of the $(2, 2L)$ COS Ciphers

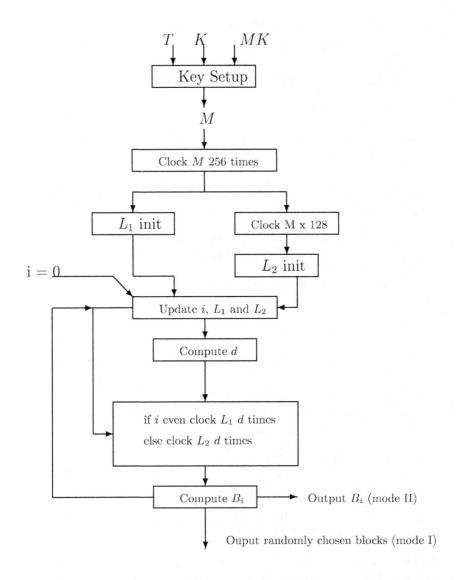

On Rabin-Type Signatures*

Marc Joye[1] and Jean-Jacques Quisquater[2]

[1] Gemplus Card International, Card Security Group
Parc d'Activités de Gémenos, B.P. 100, 13881 Gémenos Cedex, France
marc.joye@gemplus.com
http://www.geocities.com/MarcJoye/
[2] UCL Crypto Group, Université catholique de Louvain
Place du Levant 3, 1348 Louvain-la-Neuve, Belgium
jjq@dice.ucl.ac.be
http://www.uclcrypto.org/

Abstract. This paper specializes the signature forgery by Coron, Naccache and Stern (1999) to Rabin-type systems. We present a variation in which the adversary may derive the private keys and thereby forge the signature on *any* chosen message. Further, we demonstrate that, contrary to the RSA, the use of larger (even) public exponents does not reduce the complexity of the forgery. Finally, we show that our technique is very general and applies to any Rabin-type system designed in a unique factorization domain, including the Williams' M^3 scheme (1986), the cubic schemes of Loxton et al. (1992) and of Scheidler (1998), and the cyclotomic schemes (1995).

Keywords. Rabin-type systems, digital signatures, signature forgeries, factorization.

1 Introduction

In this paper, we specialize the signature forgery of Coron, Naccache and Stern [8] to Rabin-type systems [19,24]. We present a variation in which the adversary may derive the private keys and thereby forge the signature on *any* chosen message. Further, we demonstrate that, contrary to the RSA, systems using larger (even) public exponents are *equally* susceptible to the presented forgery. We also show that our technique is very general and applies to any Rabin-type systems designed in a unique factorization domain, including the Williams' M^3 scheme [27], the cubic schemes of Loxton et al. [16] and of Scheidler [20], and the cyclotomic schemes [21].

As an application, we analyze the implications of our forgery against the PKCS #1 standard [4]. Finally, and of independent interest, we propose a *generic* technique (i.e., applicable to *any* encoding message method) that reduces the overall complexity of a forgery from n to $\sqrt{n}$.

---

* A working draft of this work was presented at the ISO/IEC JTC1/SC27/WG2 meeting in August 1999.

B. Honary (Ed.): Cryptography and Coding 2001, LNCS 2260, pp. 99–113, 2001.
© Springer-Verlag Berlin Heidelberg 2001

The rest of this paper is organized as follows. We first begin by a brief presentation of Rabin-type systems. Next, in Section 3, we review the Coron-Naccache-Stern forgery and turn it into an universal forgery against Rabin-type signature schemes. In Section 4, we generalize the forgery to higher exponents and higher degree schemes. We apply it to PKCS #1 encoding method in Section 5. We also present a generic technique for reducing the complexity of the forgery. Finally, we conclude in Section 6.

2   Rabin-Type Systems

In this section, following the IEEE/P1363 specifications for public-key cryptography [2] (see also [17, Chapter 11]), we present a modified version of the Rabin-Williams signature scheme [19,24]. The scheme consists of three algorithms: the setup, the signature and the verification. For setting up the system, each user generates a pair of public/private keys. The private key is used to sign messages with the signature algorithm. Using the corresponding public key, a signature can then be verified and the signed message recovered with the verification algorithm.

setup: Generate two primes p, q such that $p \equiv 3 \pmod 8$ and $q \equiv 7 \pmod 8$ and compute $n = pq$. Define an appropriate "representation" function $R :$ $\mathcal{M} \to \mathcal{M}_R : m \mapsto \tilde{m} = R(m)$, where $\mathcal{M}$ is the set of valid messages and $\mathcal{M}_R = \{\tilde{m} = R(m) \in (\mathbb{Z}/n\mathbb{Z})^* : \tilde{m} \equiv 6 \pmod{16}\}$ is the set of message representatives. The public key is n and the private key is $d = (n-p-q+5)/8$.

signature: Compute $\tilde{m} = R(m)$ and $\hat{m}$ given by

$$\hat{m} = \begin{cases} \tilde{m} \bmod n & \text{if } (\tilde{m}|n) = 1 \\ \tilde{m}/2 \bmod n & \text{if } (\tilde{m}|n) = -1 \end{cases} .$$

The signature on message m is $s = \hat{m}^d \bmod n$.

verification: Compute $m' = s^2 \bmod n$. Then, take

$$\tilde{m} = \begin{cases} m' & \text{if } m' \equiv 6 \pmod 8 \\ 2m' & \text{if } m' \equiv 3 \pmod 8 \\ n - m' & \text{if } m' \equiv 7 \pmod 8 \\ 2(n - m') & \text{if } m' \equiv 2 \pmod 8 \end{cases} .$$

If $\tilde{m} \in \mathcal{M}_R$ then the signature is accepted and message m is recovered from $\tilde{m}$.

Remark 1. A signature scheme with appendix can also be defined along these lines. In that case, the set of messages representatives is given by $\mathcal{M}_R = \{\tilde{m} \in (\mathbb{Z}/n\mathbb{Z})^* : \tilde{m} \equiv 10 \pmod{16}\}$.

Remark 2. The recommended function R for message encoding is the international standard ISO/IEC 9796 [3]. Another possible encoding is specified in PKCS #1 v2.0 [4]; this latter, however, only covers signature schemes with appendix.

Remark 3. The presented scheme supposes $p \equiv 3 \pmod 8$ and $q \equiv 7 \pmod 8$. Similar schemes with form-free primes may be found in [26,13].

Noticing that $2d = (p-1)(q-1)/4+1$, the correctness of the method follows from the next lemma; moreover, it uses the fact that if $n \equiv 5 \pmod 8$ then $(-z|n) = (z|n) = -(2z|n)$.

Lemma 1. *Let $n = pq$, where p, q are distinct primes and $p, q \equiv 3 \pmod 4$. If $(z|n) = 1$, then*

$$z^{(p-1)(q-1)/4} \equiv \pm 1 \pmod n .$$

Proof. See [24, Lemma 1]. □

3 Signature Forgeries

This section reviews the Coron-Naccache-Stern forgery [8] when applied to the Rabin-Williams scheme. In the second part, we modify it into an universal forgery so that the signature on any message can be obtained without knowing the private key.

3.1 Coron-Naccache-Stern Forgery

As aforementioned, for each message representative $\tilde{m}_i$, $\hat{m}_i = \tilde{m}_i$ if $(\tilde{m}_i|n) = 1$ and $\hat{m}_i = \tilde{m}_i/2$ if $(\tilde{m}_i|n) = -1$; the corresponding signature is then given by $s_i = \hat{m}_i{}^d \bmod n$. (Note here that $(\hat{m}_i|n) = 1$.)

Suppose that an adversary has collected several pairs $(\hat{m}_i, s_i)$ such that the $\hat{m}_i$'s are smooth (modulo n). More precisely, suppose she knows

$$\hat{m}_i \equiv (-1)^{v_{0,i}} \prod_{1 \le j \le B} p_j{}^{v_{j,i}} \pmod n , \tag{1}$$

where $p_1 < \cdots < p_B$ are prime and $v_{j,i} \in \mathbb{Z}$, and $s_i = \hat{m}_i{}^d \bmod n$, for $1 \le i \le \ell$. Then she can forge the signature on a message m_τ, provided that $\hat{m}_\tau$ is smooth (modulo n), as follows.[1,2]

---

[1] It is here essential to note that the smoothness requirement must only be satisfied modulo n. For example, 197 is prime in $\mathbb{Z}$ but is 3-smooth as an element of $(\mathbb{Z}/437\mathbb{Z})$, i.e., $197 \equiv 2^9 \cdot 3^{-3} \pmod{437}$.

[2] In [8], the authors only consider positive "messages" $\hat{m}_i$. We slightly generalize their presentation by introducing the term $(-1)^{v_{0,i}}$ in Eq. (1).

To each $\widehat{m}_i$, she associates the B-tuple $\vec{V}_i$ given by $\vec{V}_i = (v_{1,i}, \dots, v_{B,i})$. Let $\vec{V}_\tau$ denote the B-tuple corresponding to $\widehat{m}_\tau$. If there exist integers β_i such that $\vec{V}_\tau \equiv \sum_{1 \le i \le \ell} \beta_i \vec{V}_i \pmod{4}$, then

$$v_{j,\tau} = \sum_{1 \le i \le \ell} \beta_i v_{j,i} - 4\gamma_j \quad (1 \le j \le B) \tag{2}$$

for some $\gamma_j \in \mathbb{Z}$.

Hence, the signature on message m_τ can be expressed as

$$s_\tau := \widehat{m}_\tau{}^d \bmod n \equiv \left[(-1)^{v_{0,\tau}} \prod_{1 \le j \le B} p_j{}^{v_{j,\tau}} \right]^d \quad \text{(from Eq. (1))}$$

$$\equiv \prod_{1 \le j \le B} p_j{}^{v_{\tau,j}\, d} \quad \text{(since } d \text{ is even)}$$

$$\equiv \prod_{1 \le i \le \ell} \prod_{1 \le j \le B} p_j{}^{(\beta_i v_{j,i} - 4\gamma_j)d} \quad \text{(from Eq. (2))}$$

$$\equiv \prod_{1 \le i \le \ell} s_i{}^{\beta_i} \prod_{1 \le j \le B} p_j{}^{-2\gamma_j} \pmod{n} \ . \tag{3}$$

The only difference with [8] is that we use the weaker relation $p_j{}^{4d} \equiv p_j{}^2 \pmod{n}$. (The relation $p_j{}^{2d} \equiv \pm p_j \pmod{n}$ only holds when $(p_j|n) = 1$; see Lemma 1.)

3.2   Universal Forgery

We will now show that we can do much better than forging the signature on a smooth $\widehat{m}_\tau$, namely, forging the signature on *any* chosen $\widehat{m}_\tau$. We need the following proposition.

Proposition 1. *Let $n = pq$, where p, q are distinct primes and $p, q \equiv 3 \pmod 4$ and let $d = (n - p - q + 5)/8$. If $(z|n) = -1$, then*

$$\gcd\!\left(z^{2d} \mp z \pmod{n}, n\right) = p \text{ or } q \ .$$

Proof. Since $p, q \equiv 3 \pmod 4$, both $(p-1)/2$ and $(q-1)/2$ are odd. Hence, $z^{2d} \equiv z^{(p-1)(q-1)/4} z \equiv (z|p)^{(q-1)/2} z \equiv (z|p) z \pmod p$, and similarly $z^{2d} \equiv (z|q) z \pmod q$. Noting that $(z|p) = -(z|q) = \pm 1$, the proposition is proved. $\square$

The above proposition suggests that if an adversary can derive the signature on an $\widehat{m}_\tau$ such that $(\widehat{m}_\tau|n) = -1$, then she can factor the modulus by computing $\gcd\!\left(s_\tau{}^2 - \widehat{m}_\tau \pmod{n}, n\right)$. This, however, is *not* possible. Define

$$\mathcal{J}(n, B) = \left\{ 1 \le j \le B : p_j \text{ is prime and } \left(\frac{p_j}{n}\right) = -1 \right\} \ . \tag{4}$$

Since $(\widehat{m}_i|n) = 1$ (cf. beginning of §3.1), we must have

$$\sum_{j \in \mathcal{J}(n,B)} v_{j,i} \equiv 0 \pmod 2 . \tag{5}$$

So, any linear combination, as done in Eq. (2), will always yield $\sum_{j \in \mathcal{J}(n,B)} v_{j,\tau} \equiv 0 \pmod 2$, resulting in $(\widehat{m}_\tau|n) = 1$.

But there is another way to consider Proposition 1: If the adversary is able to obtain the signature on $\widehat{m}_\tau = \pm \widehat{r}_\tau{}^2$ for some $\widehat{r}_\tau$ such that $(\widehat{r}_\tau|n) = -1$ (note here that $(\widehat{m}_\tau|n) = (\pm 1|n) \cdot (\widehat{r}_\tau{}^2|n) = 1 \cdot 1 = 1$), then

$$\gcd(s_\tau - \widehat{r}_\tau \pmod n), n) \tag{6}$$

will give a non-trivial factor of n. To make this feasible, in addition to verify Eq. (2), the components of $\vec{V}_\tau$ must satisfy

$$v_{j,\tau} \equiv 0 \pmod 2 \quad \text{for all } 1 \leq j \leq B \tag{7}$$

and

$$\sum_{j \in \mathcal{J}(n,B)} v_{j,\tau} \equiv 2 \pmod 4 . \tag{8}$$

The first condition ensures that the "message" corresponding to $\vec{V}_\tau$ is a square (i.e., $\widehat{m}_\tau = \pm \widehat{r}_\tau{}^2$), while the second condition ensures that $(\widehat{r}_\tau|n) = -1$. Replacing $v_{j,\tau}$ by Eq. (2), Eqs (7) and (8) can respectively be rewritten as

$$\sum_{1 \leq i \leq \ell} \beta_i v_{j,i} \equiv 0 \pmod 2 \quad \text{for all } 1 \leq j \leq B \tag{9}$$

and, defining $2\zeta_i = (\sum_{j \in \mathcal{J}(n,B)} v_{j,i}) \bmod 4 \ (\in \{0,2\} \text{ from Eq. (5)})$,

$$\sum_{j \in \mathcal{J}(n,B)} \sum_{1 \leq i \leq \ell} \beta_i v_{j,i} \equiv \sum_{1 \leq i \leq \ell} \beta_i \sum_{j \in \mathcal{J}(n,B)} v_{j,i} \equiv \sum_{1 \leq i \leq \ell} \beta_i 2\zeta_i \equiv 2 \pmod 4$$

$$\Longleftrightarrow \sum_{1 \leq i \leq \ell} \beta_i \zeta_i \equiv 1 \pmod 2 . \tag{10}$$

To sum up, an adversary can recover the secret factorization of n by carrying out the following steps:

(I) For $1 \leq i \leq \ell$, write $\widehat{m}_i \equiv (-1)^{v_{0,j}} \prod_{1 \leq j \leq B} p_j{}^{v_{j,i}} \pmod n$;

(II) Define $\vec{V}_i = (v_{1,i}, \dots, v_{B,i})$ and $\zeta_i = \dfrac{(\sum_{j \in \mathcal{J}(n,B)} v_{j,i}) \bmod 4}{2}$;

(III) Find $\beta_1, \dots, \beta_\ell$ such that

a) for $1 \leq j \leq B$, $\sum_{1 \leq i \leq \ell} \beta_i v_{j,i} \equiv 0 \pmod 2$;

b) $\displaystyle\sum_{1\le i\le\ell} \beta_i\,\zeta_i \equiv 1 \pmod 2$;

(IV) Compute $\vec{V}_\tau = (v_{1,\tau},\dots,v_{B,\tau}) \in (\mathbb{Z}/4\mathbb{Z})^B$,

$$\text{where } v_{j,\tau} = \sum_{1\le i\le\ell} \beta_i\,v_{j,i} - 4\gamma_j \text{ for some } \gamma_j\in\mathbb{Z};$$

(V) Set $\displaystyle\widehat{r}_\tau = \prod_{1\le j\le B} p_j{}^{v_{j,\tau}/2} \pmod n$;

(VI) Compute $\displaystyle s_\tau = \prod_{1\le i\le\ell} s_i{}^{\beta_i} \prod_{1\le j\le B} p_j{}^{-2\gamma_j} \bmod n$;

(VII) Recover the factors of n by computing $\gcd(s_\tau - \widehat{r}_\tau \pmod n, n)$.

Example 1. Here is a "toy" example to illustrate the forgery. Let $p = 8731\ (\equiv 3 \pmod 8)$ and $q = 3079\ (\equiv 7 \pmod 8)$ yielding a modulus $n = 26882749$. So, the private exponent is given by $d = 3358868$.
We consider p_4-smooth message representatives $\widetilde{m}_i \in \mathcal{M}_R$. We have $\mathcal{J}(n,4) = \{2,7\}$. Suppose, we are given:

$\widetilde{m}_i$	$\widehat{m}_i \pmod n$	s_i	$\vec{V}_i$	ζ_i
70	$70\ (= 2\cdot 5\cdot 7)$	8417525	$(1,0,1,1)$	1
294	$147\ (= 3\cdot 7^2)$	11480098	$(0,1,0,2)$	1
486	$243\ (= 3^5)$	16287310	$(0,5,0,0)$	0
630	$630\ (= 2\cdot 3^2\cdot 5\cdot 7)$	1630174	$(1,2,1,1)$	1

Conditions (III-a) and (III-b) yield

$$\begin{cases} \beta_1 + \beta_4 \equiv 0 \pmod 2 \\ \beta_2 + \beta_3 \equiv 0 \pmod 2 \\ \beta_1 + \beta_2 + \beta_4 \equiv 1 \pmod 2 \end{cases} \Longleftrightarrow \begin{cases} \beta_1 \equiv \beta_4 \pmod 2 \\ \beta_2 \equiv \beta_3 \equiv 1 \pmod 2 \end{cases}.$$

Taking $\beta_1 = \beta_4 = 0$ and $\beta_2 = \beta_3 = 1$, we have $\vec{V}_\tau = (0,2,0,2)$, which corresponds to $\widehat{m}_\tau = 3^2\cdot 7^2 = 21^2$ whose signature is given by $s_\tau = s_2{}^1 s_3{}^1 3^{-2} \bmod n = 8076196$. So, the factorization of n is obtained by computing $\gcd(8076196 - 21, n) = 8731\ (= p)$. $\diamond$

We will see below (Algorithm 1) a simple method to compute a solution $(\beta_1,\dots,\beta_\ell) \in (\mathbb{Z}/2\mathbb{Z})^\ell$. This is a slight modification of Algorithm 2.3.1 in [7, pp. 56–57] (see also [12, Algorithm N, pp. 425–426]).
Using matrix notations, Conditions (III-a) and (III-b) can be rewritten as

$$\underbrace{\begin{pmatrix} v_{1,1} & \cdots & v_{1,\ell} & 0 \\ \vdots & & \vdots & \vdots \\ v_{B,1} & \cdots & v_{B,\ell} & 0 \\ \zeta_1 & \cdots & \zeta_\ell & 1 \end{pmatrix}}_{\substack{(\bmod 2) \\ := \mathbf{U}}} \begin{pmatrix} \beta_1 \\ \vdots \\ \beta_\ell \\ 1 \end{pmatrix} \equiv \vec{0} \pmod 2. \tag{11}$$

So, the problem is reduced to find a vector $\vec{K} = (\kappa_1, \ldots, \kappa_\ell, \kappa_{\ell+1}) \in \ker \mathbf{U}$ (i.e., the kernel of matrix $\mathbf{U}$), whose last coordinate, $\kappa_{\ell+1}$, is equal to 1. In that case, we have $\beta_i = \kappa_i$, $1 \leq i \leq \ell$.

Algorithm 1. This algorithm computes a solution (if any) $\beta_1, \ldots, \beta_\ell$ satisfying Eq. (11). We let $u_{r,c}$ denote the entry (modulo 2) at row r and column c of matrix $\mathbf{U}$.

1. [initialization] Set $c \leftarrow 1$ and for $1 \leq j \leq B+1$, set $t_j \leftarrow 0$.
2. [scanning] If there is some r in the range $1 \leq r \leq B+1$ such that $u_{r,c} = 1$ and $t_r = 0$, then go to Step 3. Otherwise, go to Step 5.
3. [elimination] For all $j \neq r$, if $u_{j,c} = 1$, then add (modulo 2) row r to row j. Set $t_r \leftarrow c$.
4. [loop] If $c \leq \ell$, then set $c \leftarrow c+1$ and go to Step 2.
5. [kernel] Evaluate the vector $\vec{K} = (\kappa_1, \ldots, \kappa_{\ell+1})$ defined by

$$\kappa_i = \begin{cases} u_{j,c} & \text{if } t_j = i > 0 \\ 1 & \text{if } i = c \\ 0 & \text{otherwise} \end{cases}.$$

   If $\kappa_{\ell+1} = 0$ and $c \leq \ell$, then set $c \leftarrow c+1$ and go to Step 2.
6. [output] If $\kappa_{\ell+1} = 1$, then output $\beta_i \leftarrow \kappa_i$ for all $1 \leq i \leq \ell$; otherwise, output no solution.

Example 1 (cont'd). If we apply the previous algorithm to Example 1, matrix $\mathbf{U}$ is given by

$$\mathbf{U} = \begin{pmatrix} \boxed{1} & 0 & 0 & 1 & 0 \\ 0 & 1 & 1 & 0 & 0 \\ 1 & 0 & 0 & 1 & 0 \\ 1 & 0 & 0 & 1 & 0 \\ 1 & 1 & 0 & 1 & 1 \end{pmatrix}.$$

We then successively obtain for $c = 1, 2, 3$

$$\mathbf{U} = \begin{pmatrix} \boxed{1} & 0 & 0 & 1 & 0 \\ 0 & \boxed{1} & 1 & 0 & 0 \\ 0 & 0 & 0 & 0 & 0 \\ 0 & 0 & 0 & 0 & 0 \\ 0 & 1 & 0 & 0 & 1 \end{pmatrix}, \begin{pmatrix} \boxed{1} & 0 & 0 & 1 & 0 \\ 0 & \boxed{1} & 1 & 0 & 0 \\ 0 & 0 & 0 & 0 & 0 \\ 0 & 0 & 0 & 0 & 0 \\ 0 & 0 & \boxed{1} & 0 & 1 \end{pmatrix}, \begin{pmatrix} \boxed{1} & 0 & 0 & 1 & 0 \\ 0 & \boxed{1} & 0 & 0 & 1 \\ 0 & 0 & 0 & 0 & 0 \\ 0 & 0 & 0 & 0 & 0 \\ 0 & 0 & \boxed{1} & 0 & 1 \end{pmatrix}$$

after the elimination step. We also have $t_1 = 1$, $t_2 = 2$, $t_5 = 3$ (which is indicated by the boxes ($t_r = c$)) and $t_3 = t_4 = 0$. At this point, we have $c = 4$ and the kernel step yields the vector $\vec{K} = (u_{1,4}, u_{2,4}, u_{5,4}, 1, 0) = (1, 0, 0, 1, 0)$. Since $\kappa_5 = 0$, we increment c, $c = 5$, and go to the scanning step. Then, since $t_2 = t_5 = 0$ (note that $u_{c,2} = u_{c,5} = 1$), we directly go to the kernel step and obtain the new vector $\vec{K} = (u_{1,5}, u_{2,5}, u_{5,5}, 0, 1) = (0, 1, 1, 0, 1)$. So, we finally find $\beta_1 = 0$, $\beta_2 = 1$, $\beta_3 = 1$ and $\beta_4 = 0$.   ◇

3.3   Improvements

The methods we have presented so far are subject to numerous possible improvements. We just mention two of these.

In Eq. (1), we consider only "messages" $\widehat{m}_i$ whose largest prime factor (modulo n) is p_B. As for modern factorization methods, a substantial speed-up can be obtained by also considering the $\widehat{m}_i$'s which are p_B-smooth except for one or two factors [15]. Another speed-up can be obtained by using structured Gaussian elimination to solve Eq. (11); see [18] for an efficient variation directly applicable to our case.

4   Generalizations

4.1   Higher Exponents

The signature scheme presented in Section 2 can be generalized to other even public exponents besides $e = 2$. Define $\Lambda = \text{lcm}[(p-1)/2, (q-1)/2]$. It suffices to choose e relatively prime to Λ, the corresponding private exponent d is then given according to $ed \equiv 1 \pmod{\Lambda}$ (see [24]). The scheme remains exactly the same except that $m' = s^2 \bmod n$ must be replaced by $m' = s^e \bmod n$ in the verification stage.

In that setting, Proposition 1 becomes

Proposition 2. *Let $n = pq$, where p, q are distinct primes and $p, q \equiv 3 \pmod 4$ and let e, d such that e is even, $\gcd(e, \Lambda) = 1$ and $ed \equiv 1 \pmod{\Lambda}$. If $(z|n) = -1$, then*

$$\gcd\left(z^{ed} \mp z \pmod n, n\right) = p \text{ or } q \ .$$

Proof. From $ed \equiv 1 \pmod{\Lambda}$, we deduce that $ed \equiv 1 \pmod{(p-1)/2}$ and so there exists $\gamma \in \mathbb{Z}$ such that $ed = \gamma \frac{p-1}{2} + 1$. Further, since $p \equiv 3 \pmod 4$, $(p-1)/2$ is odd. Hence, γ must be odd since ed is even. Consequently, $z^{ed} \equiv z^{\gamma(p-1)/2} z \equiv (z|p)^\gamma z \equiv (z|p) z \equiv \pm z \pmod p$ and similarly $z^{ed} \equiv (z|q) z \equiv -(z|p) z \equiv \mp z \pmod q$, which completes the proof.   □

Since e is even, we can write $e = 2e_1$. It is here worth remarking that anyone can raise an element to the $e_1{}^{\text{th}}$ power (modulo n). So, if an adversary follows Steps (I)–(VI) as described in Section 3, she can recover the factors of n by computing

$$\gcd(S_\tau - \widehat{r}_\tau \pmod n, n) \ , \tag{12}$$

where

$$S_\tau := s_\tau{}^{e_1} \bmod n \equiv (\widehat{m}_\tau{}^d)^{e_1} \equiv \prod_{1 \le i \le \ell} \prod_{1 \le j \le B} p_j{}^{(\beta_i v_{j,i} - 4\gamma_j)de_1}$$

$$\equiv \prod_{1 \le i \le \ell} s_i{}^{\beta_i e_1} \prod_{1 \le j \le B} p_j{}^{-2\gamma_j} \pmod n \ .$$

The forgery is thus no more expensive against a scheme with a large public exponent e than against the basic scheme with $e = 2$.

4.2   Higher Degree Schemes

Rabin-type schemes can be developed in any unique factorization domain. In [27], Williams presents a scheme, called M^3, with public exponent 3 using arithmetic in a quadratic number field. This scheme was later extended to cyclotomic fields by Scheidler and Williams in [21] where they also give a scheme with a public exponent 5. In [20], Scheidler modifies Williams' M^3 scheme so that it works with a larger class of primes. Finally, in [16], Loxton et al. give another cubic scheme; the main difference with [27] being its easy geometrical interpretation. In this paragraph, we will stick on this latter scheme because it most resembles the Rabin-Williams scheme presented in Section 2.

The scheme of Loxton et al. uses the ring of Eisenstein integers, namely $\mathbb{Z}[\omega]$, where $\omega = (-1 + \sqrt{-3})/2$ is a primitive cube root of unity. Its correctness relies of the following lemma (compare it with Lemma 1).

Lemma 2. *Let* $n = pq$, *where* p, q *are distinct primes in* $\mathbb{Z}[\omega]$, $3 \nmid Nn$ *and* $Nq \equiv 2\,Np - 1 \pmod 9$. *If* $(z|n)_3 = 1$, *then*

$$z^{(Np-1)(Nq-1)/9} \equiv \left(\frac{z}{p}\right)_3^{(2/3)(Np-1)} \pmod n .$$

Proof. See [16, Lemma 1]. □

Essentially, that scheme suggests to generate two primes $p, q \in \mathbb{Z}[\omega]$ such that $p \equiv 8 + 6\omega \pmod 9$ and $q \equiv 5 + 6\omega \pmod 9$ and to compute $n = pq$. The public key is n and the private key is $d = [(Np-1)(Nq-1) + 9]/27$, where N denotes the norm. A message m is signed by computing $\tilde{m} = R(m)$ (for an appropriate function R) and $\hat{m} = (1-\omega)^{3-t}[(1-\omega)\tilde{m} + 1]$, where $\omega^t = (\tilde{m}|n)_3$; the signature is $s = \hat{m}^d \bmod n$. The signature is then verified by cubing s, and if accepted, the message m is recovered.

We need an analogue of Proposition 1.

Proposition 3. *Let* $n = pq$, *where* p, q *are primes in* $\mathbb{Z}[\omega]$, $Np \equiv 7 \pmod 9$ *and* $Nq \equiv 4 \pmod 9$, *and let* $d = [(Np-1)(Nq-1)+9]/27$. *If* $(z|n)_3 = \omega$ *or* ω^2, *then*

$$\gcd\bigl(z^{3d} - \omega^k z \pmod n, n\bigr) = p \text{ or } q ,$$

for some $k \in \{0, 1, 2\}$.

Proof. We first note that $(Np-1)/3 \equiv 2 \pmod 3$ and $(Nq-1)/3 \equiv 1 \pmod 3$. So, $z^{3d} \equiv z^{(Np-1)(Nq-1)/9} z \equiv (z|p)_3^{(Nq-1)/3} z \equiv (z|p)_3 z \pmod p$, and similarly $z^{3d} \equiv (z|q)_3^{(Np-1)/3} z \equiv (z|q)_3^2 z \pmod q$. The proposition now follows by observing that $(z|p)_3 \neq (z|q)_3^2$ because, by hypothesis, $(z|n)_3 = (z|p)_3 (z|q)_3 = \omega$ or ω^2. □

From this proposition, we can mimic the forgery presented against the Rabin-Williams scheme. The adversary has to find $\beta_1, \ldots, \beta_\ell$ such that (a) for $1 \leq j \leq B$, $\sum_{1 \leq i \leq \ell} \beta_i \, v_{j,i} \equiv 0 \pmod 3$; and (b) $\sum_{1 \leq i \leq \ell} \beta_i \, \zeta_i \equiv 1, 2 \pmod 3$, where $3\zeta_i := (\sum_{i \in \mathcal{J}(n,B)} v_{j,i}) \bmod 9$ and $\mathcal{J}(n, B) := \{1 \leq j \leq B : p_j \text{ is prime in } \mathbb{Z}[\omega]$ and $(p_j | n)_3 = \omega \text{ or } \omega^2\}$. She then computes $\vec{V}_\tau = (v_{1,\tau}, \ldots, v_{B,\tau}) \in (\mathbb{Z}/9\mathbb{Z})^B$, where $v_{j,\tau} = \sum_{1 \leq i \leq \ell} \beta_i \, v_{j,i} - 9\gamma_j$ (for some $\gamma_j \in \mathbb{Z}$), $\hat{r}_\tau = \prod_{1 \leq j \leq B} p_j^{v_{j,\tau}/3}$ (mod n) and $s_\tau = \prod_{1 \leq i \leq \ell} s_i^{\beta_i} \prod_{1 \leq j \leq B} p_j^{-3\gamma_j} \bmod n$. Finally, by computing

$$\gcd(s_\tau - \omega^k \hat{r}_\tau \pmod n), n) \tag{13}$$

for some $k \in \{0, 1, 2\}$, she finds the factors of n.

5   Applications

As an application, we will analyze the consequences of the previously described forgeries (Section 3) when the Rabin-Williams signature scheme is employed with the PKCS #1 v2.0 message encoding method as specified in [4]. We note, however, that the PKCS #1 standard is not expressly intended for use with the Rabin-Williams scheme but rather with the plain RSA scheme.

In the second part, we will present an algorithm which reduces the complexity of the forgery from n to $\sqrt{n}$. This algorithm is not restricted to PKCS #1: it remains applicable *whatever the employed encoding message method.*

5.1   PKCS #1 Encoding Method

We only briefly review the PKCS #1 message encoding method and refer the reader to [4] for details. PKCS #1 supports signature schemes with appendix. (In such a scheme, the message must accompany the signature in order to verify the validity of the signature.) The set of message representatives is thus given by $\mathcal{M}_R = \{\tilde{m} \in (\mathbb{Z}/n\mathbb{Z})^* : \tilde{m} \equiv 10 \pmod{16}\}$ (cf. Remark 1).

Let m be the message being encoded into the message representative $\tilde{m}$. First, a hash function is applied to m to produce the hash value $H = \text{Hash}(m)$. Next, the hash algorithm identifier and the hash value H are combined into an ASN.1 value and DER-encoded (see [11] for the relevant definitions). Let T denote the resulting DER-encoding. Similarly to what is done in ISO 9796 [3], we concatenate the octet $\mathtt{0A_{16}}$ to obtain the data string $D = T \| \mathtt{0A_{16}}$ (the reason is to ensure that D, viewed as an integer, is congruent to 10 modulo 16). The encoded message EM is then formed by concatenating the block-type octet BT, the padding string PS, the $\mathtt{00_{16}}$ octet and the data string D; or schematically,

$$EM = \mathtt{00_{16}} \| BT \| PS \| \mathtt{00_{16}} \| D , \tag{14}$$

where BT is a single octet containing the value $\mathtt{00_{16}}$ or $\mathtt{01_{16}}$. Let $\|n\|$ and $\|D\|$ respectively denote the octet-length of modulus n and D. When $BT = \mathtt{00_{16}}$ then PS consists of $\|n\| - \|D\| - 3$ octets having value $\mathtt{00_{16}}$; when $BT = \mathtt{01_{16}}$ then PS consists of $\|n\| - \|D\| - 3$ octets having value $\mathtt{FF_{16}}$.

The message representative $\tilde{m}$ is the integer representation of EM.

Remark 4. Three hash functions are recommended: SHA-1 [1], MD2 [5], and MD5 [6]. The default function is SHA-1; MD2 and MD5 are only recommended for compatibility with existing applications based on PKCS #1 v1.5.

In what follows, we will assume that SHA-1 is the used hash function. In that case, the data string D consists of $(15 + 20 + 1) = 36$ octets ($= 288$ bits).

Type 0 encoding. With type 0, i.e., when $BT = 00_{16}$, the message representatives, $\widetilde{m}_i$, are 288-bit long, *whatever the size of the modulus.* We have seen that the effectiveness of our forgeries is related to the smoothness (modulo n) of the $\widehat{m}_i$'s that appear in Eq. (1), where $\widehat{m}_i = \widetilde{m}_i$ or $\widetilde{m}_i/2$ according to the value of $(\widetilde{m}_i|n)$. So, when $BT = 00_{16}$, each $\widehat{m}_i$ is at most a 288-bit integer and we can hope that they factor (in $\mathbb{Z}$) into small primes.

Type 1 encoding. Type 1 encoding, i.e., $BT = 01_{16}$, is the recommended way to encode a message. In that case, the message representatives are longer. For a 1024-bit modulus, these are 1016-bit integers (the leading octet in EM is 00_{16}). Therefore, the $\widehat{m}_i$'s happen unlikely to be smooth. But, what we ultimately need is to find smooth $\widehat{m}_i$'s *modulo* n, that is, we need to find an $\widehat{M}_i \equiv \widehat{m}_i \pmod{n}$ so that $\widehat{M}_i$ is smooth as an element in $\mathbb{Z}/n\mathbb{Z}$ (cf. Footnote (^{1})). As already noted in [8], if the 1024-bit modulus has the special form

$$n = 2^{1023} \pm t, \tag{15}$$

then it is easy to find an $\widehat{M}_i$ whose magnitude is comparable to that of t. Indeed, when $BT = 01_{16}$, the padding string is formed with $(128 - 36 - 3) = 89$ octets having value FF_{16}. Therefore, considering the data string D_i as a 288-bit integer, we can write from Eq. (14), $\widetilde{m}_i = \{(2^8)^{89} + [(2^8)^{89} - 1]\}(2^8)^{37} + D_i$. Hence, $\widehat{m}_i = (2^{713} - 1)2^{296} + D_i$ or $(2^{713} - 1)2^{295} + D_i/2$ according to $(\widetilde{m}_i|n) = 1$ or -1. So, defining $M_i := 2^{g_i} \widehat{m}_i - n$ and setting $\widehat{M}_i \equiv 2^{-g_i} M_i \equiv \widehat{m}_i \pmod{n}$, an appropriate choice for g_i will remove the term $2^{713+296}$ (resp. $2^{713+295}$). Namely, we set

$$M_i = \begin{cases} 2^{14}\,\widehat{m}_i - n = 2^{14}\,D_i - 2^{310} \mp t & \text{if } (\widetilde{m}_i|n) = 1 \\ 2^{15}\,\widehat{m}_i - n = 2^{15}\,D_i - 2^{310} \mp t & \text{if } (\widetilde{m}_i|n) = -1 \end{cases}. \tag{16}$$

Hence, since a special modulus of the form (15) with a square-free t as small as 400 bits offers the same security as a regular 1024-bit modulus [14], the M_i's as given in Eq. (16) can already be as small as 400-bit integers. Consequently, hoping that the M_i's factor into small integers (in $\mathbb{Z}$),^{3} $\widehat{M}_i \equiv 2^{-g_i} M_i \equiv \widehat{m}_i$ (mod n) will be smooth as elements of $\mathbb{Z}/n\mathbb{Z}$.

---

^{3} The probability that a 400-bit integer is smooth is relatively small; but see § 3.3 for some possible alternatives.

5.2   Arbitrary Encoding

For general ("form-free") message encoding methods and moduli, the message representatives have roughly the same length as the modulus. We now present a generic method which on input an arbitrary message representative outputs a "message equivalent" whose length has the size of $\sqrt{n}$.

As in [8], we observe that if we can find integers a_i and b_i so that a_i is smooth and

$$M_i := a_i \, \widehat{m}_i - b_i \, n \tag{17}$$

is smooth as well, then $\widehat{M}_i :\equiv a_i^{-1} M_i \equiv \widehat{m}_i \pmod{n}$ is also smooth as an element of $\mathbb{Z}/n\mathbb{Z}$.

Explicitly, let $a_i = (-1)^{u_{0,i}} \prod_{1 \leq j < B} p_j{}^{u_{j,i}}$ and $M_i = (-1)^{w_{0,i}} \prod_{1 \leq j < B} p_j{}^{w_{j,i}}$ be the prime factorizations of a_i and M_i; then, we have

$$\widehat{M}_i \equiv \widehat{m}_i \equiv (-1)^{v_{0,i}} \prod_{1 \leq j \leq B} p_j{}^{v_{j,i}} \pmod{n}, \tag{18}$$

where $v_{j,i} = u_{j,i} - w_{j,i}$, as required.

A related problem has been addressed by de Jonge and Chaum in [9, §3.1] (see also [10]): they describe a method to find *small* integers a_i and M_i satisfying Eq. (17). The "Pigeonhole Principle" (e.g., see [22, Chapter 30]) quantifies how small can be a_i and M_i.

Proposition 4 (Pigeonhole Principle). *Let two integers* $n, \widehat{m} \in \mathbb{Z}$. *For any positive integer* $A < n$, *there exist integers* a^* *and* b^* *such that*

$$0 < |a^*| \leq A \quad and \quad |a^* \widehat{m} - b^* n| < \lceil n/A \rceil .$$

Proof. Consider the $(A + 1)$ "pigeons" given by the numbers $P_a := a \, \widehat{m} \bmod n$ $(0 \leq a \leq A)$, i.e., each pigeon is an integer between 0 and $(n - 1)$. We now form the A pigeonholes given by the integer intervals

$$\mathcal{I}_a = \begin{cases} [(a - 1)\lceil n/A \rceil, a\lceil n/A \rceil[& \text{for } 1 \leq a \leq (A - 1) \\ [(A - 1)\lceil n/A \rceil, n - 1] & \text{for } a = A \end{cases} .$$

Since there are more pigeons than pigeonholes, two pigeons are sitting in the same hole. Suppose these are pigeons P_x and P_y and that they are in hole $\mathcal{I}_a$; in other words, $P_x, P_y \in \mathcal{I}_a \iff |P_x - P_y| < \lceil n/A \rceil \iff |(x - y)\widehat{m} \bmod n| < \lceil n/A \rceil$. Noting that $0 \leq x, y \leq A$ and $x \neq y$, we set $a^* = x - y$ and so $0 < |a^*| \leq A$. Hence, $|a^* \widehat{m} \bmod n| < \lceil n/A \rceil$ and the lemma follows by setting $b^* = \lfloor a^* \widehat{m}/n \rfloor$. $\qquad \square$

In particular, taking $A = \lceil \sqrt{n} \rceil$, there exist $a_i, b_i \in \mathbb{Z}$ such that $|a_i| \leq \lceil \sqrt{n} \rceil$ and $|M_i| \leq \lceil n/\lceil \sqrt{n} \rceil \rceil - 1 = \lfloor n/\lceil \sqrt{n} \rceil \rfloor \leq \lfloor \sqrt{n} \rfloor$. This means that for a given 1024-bit modulus n and *any* $\widehat{m}_i$ (corresponding to the message representative $\widehat{m}_i)_i$), there are integers a_i and b_i such both a_i and $M_i = a_i \, \widehat{m}_i - b_i \, n$ are 512-bit integers, *whatever the message encoding method*. It remains to explain how to find a_i and b_i. A simple means is given by the extended Euclidean algorithm [12, Algorithm X, p. 325].

Algorithm 2. On inputs n and $\widehat{m}_i \in \mathbb{Z}/n\mathbb{Z}$, this algorithm computes a solution a_i, b_i, M_i satisfying Eq. (17) so that $|a_i|$ and $|M_i|$ are $\leq \lceil \sqrt{n} \rceil$. It makes use of a vector (u_1, u_2, u_3) such that $u_1 \widehat{m}_i - u_2 n = u_3$ always holds.

1. [initialization] Set $(u_1, u_2, u_3) \leftarrow (0, -1, n)$ and $(v_1, v_2, v_3) \leftarrow (1, 0, \widehat{m}_i)$.
2. [Euclid] Compute $Q \leftarrow \lfloor u_3/v_3 \rfloor$ and $(t_1, t_2, t_3) \leftarrow (u_1, u_2, u_3) - (v_1, v_2, v_3)Q$. Then set $(u_1, u_2, u_3) \leftarrow (v_1, v_2, v_3)$ and $(v_1, v_2, v_3) \leftarrow (t_1, t_2, t_3)$.
3. [loop] If $u_3 > \lceil \sqrt{n} \rceil$, then return to Step 2.
4. [output] Output $a_i \leftarrow u_1$, $b_i \leftarrow u_2$ and $M_i \leftarrow u_3$.

Example 2. Using the modulus of Example 1 (i.e., $n = 26882749$), suppose we are given a message m_i whose encoding is $\widetilde{m}_i = 26543210 \ (\equiv 10 \pmod{16})$. Since $(\widetilde{m}_i | n) = -1$, we have $\widehat{m}_i = \widetilde{m}_i/2 = 13271605 = 5 \cdot 79 \cdot 33599$. This $\widehat{m}_i$ is not smooth, we thus apply Algorithm 2 and obtain $M_i = a_i \widehat{m}_i + b_i n$ with $a_i = 1821$, $b_i = 899$, $M_i = 1354$. We have $a_i = 3 \cdot 607$ and $M_i = 2 \cdot 677$. Hence, $\widehat{m}_i \equiv 2 \cdot 3^{-1} \cdot 607^{-1} \cdot 677 \pmod{n}$.

$u_1 \ (a_i)$	$u_2 \ (b_i)$	$u_3 \ (M_i)$
1	0	13271605
−2	−1	339539
79	*39*	*29584*
−871	−430	14115
1821	*899*	*1354*
−19081	−9420	575
39983	19739	204
−99047	−48898	167
139030	68637	37
−655167	−323446	19
794197	392083	18
−1449364	−715529	1

Note that Algorithm 2 does not always give the best possible solution. Considering all the steps of the extended Euclidean algorithm (see above), the best solution for our example is given by $a_i = 79$ and $M_i = 29584 = 2^4 \cdot 43^2$ which yields $\widehat{m}_i \equiv 2^4 \cdot 43^2 \cdot 79^{-1} \pmod{n}$. $\diamondsuit$

6   Conclusion

This paper presented a specialized version of the Coron-Naccache-Stern signature forgery. It applies to *any* Rabin-type signature scheme and to *any* (even) public verification exponent. Furthermore, contrary to the case of RSA, the forgery is *universal*: it yields the value of the private key.

References

1. FIPS 180-1. Secure Hash Standard. Federal Information Processing Standards Publication 180-1, U.S. Department of Commerce, April 1995.
2. IEEE Std 1363-2000. IEEE Standard Specifications for Public-Key Cryptography. IEEE Computer Society, August 29, 2000.
3. ISO/IEC 9796. Information technology – Security techniques – Digital signature scheme giving message recovery, 1991.
4. PKCS #1 v2.0. RSA cryptography standard. RSA Laboratories, October 1, 1998. Available at http://www.rsasecurity.com/rsalabs/pkcs/.
5. RFC 1319. The MD2 message digest algorithm. Internet Request for Comments 1321, Burt Kaliski, April 1992. Available at http://www.ietf.org/rfc/rfc1319.txt.
6. RFC 1321. The MD5 message digest algorithm. Internet Request for Comments 1321, Ronald Rivest, April 1992. Available at http://www.ietf.org/rfc/rfc1321.txt.
7. Henri Cohen. *A Course in Computational Algebraic Number Theory*, volume 138 of *Graduate Texts in Mathematics*. Springer-Verlag, 1993.
8. Jean-Sébastien Coron, David Naccache, and Julien P. Stern. On RSA padding. In M. Wiener, editor, *Advances in Cryptology – CRYPTO '99*, volume 1666 of *Lecture Notes in Computer Science*, pages 1–18. Springer-Verlag, 1999.
9. Wiebren de Jonge and David Chaum. Attacks on some RSA signatures. In H. C. Williams, editor, *Advances in Cryptology – CRYPTO '85*, volume 218 of *Lecture Notes in Computer Science*, pages 18–27, 1986.
10. Marc Girault, Philippe Toffin, and Brigitte Vallée. Computation of approximate L-th root modulo n and application to cryptography. In S. Goldwasser, editor, *Advances in Cryptology – CRYPTO '88*, volume 403 of *Lecture Notes in Computer Science*, pages 110–117, 1990.
11. Burton S. Kaliski Jr. A layman's guide to a subset of ASN.1, BER, and DER. RSA Laboratories Technical Note, RSA Laboratories, November 1993. Available at http://www.rsasecurity.com/rsalabs/pkcs/.
12. Donald E. Knuth. *The Art of Computer Programming, v. 2. Seminumerical Algorithms*. Addison-Wesley, 2nd edition, 1981.
13. Kaoru Kurosawa, Toshiya Itoh, and Masashi Takeuchi. Public key cryptosystem using a reciprocal number with the same intractability as factoring a large number. *Cryptologia*, 12(4):225–233, 1988.
14. Arjen K. Lenstra. Generating RSA moduli with a predetermined portion. In K. Ohta and D. Pei, editors, *Advances in Cryptology – ASIACRYPT '98*, volume 1514 of *Lecture Notes in Computer Science*, pages 1–10. Springer-Verlag, 1998.
15. Arjen K. Lenstra and Mark S. Manasse. Factoring with two large primes. *Mathematics of Computation*, 63:785–798, 1994.
16. J. H. Loxton, David S. Khoo, Gregory J. Bird, and Jennifer Seberry. A cubic RSA code equivalent to factorization. *Journal of Cryptology*, 5(2):139–150, 1992.
17. Alfred J. Menezes, Paul C. van Oorschot, and Scott A. Vanstone. *Handbook of Applied Cryptography*. CRC Press, 1997.
18. Peter L. Montgomery. A block Lanczos algorithm for finding dependencies over GF(2). In L. C. Guillou and J.-J. Quisquater, editors, *Advances in Cryptology – EUROCRYPT '95*, volume 921 of *Lecture Notes in Computer Science*, pages 106–120, 1995.

19. Michael O. Rabin. Digitized signatures and public-key functions as intractable as factorization. Technical Report LCS/TR-212, M.I.T. Lab. for Computer Science, January 1979.

20. Renate Scheidler. A public-key cryptosystem using purely cubic fields. *Journal of Cryptology*, 11(2):109–124, 1998.

21. Renate Scheidler and Hugh C. Williams. A public-key cryptosystem utilizing cyclotomic fields. *Designs, Codes and Cryptography*, 6:117–131, 1995.

22. Joseph H. Silverman. *A Friendly Introduction to Number Theory*. Prentice-Hall, 1997.

23. Robert D. Silverman and David Naccache. Recent results on signature forgery, April 1999. Available at `http://www.rsasecurity.com/rsalabs/bulletins/sigforge.html`.

24. Hugh C. Williams. A modification of the RSA public-key encryption procedure. *IEEE Transactions on Information Theory*, IT-26(6):726–729, 1980.

25. _____. Some public-key crypto-functions as intractable as factorization. In G. R. Blakley and D. Chaum, editors, *Advances in Cryptology – Proceedings of CRYPTO 84*, volume 196 of *Lecture Notes in Computer Science*, pages 66–70. Springer-Verlag, 1986.

26. _____. Some public-key crypto-functions as intractable as factorization. *Cryptologia*, 9(3):223–237, 1985. An extended abstract appears in [25].

27. _____. An M^3 public key encryption scheme. In H. C. Williams, editor, *Advances in Cryptology – CRYPTO '85*, volume 218 of *Lecture Notes in Computer Science*, pages 358–368. Springer-Verlag, 1986.

Strong Adaptive Chosen-Ciphertext Attacks with Memory Dump (or: The Importance of the Order of Decryption and Validation)

Seungjoo Kim[1], Jung Hee Cheon[2], Marc Joye[3], Seongan Lim[1],
Masahiro Mambo[4], Dongho Won[5], and Yuliang Zheng[6]

[1] KISA (Korea Information Security Agency),
4Fl, IT Venture Tower W/B, 78, Garag-Dong. Songpa-Gu, Seoul, Korea 138-803
{skim, seongan}@kisa.or.kr
http://www.crypto.re.kr/ or http://www.crypto.ac/
[2] ICU (Information and Communications Univ.),
58-4 Hwaam-Dong, Yusong-Gu, Taejon 305-732, Korea
jhcheon@icu.ac.kr - http://vega.icu.ac.kr/~jhcheon/
[3] Gemplus Card International, Card Security Group,
Parc d'Activités de Gémenos, B.P. 100, 13881 Gémenos, France
marc.joye@gemplus.com − http://www.geocities.com/MarcJoye/
[4] Graduate School of Information Sciences, Tohoku University,
Kawauchi Aoba Sendai, 980-8576 Japan
mambo@icl.isc.tohoku.ac.jp − http://www.icl.isc.tohoku.ac.jp/~mambo/
[5] Sungkyunkwan University,
300 Chunchun-dong, Suwon, Kyunggi-do, 440-746, Korea
dhwon@dosan.skku.ac.kr − http://dosan.skku.ac.kr/~dhwon/
[6] UNC Charlotte,
9201 University City Blvd, Charlotte, NC 28223
yzheng@uncc.edu − http://www.sis.uncc.edu/~yzheng/

Abstract. This paper presents a new type of powerful cryptanalytic attacks on public-key cryptosystems, extending the more commonly studied adaptive chosen-ciphertext attacks. In the new attacks, an adversary is not only allowed to submit to a decryption oracle (valid or invalid) ciphertexts of her choice, but also to emit a "dump query" prior to the completion of a decryption operation. The dump query returns intermediate results that have not been erased in the course of the decryption operation, whereby allowing the adversary to gain vital advantages in breaking the cryptosystem.

We believe that the new attack model approximates more closely existing security systems. We examine its power by demonstrating that most existing public-key cryptosystems, including OAEP-RSA, are vulnerable to our extended attacks.

Keywords. Encryption, provable security, chosen-ciphertext security, ciphertext validity, OAEP-RSA, ElGamal encryption.

B. Honary (Ed.): Cryptography and Coding 2001, LNCS 2260, pp. 114–127, 2001.
© Springer-Verlag Berlin Heidelberg 2001

1   Introduction

Provably security is gaining more and more popularity. Only known to theoreticians a decade ago, provable security becomes now a standard attribute of any cryptographic scheme.

A scheme is said provably secure if the insecurity of the scheme implies that some widely-believed intractable problem is solvable. The security of a scheme is measured as the ability to resist an adversarial goal in a given adversarial model. The standard security notion for public-key encryption schemes is *indistinguishability under adaptive chosen-ciphertext attacks* (IND-CCA2) [2].

This paper introduces a *stronger* attack. In addition to having access to a decryption oracle, the adversary can perform a *"memory dump"* of the decryption oracle at any time; the sole restriction is that secret data (e.g., private keys) are inaccessible. We believe that this new attack model approximates more closely existing security systems, many of which are built on such operating systems as Unix and Windows where a reasonably privileged user can interrupt the operation of a computing process and inspect its intermediate results at ease.

We show that many IND-CCA2 encryption schemes are vulnerable within our extended attack scenario. Examples include OAEP-RSA and several ElGamal variants. A problem in almost all those schemes is that the validity of the ciphertext cannot be checked until *after* all the decryption is completed. By emitting a dump query prior to the validity checking, the adversary may get access to some secret information resulting from the *raw* decryption of an (invalid) ciphertext and thereby may weaken the scheme.

We hope that in the future cryptographers will analyze the security of their schemes in our extended setting. We note that the scope of our attacks is very broad and may apply to other cryptographic primitives as well (e.g., digital signatures). In the same vein as in [2], it may also be interesting to see what are the different implications and separations implied by our attacks when considering various adversarial goals.

ORGANIZATION. The rest of this paper is organized as follows. In the next section, we review current security notions for encryption schemes and introduce the new notion of *strong* chosen-ciphertext security. Applicative examples of our attack scenario are also provided. Next, in Section 3, we apply our attack model to the celebrated OAEP-RSA and show that it is insecure under our extended setting. Similar attacks against various ElGamal variants are also presented. Finally, we conclude in Section 4.

2   Strong Adaptive Chosen-Ciphertext Security

2.1   Security Notions

Indistinguishability of encryptions, defined by Goldwasser and Micali [14], captures the intuition that an adversary should not be able to obtain any partial information (but its length) about a message given its encryption. In their paper, Goldwasser and Micali study the case of completely *passive* adversaries, i.e., adversaries can only eavesdrop.[1] They do not consider adversaries injecting messages into a network or otherwise influencing the behavior of parties in the network.

To deal with active attacks, Naor and Yung [18] introduced security against *chosen-ciphertext attacks*. They model such an attack by allowing the adversary to get access to a "decryption oracle" and obtain decryptions of her choice. When the security goal is indistinguishability, the attack of [18] runs as follows. During the first stage, called *find stage*, the adversary (viewed as a polynomial-time machine) has access to a decryption oracle. At the end of the stage, the adversary generates two (equal-length) plaintexts m_0 and m_1. In the second stage, called *guess stage*, the adversary receives the encryption of m_b, say c_b, with b randomly drawn from $\{0,1\}$. The attack is successful if the adversary recovers the value b or, equivalently, if the adversary distinguishes the encryption of m_0 from that of m_1.

Rackoff and Simon [21] later generalized this attack by allowing the adversary to have still access to the decryption oracle after having obtained the target ciphertext c_b. Since the adversary could simply submit the target ciphertext itself to the decryption oracle, Rackoff and Simon restrict the adversary's behavior by not allowing her to probe the decryption oracle with c_b.

In summary, chosen-ciphertext attacks can be classified in two categories:

- (*Static Chosen-Ciphertext Attacks* [18]) The adversary has access to the decryption oracle uniquely prior to obtaining the target ciphertext.[2]
- (*Adaptive Chosen-Ciphertext Attacks* [21]) Not only can the adversary get access to the decryption oracle during the find stage but also during the guess stage. The only restriction is not to submit the target ciphertext itself to the decryption oracle.

As explicitly pointed out in [16], we stress that the adversary may query the decryption oracle with invalid ciphertexts. Although seemingly useless, such attacks are not innocuous. The decryption oracle can for example be used to learn whether a chosen ciphertext is valid or not. From this single bit of information and by iterating the process, Bleichenbacher successfully attacked several implementations of protocols based on PKCS #1 v1.5 [5]. More recently, Manger pointed out the importance of preventing an attacker from distinguishing between rejections at the various steps of the decryption algorithm, say, using timing analysis [17]. The lesson is that implementors must ensure that the reasons for which a ciphertext is rejected are hidden from the outside world.

---

[1] We note, however, that in the public-key setting, an adversary can always mount a chosen-plaintext attack since encryption is public, by definition.

[2] In the past, this attack has also been called "lunch-time attack" or "midnight attack".

2.2   Chosen-Ciphertext Attacks with Memory Dump

Now we consider more powerful adversaries who not only can obtain the plaintexts corresponding to chosen ciphertexts, but can also invade a user's computer and read the contents of its memory (i.e., non-erased internal data). In [24], Shoup introduced the *strong adaptive corruption model*, i.e., a memory dump attack combined with forward secrecy in the context of key exchange protocols. We apply this notion to the security of encryption algorithms and consider strong adaptive chosen-ciphertext security, i.e., a memory dump attack combined with chosen-ciphertext security.[3]

Definition 1 (Strong Chosen-Ciphertext Query). *Let k be a security parameter that generates matching encryption/decryption keys (e, d) for each user in the system. A* strong chosen-ciphertext query *is a process which, on input 1^k and e, obtains either*

- *the plaintext (relatively to d) corresponding to a chosen ciphertext; or*
- *an indication that the chosen ciphertext is invalid; or*
- *non-erased internal states of the decryption oracle decrypting the submitted ciphertext, when a "dump" query is submitted.*

In the case the decryption oracle would like to prevent the exposure of an internal state, or part of it, it can take the operation

$$\texttt{atomic\_action}[\cdots] \ .$$

The term *"atomic"* means that the action must be executed without external interruptions (e.g., memory core dump, system crash, user signal to kill the transaction, and so on [26]). After the execution being successfully completed, the internal variables used by atomic process are erased.

Most cryptosystems proposed so far do not specify which steps in the decryption process need to be protected from external probes. Such an implementation matter can be actually viewed as a cryptographic design matter: the designer of a cryptosystem *should* express the steps in the decryption process that should be protected by an "atomic action". Of course, the best design is a cryptosystem wherein only a single operation (using the private key) is executed atomically, in the above sense.

To fix the ideas, suppose that an adversary attacks the plain RSA algorithm, as originally described in [22]. Let n denote the RSA modulus and let (e, d) denote the pair of encryption/decryption keys. If the (plain) RSA decryption, $m = c^d \bmod n$, is not performed atomically then an appropriate "dump" query may reveal the values of m, c, d and n as they are, during the course of the decryption, available somewhere in the memory.

---

[3] Similarly, for digital signature schemes, stronger security notions can be defined such as, for example, *existential unforgeability* under *adaptive chosen-message attacks with memory dump*.

However, by using the operation $\mathtt{atomic\_action}[m \leftarrow c^d \bmod n]$, which is made up of the following six unit operations

$$\mathtt{atomic\_action} \begin{bmatrix} 1: & \mathtt{MEM}[1] \leftarrow n \\ 2: & \mathtt{MEM}[2] \leftarrow c \\ 3: & \mathtt{MEM}[3] \leftarrow d \\ 4: & \mathtt{MEM}[4] \leftarrow \mathtt{MEM}[2]^{\mathtt{MEM}[3]} \bmod \mathtt{MEM}[1] \\ 5: & m \leftarrow \mathtt{MEM}[4] \\ 6: & \texttt{erase internal memory states } \mathtt{MEM}[1], \mathtt{MEM}[2], \\ & \mathtt{MEM}[3] \text{ and } \mathtt{MEM}[4] \end{bmatrix},$$

the oracle has only information on n and c (which are publicly known), and on m from a "dump" query. In particular, remark that the value d ($\mathtt{MEM}[3]$) is inaccessible since it is erased from the memory (cf. Step 6).

In our model, oracle's operations using private key are always executed atomically. This restriction is the weakest possible: allowing the adversary to have access to user's private key (e.g., d in the above example) has the same effect as allowing the adversary to submit the target ciphertext itself to the decryption oracle. Note here that the power of a strong adaptive chosen-ciphertext attack deeply depends on the amount of (unit) operations contained inside a atomic action, so we can consider this amount of operations as one of the security evaluation criteria of a given cryptosystem.

Definition 2 (Strong [Static/Adaptive] Chosen-Ciphertext Attack). *A strong static chosen-ciphertext attack consists of the following scenario:*

1. *On input a security parameter k, the key generation algorithm $\mathcal{K}$ is run, generating a public key and a private key for the encryption algorithm $\mathcal{E}$. The adversary of course obtains the public key, but the private key is kept secret.*
2. *[Find stage] The adversary makes polynomially (in k) many strong chosen-ciphertext queries (as in Definition 1) to a decryption oracle.*
   (The adversary is free to construct the ciphertexts in an arbitrary way —it is certainly not required to compute them using the encryption algorithm.)
3. *The adversary prepares two messages m_0, m_1 and gives these to an encryption oracle. The encryption oracle chooses $b \in_R \{0,1\}$ at random, encrypts m_b, and gives the resulting "target ciphertext" c' to the adversary. The adversary is free to choose m_0 and m_1 in an arbitrary way, except that they must be of the same length.*
4. *[Guess stage] The adversary outputs $b' \in \{0,1\}$, representing its "guess" on b.*

In a strong adaptive chosen-ciphertext attack, the adversary has still access to the decryption oracle after having received the target ciphertext: a second series of polynomially (in k) many strong chosen-ciphertext queries may be run. The unique restriction is not to probe the decryption oracle with the target ciphertext c'.

The success probability in the previous attack scenario is defined as

$$\Pr[b' = b] \ .$$

Here, the probability is taken over coin tosses of the adversary, the key generation algorithm $\mathcal{K}$ and the encryption algorithm $\mathcal{E}$, and $(m_0, m_1) \in M^2$ where M is the domain of the encryption algorithm $\mathcal{E}$.

Definition 3 (Strong [Static/Adaptive] Chosen-Ciphertext Security).
An encryption scheme is secure if every strong (static/adaptive) chosen-ciphertext attack (as in Definition 2) *succeeds with probability at most negligibly greater than 1/2.*

2.3   Real-World Applications

When our strong adaptive chosen-ciphertext attack is mounted, the adversary may learn (i.e., "memory core-dump") the entire internal state of the decryption oracle, excluding data that has been explicitly erased.

It may be useful to illustrate the definition of security with a simple example. Consider the case of a security-enhanced electronic mail system where a public-key cryptosystem is used to encrypt messages passed among users. It is a common practice for an electronic mail user to include the original message s/he received into a reply to the message. This practice provides an avenue for chosen-ciphertext attacks, as an adversary can send a ciphertext to a target user and expect the user to send back the corresponding plaintext as part of the reply. For instance, a reply to a message may be as follows [28].

```
(original message)
> ......
> Hi, is Yum-Cha still on tonight ?
> ......

(reply to the message)
......
Yes, it's still on. I've already made the bookings.
......
```

Now suppose one step further that, via computer viruses, the adversary can modify the target user's electronic mail system to core-dump (i.e., automatically write the exact contents of the memory to a file) and send back the core-dump file in a stealthy way. This is a concrete example of our attack model. Another concrete example is when user's computer is crashed suddenly, internal secret information may then remain unerased.

We quote the following three articles as a basis of thinking about the relevance of security against strong adaptive chosen-ciphertext attack in the real life.

How your privacy is caught in the Net by Duncan Campbell [7]:

"(...) Hackers and government agencies are hard at work designing information stealing viruses. Six months ago, two of them popped up in the same week. *"Caligula"* was aimed at users who installed a privacy-protection system called PGP. Once it infected a computer, it looked for a file holding the secret keys to PGP. Then, automatically and silently, it transmitted the file to the hacker's Internet site. *Caligula* could have taken any information it wanted. Another virus called *"Picture"* was aimed at the America On-line (AOL) Internet service. *Picture* collected AOL users' passwords and log-in data, and sent them to a web site in China. Three months later, *"Melissa"* appeared. It automatically read users' address books, and used the information to mail itself to all their friends and contacts.

According to Roger Thompson, director of anti-virus security consulting firm ICSA, information theft is a price to be paid for the advent of the Internet. (...)

According to former ASIO deputy director Gerald Walsh, who proposed the new powers: *The introduction of other commands, such as diversion, copy, send, (or) dump memory to a specified site, would greatly enhance criminal investigations. (...)*".

Phone.com takes aim at WAP security hole in eWEEK [11]:

"(...) In current WAP transmissions, data must use two security protocols —WTLS during the wireless part of the journey and SSL once the data hits the wires. There is a split second when the data must decrypt and re-encrypt to switch from one protocol to the other. *A security flaw could occur if someone was able to crash the machine in the split second between decryption and re-encryption, causing a memory dump to the disk. (...)*".

Memory reconstruction attack in RSA Official Guide to Cryptography [6]:

"(...) Often, sensitive material is not stored on hard drives but does appear in a computer's memory. For example, when the program you're running allocates some of the computer's memory, the OS tags that area of memory as unavailable, and no one else can use it or see it. When you're finished with that area of memory, though, many operating systems and programs simply "free" it —marking it as available— without overwriting it. This means that anything you put into that memory area, even if you later "deleted" it, is still there. *A memory reconstruction attack involves trying to examine all possible areas of memory. The attacker simply allocates the memory you just freed and sees what's left there.*

A similar problem is related to what is called "virtual memory." The memory managers in many operating systems use the hard drive as virtual memory, temporarily copying to the hard drive any data from memory that has been allocated but is momentarily not being used. When that information is needed again, the memory manager swaps the current virtual memory for the real memory. *In August 1997, The New York Times published a report about an individual using simple tools to scan his hard drive. In the swap space, he found the password he used for a popular security application. (...)*".

3 On the Power of Strong Adaptive Chosen-Ciphertext Attacks

For the last few years, many new schemes have been proposed with provable security against chosen-ciphertext attacks. Before 1994, only theoretical (i.e., not very practical) schemes were proposed. Then Bellare and Rogaway [4] came up with the *random oracle model* [3] and subsequently designed in [4] a generic padding, called OAEP (Optimal Asymmetric Encryption Padding), to transform a one-way (partially) trapdoor *permutation* into a chosen-ciphertext secure cryptosystem. Other generic paddings, all validated in the random oracle model, were later given by Fujisaki and Okamoto [12] (improved in [13]), by Pointcheval [20], and by Okamoto and Pointcheval [19]. The first practical cryptosystem with provable security in the *standard model* is due to Cramer and Shoup [8]. They present an extended ElGamal encryption provably secure under the *decisional* Diffie-Hellman problem.

In this section, we demonstrate that most of "decrypt-then-validate"-type cryptosystems with provable security (including OAEP-RSA and most ElGamal variants) can be broken under our strong adaptive chosen-ciphertext attacks.

3.1 OAEP-RSA

We give here a brief overview of OAEP-RSA and refer the reader to [4] for details. The decryption phase of OAEP-RSA is divided into three parts: (i) RSA decryption, (ii) validation, and (iii) output.

Let $n = pq$ denote an RSA modulus, which is the product of two large primes p and q. Furthermore, let e and d, satisfying $ed \equiv 1 \pmod{\mathrm{lcm}(p - 1, q - 1)}$, respectively denote the public encryption exponent and the private decryption exponent. We assume a hash function $H : \{0,1\}^{k_m+k_1} \to \{0,1\}^{k_0}$ and a "generator" function $G : \{0,1\}^{k_0} \to \{0,1\}^{k_m+k_1}$, where $k_m + k_0 + k_1$ is the bit-length of n. The public parameters are $\{n, e, G, H\}$ and the secret parameters are $\{d, p, q\}$.

A k_m-bit plaintext message m is encrypted through OAEP-RSA as

$$c = (s\|t)^e \bmod n \quad \text{with} \quad s = m0^{k_1} \oplus G(r) \text{ and } t = r \oplus H(s)$$

for a random $r \in \{0,1\}^{k_0}$. Given a ciphertext c, m is recovered as:

(i) | RSA decryption |

   - $s\|t = \texttt{atomic\_action}[c^d \bmod n]$;
   - $z = s \oplus G(t \oplus H(s))$.

(ii) | Validation |

   If the last k_1 bits of z are not 0 then
   - Erase the internal data; and
   - Return "Invalid Ciphertext".

(iii) | Output |

   - Return the first k_m bits of z.

The attack. The problem with OAEP-RSA resides in that the validity test cannot be performed (i.e., the adversary cannot be detected) until *after* the RSA decryption is completed [25]. Thus an attacker can freely mount a strong adaptive chosen-ciphertext attack and extract the partial information from internal data in the decryption oracle's memory.

For example, consider the following game played by a strong chosen-ciphertext adversary. Let $c = (s\|t)^e \bmod n$ denote the target ciphertext. First, the adversary chooses a random r and forms the (invalid) ciphertext $c' = c \cdot r^e \bmod n$. Then she submits c' to the decryption oracle. Just after the decryption oracle computes $w' := \texttt{atomic\_action}[c'^d \bmod n]$ (note that $w' \equiv c^d \cdot r \pmod{n}$) but before it detects and rejects the invalid ciphertext c', the adversary submits a "dump" query, and then obtains the internal data w'. From w', she computes $s\|t = \frac{w'}{r} \bmod n$ and $z = s \oplus G(t \oplus H(s))$. Finally, the adversary recovers m as the first k_m bits of z, so it is easy to find a correct guess by using the recovered m.

3.2   Other Provably Secure Cryptosystems

Similarly to the plain RSA cryptosystem, the plain ElGamal cryptosystem [10] is malleable. We recall the attack hereafter.

Let $\mathcal{G} = \langle g \rangle$ be the cyclic group generated by g. The public encryption key is $X = g^x$ and the private decryption key is x. A message m, considered as an element of $\mathcal{G}$, is encrypted as $(c_1, c_2) = (g^y, X^y \cdot m)$ for some random integer y. Given (c_1, c_2), plaintext message m is recovered as $m = c_1^{-x} \cdot c_2$.

If (c_1, c_2) is the target ciphertext, then an adversary can compute the ciphertext $(c_1', c_2') = (g^r \cdot c_1, X^r \cdot c_2)$ for some random r. Next, by submitting (c_1', c_2') to the decryption oracle, she recovers $m = m' = \texttt{atomic\_action}[c_1'^{-x}] \cdot c_2'$ from a "dump" query.

Several variants of the basic ElGamal scheme were proposed in order to make it secure against chosen-ciphertexts attacks. We review some of them in the chronological order of their appearance and analyze their resistance under our strong chosen-ciphertext attacks. Here, for simplicity, the same notation G, H are used for several schemes, but functions G and H may have different domain and image from those used in OAEP-RSA of Section 3.1.

Zheng and Seberry I [28]
- encryption: $(c_1, c_2) = (g^y, G(X^y) \oplus (m\|H(m)))$
- decryption: 1) $m\|t = G(\texttt{atomic\_action}[c_1^x]) \oplus c_2$
   2) if $t = H(m)$ then output m else output $\texttt{reject}$
- attack: 1) set $(c_1', c_2') = (c_1, c_2 \oplus r)$ for a random r
   2) recover m from $(m'\| \cdots) \oplus r = m\| \cdots$

Zheng and Seberry II [28]
- encryption: $(c_1, c_2, c_3) = (g^y, H_s(m), z \oplus m)$ with $z\|s = G(X^y)$
- decryption: 1) $z\|s = G(\texttt{atomic\_action}[c_1^x])$
   2) $m = z \oplus c_3$
   3) if $c_2 = H_s(m)$ then output m else output $\texttt{reject}$

- attack: 1) set $(c_1', c_2', c_3') = (c_1, c_2', c_3)$ with $c_2' \neq c_2$
  2) recover $m = m'$

Zheng and Seberry III [28]

- encryption: $(c_1, c_2, c_3, c_4) = (g^y, g^k, \frac{H(m) - yr}{k} \pmod{\#\mathcal{G}}, z \oplus m)$
  with $r = X^{y+k}$ and $z = G(r)$
- decryption: 1) $r = \texttt{atomic\_action}[(c_1 c_2)^x]$
  2) $m = G(r) \oplus c_4$
  3) if $g^{H(m)} = c_1{}^r \cdot c_2{}^{c_3}$ then output m else output $\texttt{reject}$
- attack: 1) set $(c_1', c_2', c_3', c_4') = (c_1, c_2, c_3', c_4)$ with $c_3' \neq c_3$
  2) recover $m = m'$

Tsiounis and Yung [27]

- encryption: $(c_1, c_2, c_3, c_4) = (g^y, X^y \cdot m, g^k, y \cdot H(g, c_1, c_2, c_3) + k)$
- decryption: 1) if $g^{c_4} \neq c_1{}^{y \cdot H(g, c_1, c_2, c_3)} c_3$ then output $\texttt{reject}$
  2) output $m = \texttt{atomic\_action}[c_1{}^{-x}] \cdot c_2$

Cramer and Shoup [8]

- encryption: $(c_1, c_2, c_3, c_4) = (g_1{}^s, g_2{}^s, X_1{}^s \cdot m, X_2{}^s \cdot X_3{}^{s \cdot H(c_1, c_2, c_3)})$
  with $X_1 = g_1{}^z$, $X_2 = g_1{}^{x_1} \cdot g_2{}^{x_2}$ and $X_3 = g_1{}^{y_1} \cdot g_2{}^{y_2}$
- decryption: 1) $\alpha = H(c_1, c_2, c_3)$
  2) $v = \texttt{atomic\_action}[c_1{}^{x_1 + y_1 \alpha} \cdot c_2{}^{x_2 + y_2 \alpha}]$
  3) if $c_4 = v$ then output $m = \texttt{atomic\_action}[c_1{}^{-z}] \cdot c_3$
  else output $\texttt{reject}$

Fujisaki and Okamoto [12]

- encryption: $(c_1, c_2) = (g^{H(m\|s)}, (m\|s) \oplus X^{H(m\|s)})$
- decryption: 1) $m\|s = \texttt{atomic\_action}[c_1{}^x] \oplus c_2$
  2) if $c_1 = g^{H(m\|s)}$ then output m else output $\texttt{reject}$
- attack: 1) set $(c_1', c_2') = (c_1, c_2 \oplus r)$ for a random r
  2) recover m from $(m'\| \cdots) \oplus r = m\| \cdots$

Fujisaki and Okamoto [13]

- encryption: $(c_1, c_2, c_3) = (g^{H(s,m)}, X^{H(s,m)} \cdot s, \mathcal{E}_{G(s)}^{\mathsf{sym}}(m))$
- decryption: 1) $s = \texttt{atomic\_action}[c_1{}^{-x}] \cdot c_2$
  2) $m = \mathcal{D}_{G(s)}^{\mathsf{sym}}(c_3)$
  3) if $c_2 = X^{H(s,m)} \cdot s$ then output m else output $\texttt{reject}$
- attack: 1) set $(c_1', c_2', c_3') = (g^r \cdot c_1, X^r \cdot c_2, c_3)$ for a random r
  2) recover $m = m'$

Pointcheval [20]

- encryption: $(c_1, c_2, c_3) = (g^{H(m\|s)}, X^{H(m\|s)} \cdot k, (m\|s) \oplus G(k))$
- decryption: 1) $m\|s = G(\texttt{atomic\_action}[c_1{}^{-x}] \cdot c_2) \oplus c_3$
  2) if $c_1 = g^{H(m\|s)}$ then output m else output $\texttt{reject}$
- attack: 1) set $(c_1', c_2', c_3') = (c_1, c_2, c_3 \oplus r)$ for a random r
  2) recover m from $(m'\| \cdots) \oplus r = m\| \cdots$

Baek, Lee, and Kim [1]

- encryption: $(c_1, c_2) = (g^{H(m\|s)}, (m\|s) \oplus G(X^{H(m\|s)}))$
- decryption: 1) $m\|s = G(\texttt{atomic\_action}[c_1{}^x]) \oplus c_2$
  2) if $c_1 = g^{H(m\|s)}$ then output m else output $\texttt{reject}$
- attack: 1) set $(c_1', c_2') = (c_1, c_2 \oplus r)$ for a random r
  2) recover m from $(m'\| \cdots) \oplus r = m\| \cdots$

Schnorr and Jakobsson [23]

- encryption: $(c_1, c_2, c_3, c_4) = (g^y, G(X^y) + m, H(g^s, c_1, c_2), s + c_3 \cdot y)$
- decryption: 1) if $c_3 \neq H(g^{c_4} \cdot c_1^{-c_3}, c_1, c_2)$ then output `reject`
  2) output $m = c_2 - G(\texttt{atomic\_action}[c_1^x])$

Okamoto and Pointcheval [19]

- encryption: $(c_1, c_2, c_3, c_4) = (g^y, X^y \oplus R, \mathcal{E}^{\mathsf{sym}}_{G(R)}(m), H(R, m, c_1, c_2, c_3))$
- decryption: 1) $R = \texttt{atomic\_action}[c_1^x] \oplus c_2$
  2) $m = \mathcal{D}^{\mathsf{sym}}_{G(R)}(c_3)$
  3) if $c_4 = H(R, m, c_1, c_2, c_3)$ then output m
  else output `reject`
- attack: 1) set $(c_1', c_2', c_3', c_4') = (c_1, c_2, c_3, c_4')$ with $c_4' \neq c_4$
  2) recover $m = m'$

Table 1. Analysis of several ElGamal variants.

ElGamal variant	Type	Attack
Zheng and Seberry I, II, III [28]	Decrypt-then-validate	Yes
Tsiounis and Yung [27]	Validate-then-decrypt	No
Cramer and Shoup [8]	Validate-then-decrypt	No
Fujisaki and Okamoto [12]	Decrypt-then-validate	Yes
Fujisaki and Okamoto [13]	Decrypt-then-validate	Yes
Pointcheval [20]	Decrypt-then-validate	Yes
Baek, Lee, and Kim [1]	Decrypt-then-validate	Yes
Schnorr and Jakobsson [23]	Validate-then-decrypt	No
Okamoto and Pointcheval [19]	Decrypt-then-validate	Yes

Table 1 summarizes the cryptographic characteristics of the previously described schemes. According to the table, the "decrypt-then-validate"-type schemes are all susceptible to our extended attacks. This is certainly the case when a component c_i in the ciphertext is especially dedicated to the validity test (e.g., as in [28, II and III] or [19]); in that case, it suffices to probe the decryption oracle with the target ciphertext where component c_i is replaced by an arbitrary component $c_i' \neq c_i$.

Remark that the attacks we described are very powerful. Even if the *whole* decryption operation (excluding the validation checking) is performed through an `atomic_action`, our attacks are still successful. This, however, does not mean that it is impossible to construct a "decrypt-then-validate"-type scheme secure against strong adaptive chosen-ciphertext attacks. For example, we were not able to find an attack on the following modification of Baek, Lee, and Kim scheme wherein the $\oplus$ operator is replaced by a pair of symmetric encryption/decryption algorithms $(\mathcal{E}^{\mathsf{sym}}_K, \mathcal{D}^{\mathsf{sym}}_K)$ like DES.

- encryption: $(c_1, c_2) = (g^{H(m\|s)}, \mathcal{E}^{\mathsf{sym}}_{G(X^{H(m\|s)})}(m\|s))$
- decryption: 1) $m\|s = \mathtt{atomic\_action}[\mathcal{D}^{\mathsf{sym}}_{G(c_1{}^x)}(c_2)]$
    2) if $c_1 = g^{H(m\|s)}$ then output m else output $\mathtt{reject}$

(Note that we did not prove the security of the scheme.)

This modified scheme is nevertheless less satisfactory than a scheme wherein only the private-key operations in the decryption process are run "atomically" such as in $m\|s = \mathcal{D}^{\mathsf{sym}}_{G(\mathtt{atomic\_action}[c_1{}^x])}(c_2)$. Unfortunately, this latter scheme is insecure. Submitting $(c_1', c_2') = (c_1, c_2')$ with $c_2' \neq c_2$ to the decryption oracle, the adversary obtains the value of $R := \mathtt{atomic\_action}[c_1{}^x]$ from a "dump" query and therefore recovers m as the most significant $|m|$-bits of $\mathcal{D}^{\mathsf{sym}}_{G(R)}(c_2)$.

The ElGamal variants by Tsiounis and Yung [27] and by Schnorr and Jakobsson [23] are actually *signed* encryption schemes. The security proofs can be extended to the "strong adaptive chosen-ciphertext attack" scenario since the validity is checked prior to and separately from the decryption operation itself.

For the scheme by Cramer and Shoup [8], things are slightly different. Some secret data are involved in the validity test. As a consequence, not only the operations using the private decryption-key (i.e., z) but also the operations using the private "validation-keys" (i.e., (x_1, x_2) and (y_1, y_2)) need to be "wrapped up" in an $\mathtt{atomic\_action}$. Thus, from our evaluation criteria, we can say that this latter scheme needs more protection than the schemes of Tsiounis and Yung, and of Schnorr and Jakobsson.

4 Conclusion

Many security problems are often viewed as "implementation errors". We believe that those could be more fruitfully viewed as cryptographic design errors. In this paper we presented a new security model for encryption algorithms and analyzed the security of several algorithms provably secure against (standard) adaptive chosen-ciphertext attacks. Amongst other things, we showed that, in view of higher-level application programs, "validate-then-decrypt"-type schemes generally better behave facing our extended attacks.

References

1. J. Baek, B. Lee, and K. Kim, "Secure length-saving ElGamal encryption under the computational Diffie-Hellman assumption", *Information Security and Privacy (ACISP 2000)*, volume 1841 of *Lecture Notes in Computer Science*, pages 49–58, Springer-Verlag, 2000.
2. M. Bellare, A. Desai, D. Pointcheval, and P. Rogaway, "Relations among notions of security for public-key encryption schemes", *Advances in Cryptology – CRYPTO '98*, volume 1462 of *Lecture Notes in Computer Science*, pages 26–45, Springer-Verlag, 1998.
3. M. Bellare and P. Rogaway, "Random oracles are practical: A paradigm for designing efficient protocols", *First ACM Conference on Computer and Communications Security*, pages 62–73, ACM Press, 1993.

4. M. Bellare and P. Rogaway, "Optimal asymmetric encryption", *Advances in Cryptology – EUROCRYPT '94*, volume 950 of *Lecture Notes in Computer Science*, pages 92–111, Springer-Verlag, 1995.

5. D. Bleichenbacher, "A chosen ciphertext attack against protocols based on the RSA encryption standard PKCS #1", *Advances in Cryptology – CRYPTO '98*, volume 1462 of *Lecture Notes in Computer Science*, pages 1–12, Springer-Verlag, 1998.

6. S. Burnett and S. Paine, "RSA Security's official guide to cryptography", *RSA Press*, 2001.

7. D. Campbell "How your privacy is caught in the Net", http://www.theage.com.au/daily/990808/news/specials/news1.html, 8 August 1999.

8. R. Cramer and V. Shoup, "A practical public key cryptosystem provably secure against adaptive chosen ciphertext attack", *Advances in Cryptology – CRYPTO '98*, volume 1462 of *Lecture Notes in Computer Science*, pages 13–25, Springer-Verlag, 1998.

9. O. Dolev, C. Dwork, and M. Naor, "Non-malleable cryptography", *23rd ACM Annual Symposium on the Theory of Computing*, pages 542–552, ACM Press, 1991.

10. T. ElGamal, "A public key cryptosystems and a signature schemes based on discrete logarithms", *IEEE Transactions on Information Theory*, **IT-31**(4):469–472, 1985.

11. eWEEK, "Phone.com takes aim at WAP security hole", http://news.zdnet.co.uk/story/0,,s2081576,00.html, 23rd September 2000.

12. E. Fujisaki and T. Okamoto, "How to enhance the security of public-key encryption at minimum cost", *Public Key Cryptography*, volume 1560 of *Lecture Notes in Computer Science*, pages 53–68, Springer-Verlag, 1999.

13. E. Fujisaki and T. Okamoto, "Secure integration of asymmetric and symmetric encryption schemes", *Advances in Cryptology – CRYPTO '99*, volume 1666 of *Lecture Notes in Computer Science*, pages 537–544, Springer-Verlag, 1999.

14. S. Goldwasser and S. Micali, "Probabilistic encryption", *Journal of Computer and System Sciences*, **28**:270–299, 1984.

15. G. Itkis and L. Reyzin, "Forward-secure signatures with optimal signing and verifying", *Advances in Cryptology – CRYPTO 2001*, volume 2139 of *Lecture Notes in Computer Science*, pages 332–354, Springer-Verlag, 2001.

16. M. Joye, J.-J. Quisquater, and M. Yung, "On the power of misbehaving adversaries", *Topics in Cryptology – CT-RSA 2001*, volume 2020 of *Lecture Notes in Computer Science*, pages 208–222, Springer-Verlag, 2001.

17. J. Manger, "A chosen ciphertext attack on RSA Optimal Asymmetric Encryption Padding (OAEP) as standardized in PKCS #1", *Advances in Cryptology – CRYPTO 2001*, volume 2139 of *Lecture Notes in Computer Science*, pages 230–238, Springer-Verlag, 2001.

18. M. Naor and M. Yung, "Public-key cryptosystems provably secure against chosen ciphertext attacks", *22nd Annual ACM Symposium on Theory of Computing*, pages 427–437, ACM Press, 1990.

19. T. Okamoto and D. Pointcheval, "REACT: Rapid enhanced-security asymmetric cryptosystem transform", *Topics in Cryptology – CT-RSA 2001*, volume 2020 of *Lecture Notes in Computer Science*, pages 159–175, Springer-Verlag, 2001.

20. D. Pointcheval, "Chosen-ciphertext security for any one-way cryptosystem", *Public Key Cryptography*, volume 1751 of *Lecture Notes in Computer Science*, pages 129–146, Springer-Verlag, 2000.

21. C. Rackoff and D. Simon, "Noninteractive zero-knowledge proof of knowledge and chosen ciphertext attack", *Advances in Cryptology – CRYPTO '91*, volume 576 of *Lecture Notes in Computer Science*, pages 433–444, Springer-Verlag, 1992.

22. R.L. Rivest, A. Shamir, and L.M. Adleman, "A method for obtaining digital signatures and public-key cryptosystems", *Communications of the ACM*, **21**(2):120–126, 1978.

23. C.P. Schnorr and M. Jakobsson, "Security of Signed ElGamal Encryption", *Advances in Cryptology – ASIACRYPT 2000*, volume 1976 of *Lecture Notes in Computer Science*, pages 73–89, Springer-Verlag, 2000.

24. V. Shoup, "On formal models for secure key exchange", version 4, *Revision of IBM Research Report RZ 3120 (April 1999)*, November 15, 1999.

25. V. Shoup and R. Gennaro, "Securing threshold cryptosystems against chosen ciphertext attack", *Advances in Cryptology – EUROCRYPT '98*, volume 1403 of *Lecture Notes in Computer Science*, pages 1–16, Springer-Verlag, 1998.

26. A. Silberschatz, J. Peterson, and P. Galvin, *Operating system concepts*, Third edition, Addison-Wesley Publishing Company.

27. Y. Tsiounis and M. Yung, "On the security of ElGamal-based encryption", *Public Key Cryptography*, volume 1431 of *Lecture Notes in Computer Science*, pages 117–134, Springer-Verlag, 1998.

28. Y. Zheng and J. Seberry, "Immunizing public key cryptosystems against chosen ciphertext attacks", *IEEE Journal on Selected Area in Communications*, **11**(5):715–724, 1993.

Majority-Logic-Decodable Cyclic Arithmetic-Modular AN-Codes in 1, 2, and L Steps[*]

F. Javier Galán-Simón[1], Edgar Martínez-Moro[2], and Juan G. Tena-Ayuso[3]

[1] Dpto. Organización y Gestión de Empresas,
Universidad de Valladolid. Valladolid, 47002 Spain
javi@emp.uva.es
[2] Dpto. Matemática Aplicada Fundamental,
Universidad de Valladolid. Valladolid, 47002 Spain
edgar.martinez@ieee.org
[3] Dpto. Álgebra, Geometría y Topología,
Universidad de Valladolid. Valladolid, 47002 Spain
tena@agt.uva.es

Abstract. We generalize to any base $r \geq 2$ the Majority-Logic-Deco-dification Algorithms already considered for $r = 2$ by Chin-Long Chen, Robert T. Chien and Chao-Kai Liu [2]. The codes considered are generated by $\phi_n(r)$ where $\phi_n(x)$ is the nth-cyclotomic polynomial associated to the polynomial $x^n - 1$. Hong Decodification Algorithm [7] is also applicable to these codes, but achieves quite higher computational complexity.

1 Introduction

Arithmetic-Modular Codes were first proposed by Diamond [4]. These codes are useful on their own for error control in digital arithmetic and recently such procedures for eliminating errors in hardware arithmetic have been revealed of great cryptographic importance (See [1]). The arithmetic operations considered (in particular addition) are carried out with numbers represented the number system in base r ($r \in \mathbb{N}, r \geq 2$). One single error in addition can cause many faulty digits in the final representation information arising from the carrying in the operation.

An Arithmetic-Modular AN-code of length n, with modulus m and generator A is just the set $C(A, B) = \{AN \mid N \in \mathbb{N}, 0 \leq N < B\}$. where $m = A \cdot B$. If $m = r^n - 1$ (resp. $m = r^n + 1$) then $C(A, B)$ is a subgroup of $\mathbb{Z}/(m)$ and it is called a Cyclic (resp. Negacyclic) AN-Code [3].

If an error pattern E occurs then, $R = AN + E$ is the received information where AN stands for the correct codeword. Firstly we might find a suitable

---

[*] All three authors are supported by Junta de Castilla y León project "Construccio-nes criptográficas basadas en códigos correctores". Second one is also supported by Dgicyt PB97-0471.

B. Honary (Ed.): Cryptography and Coding 2001, LNCS 2260, pp. 128–137, 2001.

distance function since Hamming metric used in the Information Transmission Codes cannot be applied in the theory of arithmetic-modular codes [3,5,8,10].

For $x \in \mathbb{Z}$ we will denote the modular weight of x by $w_m(x)$ and it is defined as the minimal number of nonzero digits in any representation $x \equiv \sum_{i=0}^{\infty} c_i r^i$ mod m where the integers c_i are such that $|ci| < r$ for all i, r is the chosen base and $c_i = 0$ if $i \geq i_0$ for a fixed index i_0 (see [5,11,10]).

The minimum distance of an arithmetic-modular AN-code in base r is defined as (see [10,11]):

$$d = \min\{d(x, y) = w_m(x - y) \mid x, y \in C(A, B), x \neq y\}$$
$$= \min\{w_m(x) \mid x \in C(A, B), x \neq 0\} \tag{1}$$

As usual, a code where the distance between two any different words is at least $2t+1$ is called a t-error correcting code. The Correcting Capacity is the maximum t for which $d \geq 2t + 1$, that is $t = \left[\frac{d-1}{2}\right]$. Thus a t-error-correcting AN- code with minimum distance d detects every error pattern E with $w_m(E) < d - 1$ and corrects it for $w_m(E) < t$ (see [10]).

In order to compute the minimum distance we need the definition of Non Adjacent Form (NAF) and Cyclic Nonadjacent Form (CNAF), as well as the study of the exceptional cases ($x \in \mathbb{Z}$ is exceptional if $x \equiv 0$ mod m but $(r + 1)x \not\equiv 0$ mod m). These algorithms and the definition and properties of the syndrome of an integer can be found in [5] and [8].

In the theory of Arithmetic-Modular AN-Codes, the main problem still unsolved is finding good decoding algorithms. In all the AN-codes considered in [2] it is possible to accomplish the decodification using majority logic. This method was already used by Chin-Long Chen, Robert T. Chien and Chao-Kai Liu in the binary case $r = 2$. In this paper we will generalize this problem to any base $r > 2$ for one, two and L steps [6].

2   Majority-Logic-Decodable Cyclic Arithmetic-Modular AN-Codes (1 Step)

We consider the cyclic Arithmetic-Modular AN-Code $C(A, B)$ generated by $A = \frac{r^n - 1}{r^{n_2} - 1}$ and having information rank $B = r^{n_2} - 1$, where the code length n verifies that $n = n_1 \cdot n_2$. We suppose $n_1 \geq 3$ and $n_2 \geq 2$. The minimum modular distance of $C(A, B)$ is $d = n_1$, in fact:

$$w_m(A) = w_m(1 + r^{n_2} + r^{2n_2} + \cdots + r^{(n_1-1)n_2}) = n_1$$

Let $N \in \mathbb{N}$ be such that $1 < N < B = r^{n_2} - 1$ and its CNAF $N \equiv c_0 + c_1 r + \cdots + c_{n_2-1} r^{n_2-1}$ mod m. If we denote $N = \underline{c_0\, c_1\, \ldots\, c_{n_2-1}}$, then the CNAF of AN is:

$$AN \equiv c_0 + c_1 r + \cdots + c_{n_2-1} r^{n_2-1} + c_0 r^{n_2} + c_1 r^{n_2+1} + \cdots + c_{n_2-1} r^{2n_2-1}$$
$$+ \cdots + c_0 r^{(n_1-1)n_2} + c_1 r^{(n_1-1)n_2+1} + \cdots + c_{n_2-1} r^{n-1} \quad \text{mod } m \tag{2}$$

therefore:

$$AN = \frac{c_0 \, c_1 \, \cdots \, c_{n_2-1}}{1 \star n_2} \quad \frac{c_0 \, c_1 \, \cdots \, c_{n_2-1}}{2 \star n_2} \quad \cdots \quad \frac{c_0 \, c_1 \, \cdots \, c_{n_2-1}}{n_1 \star n_2} \qquad (3)$$

where $x \star y$ means block number x with y elements. Accordingly $C(A,B)$ detects all error pattern E with modular weight $w_m(E) \le n_1 - 1$ and corrects all error pattern with modular weight $w_m(E) \le t = \left[\frac{n_1-1}{2} \right]$.

We recall R is the received information, i.e., $R = AN + E$ with $w_m(E) < t$ and AN the codeword that decodifies R then E and N are the values we want to calculate. The proof of next lemma becomes straightforward.

Lemma 1. *Consider a code $C(\frac{r^n-1}{r^{n_2}-1}, r^{n_2}-1)$ and let $R = AN+E$. Then $N = 0$ if and only if $w_m(R) < t = \left[\frac{n_1-1}{2} \right]$.*

If N is an integer, $0 \le N < B$ and we consider the usual expression in base r, $0 \le a_i < r, i = 0, 1, 2, \ldots, n_2 - 1$, i.e., $N = a_0 \, a_1, \ldots a_{n_2-1}$. (Note that digits a_i are used in the standard r-ary expression and digits c_i stand for the CNAF). Then we have:

$$AN = \frac{a_0 \, a_1 \, \cdots \, a_{n_2-1}}{1 \star n_2} \quad \frac{a_0 \, a_1 \, \cdots \, a_{n_2-1}}{2 \star n_2} \quad \cdots \quad \frac{a_0 \, a_1 \, \cdots \, a_{n_2-1}}{n_1 \star n_2} \qquad (4)$$

i.e., n_1 identical blocks, each of them with n_2 elements.

Lemma 2. *A single error E in any position of the expression in (4) of any nonzero word of $C(A,B)$ can modify at most n_2 consecutive digits of that word.*

Proof. Assuming that $0 < N < r^{n_2} - 1$, there is, at least, a nonzero digit a_i and, at least a digit a_j such that $a_j < r - 1$. This allows the definition of h and j as follows:

$$h = \min \{h' \in \{0, 1, \ldots, n_2 - 1\} \mid a_{h'} < r - 1\}$$
$$j = \min \{j' \in \{0, 1, \ldots, n_2 - 1\} \mid a_{j'} > 0\} \qquad (5)$$

Since $w_m(E) = 1$ we can write $E \equiv b_i r^{kn_2+i} \mod m$ where $0 < |b_i| < r$ and the digit k shows us the block where the error E is affecting.

$$R = \frac{a_0 \, a_1 \, \cdots \, a_{n_2-1}}{1 \star n_2} \cdots \frac{a_0 \, a_1 \, \cdots \, (a_i + b_i) \, \cdots \, a_{n_2-1}}{k \star n_2} \cdots \frac{a_0 \, a_1 \, \cdots \, a_{n_2-1}}{n_1 \star n_2} \qquad (6)$$

According to the sign of the digit b_i we are able to distinguish two cases:

1. **Case 1:** $b_i > 0$

   If $a_i + b_i < r$ then (6) is the usual expression of R in base r and the error E will modify the word AN only in one digit, the one in the position $kn_2 + i$. But if $a_i + b_i \ge r$, the block k in (6) will be:

   $$\frac{a_0 \, a_1 \, \cdots \, (a_i + b_i - r) \, (a_{i+1} + 1) \, \cdots \, a_{n_2-1}}{k \star n_2} \qquad (7)$$

so that if $a_{i+1} + 1 < r$ it can be considered as the usual expression of R in base r and will not modify the following digits, otherwise (i.e. if $a_{i+1} + 1 = r$) the block k will be

$$\frac{a_0 \, a_1 \, \ldots \, (a_i + b_i - r) \, 0 \, (a_{i+2} + 1) \, \ldots \, a_{n_2 - 1}}{k \star n_2} \qquad (8)$$

and so on.

This process of "carry propagation" will finish at the first digit a_j such that $a_j < r - 1$; Strictly speaking, at the digit h either belonging to the block k (if $i < h$) or to the block $k + 1$ (if $i \geq h$). Therefore it can never corrupt more than n_2 consecutive digits from the position $kn_2 + i$.

2. **Case 2:** $b_i < 0$

If $a_i + b_i \geq 0$ the usual expression of R in base r is (6). (Notice that we have $a_i + b_i < r$ for $b_i < 0$) The error E will have modified only one digit in AN. But if $a_i + b_i < 0$ then the block k of (6) will be:

$$\frac{a_0 \, a_1 \, \ldots \, (a_i + b_i + r) \, (a_{i+1} - 1) \, \ldots \, a_{n_2 - 1}}{k \star n_2} \qquad (9)$$

So, if $a_{i+1} - 1 \geq 0$ it can never corrupt subsequent digits but if $a_{i+1} - 1 < 0$ the block k will be:

$$\frac{a_0 \, a_1 \, \ldots \, (a_i + b_i + r) \, (a_{i+1} - 1 + r) \, (a_{i+2} - 1) \, \ldots \, a_{n_2 - 1}}{k \star n_2} \qquad (10)$$

and so on.

At this stage, the process of "carry propagation" will be finished when the first digit $a_j > 0$ is found. This process will finish in the digit a_j of the block k (if $i < j$) or of the block $k + 1$ (if $i \geq j$) at most. Consequently it can never modify more than n_2 consecutive digits of the usual expression of the codeword AN. Having in mind that block $k + 1$ is the same that block 1 when $k = n_1$ the proof of the lemma is complete.$\square$

Lemma above provides the tools necessary for proving the theorem and the decodification algorithm below. It is only necessary to take into account that $n_1 - t > \frac{n_1 + 1}{2}$.

Theorem 1. *Let $C(A, B)$ be a cyclic arithmetic-modular AN-code generated by $A = \frac{r^n - 1}{r^{n_2} - 1}$ where the length n of the code verifies that $n = n_1 n_2$ and let $E \in \mathbb{Z}$, $w_m(E) < t = \left[\frac{n_1 - 1}{2}\right]$ where $R = AN + E$ is the received information. We will consider the digits of positions $k, n_2 + k, 2n_2 + k, \ldots, (n_1 - 1)n_2 + k$ in the usual r-ary expansion of R. Then at least $n_1 - t$ of those digits are equal, and they correspond to those in the same position as in the r-ary expansion of the codeword AN. This stands for $k = 0, 1, 2, \ldots, n_2 - 1$.*

The proof follows directly from lemma above. This result allow us to state the following decoding algorithm:

Algorithm 1 (Majority-Logic Decoding (One step))

- **If** $w_m(R) \leq t$ *then* **Return** $N = 0$ *and* $R = E$ **fi**
- **If** $w_m(R) > t$ *then* **do**
  For $k = 0, \ldots, n_2 - 1$ **do**
  Consider the positions $k, n_2 + k, 2n_2 + k, \ldots, (n_1 - 1)n_2 + k$

$$\underbrace{* \, * \, \ldots \, * \, \overbrace{d_1}^{k} \, * \, \ldots \, *}_{1 \star n_2} \quad \underbrace{* \, * \, \ldots \, * \, \overbrace{d_2}^{k} \, * \, \ldots \, *}_{2 \star n_2} \quad \cdots \quad \underbrace{* \, * \, \ldots \, * \, \overbrace{d_{n_1}}^{k} \, * \, \ldots \, *}_{n_1 \star n_2}$$

Take as the t_k "true value" of a_k the most repeated value among the values in the multiset $\{d_1, \ldots, d_{n_1}\}$.
od
return $N = \underline{t_0 \, t_1 \, \ldots \, t_{n_2-1}}$ **fi**

Example 1. $r = 3, n_1 = 5, n_2 = 4, n = 20, A = \frac{3^{20}-1}{3^4-1} = 43.584.805$
$B = 3^4 - 1, m = AB = 3^{20} - 1 = 3.486.784.401, d = n_1 = 5$ and $t = 2$.
  Let $R = 3.094.403.084$ be the received information with $w_m(R) = 5 > t$. Its 3-ary expression is:

$$R = \frac{2\,2\,1\,0}{1 \star 4} \, \frac{0\,0\,2\,2}{2 \star 4} \, \frac{2\,2\,2\,1}{3 \star 4} \, \frac{2\,2\,1\,2}{4 \star 4} \, \frac{2\,2\,1\,2}{5 \star 4}$$
$$\downarrow \text{decoding algorithm}$$
$$N = \underline{2\,2\,1\,2}$$

Therefore $N = 71$ and $E = R - AN = -118.071 = 3^3 - 2 \cdot 3^{10}$. Note that $w_m(3^3 - 2 \cdot 3^{10}) = 2 < t$.

3 Majority-Logic-Decodable Cyclic Arithmetic-Modular AN-Codes (L Steps)

Let n be a positive integer such that $n = \prod_{i=1}^{L} p_i^{s_i}$, where the distinct primes p_i satisfy the condition $p_i \geq 2^{i-1}(p_1 - 1) + 1$, for $i = 2, 3, \ldots, L$ and $p_1 > 2$ in order to get $w_m(E) > 0$. It is a well known fact [9] that the polynomial $x^n - 1$ can be expressed as the product of cyclotomic polynomials over the rational field, i.e., $x^n - 1 = \prod_{d|n} \phi_d(x)$. In particular, we can write $r^n - 1 = \prod_{d|n} \phi_d(r)$. Note also that although $\phi_d(x)$ is irreducible over $\mathbb{Q}[x]$, $\phi_d(r)$ might not be a prime number. One can consider the cyclic arithmetic-modular AN-code $C(A, B)$ with generator $A = \phi_n(r)$ and rank of information $B = \frac{r^n-1}{A} = \prod_{d|n, d \neq n} \phi_d(r)$.
  We will show that the AN-code corrects all error patterns E with $w_m(E) \leq \frac{p_1-1}{2}$. This will lead us to the conclusion that $d = p_1$, and to a Majority-Logic Decodification Algorithm of such a code in L steps.

Let $R = AN + E$ be the received information where $w_m(E) < \frac{p_1-1}{2}$. Multiplying both sides of the equation by $\prod_{i=1}^{L-1}\left(r^{\frac{n}{p_i}} - 1\right)$ we obtain

$$R \prod_{i=1}^{L-1}\left(r^{\frac{n}{p_i}} - 1\right) = (AN + E) \prod_{i=1}^{L-1}\left(r^{\frac{n}{p_i}} - 1\right)$$

$$= A \prod_{i=1}^{L-1}\left(r^{\frac{n}{p_i}} - 1\right) N + E \prod_{i=1}^{L-1}\left(r^{\frac{n}{p_i}} - 1\right) \tag{11}$$

We denote $R_L =: \left| R \prod_{i=1}^{L-1}\left(r^{\frac{n}{p_i}} - 1\right) \right|_m$ where $\mod m = r^n - 1$, and we define the integers:

$$A_L = \frac{r^n - 1}{r^{\frac{n}{p_L}} - 1}, \quad B_L = r^{\frac{n}{p_L}} - 1, \quad K_L = \frac{B}{B_L}, \quad M_L = \frac{\prod_{i=1}^{L}\left(r^{\frac{n}{p_i}} - 1\right)}{B}.$$

It is not difficult to show that:

$$A_L = A K_L, \quad \prod_{i=1}^{L-1}\left(r^{\frac{n}{p_i}} - 1\right) = K_L M_L, \quad \left(r^{\frac{n}{p_{L-1}}} - 1\right) |R_L$$

Replacing in equation (11) we obtain:

$$R \prod_{i=1}^{L-1}\left(r^{\frac{n}{p_i}} - 1\right) = A K_L M_L N + E \prod_{i=1}^{L-1}\left(r^{\frac{n}{p_i}} - 1\right)$$

$$= A_L M_L N + E \prod_{i=1}^{L-1}\left(r^{\frac{n}{p_i}} - 1\right) \tag{12}$$

therefore if $M =: M_L N$ and $E_L =: \left| E \prod_{i=1}^{L-1}\left(r^{\frac{n}{p_i}} - 1\right) \right|_m$, it follows that

$$R_L = |A_L| M|_{B_L} + E_L|_m \tag{13}$$

Lemma 3. $W_m(E_L) \leq \frac{p_L-1}{2}$

Proof.

$$w_m(E_L) = w_m \left(E \left(r^{\frac{n}{p_1}} - 1\right)\left(r^{\frac{n}{p_2}} - 1\right) \dots \left(r^{\frac{n}{p_{L-1}}} - 1\right)\right)$$

$$\leq 2 w_m \left(E \left(r^{\frac{n}{p_2}} - 1\right) \dots \left(r^{\frac{n}{p_{L-1}}} - 1\right)\right)$$

$$\leq 2^2 w_m \left(E \left(r^{\frac{n}{p_3}} - 1\right) \dots \left(r^{\frac{n}{p_{L-1}}} - 1\right)\right) \leq \dots$$

$$\leq 2^{L-1} w_m(E) \leq 2^{L-1} \frac{p_L - 1}{2}$$

$$= \frac{\left(2^{L-1}(p_1 - 1) + 1\right) - 1}{2} \leq \frac{p_L - 1}{2}$$

Note that the condition $p_L \geq 2^{L-1}(p_1 - 1) + 1$ has been used. $\square$

In addition, we have $R_L = |A_L|M|_{B_L} + E_L|_m$ where $w_m(E_L) \leq \frac{p_L-1}{2}$ and $d_{\min}C(A_L, B_L) = p_L$. Using the algorithm 1 in previous section we will be able to find N_L, where $0 \leq N_L < B$ and E'_L such that $R_L = A_L N_L + E'_L$ and

$$E'_L \equiv E \prod_{i=1}^{L-1} \left(r^{\frac{n}{p_i}} - 1 \right) \quad \bmod m \tag{14}$$

From equation above it follows that there will be an integer b_{L-1} such that:

$$E'_L = E \prod_{i=1}^{L-1} \left(r^{\frac{n}{p_i}} - 1 \right) + b_{L-1} \left(r^n - 1 \right)$$

$$\Downarrow \tag{15}$$

$$\frac{E'_L}{r^{\frac{n}{p_{L-1}}} - 1} = E \prod_{i=1}^{L-2} \left(r^{\frac{n}{p_i}} - 1 \right) + b_{L-1} \frac{r^n - 1}{r^{\frac{n}{p_{L-1}}} - 1}$$

If we define

$$A_{L-1} = \frac{r^n - 1}{r^{\frac{n}{p_{L-1}}} - 1}, \quad B_{L-1} = r^{\frac{n}{p_{L-1}}} - 1, \quad R_{L-1} = \left| \frac{E'_L}{r^{\frac{n}{p_{L-1}}} - 1} \right|_m \quad \text{and}$$

$$E_{L-1} = \left| E \prod_{i=1}^{L-2} \left(r^{\frac{n}{p_i}} - 1 \right) \right|_m$$

reducing (15)   mod m we obtain

$$R_{L-1} = |A_{L-1}|b_{L-1}|_m + E_{L-1}|_m$$

Bearing in mind that $p_{L-1} \geq 2^{L-2}(p_1 - 1) + 1$ and following the previous procedures we can apply the algorithm 1 by using the AN-code $C(A_{L-1}, B_{L-1})$ whose minimum distance is $d = p_{L-1}$. By a recurrent process we will obtain in the $L-1$ step the following relations:

$$R_2 = A_2 N_2 + E'_2 \text{ where } A_2 = \frac{r^n - 1}{r^{\frac{n}{p_2}} - 1}, B_2 = r^{\frac{n}{p_2}} - 1,$$

$$E'_2 \equiv E(r^{\frac{n}{p_1}} - 1) \quad \bmod m \quad \text{and} \quad (r^{\frac{n}{p_1}} - 1)|E'_2$$

Therefore $\frac{E'_2}{r^{\frac{n}{p_1}} - 1} = E + b_1 \frac{r^n - 1}{r^{\frac{n}{p_1}} - 1}$ where $b_1 \in \mathbb{Z}$. If we let $A_1 = \frac{r^n - 1}{r^{\frac{n}{p_1}} - 1}$, $B_1 = r^{\frac{n}{p_1}} - 1$, $R_1 = \left| \frac{E'_2}{r^{\frac{n}{p_1}} - 1} \right|_m$ and $E_1 = |E|_m$; then $R_1 = |A_1|b_1|_m + E_1|_m$ with $w_m(E_1) = w_m(E) \leq \frac{p_L-1}{2}$.

Thus if we apply the algorithm 1 on the section above to R_1 using the AN-code $C(A_1, B_1)$, whose minimum distance is p_1 ,there will be $N_1 \in \mathbb{Z}$, $0 \leq N_1 < B$ and E'_1 such that $R_1 = A_1 N_1 + E'_1$ where $E'_1 \equiv E \bmod m$, and this is the error that had been made. Hence it follows this L-steps algorithm:

Algorithm 2 (Majority-Logic Decoding Algorithm(L steps))

- **Input** R,
- **If** $w_m(R) \leq \frac{p_L - 1}{2}$ then **Return** $N = 0$ and $R = E$ **fi**
- **If** $w_m(R) > \frac{p_L - 1}{2}$ then **do**

  Compute $R_L = \left| R \prod_{i=1}^{L-1} \left(r^{\frac{n}{p_i}} - 1 \right) \right|_m$

  For $s = L, \ldots, 1$ *stepsize=-1* **do**
  Make majority decision (algorithm 1) on R_s *in* $C(A_s, B_s)$ *to get* $R_s = A_s N_s + E'_s$ *where* $0 \leq N_s < B_s$

  $$R_{s-1} = \left| \frac{E'_s}{\left(r^{\frac{n}{p_{s-1}}} - 1 \right)} \right|_m$$

  od
  return $E \equiv E'_1 \mod m$ *and* $N = \frac{R-E}{A}$ **fi**

Corollary 1. *The minimum distance of the Cyclic AN-code generated by* $A = \phi_n(r)$ *is* $d = p_1$, *where the length* $n = \prod_{i=1}^{L} p_i^{s_i}$ *and the distinct primes* p_i *verify the condition* $p_i \geq 2^{i-1}(p_1 - 1) + 1$, *for* $i = 2, \ldots, L$ *and* $p_1 > 2$.

Proof. We have shown that $C(A, B)$ may correct all error pattern E with $w_m(E) \leq \frac{p_L - 1}{2}$. From this fact we obtain that $d \geq 2\left(\frac{p_1 - 1}{2}\right) + 1 = p_1$ (see [10]). Now we consider the integer $N = \frac{r^n - 1}{(r^{\frac{n}{p_1}} - 1)A}$. It follows that $w_m(AN) = w_m\left(\frac{r^n - 1}{r^{\frac{n}{p_1}} - 1}\right) = p_1$ because the CNAF of $\frac{r^n - 1}{r^{\frac{n}{p_1}} - 1} = 1 + r^{\frac{n}{p_1}} + r^{\frac{2n}{p_1}} + \cdots + r^{\frac{(p_1 - 1)n}{p_1}}$ has p_1 nonzero digits. Thus $d \leq p_1$ and hence $d = p_1$. $\square$

4   Particular Case: $L = 2$

This algorithm is also true if $n = \prod_{i=1}^{L} n_i^{s_i}$ is a decomposition of n in relatively prime integers with the additional condition $n_i \geq 2^{i-1}(n_1 - 1) + 1$ for $i = 2, \ldots, L$ and $n_1 > 1$. If $L = 2$, $n = n_1 n_2$ where $n_1, n_2 \in \mathbb{N}$, $gcd(n_1, n_2) = 1$ and $n_2 \geq 2n_1 + 1$. The generator of the arithmetic- modular AN-code $C(A, B)$ will be $A = \frac{(r^n - 1)(r - 1)}{(r^{n_1} - 1)(r^{n_2} - 1)}$.

Example 2. $r = 4, n_1 = 3, n_2 = 8, n = 24$,

$$A = \frac{(4^{24} - 1)(4 - 1)}{(4^3 - 1)(4^8 - 1)} = 204.525.373, \quad B = \frac{(4^3 - 1)(4^8 - 1)}{4 - 1} = 1.376.235,$$

$$m = AB = 4^{24} - 1 = 281.474.976.710.655, \quad d = n_1 = 3 \text{ and } t = 1.$$

Let $R = 4.704.075.387$ be the received information with $w_m(R) = 11 > t$.

- **Step 1** Compute $R_2 = |R(4^8 - 1)|_m = 26.806.603.776.390$

$$A_2 = \frac{4^{24} - 1}{4^3 - 1} = 4.467.856.773.185, \qquad B_2 = 4^3 - 1 = 63.$$

The expression of R_2 in base 4 is:

$$R_2 = \frac{2\,1\,0\ \ 2\,1\,0\ \ 0\,2\,0\ \ 2\,1\,0\ \ 2\,1\,2\ \ 1\,1\,0\ \ 2\,1\,0\ \ 2\,1\,0}{1\star 3\ \ 2\star 3\ \ 3\star 3\ \ 4\star 3\ \ 5\star 3\ \ 6\star 3\ \ 7\star 3\ \ 8\star 3}$$
$$\downarrow \text{ decoding algorithm 1}$$
$$N_2 = \underline{2\,1\,0}$$

$N_2 = (2 + 1 \cdot 4 + 0 \cdot 4^2) = 6$ and $E_2' = R_2 - A_2 N_2 = -536.862.720$

- **Step 2** Compute $R_1 = \left| \frac{E_2'}{4^8 - 1} \right|_m = 281.474.976.702.463.$

$$A_1 = \frac{4^{24} - 1}{4^8 - 1} = 4.295.032.833, \qquad B_1 = 4^8 - 11 = 65.535$$

Therefore, expressing R_1 in base 4:

$$R_1 = \frac{3\,3\,3\,3\,3\,3\,1\,3\ \ 3\,3\,3\,3\,3\,3\,3\,3\ \ 3\,3\,3\,3\,3\,3\,3\,3}{1\star 8 \qquad\quad 2\star 8 \qquad\quad 3\star 8}$$
$$\downarrow \text{ decoding algorithm 1}$$
$$N_1 = \underline{3\,3\,3\,3\,3\,3\,3\,3}$$

$N_1 = 65.535$ and $E_1' = R_1 - A_1 N_1 = -8.192 = -2 \cdot 4^6$, hence our algorithm results:

$$N = \frac{R - E}{A} = 23 \qquad E \equiv -2 \cdot 4^6 \mod m$$

References

1. D. Boneh, R.A. DeMillo, and R.J. Lipton. On the importance of eliminating errors in cryptographic computations. *Journal of Cryptology*, 14:101–119, 2001.
2. C. L. Chen, R. T. Chien, and C. K. Liu. On majority-logic-decodable arithmetic codes. *IEEE Trans. Information Theory*, IT-19:678–682, 1973.
3. R. T. Chien, S. J. Hong, and F. P. Preparata. Some results in the theory of arithmetic codes. *Information and Control*, 19:246–264, 1971.
4. Joseph M. Diamond. Checking codes for digital computers. *IRE*, 1954.
5. Fco. Javier Galán-Simón. Pesos aritméticos y modulares en los códigos aritméticos. Master thesis, Facultad de Ciencias, Universidad de Valladolid, 1990.
6. Fco. Javier Galán-Simón. *Detección y corrección de errores en códigos aritméticos y modulares*. PhD thesis, Facultad de Ciencias, Universidad de Valladolid, 1993.
7. S. J. Hong *On bounds and implementation of arithmetic codes*, Procedings of National Electronics Cooference, R-437, Univ. of Illinois, Urbana. 1976.
8. Antoine Lobstein. On modular weights in arithmetic codes. In *Coding theory and applications (Cachan, 1986)*, pages 56–67. Springer, Berlin, 1988.

9. Félix López Fernández-Asenjo and Juan Tena Ayuso. *Introducción a la teoría de números primos*. Universidad de Valladolid Instituto de Ciencias de la Educación, Valladolid, 1990.

10. T. R. N. Rao. *Error coding for arithmetic processors*. Academic Press, New York-London, 1974.

11. J. H. van Lint. *Introduction to coding theory*. Springer-Verlag, Berlin, 1999.

Almost-Certainly Runlength-Limiting Codes

David J.C. MacKay

Department of Physics, University of Cambridge
Cavendish Laboratory, Madingley Road,
Cambridge, CB3 0HE, United Kingdom.
mackay@mrao.cam.ac.uk
http://wol.ra.phy.cam.ac.uk/mackay/

Abstract. Standard runlength-limiting codes – nonlinear codes defined by trellises – have the disadvantage that they disconnect the outer error-correcting code from the bit-by-bit likelihoods that come out of the channel. I present two methods for creating transmissions that, with probability extremely close to 1, both are runlength-limited and are codewords of an outer linear error-correcting code (or are within a very small Hamming distance of a codeword). The cost of these runlength-limiting methods, in terms of loss of rate, is significantly smaller than that of standard runlength-limiting codes. The methods can be used with any linear outer code; low-density parity-check codes are discussed as an example.

The cost of the method, in terms of additional redundancy, is very small: a reduction in rate of less than 1% is sufficient for a code with blocklength 4376 bits and maximum runlength 14.

This paper concerns noisy binary channels that are also constrained channels, having maximum runlength limits: the maximum number of consecutive 1s and/or 0s is constrained to be r. The methods discussed can also be applied to channels for which certain other long sequences are forbidden, but they are not applicable to channels with minimum runlength constraints such as maximum transition-run constraints.

I have in mind maximum runlengths such as $r = 7$, 15, or 21. Such constraints have a very small effect on the capacity of the channel. (The capacity of a noiseless binary channel with maximum runlength r is about $1 - 2^{-r}$.)

There are two simple ways to enforce runlength constraints. The first is to use a nonlinear code to map, say, 15 data bits to 16 transmitted bits [8]. The second is to use a linear code that is guaranteed to enforce the runlength constraints. The disadvantage of the first method is that it separates the outer error-correcting code from the channel: soft likelihood information may be available at the channel output, but once this information has passed through the inner decoder, its utility is degraded. The loss of bit-by-bit likelihood information can decrease the performance of a code by about 2 dB [3]. The second method may be feasible, especially if low-density parity-check codes are used, since they are built out of simple parity constraints, but it only gives a runlength limit r smaller than 16 if the outer code's rate is smaller than the rates that are conventionally required for magnetic recording (0.9 or so) [7].

B. Honary (Ed.): Cryptography and Coding 2001, LNCS 2260, pp. 138–147, 2001.

I now present two simple ideas for getting the best of both worlds. The methods presented involve only a small loss in communication rate, *and* they are compatible with the use of linear error-correcting codes. The methods do not give an absolute guarantee of reliable runlength-limited communication; rather, as in a proof of Shannon's noisy channel coding theorem, we will be able to make a statement like 'with probability $1 - 10^{-20}$, this method will communicate reliably and satisfy the $r = 15$ runlength constraint'. This philosophy marries nicely with modern developments in magnetic recording, such as (a) the use of low-density parity-check codes, which come without cast-iron guarantees but work very well empirically [7]; and (b) the idea of digital fountain codes [1], which can store a large file on disc by writing thousands of packets on the disc, each packet being a random function of the original file, and the original file being recoverable from (almost) any sufficiently large subset of the packets – in which case occasional packet loss is unimportant. The ideas presented here are similar to, but different from, those presented by Immink [4,5,6], Deng and Herro [2], and Markarian *et al.* [9].

1   Linear Method

The first idea is a method for producing a codeword of the linear outer code that, with high probability, satisfies the runlength-limiting constraints.

We assume that a good outer (N, K) linear error-correcting code has been chosen, and that it is a systematic code. We divide the K source bits into K_u user bits and $K_r > \log_2 M$ additional special source bits that we will set so as to satisfy the runlength constraints. The code has $M = N - K$ parity bits. We choose the special source bits such that all M of the parity bits are influenced by at least one of the K_r special bits. When transmitting the codeword we order the bits as shown below, such that the $K_r + M$ special bits and parity bits appear uniformly throughout the block. We call these $K_r + M$ bits the pad bits.

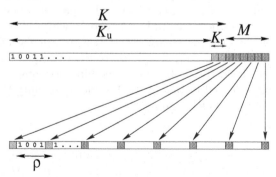

We modify the code by adding an offset **o** to all codewords. The random vector **o** is known to the sender and receiver and satisfies the runlength constraints. As we will see later, it may be useful for **o** to change from block to block; for example, in magnetic recording, it might depend on a seed derived from the sector number.

The method we will describe is appropriate for preventing runs of length greater than r for any r greater than or equal to ρ, the distance between pad bits. Two examples of the code parameters are given in table 1. R is the overall code rate from the user's perspective. We intend the number of special source bits

Table 1. Some feasible parameter settings.

K_{u}	4096	4096
K_{r}	20	24
M	296	432
N	4412	4552
$R = K_{\mathrm{u}}/N$	0.928	0.9
$\rho = N/(M + K_{\mathrm{r}})$	14	10

K_{r} to be very small compared with the blocklength. For comparison, a standard rate 15/16 runlength-limiting inner code would correspond to $K_{\mathrm{r}} \simeq 256$ bits. If instead we use $K_{\mathrm{r}} = 24$ then we are using one tenth the number of bits to achieve the runlength constraint.

The idea of the linear runlength-limiting method is that, once the user bits have been set, there are very likely to be codewords that satisfy the runlength constraints among the $2^{K_{\mathrm{r}}}$ codewords corresponding to the $2^{K_{\mathrm{r}}}$ possible settings of the special source bits.

Encoding Method

We write the user data into the K_{u} bits and note which, if any, of the $K_{\mathrm{r}} + M$ pad bits are forced by the runlength constraints to take on a particular state. [If the maximum runlength, r, is greater than the spacing between pad bits, ρ, then there may be cases where we are free to choose which of two adjacent pad bits to force. We neglect this freedom in the following calculations, noting that this means the probability of error will be overestimated.]

For a given user block, the probability, averaging over offset vectors $\mathbf{o}$, that a particular pad bit is forced to take state 0 is the probability that it lies in or adjacent to a run of r 1s, which is approximately

$$\beta \equiv r2^{-r}. \tag{1}$$

The probability that a particular pad bit is forced to take state 1 is also β. The expected number of forced pad bits in a block is thus $2\beta(K_{\mathrm{r}} + M)$. Table 2 shows that, for $r \geq 14$, it will be rare that more than one or two pad bits are forced.

Having identified the forced bits, we use linear algebra to find a setting of the K_{r} special bits such that the corresponding codeword satisfies the forced bits, or, in the rare cases where no such codeword exists, to find a codeword that violates the smallest number of them.

Table 2. Examples of the expected number of forced pad bits, $2\beta(M + K_r)$.

r	14	16	20
β	8×10^{-4}	2×10^{-4}	2×10^{-5}
$K_r + M$	316	316	316
$2\beta(M + K_r)$	0.5	0.15	0.01

1.1 Probability of Failure

The probability that this scheme will fail to find a satisfactory codeword depends on the details of the outer code. We first make an estimate for the case of a low-density parity-check code; later, we will confirm this estimate by numerical experiments on actual codes.

Let the columns of the parity-check matrix corresponding to the $K_r + M$ pad bits form a submatrix $\mathbf{F}$. In the case of a regular low-density parity-check code, this binary matrix will be sparse, having column weight 4, say, or about 8 in the case of a code originally defined over $GF(16)$. If the code is an irregular low-density parity-check code, these columns might be chosen to be the higher weight columns; we will see that this would reduce the probability of error.

Consider one row of $\mathbf{F}$ of weight w. If the w corresponding pad bits are *all* constrained, then there is a probability of $1/2$ that the parity constraint corresponding to this row will be violated. In this situation, we can make a codeword that violates one of the w runlength constraints and satisfies the others. The probability of this event happening is $(2\beta)^w$. For every row of $\mathbf{F}$, indeed *for every non-zero codeword of the dual code* corresponding to $\mathbf{F}$, there is a similar event to worry about. The expected number of runlength constraints still violated by the best available codeword is thus roughly

$$\sum_w \frac{1}{2} A_{\mathbf{F}}(w)(2\beta)^w, \tag{2}$$

where $A_{\mathbf{F}}(w)$ is the weight enumerator function of the code whose generator matrix is $\mathbf{F}$. For small β, this expectation is dominated by the low-weight words of $\mathbf{F}$, so, if there are M words of lowest weight $w_{\min}$, the expected number of violations is roughly

$$\frac{1}{2} M (2\beta)^{w_{\min}} \tag{3}$$

Table 3 shows this expected number for $w_{\min} = 4$ and 8.

For example, assuming $w_{\min} = 8$ (which requires a matrix $\mathbf{F}$ whose columns have weight 8 or greater), for a maximum runlength r of 14 or more, we can get the probability of failure of this method below 10^{-20}.

What the above analysis has not pinned down is the relationship between the column weight of $\mathbf{F}$ and $w_{\min}$. We now address this issue, assuming $\mathbf{F}$ is a low-density parity check matrix. If $\mathbf{F}$ has a row of weight w, then the dual code has a word of weight w. Any linear combination of rows also gives a dual

Table 3. The expected number of violations for $w_{\min} = 4$ and 8.

r	14	16	20
M	296	296	296
β	8×10^{-4}	2×10^{-4}	2×10^{-5}
$\frac{1}{2}M(2\beta)^4$	1×10^{-9}	8×10^{-12}	3×10^{-16}
$\frac{1}{2}M(2\beta)^8$	1×10^{-20}	5×10^{-25}	7×10^{-34}

codeword. Is it likely that the dual code has words of lower weight than the already sparse rows that make up the parity check matrix? It would be nice to know it is not likely, because then we could approximate the expected number of violations (2) by

$$\frac{1}{2} \sum_w g(w)(2\beta)^w, \tag{4}$$

where $g(w)$ is the number of rows of $\mathbf{F}$ that have weight w. However, as $\mathbf{F}$ becomes close to square, it becomes certain that linear combinations of those low-weight rows will be able to make even lower weight dual words. The approximation (4) would then be a severe underestimate.

We now test these ideas empirically.

1.2   Explicit Calculation of the Probability of Conflict

I took a regular low-density parity-check code with blocklength $N = 4376$ bits and $M = 282$ (the true number of independent rows in the parity check matrix was 281). The column weight was $j = 4$. In four experiments I allocated $K_r = 11, 21, 31, 41$ of the source bits to be special bits and found experimentally the conditional probability, given w, that a randomly selected set of w pad bits constrained by the runlength constraints would conflict with the code constraints.

[Method: given w randomly chosen pad bits, Gaussian elimination was attempted to put the generator matrix into systematic form with respect to the chosen bits. This was repeated for millions of choices of the pad bits, and the probability of failure of Gaussian elimination was estimated from the results. The actual probability of failure will be smaller than this quantity by a factor between 1 and 2, because it is possible, even though the pad bits are not independent, that the runlength constraints will be compatible with the code constraints.]

Under a union-type approximation (like (4)) that only counts the dual codewords that are rows of $\mathbf{F}$, we would predict this conditional probability to be

$$P(\text{conflict}|w) \simeq \sum_{w'} g(w')\frac{\binom{N-w'}{w-w'}}{\binom{N}{w}}. \tag{5}$$

The empirically-determined conditional probability of failure, as a function of the weight w of the constraint, is shown in figure 1, along with the approximation (5), shown by the lines without datapoints.

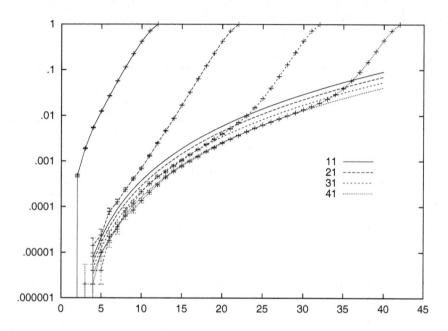

Fig. 1. Probability of failure of the linear runlength-limiting method as a function of number of constrained pad bits, w. The four curves with points and error bars show empirical results for $K_r = 11$, 21, 31, and 41. The four lines show the approximation (5). All systems derived from the same regular low-density parity-check code with $M = 281$ constraints and column weight $j = 4$. The runlength limits for these four cases are $r = N/(K_r+M) = 15$, 14.5, 14, and 13.6.

It can be seen that for $K_r = 11$ and 21, the approximation is an underestimate, but for $K_r = 31$ and 41, it gives a snug fit in the important area (*i.e.*, low w). From the empirical results we can also deduce the probability of failure, which is

$$P(\text{conflict}) = \sum_w P(w)P(\text{conflict}|w) \tag{6}$$

$$= \binom{N}{w}(2\beta)^w(1-2\beta)^{N-w}P(\text{conflict}|w). \tag{7}$$

Plugging in $\beta = 8 \times 10^{-4}$ (corresponding to constrained length $r = 14$), we find that for $K_r = 21$, 31, and 41, the probability of a conflict is about 10^{-15}. We will discuss how to cope with these rare failures below.

1.3   Further Experiments

I also explored the dependence on column weight by making three codes with identical parameters N, M, K_r, but different column weights: $j = 3$, $j = 4$, and $j \simeq M/2$ (a random code). Figure 2 shows that for $j = 4$, the failure probability, at small w, is quite close to that of a random code.

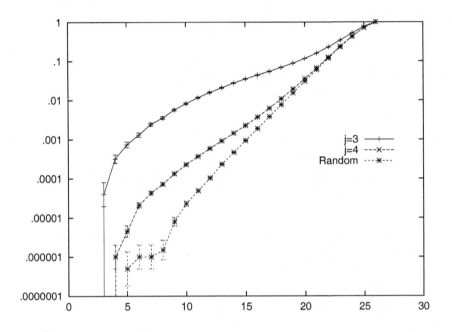

Fig. 2. Probability of failure of the linear runlength-limiting method for various column weights. The code parameters were $K_r = 25$, $M = 220$.

1.4   How to Make the Probability of Failure Even Smaller

We showed above that our runlength-limiting method can have probability of failure about 10^{-15}. What if this probability of failure is too large? And what should be done in the event of a failure? I can suggest two simple options. First, in discdrive applications, if the offset vector **o** is a random function of the sector number where the block is being written, we could have an emergency strategy: when the optimally encoded block has runlength violations, leave a pointer in the current sector and write the file in another sector, where the coset **o** is different. This strategy would incur an overhead cost at write-time on the rare occasions where the writer has to rearrange the blocks on the disc.

The second option is even simpler, as described in the following section.

2 Nonlinear Method

If the linear method fails to satisfy all the runlength constraints, a really dumb option is to modify the corresponding pad bits so that the runlength constraints *are* satisfied, and transmit the modified word, which will no longer be a codeword of the outer error-correcting code. *As long as the number of flipped bits is not too great, the decoder of the outer code will be able to correct these errors.* The average probability of a bit's being flipped is raised by a very small amount compared with a typical noise level of 10^{-4}. The probability distribution of the number of flipped bits depends on the details of the code, but for a low-density parity-check code whose graph has good girth properties, we'd expect the probability of t flips to scale roughly as

$$p_0 p_1^{t-1}, \tag{8}$$

where $p_0 = \frac{1}{2}(2\beta)^{w_{\min}}$ and $p_1 = \frac{1}{2}(2\beta)^{w_{\min}-1}$, and no worse than

$$\binom{K_r + M}{t} p_0 p_1^{t-1}, \tag{9}$$

which is roughly

$$\frac{(K_r + M)^t}{t!}(2\beta)^{t(w_{\min}-1)} \tag{10}$$

For $w_{\min} = 4$, $K_r + M = 316$, and $t = 6$, for example, the probability of t errors would be roughly as shown in table 4. Thus as long as the outer code is

Table 4.

r	14	16	20
$\frac{(K_r+M)^t}{t!}(2\beta)^{6\times 3}$	2×10^{-44}	4×10^{-54}	5×10^{-74}

capable of correcting substantially more than 6 errors, the probability of failed transmission using this scheme is very low indeed.

2.1 Use of the Nonlinear Method Alone

For some applications, the dumb nonlinear scheme by itself might suffice. At least in the case of a low-density parity-check code, it is simple to modify the decoder to take into account the fact that the pad bits are slightly less reliable than the user bits. [We could even include in the belief propagation decoding algorithm the knowledge that the pad bit is least reliable when it sits in the middle of run.]

Let the pad bits be the parity bits, *i.e.*, let $K_r = 0$, use a random offset vector $\mathbf{o}$, and flip whichever pad bits need to be flipped to satisfy the runlength constraints. The probability of a pad bit's being flipped is β, which was given in table 2. If the ambient noise level is a bit-flip probability of 10^{-3}, then for runlength constraints r greater than or equal to 16, the increase in noise level for the pad bits (from 10^{-3} to $10^{-3} + \beta$) is small enough that the average effect on performance will be small. For a t-error-correcting code, this dumb method would suffer an absolutely uncorrectable error (*i.e.*, one that cannot be corrected at any noise level) with probability about $M^{(t+1)}\beta^{(t+1)}/(t+1)!$. For $r = 16$ and $M = 316$, this probability is shown in table 5.

Table 5. Probability of an absolutely uncorrectable error for the nonlinear method and a t-error-correcting code with $r = 16$ and $M = 316$.

t	5	9	19
$\frac{M^{(t+1)}\beta^{(t+1)}}{(t+1)!}$	2×10^{-10}	2×10^{-18}	2×10^{-41}

Thus, the dumb method appears to be feasible, in that it can deliver a failure probability smaller than 10^{-15} with a $t{=}9$-error-correcting code. If it is coupled with the trick of having the offset **o** vary from sector to sector, then it appears to offer a cheap and watertight runlength limiting method, even with a weaker outer code: on the rare occasions when more than, say, 5 bits need to be flipped, we move the data to another sector; this emergency procedure would be used of order once in every billion writes. The only cost of this method is a slight increase in the effective noise level at the decoder.

3   Discussion

In an actual implementation it would be a good idea to compute the weight enumerator function of the dual code defined by **F** and ensure that it has the largest possible minimum distance.

The ideas in this paper can be glued together in several ways.

- Special bits: various values of K_r can be used, including $K_r = 0$ (*i.e.*, use the nonlinear method alone). The experiments suggest that increasing beyond $K_r = 20$ or 30 gives negligible decrease in the probability of conflict.
- Nonlinear bitflipping. This feature could be on or off.
- Variable offset vector **o**. If the offset vector can be pseudorandomly altered, it is very easy to cope with rare cases where either of the above methods fails.

If the variable offset vector is available, then either of the two ideas – the linear method or the nonlinear method – should work fine by itself. Otherwise, a belt-and-braces approach may be best, using the linear and nonlinear methods together.

Acknowledgements. This work was supported by the Gatsby Foundation and by a partnership award from IBM Zürich research laboratory. I thank Brian Marcus, Steven McLaughlin, Paul Siegel, Jack Wolf, Andreas Loeliger, Kees Immink, and Bane Vasic for helpful discussions, and Evangelos Eleftheriou of IBM Zürich for inviting me to the 1999 IBM Workshop on Magnetic Recording.

References

1. John Byers, Michael Luby, Michael Mitzenmacher, and Ashu Rege. A digital fountain approach to reliable distribution of bulk data. In *Proceedings of ACM SIG-COMM '98, September 2-4, 1998*, 1998.
2. R. H. Deng and M. A. Herro. DC-free coset codes. *IEEE Trans. Inf. Th.*, 34:786–792, 1988.
3. R. G. Gallager. *Low Density Parity Check Codes*. Number 21 in Research monograph series. MIT Press, Cambridge, Mass., 1963.
4. K. A. S. Immink. Constructions of almost block-decodable runlength-limited codes. *IEEE Transactions on Information Theory*, 41(1), January 1995.
5. K. A. S. Immink. A practical method for approaching the channel capacity of constrained channels. *IEEE Trans. Inform. Theory*, 43(5):1389–1399, Sept 1997.
6. K. A. S. Immink. Weakly constrained codes. *Electronics Letters*, 33(23), Nov. 1997.
7. D. J. C. MacKay and M. C. Davey. Evaluation of Gallager codes for short block length and high rate applications. In B. Marcus and J. Rosenthal, editors, *Codes, Systems and Graphical Models*, volume 123 of *IMA Volumes in Mathematics and its Applications*, pages 113–130. Springer-Verlag, New York, 2000.
8. B. H. Marcus, P. H. Siegel, and J. K. Wolf. Finite-state modulation codes for data storage. *IEEE Journal on Selected Areas in Communication*, 10(1):5–38, January 1992.
9. G. S. Markarian, M. Naderi, B. Honary, A. Popplewell, and J. J. O'Reilly. Maximum likelihood decoding of RLL-FEC array codes on partial response channels. *Electronics Letters*, 29(16):1406–1408, 1993.

Weight vs. Magnetization Enumerator for Gallager Codes

Jort van Mourik[1], David Saad[1], and Yoshiyuki Kabashima[2]

[1] Neural Computing Research Group, Aston University, Birmingham B4 7ET, UK
[2] Department of Computational Intelligence and Systems Science, Tokyo Institute of
Technology, Yokohama 2268502, Japan

Abstract. We propose a method to determine the critical noise level for
decoding Gallager type low density parity check error correcting codes.
The method is based on the magnetization enumerator ($\mathcal{M}$), rather than
on the weight enumerator ($\mathcal{W}$) presented recently in the information
theory literature. The interpretation of our method is appealingly simple,
and the relation between the different decoding schemes such as typical
pairs decoding, MAP, and finite temperature decoding (MPM) becomes
clear. Our results are more optimistic than those derived via the methods
of information theory and are in excellent agreement with recent results
from another statistical physics approach.

1   Introduction

Triggered by active investigations on error correcting codes in both of information theory (IT) [1,2,3] and statistical physics (SP) [4,5] communities, there is a
growing interest in the relationship between IT and SP. As the two communities
investigate similar problems, one may expect that standard techniques known
in one framework would bring about new developments in the other, and vice
versa. Here we present a direct SP method to determine the critical noise level
for Gallager type low density parity check codes which allows us to focus on the
differences between the various decoding criteria and their approach for defining the critical noise level for which decoding, using Low Density Parity Check
(LDPC) codes, is theoretically feasible.

2   Gallager Code

In a general scenario, the N dimensional Boolean message $\boldsymbol{s}^o \in \{0,1\}^N$ is encoded to the $M(>N)$ dimensional Boolean vector $\boldsymbol{t}^o$, and transmitted via a noisy
channel, which is taken here to be a Binary Symmetric Channel (BSC) characterized by an independent flip probability p per bit; other transmission channels may
also be examined within a similar framework. At the other end of the channel, the
corrupted codeword is decoded utilizing the structured codeword redundancy.

The error correcting code that we focus on here is Gallager's linear code [6].
Gallager's code is a low density parity check code defined by the a binary ($M-$

B. Honary (Ed.): Cryptography and Coding 2001, LNCS 2260, pp. 148–157, 2001.

$N)\times M$ matrix $\mathbf{A} = [\mathbf{C}_1|\mathbf{C}_2]$, concatenating two very sparse matrices known to both sender and receiver, with the $(M-N)\times(M-N)$ matrix $\mathbf{C}_2$ being invertible. The matrix $\mathbf{A}$ has K non-zero elements per row and C per column, and the code rate is given by $R=1-C/K=1-N/M$. Encoding refers to multiplying the original message $\boldsymbol{s}^o$ with the $(M\times N)$ matrix $\mathbf{G}^T$ (where $\mathbf{G}=[\mathbb{1}_N|\mathbf{C}_2^{-1}]$), yielding the transmitted vector $\boldsymbol{t}^o$. Note that all operations are carried out in (mod 2) arithmetic. Upon sending $\boldsymbol{t}^o$ through the binary symmetric channel (BSC) with noise level p, the vector $\boldsymbol{r} = \boldsymbol{t}^o + \boldsymbol{n}^o$ is received, where $\boldsymbol{n}^o$ is the true noise.

Decoding is carried out by multiplying $\boldsymbol{r}$ by $\mathbf{A}$ to produce the syndrome vector $\boldsymbol{z} = \mathbf{A}\boldsymbol{r}$ $(= \mathbf{A}\boldsymbol{n}^o$, since $\mathbf{A}\mathbf{G}^T = \mathbf{0})$. In order to reconstruct the original message $\boldsymbol{s}^o$, one has to obtain an estimate $\boldsymbol{n}$ for the true noise $\boldsymbol{n}^o$. First we select all $\boldsymbol{n}$ that satisfy the parity checks $\mathbf{A}\boldsymbol{n} = \mathbf{A}\boldsymbol{n}^o$:

$$\mathcal{I}_{\mathrm{pc}}(\mathbf{A},\boldsymbol{n}^o) \equiv \{\boldsymbol{n} \mid \mathbf{A}\boldsymbol{n} = \boldsymbol{z}\}, \text{ and } \mathcal{I}_{\mathrm{pc}}^{\mathrm{r}}(\mathbf{A},\boldsymbol{n}^o) \equiv \{\boldsymbol{n} \in \mathcal{I}_{\mathrm{pc}}(\mathbf{A},\boldsymbol{n}^o)|\boldsymbol{n}\neq\boldsymbol{n}^o\}, \quad (1)$$

the (restricted) parity check set. Any general decoding scheme then consists of selecting a vector $\boldsymbol{n}^*$ from $\mathcal{I}_{\mathrm{pc}}(\mathbf{A},\boldsymbol{n}^o)$ on the basis of some noise statistics criterion. Upon successful decoding $\boldsymbol{n}^0$ will be selected, while a decoding error is declared when a vector $\boldsymbol{n}^* \in \mathcal{I}_{\mathrm{pc}}^{\mathrm{r}}(\mathbf{A},\boldsymbol{n}^o)$ is selected. An measure for the error probability is usually defined in the information theory literature [7] as

$$P_e(p) = \left\langle \Delta\left(\exists\, \boldsymbol{n} \in \mathcal{I}_{\mathrm{pc}}^{\mathrm{r}}(\mathbf{A},\boldsymbol{n}^o) : w(\boldsymbol{n}) \leq w(\boldsymbol{n}^o) \mid \boldsymbol{n}^o\right)\right\rangle_{\mathbf{A},\boldsymbol{n}^o}, \quad (2)$$

where $\Delta(\cdot)$ is an indicator function returning 1 if there exists a vector $\boldsymbol{n} \in \mathcal{I}_{\mathrm{pc}}^{\mathrm{r}}(\mathbf{A},\boldsymbol{n}^o)$ with lower weight than that of the given noise vector $\boldsymbol{n}^o$. The weight of a vector is the average sum of its components $w(\boldsymbol{n}) \equiv \frac{1}{M}\sum_{j=1}^{M} n_j$. To obtain the error probability, one averages the indicator function over all $\boldsymbol{n}^o$ vectors drawn from some distribution and the code ensemble $\mathbf{A}$ as denoted by $\langle\cdot\rangle_{\mathbf{A},\boldsymbol{n}^o}$.

Carrying out averages over the indicator function is difficult, and the error probability (2) is therefore upper-bounded by averaging over the *number* of vectors $\boldsymbol{n}$ obeying the weight condition $w(\boldsymbol{n}) \geq w(\boldsymbol{n}^o)$. Alternatively, one can find the average number of vectors with a given weight value w from which one can construct a complete weight distribution of noise vectors $\boldsymbol{n}$ in $\mathcal{I}_{\mathrm{pc}}^{\mathrm{r}}(\mathbf{A},\boldsymbol{n}^o)$. From this distribution one can, in principle, calculate a bound for P_e and derive critical noise values above which successful decoding cannot be carried out.

A natural and direct measure for the average number of states is the entropy of a system under the restrictions described above, that can be calculated via the methods of statistical physics.

It was previously shown (see e.g. [4] for technical details) that this problem can be cast into a statistical mechanics formulation, by replacing the field ($\{0,1\}$, $+\mathrm{mod}(2)$) by ($\{1,-1\}$, $\times$), and by adapting the parity checks correspondingly. The statistics of a noise vector $\boldsymbol{n}$ is now described by its magnetization $m(\boldsymbol{n}) \equiv \frac{1}{M}\sum_{j=1}^{M} n_j$, ($m(\boldsymbol{n}) \in [1,-1]$), which is inversely linked to the vector weight in the $[0,1]$ representation. With this in mind, we introduce the conditioned magnetization enumerator, for a given code and noise, measuring the noise vector magnetization distribution in $\mathcal{I}_{\mathrm{pc}}^{\mathrm{r}}(\mathbf{A},\boldsymbol{n}^o)$

$$\mathcal{M}_{\mathbf{A},\boldsymbol{n}^o}(m) \equiv \frac{1}{M} \ln \left[\mathop{\mathrm{Tr}}_{\boldsymbol{n} \in \mathcal{I}_{\mathrm{pc}}^{\mathrm{r}}(\mathbf{A},\boldsymbol{n}^o)} \delta(m(\boldsymbol{n})-m) \right]. \tag{3}$$

To obtain the *magnetization enumerator* $\mathcal{M}(m)$

$$\mathcal{M}(m) = \Big\langle \, \mathcal{M}_{\mathbf{A},\boldsymbol{n}^o}(m) \, \Big\rangle_{\mathbf{A},\boldsymbol{n}^o}, \tag{4}$$

which is the entropy of the noise vectors in $\mathcal{I}_{\mathrm{pc}}^{\mathrm{r}}(\mathbf{A},\boldsymbol{n}^0)$ with a given m, one carries out uniform explicit averages over all codes $\mathbf{A}$ with given parameters K, C, and weighted average over all possible noise vectors generated by the BSC, i.e.,

$$P(\boldsymbol{n}^o) = \prod_{j=1}^{M} \big((1-p)\, \delta(n_j^o-1) + p\, \delta(n_j^o+1)\big). \tag{5}$$

It is important to note that, in calculating the entropy, the average quantity of interest is the magnetization enumerator rather than the actual number of states. For physicists, this is the natural way to carry out the averages due to three main reasons: a) The entropy obtained in this way is believed to be *self-averaging*, i.e., its average value (over the disorder) coincides with its *typical* value. b) This quantity is *extensive* and grows linearly with the system size. c) This averaging distinguishes between *annealed* variables that are averaged or summed for a given set of *quenched* variables, that are averaged over later on. In this particular case, summation over all $\boldsymbol{n}$ vectors is carried for a *fixed* choice of code $\mathbf{A}$ and noise vector $\boldsymbol{n}^o$; averages over these variables are carried out at the next level.

One should point out that in somewhat similar calculations, we showed that this method of carrying out the averages provides more accurate results in comparison to averaging over both sets of variables simultaneously [8].

A positive magnetization enumerator, $\mathcal{M}(m) > 0$ indicates that there is an exponential number of solutions (in M) with magnetization m, for typically chosen $\mathbf{A}$ and $\boldsymbol{n}^o$, while $\mathcal{M}(m) \to 0$ indicates that this number vanishes as $M \to \infty$ (note that negative entropy is unphysical in discrete systems).

Another important indicator for successful decoding is the overlap ω between the selected estimate $\boldsymbol{n}^*$, and the true noise $\boldsymbol{n}^o$: $\omega(\boldsymbol{n},\boldsymbol{n}^o) \equiv \frac{1}{M} \sum_{j=1}^{M} n_j n_j^o$, $(\omega(\boldsymbol{n},\boldsymbol{n}^o) \in [-1,1])$, with $\omega = 1$ for successful (perfect) decoding. However, this quantity cannot be used for decoding as $\boldsymbol{n}^o$ is unknown to the receiver. The (code and noise dependent) overlap enumerator is now defined as:

$$\mathcal{W}_{\mathbf{A},\boldsymbol{n}^o}(\omega) \equiv \frac{1}{M} \ln \left[\mathop{\mathrm{Tr}}_{\boldsymbol{n} \in \mathcal{I}_{\mathrm{pc}}^{\mathrm{r}}(\mathbf{A},\boldsymbol{n}^o)} \delta(\omega(\boldsymbol{n},\boldsymbol{n}^o)-\omega) \right], \tag{6}$$

and the average quantity being

$$\mathcal{W}(\omega) = \Big\langle \, \mathcal{W}_{\mathbf{A},\boldsymbol{n}^o}(\omega) \Big\rangle_{\mathbf{A},\boldsymbol{n}^o}. \tag{7}$$

This measure is directly linked to the *weight enumerator* [3]), although according to our notation, averages are carried out distinguishing between annealed and quenched variables unlike the common definition in the IT literature. However, as we will show below, the two types of averages provide identical results *in this particular case*.

3   The Statistical Physics Approach

Quantities of the type $\mathcal{Q}(c) = \langle \mathcal{Q}_y(c) \rangle_y$, with $\mathcal{Q}_y(c) = \frac{1}{M} \ln [\mathcal{Z}_y(c)]$ and $\mathcal{Z}_y(c) \equiv \mathrm{Tr}_x \; \delta(c(x, y) - Mc)$, are very common in the SP of disordered systems; the macroscopic order parameter $c(x, y)$ is fixed to a specific value and may depend both on the disorder y and on the microscopic variables x. Although we will not prove this here, such a quantity is generally believed to be *self-averaging* in the large system limit, i.e., obeying a probability distribution $P(\mathcal{Q}_y(c)) = \delta(\mathcal{Q}_y(c) - \mathcal{Q}(c))$. The direct calculation of $\mathcal{Q}(c)$ is known as a *quenched* average over the disorder, but is typically hard to carry out and requires using the replica method [9]. The replica method makes use of the identity $\langle \ln \mathcal{Z} \rangle = \langle \; \lim_{n \to 0}[\mathcal{Z}^n - 1]/n \; \rangle$, by calculating averages over a product of partition function replicas. Employing assumptions about replica symmetries and analytically continuing the variable n to zero, one obtains solutions which enable one to determine the state of the system.

To simplify the calculation, one often employs the so-called *annealed* approximation, which consists of performing an average over $\mathcal{Q}_y(c)$ first, followed by the logarithm operation. This avoids the replica method and provides (through the convexity of the logarithm function) an upper bound to the quenched quantity:

$$Q_a(c) \equiv \frac{1}{M} \ln[\langle \mathcal{Z}_y(c) \rangle_y] \geq Q_q(c) \equiv \frac{1}{M} \langle \ln[\mathcal{Z}_y(c)] \rangle_y = \lim_{n \to 0} \frac{\langle \mathcal{Z}_y^n(c) \rangle_y - 1}{nM} . \tag{8}$$

The full technical details of the calculation will be presented elsewhere, and those of a very similar calculation can be found in e.g. [4]. It turns out that it is useful to perform the gauge transformation $n_j \to n_j n_j^o$, such that the averages over the code $\mathbf{A}$ and noise $\boldsymbol{n}^o$ can be separated, $\mathcal{W}_{\mathbf{A}, \boldsymbol{n}^o}$ becomes independent of $\boldsymbol{n}^o$, leading to an equality between the quenched and annealed results, $\mathcal{W}(m) = \mathcal{M}_a(m)|_{p=0} = \mathcal{M}_q(m)|_{p=0}$. For any finite noise value p one should multiply $\exp[\mathcal{W}(\omega)]$ by the probability that a state obeys all parity checks $\exp[-K(\omega, p)]$ given an overlap ω and a noise level p [3]. In calculating $\mathcal{W}(\omega)$ and $\mathcal{M}_{a/q}(m)$, the δ-functions fixing m and ω, are enforced by introducing Lagrange multipliers $\hat{m}$ and $\hat{\omega}$.

Carrying out the averages explicitly one then employs the saddle point method to extremize the averaged quantity with respect to the parameters introduced while carrying out the calculation. These lead, in both quenched and annealed calculations, to a set of saddle point equations that are solved either analytically or numerically to obtain the final expression for the averaged quantity (entropy).

The final expressions for the annealed entropy, under both overlap (ω) and magnetization (m) constraints, are of the form:

$$
\mathcal{Q}_a = -\frac{C}{K}\left(\ln(2)+(K{-}1)\ln(1+q_1^K)\right)+\ln\left\langle \operatorname*{Tr}_{n=\pm 1}\ e^{(n\hat{\omega}+\hat{m}n^o)}(1+nq_1^{K-1})^C\right\rangle_{n^o}
$$
$$
-\hat{\omega}\omega - \hat{m}m \ , \tag{9}
$$

where q_1 has to be obtained from the saddle point equation $\frac{\partial \mathcal{Q}_a}{\partial q_1}=0$. Similarly, the final expression in the quenched calculation, employing the simplest replica symmetry assumption [9], is of the form:

$$
\mathcal{Q}_q = -C\int dx d\hat{x}\ \pi(x)\hat{\pi}(\hat{x})\ \ln[1+x\hat{x}]+\frac{C}{K}\int\left\{\prod_{k=1}^{K} dx_k \pi(x_k)\right\}\ln\left[\frac{1}{2}\left(1+\prod_{k=1}^{K}x_k\right)\right]
$$
$$
+\int\left\{\prod_{c=1}^{C} d\hat{x}_c \hat{\pi}(\hat{x}_c)\right\}\left\langle\ln\left[\operatorname*{Tr}_{n=\pm 1}\exp(n(\hat{\omega}+\hat{m}n^o))\prod_{c=1}^{C}(1+n\hat{x}_c)\right]\right\rangle_{n^o}
$$
$$
-\hat{\omega}\omega - \hat{m}m \ . \tag{10}
$$

The probability distributions $\pi(x)$ and $\hat{\pi}(\hat{x})$ emerge from the calculation; the former represents a probability distribution with respect to the noise vector local magnetization [10], while the latter relates to a field of conjugate variables which emerge from the introduction of δ-functions while carrying out the averages (for details see [4]). Their explicit forms are obtained from the functional saddle point equations $\frac{\delta \mathcal{Q}_q}{\delta \pi(x)}$, $\frac{\delta \mathcal{Q}_q}{\delta \hat{\pi}(\hat{x})}=0$, and all integrals are from –1 to 1. Enforcing a δ-function corresponds to taking $\hat{\omega}, \hat{m}$ such that $\frac{\partial \mathcal{Q}_{a/q}}{\partial \hat{\omega}}, \frac{\partial \mathcal{Q}_{a/q}}{\partial \hat{m}}=0$, while not enforcing it corresponds to putting $\hat{\omega}, \hat{m}$ to 0. Since ω, m follow from $\frac{\partial \mathcal{Q}_{a/q}}{\partial \hat{\omega}}=0$, $\frac{\partial \mathcal{Q}_{a/q}}{\partial \hat{m}}=0$, all the relevant quantities can be recovered with appropriate choices of $\hat{\omega}, \hat{m}$.

4   Qualitative Picture

We now discuss the qualitative behaviour of $\mathcal{M}(m)$, and the interpretation of the various decoding schemes. To obtain separate results for $\mathcal{M}(m)$ and $\mathcal{W}(m)$ we calculate the results of Eqs.(9) and (10), corresponding to the annealed and quenched cases respectively, setting $\hat{\omega} = 0$ for obtaining $\mathcal{M}(m)$ and $\hat{m} = 0$ for obtaining $\mathcal{W}(\omega)$ (that becomes $\mathcal{W}(m)$ after gauging). In Fig. 1, we have qualitatively plotted the resulting function $\mathcal{M}(m)$ for relevant values of p. $\mathcal{M}(m)$ (solid line) only takes positive values in the interval $[m_-(p), m_+(p)]$; for even K, $\mathcal{M}(m)$ is an even function of m and $m_-(p) = -m_+(p)$. The maximum value of $\mathcal{M}(m)$ is always $(1-R)\ln(2)$. The true noise n^o has (with probability 1) the typical magnetization of the BSC: $m(n^o)=m_0(p)=1-2p$ (dashed-dotted line).

The various decoding schemes can be summarized as follows:

- **Maximum likelihood (MAP) decoding** - minimizes the *block error probability* [11] and consists of selecting the n from $\mathcal{I}_{pc}(\mathbf{A}, n^o)$ with the

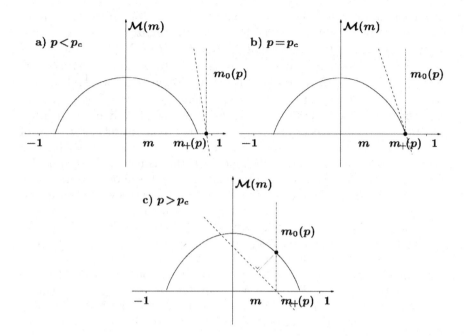

Fig. 1. The qualitative picture of $\mathcal{M}(m) \geq 0$ (solid lines) for different values of p. For MAP, MPM and typical set decoding, only the relative values of $m_+(p)$ and $m_0(p)$ determine the critical noise level. Dashed lines correspond to the energy contribution of $-\beta F$ at Nishimori's condition ($\beta = 1$). The states with the lowest free energy are indicated with •. **a)** Sub-critical noise levels $p < p_c$, where $m_+(p) < m_0(p)$, there are no solutions with higher magnetization than $m_0(p)$, and the correct solution has the lowest free energy. **b)** Critical noise level $p = p_c$, where $m_+(p) = m_0(p)$. The minimum of the free energy of the sub-optimal solutions is equal to that of the correct solution at Nishimori's condition. **c)** Over-critical noise levels $p > p_c$ where many solutions have a higher magnetization than the true typical one. The minimum of the free energy of the sub-optimal solutions is lower than that of the correct solution.

highest magnetization. Since the probability of error below $m_+(p)$ vanishes, $P(\exists n \in \mathcal{I}_{pc}^r : m(n) > m_+(p)) = 0$, and since $P(m(n^o) = m_0(p)) = 1$, the critical noise level p_c is determined by the condition $m_+(p_c) = m_0(p_c)$. The selection process is explained in Fig.1(a)-(c).

- **Typical pairs decoding** - is based on randomly selecting a n from $\mathcal{I}_{pc}$ with $m(n) = m_0(p)$ [3]; an error is declared when n^0 is not the only element of $\mathcal{I}_{pc}$. For the same reason as above, the critical noise level p_c is determined by the condition $m_+(p_c) = m_0(p_c)$.
- **Finite temperature (MPM) decoding** - An energy $-Fm(n)$ (with $F = \frac{1}{2}\ln(\frac{1-p}{p})$) according to Nishimori's condition[1] is attributed to each $n \in \mathcal{I}_{pc}$,

---

[1] This condition corresponds to the selection of an accurate prior within the Bayesian framework.

and a solution is chosen from those with the magnetization that minimizes the free energy [4]. This procedure is known to minimize the *bit error probability* [11]. Using the thermodynamic relation $\mathcal{F} = \mathcal{U} - \frac{1}{\beta}\mathcal{S}$, β being the inverse temperature (Nishimori's condition corresponds to setting $\beta = 1$), the free energy of the sub-optimal solutions is given by $\mathcal{F}(m) = -Fm - \frac{1}{\beta}\mathcal{M}(m)$ (for $\mathcal{M}(m) \geq 0$), while that of the correct solution is given by $-Fm_0(p)$ (its entropy being 0). The selection process is explained graphically in Fig.1(a)-(c). The free energy differences between sub-optimal solutions relative to that of the correct solution in the current plots, are given by the orthogonal distance between $\mathcal{M}(m)$ and the line with slope $-\beta F$ through the point $(m_0(p), 0)$. Solutions with a magnetization m for which $\mathcal{M}(m)$ lies above this line, have a lower free energy, while those for which $\mathcal{M}(m)$ lies below, have a higher free energy. Since negative entropy values are unphysical in discrete systems, only sub-optimal solutions with $\mathcal{M}(m) \geq 0$ are considered. The lowest p value for which there are sub-optimal solutions with a free energy equal to $-Fm_0(p)$ is the critical noise level p_c for MPM decoding. In fact, using the convexity of $\mathcal{M}(m)$ and Nishimori's condition, one can show that the slope $\partial \mathcal{M}(m)/\partial m > -\beta F$ for any value $m < m_o(p)$ and any p, and equals $-\beta F$ only at $m = m_o(p)$; therefore, the critical noise level for MPM decoding $p = p_c$ is identical to that of MAP, in agreement with results obtained in the information theory community [12].

The statistical physics interpretation of finite temperature decoding corresponds to making the specific choice for the Lagrange multiplier $\hat{m} = \beta F$ and considering the free energy instead of the entropy. In earlier work on MPM decoding in the SP framework [4], negative entropy values were treated by adopting different replica symmetry assumptions, which effectively result in changing the inverse temperature, i.e., the Lagrange multiplier $\hat{m}$. This effectively sets $m = m_+(p)$, i.e. to the highest value with non-negative entropy. The sub-optimal states with the lowest free energy are then those with $m = m_+(p)$.

The central point in all decoding schemes, is to select the correct solution only on the basis of its magnetization. As long as there are no sub-optimal solutions with the same magnetization, this is in principle possible. As shown here, all three decoding schemes discussed above, manage to do so. To find whether at a given p there exists a gap between the magnetization of the correct solution and that of the nearest sub-optimal solution, just requires plotting $\mathcal{M}(m)(>0)$ and $m_0(p)$, thus allowing a graphical determination of p_c. Since MPM decoding is done at Nishimori's temperature, the simplest replica symmetry assumption is sufficient to describe the thermodynamically dominant state [9]. At p_c the states with $m_+(p_c) = m_0(p_C)$ are thermodynamically dominant, and the p_c values that we obtain under this assumption are exact.

5   Critical Noise Level - Results

Some general comments can be made about the critical MAP (or typical set) values obtained via the annealed and quenched calculations. Since $\mathcal{M}_q(m) \leq \mathcal{M}_a(m)$ (for given values of K, C and p), we can derive the general inequality $p_{c,q} \geq p_{c,a}$. For all K, C values that we have numerically analyzed, for both annealed and quenched cases, $m_+(p)$ is a non increasing function of p, and p_c is unique. The estimates of the critical noise levels $p_{c,a/q}$, based on $\mathcal{M}_{a/q}$, are obtained by numerically calculating $m_{c,a/q}(p)$, and by determining their intersection with $m_0(p)$. This is explained graphically in Fig.2(a). As the results for

a)

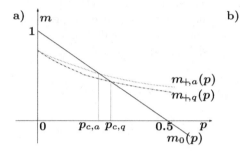

b)

(K, C)	$(6, 3)$	$(5, 3)$	$(6, 4)$	$(4, 3)$
Code rate	$1/2$	$2/5$	$1/3$	$1/4$
IT ($\mathcal{W}_a$)	0.0915	0.129	0.170	0.205
SP	0.0990	0.136	0.173	0.209
$p_{c,a}$ ($\mathcal{M}_a$)	0.031	0.066	0.162	0.195
$p_{c,q}$ ($\mathcal{M}_q$)	0.0998	0.1365	0.1725	0.2095
Shannon	0.109	0.145	0.174	0.214

Fig. 2.   a) Determining the critical noise levels $p_{c,a/q}$ based on the function $\mathcal{M}_{a/q}$, a qualitative picture. **b)** Comparison of different critical noise level (p_c) estimates. Typical set decoding estimates have been obtained via the methods of IT [3], based on having a unique solution to $\mathcal{W}(m) = K(m, p_c)$, as well as using the methods of SP [14]. The numerical precision is up to the last digit for the current method. Shannon's limit denotes the highest theoretically achievable critical noise level p_c for any code [15].

MPM decoding have already been presented elsewhere [13], we will now concentrate on the critical results p_c obtained for typical set and MAP decoding; these are presented in Fig.2(b), showing the values of $p_{c,a/q}$ for various choices of K and C are compared with those reported in the literature.

From the table it is clear that the annealed approximation gives a much more pessimistic estimate for p_c. This is due to the fact that it overestimates $\mathcal{M}$ in the following way. $\mathcal{M}_a(m)$ describes the combined entropy of n and n^o as if n^o were thermal variables as well. Therefore, exponentially rare events for n^o (i.e. $m(n^o) \neq m_0(p)$) still may carry positive entropy due to the addition of a positive entropy term from n. In a separate study [14] these effects have been taken care of by the introduction of an extra exponent; this is not necessary in the current formalism as the quenched calculation automatically suppresses such contributions. The similarity between the results reported here and those obtained in [8] is not surprising as the equations obtained in quenched calculations are similar to those obtained by averaging the upper-bound to the reliability exponent using a methods presented originally by Gallager [6]. Numerical differences between the two sets of results are probably due to the higher numerical precision here.

6   Conclusions

To summarize, we have shown that the *magnetization enumerator* $\mathcal{M}(m)$ plays a central role in determining the achievable critical noise level for various decoding schemes. The formalism based on the magnetization enumerator $\mathcal{M}$ offers a intuitively simple alternative to the weight enumerator formalism as used in typical pairs decoding [3,14], but requires invoking the replica method given the very low critical values obtained by the annealed approximation calculation. Although we have concentrated here on the critical noise level for the BSC, both other channels and other quantities can also be treated in our formalism. The predictions for the critical noise level are more optimistic than those reported in the IT literature, and are up to numerical precision in agreement with those reported in [14]. Finally, we have shown that the critical noise levels for typical pairs, MAP and MPM decoding must coincide, and we have provided an intuitive explanation to the difference between MAP and MPM decoding.

Support by Grants-in-aid, MEXT (13680400) and JSPS (YK), The Royal Society and EPSRC-GR/N00562 (DS/JvM) is acknowledged.

References

1. MacKay, D.J.C.: Good Error-correcting Codes Based on Very Sparse Matrices: IEEE Transactions on Information Theory **45** (1999) 399-431.
2. Richardson, T., Shokrollahi, A., Urbanke, R.: Design of Provably Good Low-density Parity-check Codes: IEEE Transactions on Information Theory (1999) in press.
3. Aji, S., Jin, H., Khandekar, A., MacKay, D.J.C., McEliece, R.J.: BSC Thresholds for Code Ensembles Based on "Typical Pairs" Decoding: In: Marcus, B., Rosenthal, J. (eds): Codes, Systems and Graphical Models. Springer Verlag, New York (2001) 195-210.
4. Kabashima, Y., Murayama, T., Saad, D.: Typical Performance of Gallager-type Error-Correcting Codes: Phys. Rev. Lett. **84** (2000) 1355-1358. Murayama, T., Kabashima, Y., Saad, D., Vicente, R.: The Statistical Physics of Regular Low-Density Parity-Check Error-Correcting Codes: Phys. Rev. E **62** (2000) 1577-1591.
5. Nishimori, H., Wong, K.Y.M.: Statistical Mechanics of Image Restoration and Error Correcting Codes: Phys. Rev. E **60** (1999) 132-144.
6. Gallager, R.G.: Low-density Parity-check Codes: IRE Trans. Info. Theory **IT-8** (1962) 21-28 . Gallager, R.G.: Low-density Parity-check Codes, MIT Press, Cambridge, MA. (1963).
7. Gallager, R.G.: Information Theory and Reliable Communication: Weily & Sons, New York (1968).
8. Kabashima, Y., Sazuka, N., Nakamura, K., Saad, D.: Tighter Decoding Reliability Bound for Gallager's Error-Correcting Code: Phys. Rev. E (2001) in press.
9. Nishimori, H.: Statistical Physics of Spin Glasses and Information Processing. Oxford University Press, Oxford UK (2001).
10. Opper, M., Saad, D.: Advanced Mean Field Methods - Theory and Practice. MIT Press, Cambridge MA (2001).

11. Iba, Y.: The Nishimori Line and Bayesian Statistics : Jour. Phys. A **32** (1999) 3875-3888.
12. MacKay, D.J.C.: On Thresholds of Codes: "`http://wol.ra.phy.cam.ac.uk/` .. .. `mackay/CodesTheory.html`" (2000) unpublished.
13. Vicente, R., Saad, D., Kabashima, Y.: Error-correcting Code on a Cactus - a Solvable Model: Europhys. Lett. **51**, (2000) 698-704.
14. Kabashima, Y., Nakamura, K., van Mourik, J.: Statistical Mechanics of Typical Set Decoding: cond-mat/0106323 (2001) unpublished.
15. Shannon, C.E.: A Mathematical Theory of Communication: Bell Sys. Tech. J., **27** (1948) 379-423, 623-656.

Graph Configurations and Decoding Performance

J.T. Paire, P. Coulton, and P.G. Farrell

Department of Communication Systems, Lancaster University,
Lancaster, LA1 4YR, UK
P.G.Farrell@lancaster.ac.uk

Abstract. The performance of a new method for decoding binary error correcting codes is presented, and compared with established hard and soft decision decoding methods. The new method uses a modified form of the max-sum algorithm, which is applied to a split (partially disconnected) modification of the Tanner graph of the code. Most useful codes have Tanner graphs that contain cycles, so the aim of the split is to convert the graph into a tree graph. Various split graph configurations have been investigated, the best of which have decoding performances close to maximum likelihood.

1. Introduction

This paper describes some new results arising from a continuing investigation [1, 2, 3] into the graph decoding of binary linear block error-correcting codes. Graph algorithms have been applied very successfully to the decoding of long and powerful codes, such as the low density parity check (LDPC) codes [4], where performances comparable to that of turbo codes [5] have been achieved. In the case of short or medium length codes, however, the existence of cycles or loops in a code graph leads to difficulties in applying the various decoding algorithms, because of instability or lack of convergence. Unfortunately, almost all the codes of interest (i.e., with optimum or near-optimum parameters), whatever their length, have graphs containing cycles [6]. In the case of long codes the problem does not arise, as the algorithms are stable and converge, for reasons which are still not entirely clear. In the case of short and medium length codes it is necessary to find ways of modifying the code graph and/or the decoding algorithm, so as to reduce or remove cycles and thus achieve the potential simplicity and performance of graph decoding. In [1] a method of splitting (partially disconnecting) a code graph so as to remove cycles was introduced, and further investigated in [2]. Subsequently this method was studied [3] to evaluate its performance and the effect of reconfiguring the code graph in various different ways. The main results of the study are presented in this paper. The fully disconnected sub-graph configuration has also been independently studied [7], with similar results being obtained.

B. Honary (Ed.): Cryptography and Coding 2001, LNCS 2260, pp. 158-165, 2001.

2. The Code Graph

2.1 Full Graph

The Tanner [8] or factor graph of an (n,k) code (block length n and n − k parity checks) is a bipartite graph representing the parity check relationships of the code. Each position (bit) of a codeword in the code is represented by one of a first set of n nodes in the graph. All the position nodes in a parity relationship are joined by edges to one of a second set of n − k nodes called parity nodes. If two position nodes are connected to the same pair of check nodes then there is a cycle or loop in the graph. There will be many such cycles in the graph of a useful code.

Each parity check relationship of a code is represented by a row in the parity check matrix of the code, and each non-zero symbol in the row corresponds to an edge in the graph. Minimising the number of edges connected to each parity node minimises the number of computations required in the decoding algorithm [3, 9], and may also reduce the number of cycles in the graph, sometimes removing them entirely [9]. Therefore, since a code can be specified by a number of equivalent parity check matrices, it can be important to find the minimum weight matrix from which to construct the graph. This minimum weight parity matrix corresponds to the minimum span (trellis oriented) form of the generator matrix of the code [10], in the case of several simple codes. We conjecture that this will always be the case, but do not have a general proof at present. If the graph corresponding to the minimum weigh parity matrix contains cycles, or if a graph with cycles and more than the minimum number of edges is used for other reasons, then in the case of short and medium length codes it is necessary to modify either the graph or the algorithm, or both, before it can be decoded effectively.

2.2 Split Graph

The graph modification introduced in [1, 2] and studied in [3] splits the graph so as to break all the cycles, thus creating a tree graph. This is done by disconnecting an appropriate position node in each cycle into two sub-nodes (or more if more than one cycle is being simultaneously split), so that one or more position nodes will appear as two or more sub-nodes in the modified graph. The number of edges remains the same, but the number of nodes increases. Clearly, there are a number of ways of splitting a given graph. So-called line, star and decomposed configurations [3,7] were found interesting, although the research reported in [3] did not investigate the decomposed configuration. The first two configurations generate a connected tree graph, and the third a set of n − k disconnected tree sub-graphs. For example, Figure 1 shows the Tanner graph for the Hamming (7,4,3) code (block length n = 7, dimension k = 4 and minimum distance d = 3) with minimum weight parity check matrix

$$[H] = \begin{matrix} 0\ 1\ 0\ 1\ 1\ 0\ 1 \\ 1\ 0\ 0\ 1\ 0\ 1\ 1 \\ 1\ 0\ 1\ 0\ 1\ 0\ 1 \end{matrix}$$

together with line, star and decomposed split graph configurations, where the split nodes are identified by a pair of numbers. In addition, a number of different parity check matrices have been investigated for each code. These different matrices, some minimum weight and others not, all correspond to equivalent codes (ie, codes with the same (n,k,d) parameters).

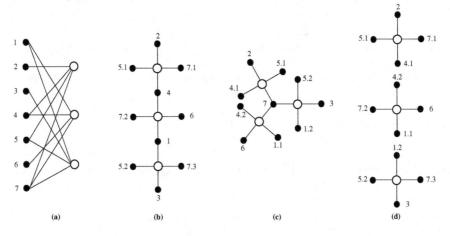

Fig. 1. Full and split graphs for the (7,4,3) code

(a): Tanner graph
(b): line configuration
(c): star configuration
(d): decomposed configuration

3. The Decoding Algorithm

The max-sum algorithm, or its equivalent the min-difference algorithm (as the code is binary) [2,11], has been used. As there are no cycles in the split graph, the algorithm can be applied in the normal manner [2, 3, 11], the computations flowing in from the leaf nodes and out again, but with a modification as to how the metrics at the split position nodes are handled. If a node has been split into s sub-nodes, then the initial metric value for this node is divided by s to determine the initial metric value for each sub-node, since otherwise the metrics of these nodes will tend to dominate in the computations. After completion of the algorithm computation, then the final metric value of the split position node is the sum of the s sub-node final metric values. In addition, it is necessary to iterate the algorithm several times until the final metric values converge. As before, at each iteration of the algorithm the new initial metrics for the s sub-nodes of a split node are determined by dividing the previous final metric at that node by s. This effectively averages the sub-node metric values before each iteration. In the case of the disconnected configuration, each iteration of the modified algorithm is applied to the n − k sub-graphs simultaneously.

4. Performance Results

A number of computer simulations were run to assess the performance of several relatively simple binary error-correcting codes using the split graph decoding technique.

4.1 The (7,4,3) Code

Figure 2 shows the performance in additive white Gaussian noise (AWGN) of the Hamming (7,4,3) code, as plots of bit error rate (log (BER)) versus energy per information bit to noise spectral density ratio (E_b/N_0 in dB). The figure allows the performance of hard-decision (minimum distance) decoding, soft-decision (Viterbi algorithm (VA) maximum likelihood trellis) decoding and split graph decoding to be compared. The minimum weight parity check matrix given above in section 2.2 was used for the split graph curve, and convergence was achieved with 5 iterations, using the line configuration in Figure 1(b) as the structure of the split graph. There was almost no difference in the performance of the code when using other minimum weight [H] matrices with line configurations, but using star configurations slightly degraded the performance (by about a tenth of a dB at an error rate of 10^{-6}).

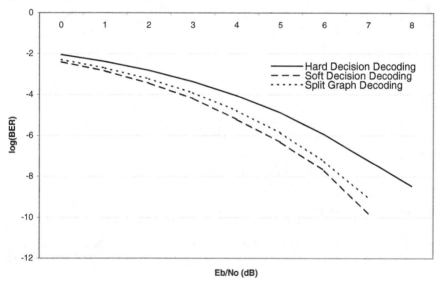

Fig. 2. Performance of the (7,4,3) code

4.2 The (9,5,3) Code

Figure 3 shows the performance in AWGN of the (9,5,3) generalised array code (GAC) [1]. Again, hard-decision, soft-decision (VA) and split graph decoding is compared. The non-minimum weight matrix

$$[H] = \begin{matrix} 0\ 1\ 0\ 1\ 1\ 0\ 1\ 0\ 1 \\ 0\ 1\ 1\ 0\ 0\ 1\ 1\ 0\ 1 \\ 1\ 0\ 1\ 0\ 0\ 1\ 0\ 1\ 1 \\ 1\ 0\ 0\ 1\ 0\ 1\ 1\ 0\ 1 \end{matrix}$$

was used, with the line configuration of Figure 4, and 7 iterations were required. As with the (7,4,3) code, the best star configuration had slightly degraded performance compared to the best line configuration .

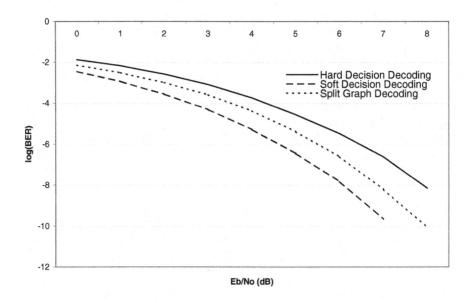

Fig. 3. Performance of the (9,5,3) code

4.3 The (16,11,4) Code

Figure 5 shows the performance in AWGN of the (16,11,4) extended Hamming code, comparing the hard-decision, soft-decision and split graph results. The following [H] matrix was used:

$$[H] = \begin{matrix} 0\ 1\ 0\ 1\ 0\ 1\ 0\ 1\ 0\ 1\ 0\ 1\ 0\ 1\ 0\ 1 \\ 1\ 1\ 0\ 0\ 1\ 1\ 0\ 0\ 0\ 0\ 1\ 1\ 0\ 0\ 1\ 1 \\ 0\ 0\ 1\ 1\ 0\ 0\ 1\ 1\ 0\ 0\ 1\ 1\ 0\ 0\ 1\ 1 \\ 1\ 0\ 0\ 1\ 1\ 0\ 0\ 1\ 1\ 0\ 0\ 1\ 1\ 0\ 0\ 1 \\ 0\ 0\ 1\ 1\ 1\ 1\ 0\ 0\ 0\ 0\ 1\ 1\ 1\ 1\ 0\ 0 \end{matrix}$$

with the line configuration of Figure 6, which gave the best results of all the line configurations investigated. 7 iterations were required. Star configurations were not investigated for this code.

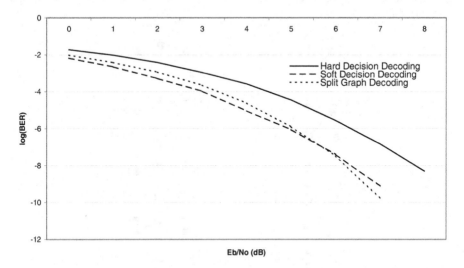

Fig. 5. Performance of the (16,11,4) code

4.4 Other Codes

In addition to the above codes, the (15,7,5) and (16,5,8) codes were studied. Unfortunately it was not possible to validate the results obtained, and it is conjectured that the parity check matrices used inadvertently contained mistakes.

5. Conclusions

The results for the (7,4,3) and (16,11,4) codes are very close (within a few tenths of a dB at BER = 10^{-6}) to their maximum likelihood (ML) performance. At very low error rates the split graph decoding algorithm, because it is an optimum symbol decoding algorithm, does better than the ML (soft decision) algorithm. The good performance of the split graph decoding algorithm for these codes is most probably because a minimum weight parity check matrix was used (in the case of the (16,11,4) code this needs to be confirmed), and because an effective split graph configuration was found. The performance of the (9,5,3) code was not as good (about 0.7 dB away from ML at BER = 10^{-6}), probably because a non-minimum weight [H] matrix was used. Therefore it seems to be important to use a minimum weight parity check matrix, and to find an optimum split graph configuration. The balanced symmetrical structures of the split graphs for the (7,4,3) and (16.11.4) codes, as against the less symmetrical graph for the 9,5,3) code, may be a clue here. Except in the case of the (7,4,3) code, where the matrix and configuration options are limited, different matrices and configurations lead to a wide range of performance results. The best star configuration results (where available) were slightly (about 0.1dB) worse than the best line configuration results, however.

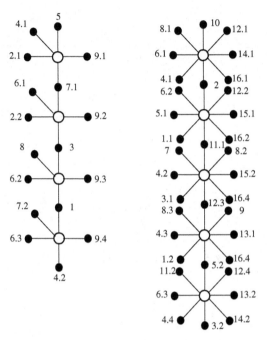

Fig. 4. Line configuration split (9,5,3) code

Fig. 6. Line configuration graph for the split graph for the (16,11,4) code

There are no results in this investigation [3] for the decomposed split graph configuration, because at the time it was thought that this configuration could not perform as well as the others. In [7], however, the authors obtained virtually similar results for star and decomposed configurations of split graphs for the (8,4,4) and (10,6,3) codes, with performances very close to those presented here. They also suggest that an approximate bound on the number of iterations required for convergence is the number of edges on the shortest path between the two nodes furthest apart in the Tanner graph before it is split, which agrees with our own observations. This welcome verification of our results confirms the potential value of the split graph decoding technique, and supports our conjecture that incorrect parity check matrices were used in the simulations for the (15,7,5) and (16,5,8) codes.

Further research is now needed to correct the mistakes in the simulations for the (15,7,5) and (16,5,8) codes, to find or confirm minimum weight [H] matrices and optimum splitting configurations for all the codes investigated so far, to prove the conjecture that the minimum span generator matrix of a code corresponds to the minimum weight parity check matrix, to investigate the performance of longer and more powerful codes using this decoding technique, and to assess the relative computational complexity of this technique compared to other effective decoding methods.

References

[1] P.G. Farrell: Graph decoding of error-control codes; DSPCS'99, Scarborough, Perth, Australia, 1-4 February, 1999.

[2] P.G. Farrell & S.H. Razavi: Graph decoding of array error-correcting codes; IMA Conf. Cryptography & Coding, Cirencester, UK, 20-22 December, 1999.

[3] J.T. Paire: Graph decoding of block error-control codes; MSc Dissertation, Lancaster University, September 2000.

[4] D.J.C. MacKay: Good error-correcting codes based on very sparse matrices; IEEE Trans Info Theory, Vol 45, No 2, pp 399-431, March 1999.

[5] C. Berrou, A. Glavieux & P. Thitimajshima: Near Shannon limit error-correcting coding and decoding: turbo-codes (1); Proc ICC'93, Geneva, Switzerland, pp1064-70, June 1993.

[6] T. Etzion, A. Trachtenberg & A. Vardy: Which codes have cycle-free Tanner graphs?; IEEE Trans Info Theory, Vol 45, No 5, pp 2173-81, Sept 1999.

[7] B. Magula & P. Farkas: On decoding of block error control codes using Tanner graphs: ISCTA'01,Ambleside, UK, 15-20 July, 2001.

[8] R.M. Tanner: A recursive approach to low-complexity codes; IEEE Trans Info Theory, Vol IT-27, No 5, pp533-547, Sept 1981.

[9] M. Esmaeili & A.K. Khandani: Acyclic Tanner graphs and maximum-likelihood decoding of linear block codes; IEE Proceedings-Communications, Vol 147, No 6, pp 322-332, Dec 2000.

[10] R.J. McEliece: On the BCJR trellis for linear block codes; IEEE Trans Info Theory, Vol IT-42, No 4, pp1072-92, July 1996.

[11] G.D. Forney: On iterative decoding and the two-way algorithm; Int. Symp. On Turbo Codes, Brest, France, Sept 1997.

A Line Code Construction for the Adder Channel with Rates Higher than Time-Sharing

P. Benachour, P.G. Farrell, and Bahram Honary

Department of Communication Systems, Lancaster University, UK

benachou@exchange.lancs.ac.uk

Abstract. In this paper, line coding schemes and their application to the multi-user adder channel are investigated. The focus is on designing line codes with higher information per channel use rates than time- sharing. We show that by combining short multi-user line codes, it is possible to devise longer coding schemes with rate sums which increase quite rapidly at each iteration of the construction. Asymptotically, there is no penalty in requiring the coding schemes to be DC-free.

1 Introduction

The multiple access adder channel (MAAC) is based on a channel model which permits simultaneous transmission by two or more users in the same bandwidth without sub-division in time, frequency or the use of orthogonal codes. This can be achieved by the use of a multi-user coding scheme (sometimes called a superimposed scheme) on the adder channel, and also leads to a significantly larger capacity. This results from the higher combined information rate sum compared, say, to time-sharing; and hence leads to a potentially more efficient system. A block diagram of a two-user *(M=2)* scheme is shown in Figure 1, where the inputs and their associated sources have independent encoders and a single decoder estimates their combined output.

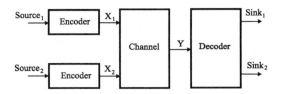

Fig. 1. 2-user Multi Access Scheme

A number of multi-user coding schemes for the MAAC have been constructed [1-4], and a useful survey of the literature on these schemes before 1980 is given in [5]. Potential practical applications of multi-user coding schemes have been investigated for local area networks (LANs), using a baseband MAAC model [6]; and for mobile

B. Honary (Ed.): Cryptography and Coding 2001, LNCS 2260, pp. 166-175, 2001.

radio systems, using a pass band MAAC model [7]. In the former case, it is advantageous to have a DC-free multi-user line coding scheme, as was first explored in [8]. The penalty for this is a fall in the combined information rate; however, it is shown in this paper that by using a code combining technique known as the "direct sum code construction" [9], this rate penalty can be reduced almost entirely as n increases. The advantage of this technique can be highlighted when compared to binary multi-user coding schemes -which are not DC-free- where the rate sum is always constant when combining codes.

The paper is organised as follows: In section 2, line codes and their properties are introduced together with an example of a 2-user coding scheme on the adder channel. Section 3 describes the code combining technique that would enable the user codes to achieve a higher rate than one information bit per channel use (time-sharing). Trellis construction for some line codes are constructed in section 4 then conclusions and further work are suggested in section 5.

2 Line Codes and Their Properties

It is usually the case that a communication link cannot transmit DC components. An example of this situation is the telephone network where AC coupling is provided by transformers. Further in base band transmissions systems, errors may occur in the channel due to interference and noise. Line coding is concerned with providing error control coding for base band transmission, together with reflecting the requirements of its medium [10]. Various line codes have been designed to meet a number of transmission requirements. In choosing a line-coding scheme for base band transmission, the following parameters are considered: spectral characteristics, bit synchronisation, error detection, bandwidth compression and noise immunity. For instance, the Manchester Code (MC) is a popular code since it is DC-free and is self-clocking. This code as well as the Coded Mark Inversion (CMI) scheme, which will be shown to form a two-user uniquely decodable scheme, are introduced and described in this section. The MC and CMI codes are widely used for high speed data transmission such as: TF-34, F-32 and CIT [11-12]. In a MC, a low-to-high level (-11) transformation during the symbol interval T, indicates a logical-zero at the encoder input, while a high-to-low transformation (1-1) indicates a logical one. This code is also known as a self-synchronising code, since the data and the clock are combined. Each encoded bit contains a transition at the midpoint of a bit period as is shown in Figure 2

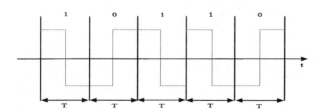

Fig. 2. Signal Waveform of the Manchester Code

In the CMI coding sequence (see Figure 3), input data 0's are encoded by (-11). On the other hand, input data 1's are encoded with opposite polarity as (-1-1) and (11) alternately.

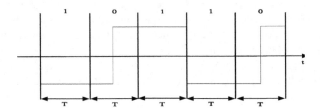

Fig. 3. Signal Waveform of the CMI Code

An effective way of achieving DC-freedom on the base band adder channel is to ensure that all the component codes are themselves DC-free, since then their sum also will be DC-free [8]. This can be done by using a balanced set of code symbols (eg, for $q=2$ {0, 1} maps to {1, -1}, for $q=3$ {0, 1, 2} maps to {1, 0, -1}), (where q is the number of encoder output levels from each user), and ensuring that all codewords either have zero disparity (ie, the algebraic sum of the codewords symbols is zero) or exist in opposite disparity pairs which are used alternately. These mappings and pairings are standard line coding techniques [13]. The encoding table (Table 1) – which will also be considered as Example 1- represents the composite codeword outputs when applied to the base band adder channel. This code is a mapping of a 2-user binary scheme presented in [1]. The original code from [1], $C_1=\{00,11\}$ and $C_2=\{00,01,10\}$, has a rate sum $R_{sum}=1.292$ information bits per channel use, which reduces to 1.0 when line coding is introduced. The reason behind the reduction in the rate sum is that binary words 11 and -1-1 must be used alternately in order to keep the disparity at *zero* and as a result are considered to be a single *effective* codeword. It is interesting to note that the CMI code can be represented as a binary convolutional code where the encoder can be described as having two states each defined by the sign of the running digital sum (RDS) [14]. Trellis constructions for these and other codes are dealt with in section 4.

Table 1. Output from a 2-user Line Coding Scheme

User$_2$\ User$_1$	-11	1-1
-11	-22	00
11	02	20
-1-1	-20	0-2

3 Construction of Multi-user Line Codes with Higher Rates

In this section, a code combining method known as the *Direct Sum Construction* is described with examples. The direct sum technique has been used previously by Chang [15] to construct multi-user coding schemes for the MAAC. He called it concatenation, and noted the advantage of being able to form longer multi-user codes whilst maintaining the same overall rate. Chang was not concerned with the DC-free case, however, and so did not note the increase in rate that occurs when constructing longer DC-free multi-user coding schemes using this method.

3.1 Definition

Given two block codes C_a C_b, with lengths n_a and n_b, having N_a and N_b codewords respectively, their *direct sum* consists of all codewords $|u|v|$ where $u \in C_a$ and $v \in C_b$ [8]. This code C_{ab} (and similarly the code C_{ba} consisting $|v|u|$) has length $n_a + n_b$ and $N_a N_b$ codewords. If both C_a and C_b are DC-free, then C_{ab} and C_{ba} are also DC-free. If $\{C_a, C_b\}$ form a 2-user coding scheme on the adder channel, then so do the pairs of codes C_{ab} and C_{ba} (*cross direct sum*), and C_{aa} and C_{bb} (*self direct sum*).

3.2 Examples

Applying the self direct sum construction to the following scheme:

Example 2:

$C_{11} = \{1\text{-}11\text{-}1,\ 1\text{-}1\text{-}11,\ \text{-}111\text{-}1,\ \text{-}11\text{-}11\}$

$C_{22} = \{1\text{-}11\text{-}1,\ 11\text{-}1\text{-}1,\ \text{-}1\text{-}111,\ 1\text{-}111/1\text{-}1\text{-}1\text{-}1,\ 111\text{-}1/\text{-}1\text{-}11\text{-}1,1111/\text{-}1\text{-}1\text{-}1\text{-}1\}$

which have 4 and 6 *effective codewords* respectively. Note that C_{11} consists of balanced (zero disparity) words only. Now $R_{11} = 0.5$ and $R_{22} = 0.646$ so $R_{1122} = 1.146$ which is greater than the rate sum $R_{12} = 1.0$ of the C_1, C_2 scheme. The reason for the higher rate is that codewords with opposite disparities in the original shorter codes form codewords with zero disparity in the longer direct sum codewords. Thus complementary pairs forming single effective codewords in C_2 become two effective codewords in the direct sum code C_{22}. This does not occur in C_{11}, as C_1 is already all balanced, so $R_{11}=R_1=0.5$. The direct sum process can be iterated to form longer and longer coding schemes with increasing rate R_{22} and overall rate R_{1122}, as shown in Table 2 for this code. In the limit as n increases, the overall rate is bounded by the overall rate of the original code when regarded as a non-DC-free code; ie, counting all codewords as effective. Thus for Example 2, the asymptotic rate is 1.292, and therefore in the limit there is no penalty in requiring the scheme to be DC-free.

Example 3:

Another DC-free uniquely decodable 2-user coding scheme (Example 3) has $n=2$, $q_1=2$, $q_2=3$, $Q=5$ and component codes

$C_1=\{1-1, -11\}$ and

$C_2=\{00, 1-1, 01/0-1, 10/-10, 11/-1-1\}$.

Here, $R_1=0.5$, because C_1 is a binary code and $R_2=0.732$, as C_2 is a ternary code which has 5 *effective codewords*. This scheme has $R_{12}=1.232$ bits/channel use, the highest rate for the parameters concerned. The iterated rates R_{22} and overall rates R_{1122} for the direct summing of the 2-user scheme of this example are also shown in Table2. Here $R_{11}=R_1$ remains constant at 0.5, and $R_{22} > R_2$ rapidly increases. The asymptotic limit on R_{1122} is given by 1.446 information bits per channel use. In practice, this limit seems to be approached fairly rapidly and It is interesting to note that this asymptotic rate is quite close to that of the best 2-user binary adder channel coding schemes found so far [3-4].

Table 2. Rate of 2-user DC-free Schemes

Length	**Example 1**		**Example 2**	
n	R_{22}	R_{1122}	R_{22}	R_{1122}
2	0.5	**1.0**	0.733	**1.233**
4	0.646	**1.146**	0.839	**1.339**
8	0.705	**1.205**	0.884	**1.384**
16	0.744	**1.244**	0.913	**1.413**
32	0.766	**1.266**	0.928	**1.428**
∞	0.792	**1.292**	0.946	**1.446**

It is also possible to use cross direct summing to construct longer DC-free multi-user codes, but it has been found that in every case considered fewer codewords are generated. The overall rate is less (sometimes significantly) than when using self direct summing. With the scheme of Example 2 for instance, $R_{1221}=1.0 = R_{12}$, so no increase in overall rate occurs in this case.

3.3 Direct Sum Codes with M>2

The direct sum construction method can be applied to multi-user DC-free coding schemes with any number of users M. In this case, a 3-user scheme exists with $n=2$, $q_1=3$, $q_2=2$, and $q_3=3$. The component codes are $C_1=\{00, -11\}$, $C_2=\{1-1,-11\}$ and $C_3=\{-11, 10/-10, 11/-1-1\}$, where C_3 has 3 effective words. The code rates are $R_1=0.315$, $R_2=0.5$ and $R_3=0.5$, and the overall rate $R_{123}=1.315$. After the first iteration of the self direct sum construction, $R_{112233}=1.616$, and the asymptotic rate as n tends to infinity is 1.732.

4 Trellis Construction of Multi-user Line Coding Schemes

It is relatively straight forward to design trellises (in general non-linear) for the component codes of a multi-user coding scheme. It is also convenient to use disparity values to label the trellis states. If the codewords of a component code are all balanced, then the trellis is that of a block code. If, on the other hand, a component code contains complementary disparity pairs of codewords, then it is equivalent to a code with memory, and has a convolutional-type code trellis. The trellis of the composite coding scheme can then be obtained by using the Shannon Product construction [8,16]. Once this composite trellis is obtained, then the soft or hard decision Viterbi Algorithm (VA) decoding can be used to recover the information bits/symbol of each user [8]. Figures 4, 5 and 6 show the trellises for code C_1, code C_2 and the composite multi-user coding scheme of Example 1, respectively.

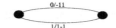

Fig. 4. Trellis for Code C_1

Fig. 5. Trellis for Code C_2

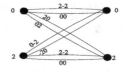

Fig. 6. Combined trellis for Example 1

Figure 7 shows the trellises for the first iteration of the self direct sum construction for the codes of Example 2 with block length $n=4$.

C_{11}

C_{22}

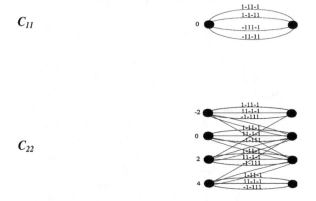

Fig. 7. Trellises for the codes in Example 2 (Direct Sum)

Various versions of the code trellises and therefore of the composite trellis, can be constructed, which vary in their spectral properties. In the case of Example 3, there is no penalty in choosing the trellis with the best low frequency roll-off, as the decoding performance is almost identical for all the trellis versions investigated. The trellis structures shown in Figures 8, 9 and 10 differ in the number of path emerging and converging to the nodes at a given depth. The nodes are labelled with disparity values at it was prove that if these value ranges are closer to zero then better low frequency roll-off can be attained.

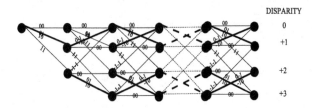

Fig. 8. Trellis Structure 1 of Code ={00,01/0-1,10/-10,11/-1-1}, the bold paths represent parallel paths

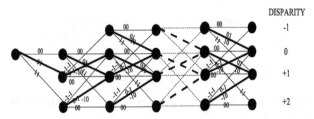

Fig. 9. Trellis Structure 2: because the Disparity range is closer to Zero, the Code has less Magnitude in its Power Spectrum at Low frequencies (nearer DC)

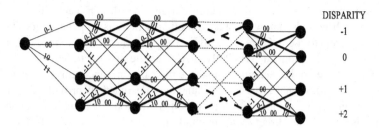

Fig. 10. Trellis Structure 3: The Structure above has 4 branches from each node, therefore making it homogeneous.

From a complexity viewpoint, it is possible to set a limit on the maximum number of nodes required for a given DC-free Code. If the maximum disparity value is d_{max}, then the minimum number of nodes per depth required in a trellis representing such a code is $N_t = 2\, d_{max}$. For this to occur, it must be possible to have a branch with disparity either $+d_{max}$ or $-d_{max}$ from every node. This in itself sets the maximum number of nodes required in order to represent a code by a trellis. The following Example in Figure 11 illustrates this

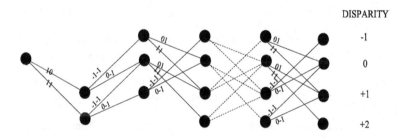

Fig. 11. Example for Code={01/0-1,11/-1-1} with $d_{max}=2$, then $N_t= 4$ Nodes.

Finally, a note on the performance of these schemes under noisy conditions. From the results of the simulations -not shown here but will be presented in the talk- that the redundancy inherent in the coding schemes enables advantage to be taken of the use of soft-decision decoding, even though the redundancy has not been structured to provide error control decoding power (the free distance of the trellis for C_2 in example 1 is only 2, for example).

5. Conclusions

In this paper, multi-user DC-free coding schemes for the MAAC have been presented and described for $M=2,3$ users. It was shown that by using the direct sum construction on short multi-user codes, it is possible to devise longer DC-free multi-user coding schemes with rate sums, which increase quite rapidly at each iteration of the construction. It was also shown that asymptotically there is no penalty in requiring the coding schemes to be DC-free. In addition, the schemes can be efficiently soft decision decoded using a relatively low complexity sectionalised trellis. The redundancy in the coding schemes enhances their performance, especially in the soft-decision case. Additional structured redundancy designed to provide specific error-control power can be achieved by finding other methods of combining codes with a greater distance. The direct sum method appears to be very effective for the construction of both DC-free and non-DC-free multi-user coding schemes with a gain for DC-free codes.

Acknowledgment. This work is supported by the Engineering and Physical Science Research Council under EPSRC Grant: GR/M37868.

References

[1]   Kasami T. and Lin S., "Coding for a Multiple access channel", IEEE Trans. On Inf. Theory, Vol. IT-22, No.2, pp 129-137, March 1976.

[2]   Chang., S. C. and Weldon, Jr., E. J. "Coding for T-user multiple access channel," IEEE Trans. Inform. Theory, vol. IT-25, no. 6, pp. 684-691, Nov. 1979.

[3]   Coebergh van den Braak P.A.B. M. and van Tilborg, H. C. A., "A class of uniquely decodable code pairs for the two-access binary adder channel," IEEE Trans. Inform. Theory, vol. IT-31, no.1, pp. 3-9, Jan. 1985.

[4]   Ahlswede, R and Balakirsky, V. B, "Construction of Uniquely Decodable Codes for the Two-user Binary Adder Channel", IEEE Trans. Inform. Theory, vol. 45 N. 1 pp. 326-330, Jan.1999.

[5]   Farrell P. G., "Survey of channel coding for multi-user systems", in *Skwirzynski, J.K.* (Ed.) : 'New concepts in multi-user communications' (Sijthoff and Noordhoff, 1981), pp. 133-159.

[6]   Bridge, P., "Collaborative Coding for Optical Fibre multi-user Channels", in *Skwirzynski, J.K.* (Ed.) "Performance limits in communication theory and practice", Kluwer, 1988, pp 99-111.

[7]   Farrell P. G., Brine A., Clark A. P., and Tait, D. J., "Collaborative Coding for Optical Fibre multi-user Channels", in *Skwirzynski, J.K.* (Ed.) "Limits of radio communication – collaborative transmission over cellular radio channels", Kluwer, 1988, pp 281-307.

[8]   Markarian G., Honary B. and Benachour, P., "Trellis decoding techniques for the binary adder channel with M users", IEE Proceedings on Communications, vol.144, 1997.

[9]   MacWilliams F. J. and Sloane N. J. A., "The Theory of Error Correcting Codes", Elsevier Science Publishers, North-Holland 1977, Chapter 2, page 76.

[10]   O'Reilly J.J., "Telecommunication Principles", Van Nostrand Reinhold (UK), 1984.

[11]   Takasaka Y., Yamashida K. and Tokahashi Y., "Two level AMI coding family for optical fibre systems", International Journal of Electronics, vol. 55, No. 1, pp. 121-131, 1983.

[12]   Cattermole K. W., O'Reilly J.J. "Problems of randomness in communications engineering", vol.2, Pentech Press, London 1984.

[13]   Bylanski, P and Ingram, D. G. W., "Digital Transmission Systems", Peter Peregrinus (IEE), Second Edition, 1980.

[14] Justesen J., "information rates and power spectra of digital codes", IEEE Trans. Inform. Theory, vol. IT- 28, No.3, pp.457-472, May 1982.

[15] Chang, S. C.,"Coding for the Multiple Access Adder channel with and without Idle Sources", Elec. Letters, Vol 31, N. 7, pp 523-4, 30th of March 1995.

[16] Sidorenko V., Markarian G. and Honary B., ''Minimal trellis design for linear codes based on the Shannon product '', IEEE Trans. Inform. Theory, vol. 42   pp. 2048-2053, Nov.1996.

The Synthesis of TD-Sequences and Their Application to Multi-functional Communication Systems

Ahmed Al-Dabbagh and Michael Darnell

Institute of Integrated Information Systems School of Electronic and Electrical
Engineering
University of Leeds
Leeds, LS2 9JT
UK
eenaa@electeng.leeds.ac.uk
M.Darnell@elec-eng.leeds.ac.uk

Abstract. The multi-functional approach to design has been previously defined. This paper emphasizes the need for such approach in order to reduce the increasing level of complexity of future communication systems. The paper then describes a class of deterministic quasi-analogue (multi-level) pseudorandom sequences with good correlation properties and the means for tailoring the sequences' probability density functions. Finally, the family of sequences is shown to offer a vehicle for coalescing a number of aspects of multi-functionality within the context of spread-spectrum communication systems.

1 Introduction

The nature of communication systems is evolving towards a higher degree of mobility and availability of service. With this drive, future communication systems will be required to offer greater flexibility, leading to an increase in complexity.

Previously, the multi-functional approach to design has been proposed as the means by which algorithmic complexity may be reduced, with no sacrifice in the overall system functionality, adaptation and performance [1]. Due to the multiplicity of functions that are expected from modern communication systems, multi-functionality has been recognised as an ideal that cannot be fully met in practice.

This paper shows that several aspects of multi-functionality may be combined into one algorithm. In particular, the various functions of the physical-layer of a spread-spectrum (SS) communication system are coalesced into one algorithm using a class of quasi-analogue, or multi-level, pseudorandom sequences, known as trajectory-derived (TD) sequences.

In section (2), the need for system flexibility and adaptivity is emphasized and the system parameters that can be made adaptable are identified. The concept of multi-functionality is then formally defined in section (3) and the various aspects of the concept are outlined.

B. Honary (Ed.): Cryptography and Coding 2001, LNCS 2260, pp. 176–190, 2001.

The synthesis procedure for TD-sequences is described in section (4) and example sequences are given. The dependency of the sequences upon initial conditions is also discused and demonstrated. Next, various conditions are stated that guarantee algorithmic determinism. A close form expression for the sequences' probability density function (PDF) is derived and compared against estimates of the PDF of the generated example sequences. It is also shown that the algorithm offers a simple mechanism for tailoring the PDF of the resultant sequences; two designs that yield sequences with Gaussian and uniform PDFs are specified and demonstrated. Formal definitions of the periodic/aperiodic real/complex sequence obtainable from TD-sequences are then given. A study of the sequences correlation properties is also made. In particular, it is shown that the sequences possess quasi-impulsive aperiodic autocorrelation functions (ACF), with the aperiodic crosscorrelation functions (CCF) between different sequences being low.

The application of the TD family of sequences to multi-functional communication systems is treated in section (5). The various aspects of an M-ary phase-shift-keying (MPSK) spread-spectrum communication system, which are deliverable by TD-sequences, are explained; a number of simulation results are presented to verify the multi-functional system concept.

Finally, concluding remarks are given in section (6).

Notation: The following symbols and notation are used: $\mathcal{Z}$ denotes the set of all integers; $\mathcal{R}$ denotes the set of all real numbers; $\mathcal{C}$ denotes the set of all complex numbers; $\mathcal{R}e\ x$ denotes the real part of x; $\mathcal{I}m\ x$ denotes the imaginary part of x; the symbol $*$ denotes complex conjugates; $\otimes$ denotes convolution; $< p, q >$ denotes the p modulo-q; the symbol $\forall$ means for all.

2   System Adaptation

Current trends show that demands on *mobility and availability of service* from communication systems will continue to rise. In order to sustain communication services under variable channel conditions, system *flexibility* has become a primary concern for system designers. System flexibility is greatly enhanced if *adaptivity* becomes the guiding theme during the process of system design. The main system parameters that can be made adaptable are [2] :

- Occupied Bandwith;
- Transmission Rate;
- Transmitted Signal Format;
- Radiated Power;
- Operating Frequency;
- Control Protocol;
- Spacial Selectivity.

3   Multi-functionality

The above list of adaptable parameters indicates that the level of sophistication of future communication systems will undoubtedly continue to increase, thus leading to systems with higher levels of complexity. Whilst computational power continues to increase by technology development, there remains a basic need to simplify system designs *without* sacrificing functionality and/or performance.

The concept of *multi-functionality* has been defined as the means [1]

> by which the various elements of channel encoding/decoding applicable to adaptive communication systems could be coalesced in a systematic manner to achieve a more powerful and elegent algorithmic structure, together with a more efficient implementation within a digital signal processing architecture.

From the definition, it is clear that multi-functionality is primarily concerned with the physical layer aspects of the communication process, which include :

- Data-Encoding/Decoding;
- Modulation/Demodulation;
- Error-Control;
- Synchronisation;
- Security;
- Multiple/Random Access.

It may be said that a *full multi-functional solution* to the communication problem in its entirety is an ideal that is not likely to be fully achieved. This can be attributed to the diversity and complexity of the tasks that must be undertaken by the physical layer. However, a system designer must strive to seek the reduction of system complexity by focusing on the techniques that achieve one particular aspect of communications, whilst simultaneously offering at least one other 'free' aspect of the system functionality.

4   Trajectory-Derived Sequences

Recently, there has been an interest in analogue-type pseudorandom sequences for application in digital data transmission systems [3]. In this section, TD-sequences are presented as an example of such sequences which offer sufficient degrees of randomness and versatility for use in multi-functional digital communications.

4.1   Sequence Synthesis

The sequences are generated via geometric ray-tracing. Here, a recti-linear path is plotted within a pre-defined, and perfectly-reflecting, enclosure. The trajectory can be visualised as the path which would be taken by, say, a small ball 'bouncing' off the inner walls of the enclosure, with no energy loss. An example configuration is taken to be a square with side 2 units, centred on the origin of the Cartesian co-ordinate system and rotated by 45 deg. A circular reflector centred on the origin is also positioned inside the enclosure. The x and y coordinates of the successive points of impact define either a complex sequence or, alternatively, two real sequences. To indicate the generation method, these sequences have been termed *trajectory-derived* sequences.

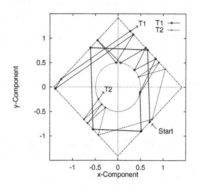

Fig. 1. TD-Sequence Synthesis and Sensitivity to Initial Conditions.

An example trajectory, T_1, was generated using the algorithm described above; figure 1 shows the enclosure and the first 13 reflections of the trajectory. The initial point was taken at $\frac{1}{\sqrt{2}}(1, -1)$ with an initial angle of 47 deg. The trajectory was generated with a computational accuracy of 6 decimal places.

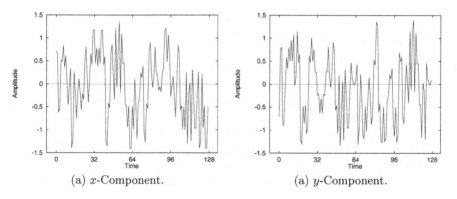

(a) x-Component.   (a) y-Component.

Fig. 2. The x and y Components of Trajectory T_1.

4.2   Sensitivity to Initial Conditions

A second trajectory, T_2, was also generated using the same enclosure configuration but with a small change in the starting x-coordinate value and the initial angle. The exact value of the starting point was $\frac{1}{\sqrt{2}}(-1.01, -1)$ and the initial angle was chosen to be 47.01 deg. Figure 1 also shows the series of reflection points resulting from the second trajectory. The plot shows the two trajectories start from almost the same point and with the same initial take off angle. However, at the 5th reflection point, the two trajectories begin to diverge and follow independent paths and by the 7th reflection, the trajectories display distinct behaviours. Thus, it is demonstrated that past points of reflection and the respective take-off angles have a major influence on the sequence characteristics. In order to ensure that the algorithm is deterministic, some form of precision control is required. A simple computational rounding procedure of decimal places suffices to ensure algorithmic determinism.

4.3   Parameters for Sequence Specification and Determinism

The structure of TD-sequences is completely determined by the following factors:

- Starting point and direction of initial path;
- Geometric enclosure configuration;
- Arithmetic precision of trajectory computation;
- Number of successive impacts;
- Choice of x, y or complex sequences.

If all these factors are specified, then the sequence produced is deterministic and can be exactly regenerated.

4.4   Derivation of the Sequence PDFs

The PDF analysis of TD-sequences can be geometrically derived by considering the average projection of the reflection points on to the x and y axes. Assuming that the trajectories are uniformly distributed within the enclosure, the probability of the reflection points being in any bin may then be expressed in terms of the total boundary and enclosure length existing in that bin.

Consider the set of trajectories falling on the four inner enclosure walls. Let $P_e(x, \Delta x)$ denote the probability of the reflection points falling on the proportion of the boundary in the bin centred at x with width Δx. Then

$$P_e(x, \Delta x) = \frac{2[\text{boundary length in the } (x, \Delta x)^{th} \text{ bin}]}{\text{total enclosure length}} \tag{1}$$

$$= \frac{(x + \frac{\Delta x}{2}) - (x - \frac{\Delta x}{2})}{2\sqrt{2}} \tag{2}$$

In a similar manner, if $P_c(x, \Delta x)$ denotes the probability of the reflection points falling on the proportion of the circular reflector in the bin centred at x with width Δx, then

$$P_c(x, \Delta x) = 2 \frac{\text{arc length in the } (x, \Delta x)^{th} \text{ bin}}{\text{cicumference of the reflector}} \tag{3}$$

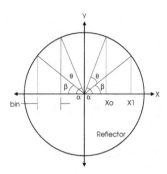

Fig. 3. Sequence PDF Computations.

Consider the reflector shown in figure 3: the arc length projected by the angle θ (in radians) is given by arc length $= r \times \theta$. Using triangulation, the angle θ, created by the points $(x_0, 0)$ and $(x_1, 0)$, may be shown to be

$$\theta(x_0, x_1) = \alpha - \beta = \begin{cases} (\frac{\sqrt{(r^2 - x_0^2)}}{r}) - \sin^{-1}(\frac{\sqrt{(r^2 - x_1^2)}}{r}) \mid x_0 \mid < \mid x_1 \mid \\ (\frac{\sqrt{(r^2 - x_1^2)}}{r}) - \sin^{-1}(\frac{\sqrt{(r^2 - x_0^2)}}{r}) \mid x_0 \mid > \mid x_1 \mid \end{cases} \tag{4}$$

Letting $x_o = x - \Delta x/2$ and $x_1 = x + \Delta x/2$ in equation 4 expresses the angle, θ, in terms of the $(x, \Delta x)^{th}$ bin and substituting the results into equation 3 yields

$$P_c(x, \Delta x) = \begin{cases} \frac{1}{\pi} \sin^{-1}(\frac{\sqrt{(r^2 - (x - \Delta x/2)^2)}}{r}) - \sin^{-1}(\frac{\sqrt{(r^2 - (x + \Delta x/2)^2)}}{r}) \mid x \mid \geq 0 \\ \frac{1}{\pi} \sin^{-1}(\frac{\sqrt{(r^2 - (x + \Delta x/2)^2)}}{r}) - \sin^{-1}(\frac{\sqrt{(r^2 - (x - \Delta x/2)^2)}}{r}) \mid x \mid < 0 \end{cases} \tag{5}$$

If they exist, the probability density functions $p_e(x)$ and $p_c(x)$ are related to the obtained probability functions for $P_e(x, \Delta x)$ and $P_c(x, \Delta x)$ by [4]

$$\int_{x - \Delta x/2}^{x + \Delta x/2} p_e(x')dx' = P_e(x, \Delta x) \Rightarrow p_e(x) = \frac{1}{2\sqrt{2}} \tag{6}$$

$$\int_{x - \Delta x/2}^{x + \Delta x/2} p_c(x')dx' = P_c(x, \Delta x) \Rightarrow p_c(x) = \frac{1}{\pi\sqrt{r^2 - x^2}} \tag{7}$$

$$\tag{8}$$

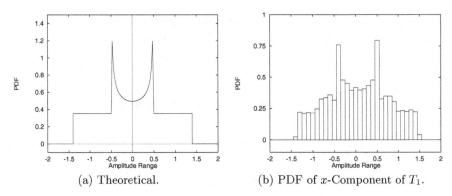

(a) Theoretical.   (b) PDF of x-Component of T_1.

Fig. 4. The PDF.

If p denotes the probability of the trajectory landing on the reflector, then the sequence density function is

$$p(x) = \begin{cases} \frac{p}{\pi\sqrt{r^2-x^2}} + \frac{1-p}{2\sqrt{2}} & |x| \geq 0 \\ \frac{1}{2\sqrt{2}} & Otherwise. \end{cases} \tag{9}$$

A plot of the theoretical PDF is shown in figure 4(a) for $r = 0.5$ and $p = 0.5$, whilst estimate of the PDF of the x component is given in figure 4(b).

4.5   The Tailoring of Sequence PDFs

The use of multiple reflectors and appropriate enclosure scaling constitutes a tool for tailoring the sequence PDF. In particular, the shape of the PDF may be controlled by the location, number and sizes of the internal circular reflectors, whilst the enclosure size determines the maximum and minimum values of the sequence. Since, each reflector offers a PDF shaped by equation 9, then equation 9 constitutes the building block for any desired PDF.

Figure 5(a) describes the enclosure design that may be used to synthesise a random sequence whose PDF is approximately Gaussian, whilst figure 5(b) gives the design that will yield pseudo-random sequences whose PDF are approximately uniform.

For the Gaussian case, a trajectory was generated using the arrangement of figure 5(a). The trajectory is specified by the starting point $(\sqrt{2}, 0.0)$, a take-off angle of 20 deg and a computational accuracy of 6. The PDF of the initial 10,000 points of reflection of the resultant x component were computed and is shown in figure 5(c). The mean and variance (the square of the standard-deviation) of the TD Gaussian sequence were estimated as 0.0 and 0.45, respectively. The theoretical PDF corresponding to a zero mean Gaussian distributed random variable with variance equal 0.45 is also plotted in figure5(c). It is found that the distribution of the generated sequence reasonably approximates the theoretical distribution.

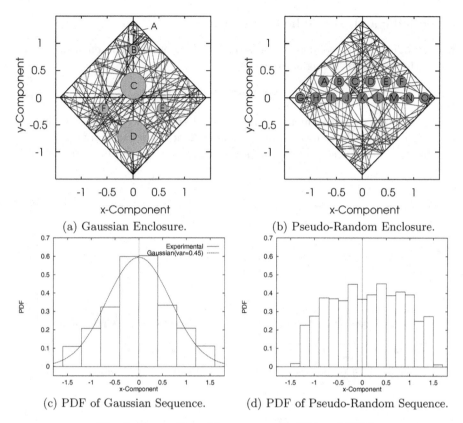

(a) Gaussian Enclosure.   (b) Pseudo-Random Enclosure.

(c) PDF of Gaussian Sequence.   (d) PDF of Pseudo-Random Sequence.

Fig. 5. Generation of Sequences with Different PDFs.

Similarly, to synthesise a pseudo-random sequence, another trajectory was generated using the enclosure of figure 5(b) and the PDF of the initial $10,000$ points of reflection of the resultant x component were computed and are shown in figure 5(d). The trajectory is specified by the starting point $(\sqrt{2}, 0.0)$, a take-off angle of $20\,\mathrm{deg}$ and a computational accuracy of 6. The obtained PDF indicates that the different sequence levels are approximately equi-probable.

4.6   Real and Complex Sequences

Given a series of consecutive ordered pairs, $T = \{(x_i, y_i)\}$, describing a unique trajectory, then a pair of real sequences, $s_1, s_2 \in \mathcal{R}$, may be defined by

$$s_1 = \mathcal{R}e\,T \qquad (10)$$

$$s_2 = \mathcal{I}m\,T \qquad (11)$$

Similarly, one complex sequence, $s \in \mathcal{C}$, may be defined by

$$s = T \qquad (12)$$

4.7   Periodic and Aperiodic Sequences

For a periodic type of sequence, the selected elements are chosen in a cyclic manner. Hence, the resulting sequence is

$$s = \{(x_i, y_i) : \forall < i + D, N >\} \tag{13}$$

where N is the required sequence period and D is the initial phase of the sequences. Real periodic sequences can be defined in a similar manner. For an aperiodic type of sequence, the period of the sequence is made infinite by letting $N \to \infty$.

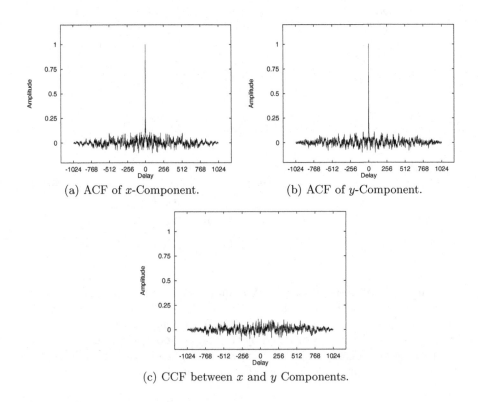

(a) ACF of x-Component.   (b) ACF of y-Component.

(c) CCF between x and y Components.

Fig. 6. Correlation of Aperiodic TD-Sequences.

4.8   Correlation Analysis

The CCF between two complex functions f_1 and f_2 is defined by [5]

$$\phi_{12}(\tau) = \int f_1(t) f_2^*(t + \tau) dt \tag{14}$$

Thus, ACF of a given function is obtained by computing $\phi_{11}(\tau)$.

A 1024-reflection trajectory, T_1, was generated and the aperiodic ACFs of the x and y components were computed; the trajectory is specified by the starting point $(-1, 0)$, a take-off angle of 47 deg and a computational accuracy of 6. The x and y components of the first 1024 trajectory reflections were employed as aperiodic sequences. Figures 6(a) and 6(b) shows the obtained results which indicate that the sequences' aperiodic ACF s are of the quasi-impulsive type. In addition, the aperiodic CCF between the two considered sequences was computed and is shown in figure 6(c). Note that the amplitude of the CCF is of the same order as the sidelobes of the sequences' ACFs.

Similar low ACF and CCF sidelobes are obtainable for the periodic and complex sequences.

5   Multi-functional System Design

In the previous section, the synthesis and properties of TD-sequences was described. The availability of many deterministic sequences, combined with their correlation and PDF properties, suggests that the sequences offer a simple mechanism for achieving multi-functionality within the context of SS communications.

5.1   Aspects of Multi-functionality

The applicability of TD-sequences to achieve different aspects of multi-functional communication systems will now be addressed.

Data-Encoding: Given that the sequence ACFs are of the quasi-impulsive type, the sequences offer the ability to encode data using the method of *sequence inversion keying* with respect tothe random data, resulting in binary communications. Extension to M-ary signalling is also possible [6].

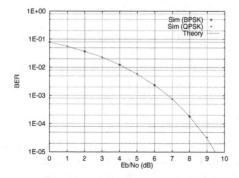

Fig. 7. Performance of TD-Sequence Based MPSK DS-SS Communication System.

A family of M-ary phase-shift-keying (MPSK) SS communication systems may be defined which makes use of TD-sequences as spreading codes [7]. In such systems, TD-sequences are encoded by two independent streams of random binary data yielding the baseband version of the information bearing signal. The two waveforms are then applied to a quadrature modulator for frequency up-conversion purposes. At the receiver, the received signal is first quadrature demodulated and the data is then recovered by the application of two matched filters which are matched to replicas of the TD-sequences used at the transmitter.

A binary phase-shift-keying (BPSK) and quaternary phase-shift-keying (QPSK) members of the family of MPSK systems described above were simulated over an additive-white Gaussian noise channel. The bit-error-rates (BER) for various values of the ratio of energy-per-bit to noise power spectral density, denoted by E_b/N_o, are plotted in figure 7 for the simulated systems, together with the theoretical BER for a BPSK modem given in [6]. The results clearly show that the use of TD-sequences as spreading codes causes no increase in the system BER.

Spectral-Spreading: The process of sequence inversion keying (or bi-orthogonal encoding in general) results in the application of many sequence digits per data symbol. Thus, the transmitted signal is made to occupy a greater bandwidth than the data. Bandwidth spreading equips the system with the ability to mitigate the effects of multi-path channels [7].

Channel Estimation: Given that the sequence ACFs are of the quasi-impulsive type, channel estimation becomes possible. However, the measured responses will be somewhat inaccurate due to the imperfection of the ACF. Inverse-filtering methods may be used to account for such imperfections [8].

Let $s(t)$ and $r(t)$ denote the test signal applied to the channel input and the corresponding channel output, respectively. Then, the channel input and output are related by the convolution integral

$$r(t) = \otimes h(t)s(t) \tag{15}$$

where $h(t)$ is the channel impulse response. Estimates of $h(t)$ may be obtained by convolving the channel output with the reciprocal of $s(t)$, which is denoted by $\bar{s}(t)$. The signal $s(t)$ and its reciprocal $\bar{s}(t)$ have the property that their cross-convolution is a perfect Dirac impulse, ie

$$\otimes s\bar{s} = \delta(t) \tag{16}$$

This measurement method of the impulse response has been termed *generalised channel identification* since it constitutes a generalisation of the classical approach in which the estimate of the channel impulse response is obtained by cross-correlating the channel output with a replica of the applied test signal [9].

The correlation-based classical method suffers from a deterministic error due to the non-zero sidelobes of the test signal ACF; this error is exactly given by

$$e(\tau) = \int h(u)g(\tau - u)du \qquad (17)$$

where $g(\tau)$ is the sidelobe of the test signal ACF at delay τ. When employed, the generalised approach completely eliminates this error [8].

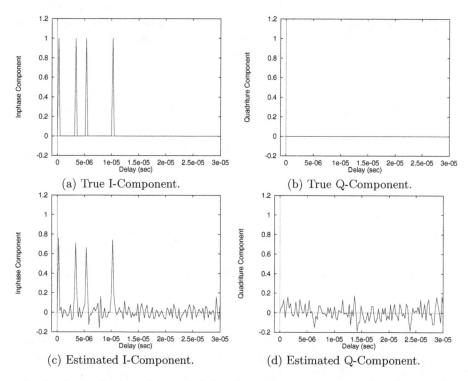

(a) True I-Component.   (b) True Q-Component.

(c) Estimated I-Component.   (d) Estimated Q-Component.

Fig. 8. Four-Path Radio Channel Estimation Using TD-Sequences ($SNR = 10dB$).

Figures 8(a) and 8(b) describes the inphase (I) and quadrature (Q) components of an artificial four-path radio channel, whilst the obtained channel estimates are given in figures 8(c) and 8(d), respectively. The results show that TD-sequences may be used effectively for the purpose of channel estimation within the context of generalised channel identification. Note that the distortion in the measured responses is solely due to the additive white Gaussian noise present in the observed channel output; the noise was added at an signal-to-noise-ratio of 20dB.

Synchronisation: Again, using the sequences ACF property means that 'free' time-synchronisation becomes available at the receiver. Combined with modulation derived synchronisation, robust symbol timing recovery may be obtained [10].

Figure 9 shows the initial period of the synchroniser output and the symbol synchronisation epoch flag. The synchroniser output consists of a rising series of impulses whose time positions correspond to symbol boundaries in the received signal. The magnitude of the impulses saturates when the synchroniser reaches the steady state part of its response, when the receiver is considered to be fully time-aligned with the received signal.

The diagram also shows that the synchronisation epoch flag changes state in an irregular fashion when the receiver is un-synchronised. However, once a degree of time synchronistion is achieved, the flag changes state at a regular time intervals, corresponding to the symbol period.

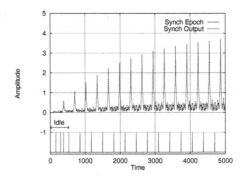

Fig. 9. Timing Recovery in the MPSK SS Communication System.

Multi-user. The low CCF between any pair of TD-sequences enables their use in a multi-user scenario, where a number of users can simultaneously access the same channel, either in a synchronous or asynchronous manner [11].

Security. The bandwidth spreading feature of the sequences, combined with their sensitivity to initial conditions, implies that communication systems employing TD-sequences as spreading codes have the following three naturally embedded security aspects.

 – *Low-Probability of Intercept (LPI):* As a consequence of spectral-spreading, the power spectral density of the transmitted signal is made to extend over a greater bandwidth than that of the data. Thus, the signal becomes hidden in the background noise, making it difficult for an interceptor to determine the presence/absence of transmission. In addition, the ability to tailor the sequences' PDF contributes to the LPI system feature.
 – *Anti-Jamming:* Spectral-spreading also leads to an increased system anti-jamming capability, if the signal presence is detected by an eavsdropper.

– *Privacy:* The dependency of TD-sequences on the initial conditions (enclosure configuration, starting points and initial angle) leads to some form of system privacy, since any small error in the specification of the initial conditions yields an entirely different trajectory. It has been established that the CCF between different TD-sequences is low; thus, it follows that the transmitted message from a communication system employing TD-sequences, can only be successfully recovered by the receiver(s) to whom the trajectory's initial conditions are *exactly* known; this knowledge acts as a form of a 'key' without which the transmitted information cannot be optimally de-spread.

6   Concluding Remarks

In this paper, the concept of multi-functionality was defined and identified as a means which leads to potential complexity reduction in future communication systems. The TD method of generating a class of quasi-analogue pseudorandom sequences was then described and shown to offer the desired properties for achieving multi-functionality within a SS communication system. In particular, the family of TD-sequences offers :

– many of sequences;
– quasi-impulsive aperiodic ACF;
– low aperiodic CCF;
– determinism if, and only if, the initial conditions are accurately known;
– a mechanisim for PDF tailoring.

These properties were exploited in the definition a multi-functional MPSK SS communication system in which TD-sequences facilitated data-encoding, spectral-spreading, channel estimation, synchronisation, multi-user capability and security. Simulation results have been presented to clarify and verify the concepts.

In essence, this paper has demonstrated that a multi-functional algorithm may be developed which fullfils many of the required tasks of the physical layer of a SS communication system.

References

1. M Darnell and B Honary, "Multi-functionality and its application in digital communication system design," in *Cryptography and Coding III*. Oxford, 1993.
2. M Darnell, "The interaction betweem radio communication system adaptation capability and the availability of channel state data," in *International Symposium on Radio Propagation, China*, Aug 1997.
3. M Darnell, "Analogue pseudorandom sequences for communication applications," in *Codes and Cyphers*, pp. 121–139. Oxford, ISBN 0-905091-03-5, 1995.
4. A Papoulis, *Probability, Random Variables and Stochastic Processes*, McGraw-Hill Book Company, 1984.
5. P Lynn, *An Introduction to the Analysis and Processing of Signals*, McMillan, ISBN: 0-333-48887-3, 3rd edition, 1994.

6. B Sklar, *Digital Communications:Fundementals and Applications*, Prentice Hall, 1988.
7. R Peterson, R Ziemer, and D Borth, *Introduction to Spread Spectrum Communications*, Prentice-Hall, ISBN: 0-02-431623-7, 1995.
8. A Al-Dabbagh and M Darnell, "The theory and application of reciprocal pairs of periodic sequences," in *Cryptography and Coding VI*, Lecture Notes in Computer Science (1355), pp. 1–16. Springer, ISBN: 3-540-63927-6, 1997.
9. S Sampei, *Applications of Digital Wireless Technologies to Global Wireless Communications*, Prentice Hall ISBN: 0-13-214272-4, 1997.
10. M Shaw and B Honary, "Modulation derived synchronisation," *Electronic Letters*, vol. 25, no. 11, pp. 750–751, May 1989.
11. A Viterbi, *CDMA: Principles of Spread Spectrum Communication*, Addison Wesley, 1995.

Improvement of the Delsarte Bound for τ-Designs in Finite Polynomial Metric Spaces

Svetla Nikova[1]* and Ventzislav Nikov[2]

[1] Department Electrical Engineering, ESAT/COSIC,
Katholieke Universiteit Leuven, Kasteelpark Arenberg 10,
B-3001 Heverlee-Leuven, Belgium
svetla.nikova@esat.kuleuven.ac.be
[2] Department of Mathematics and Informatics
Veliko Turnovo University, 5000 Veliko Turnovo, Bulgaria
vnikov@mail.com

Abstract. In this paper the problem for the improvement of the Delsarte bound for τ-designs in finite polynomial metric spaces is investigated. First we distinguish the two cases of the Hamming and Johnson Q- polynomial metric spaces and give exact intervals, when the Delsarte bound is possible to be improved. Secondly, we derive new bounds for these cases. Analytical forms of the extremal polynomials of degree $\tau + 2$ for non-antipodal PMS and of degree $\tau + 3$ for antipodal PMS are given. The new bound is investigated in the following asymptotical process: in Hamming space when τ and n grow simultaneously to infinity in a proportional manner and in Johnson space when τ, w and n grow simultaneously to infinity in a proportional manner. In both cases, the new bound has better asymptotical behavior then the Delsarte bound.

1 Introduction

Let $\mathcal{M}$ be a *polynomial metric space* (PMS) [14] with metric $d(x,y)$, standard substitution $t = \sigma(d(x,y))$ and a normalized measure μ, $\mu(\mathcal{M}) = 1$. When we describe the PMS we follow Levenshtein [14]. Any finite nonempty subset C of $\mathcal{M}$ is called a code. A code for which $\sigma(d(x,y)) \leq \sigma(d)$ $(x,y \in C)$ and d is the *minimal distance* of C is an $(\mathcal{M}, |C|, \sigma)$-code. We consider some parameters of the codes connected with their metric properties. For any $C \subset \mathcal{M}$, let with $\Delta(C)$ denote the *distance set* of C, i.e set of values of $d(x,y)$ when $x,y \in C$. For any code C the *minimal distance* of C is defined to be $d(C) = \min\limits_{x,y \in C, x \neq y} d(x,y)$ and the parameter $s(C) = |\Delta(C) \backslash \{0\}|$ characterizes the number of different distances between distinct points of C. Obviously for any code C, $d(C) = \min(\Delta(C) \backslash \{0\})$. The diameter of the whole space $\mathcal{M}$ can be defined as $D(\mathcal{M}) = \max(\Delta(\mathcal{M}) \backslash \{0\})$. Correspondingly the diameter of the code C is $D(C) = \max(\Delta(C) \backslash \{0\})$.

---

* The author was partially supported by NATO research fellowship and Concerted Research Action GOA-MEFISTO-666 of the Flemish Government

B. Honary (Ed.): Cryptography and Coding 2001, LNCS 2260, pp. 191–204, 2001.
© Springer-Verlag Berlin Heidelberg 2001

Now let us introduce the concept of a τ-design. For our purposes it is useful to consider the code C as a weighted set $C = (C, m)$, where m is a certain positive-valued function on C. We suppose that the weights $m(x)$ of elements x are normalized. Hereafter we consider a code C as a special case of a weighted set, when $m(x) \equiv 1$ for all $x \in C$. We will define a (weighted) τ-design by means of the strictly monotone real function (substitution) $\sigma(d)$ defined on the interval $[0, D(\mathcal{M})]$.

Definition 1. *A weighted set (C, m) will be referred to as a weighted τ-design in $\mathcal{M}$ with respect to the substitution $\sigma(d)$ if for any polynomial $f(t)$ in a real t of degree at most τ,*

$$\int_{\mathcal{M}} \int_{\mathcal{M}} f(\sigma(d(x,y))) d\mu(x) d\mu(y) = \frac{1}{|C|^2} \sum_{x,y \in C} f(\sigma(d(x,y))) m(x) m(y).$$

The τ-design is a special case of the weighted τ-design, when $m(x) \equiv 1$ for all $x \in C$. The maximum integer τ ($\tau \leq s(\mathcal{M})$) such that a (weighted) set C is a (weighted) τ-design is called the *strength* of C and denoted by $\tau(C)$. Suppose $\mathcal{M}$ is finite and $\Delta(C) = \{d_0, d_1, \ldots, d_n\}$, is the *distance distribution* of C. The *dual distance distribution* (so called MacWilliams transform [16, p.137]) of C is defined to be $\Delta'(C) = \{d'_0, d'_1, \ldots, d'_n\}$. The *dual distance* of the code C is the smallest i, $i = 1, \ldots, n$, such that $d'_i \neq 0$; the *dual degree* $s'(C)$ of C is the number of i, $i = 1, \ldots, n$, such that $d'_i \neq 0$. As proved in [3] $\tau(C) + 1 = d'(C)$.

The basic problem of the coding theory is the construction of the maximum (on cardinality) σ-code. Together with this problem there exists another one of constructing the minimum (on cardinality) τ-design (or equivalently a code with dual distance $d' = \tau + 1$). As it is proved in [12,15,1] this two problems are dual.

The PMS $\mathcal{M} = (\mathcal{M}, d(x, y), \mu)$ with a given substitution $\sigma(d)$ is connected with a system of orthogonal (to the measure $\nu(t)$) polynomials $\{U_i(t)\}$ of degree i, $i = 0, 1, \ldots, s(\mathcal{M})$, the so called *zonal spherical functions* (ZSF). The function $\nu(t)$ is equal to $1 - \mu(\sigma^{-1}(t))$ on the interval $[-1, 1]$. So the system $U_i(t)$ is defined by the substitution $\sigma(d)$ and measure $\mu(d)$. We can assume without loss of generality that $\sigma(d)$ is a continuous strictly decreasing function on $[0, D(\mathcal{M})]$ such that $\sigma(D(\mathcal{M})) = -1 \leq \sigma(d) \leq \sigma(0) = 1$.

A polynomial metric space $\mathcal{M}$ is called *antipodal* if for every point $x \in \mathcal{M}$ there exists a point $\bar{x} \in \mathcal{M}$ such that for any point $y \in \mathcal{M}$ we have $\sigma(d(x, y)) + \sigma(d(\bar{x}, y)) = 0$.

Since $U_i(t)$ is a system of orthogonal polynomials there exists a unique system of positive constants r_i, $i = 0, \ldots, s(\mathcal{M})$ such that

$$r_i \int_{-1}^{1} U_i(t) U_j(t) d\nu(t) = \delta_{i,j}, \quad U_i(1) = 1, \quad i = 0, 1, \ldots, s(\mathcal{M}). \tag{1}$$

The integral on the left-hand side can be considered as a Lebesgue-Stieltjes integral on $[-1, 1]$. When $\mathcal{M}$ is finite $\nu(t)$ is left continuous and has $s(\mathcal{M}) + 1$ steps at the point $t_i = \sigma(d_i)$ with positive step sizes w_i, $i = 0, 1, \ldots, s(\mathcal{M})$, $\sum_{i=0}^{s} w_i = 1$. Denote $w(t) = \nu'(t)$ if $\mathcal{M}$ is infinite. Since the measure is normalized we have

$U_0(t) \equiv 1, r_0 = 1$. The orthogonality condition (1) for a finite $\mathcal{M}$ can be rewritten respectively in the following form: $r_i \sum_{k=0}^{s(\mathcal{M})} U_i(\sigma(d_k))U_j(\sigma(d_k))w_k = \delta_{i,j}$ for $i, j = 0, 1, \ldots, s(\mathcal{M})$.

For arbitrary $a, b \in \{0, 1\}$ we define the so called *adjacent* to $\{U_k(t)\}$ system of polynomials $\{U_k^{a,b}(t)\}$ in a real t of degree k, $k = 0, 1, \ldots, s(\mathcal{M}) - \delta_{a,1} - \delta_{b,1}$ as follows.

First we define positive constants $c^{a,b}$ and the function $\nu^{a,b}(t)$. In the infinite case $d(\nu^{a,b}(t)) = c^{a,b}(1 - t)^a(1 - t)^b w(t)dt$. For the finite case $\nu^{a,b}(t)$ is a step-function and it is equal to $c^{a,b}(1 - \sigma(d_k))^a(1 + \sigma(d_k))^b w_k$ in the points $t_k = \sigma(d_k)$ $(k = 0, 1, \ldots, s(\mathcal{M}))$. The constants $c^{a,b}$ are chosen in such a way that the Lebesgue-Stieltjes measure $\nu^{a,b}$ on $[-1, 1]$ generated by the function $\nu^{a,b}(t)$ is normalized, i.e. $\int_{-1}^{1} d\nu^{a,b}(t) = c^{a,b} \int_{-1}^{1}(1 - t)^a(1 + t)^b d\nu(t) = 1$ in the infinite case. Correspondingly for the finite case we have

$$c^{a,b} \sum_{k=0}^{s(\mathcal{M})} (1 - \sigma(d_k))^a(1 + \sigma(d_k))^b w_k = 1. \tag{2}$$

Then the polynomials $U_k^{a,b}(t)$ together with the positive constants $r_k^{a,b}$ are defined uniquely by the following orthogonality relations

$$r_i^{a,b} \int_{-1}^{1} U_i^{a,b}(t)U_i^{a,b}(t)d\nu^{a,b}(t) = \delta_{i,j}$$

$$r_i^{a,b}c^{a,b} \sum_{k=0}^{s(\mathcal{M})} U_i^{a,b}(\sigma(d_k))U_j^{a,b}(\sigma(d_k))(1 - \sigma(d_k))^a(1 + \sigma(d_k))^b w_k = \delta_{i,j}. \tag{3}$$

It is easy to see that $U_0^{a,b}(t) = 1$, $U_k^{a,b}(1) = 1$, $r_0^{a,b} = 1$. When $a = b = 0$ we omit the upper indices. Denote by $-1 < t_{k,i}^{a,b} < 1$, $i = 1, \ldots, k$, the roots of the polynomial $U_k^{a,b}(t)$, $k \geq 0$, ordered in increasing order and $t_k^{a,b} = t_{k,k}^{a,b}$. Note that by the normalization $U_k^{a,b}(1) = 1$ the leading coefficient $a_{k,k}^{a,b}$ of polynomials $U_k^{a,b}(t)$ is positive and $sgn U_k^{a,b}(-1) = (-1)^k$ for $k \geq 0$. Let us introduce the notation

$$U_k^{a,b}(t) = \sum_{i=0}^{k} a_{k,i}^{a,b} t^i.$$

For any integer k, and reals x and y we have the well-known Christoffel-Darboux formulae

$$\sum_{i=0}^{k} r_i U_i(x)U_i(y) = \begin{cases} r_k m_k \dfrac{U_{k+1}(x)U_k(y) - U_k(x)U_{k+1}(y)}{x - y} & \text{if } x \neq y, \\ r_k m_k(U_{k+1}'(x)U_k(x) - U_k'(x)U_{k+1}(x)) & \text{if } x = y. \end{cases}$$

It is known that for the ZSF the following recurrence formula holds:

$$(t + m_i + c_i - 1)U_i(t) = m_i U_{i+1}(t) + c_i U_{i-1}(t), \tag{4}$$

for $i \geq 0$, where $r_{-1} = m_{-1} = 0$, $m_i = \frac{a_{i,i}}{a_{i+1,i+1}}$, $c_i = \frac{r_{i-1} m_{i-1}}{r_i}$ and $U_{-1}(t) \equiv 0$, $U_0(t) \equiv 1$. By definition

$$T_k^{a,b}(x,y) = \sum_{i=0}^{k} r_i^{a,b} U_i^{a,b}(x) U_i^{a,b}(y), \tag{5}$$

For any non-negative integer k [10,14] the following relations between the adjacent system of orthogonal polynomials hold: $U_k^{a,b+1}(t) = \frac{T_k^{a,b}(t,-1)}{T_k^{a,b}(1,-1)}$, $U_k^{a+1,b}(t) = \frac{T_k^{a,b}(t,1)}{T_k^{a,b}(1,1)}$.

We consider the *Linear Programming Theorem* due to Delsarte [3].

Theorem 1. *Let $C \subset \mathcal{M}$ be an $(\mathcal{M}, |C|, \sigma)$-code (reps. τ-design) and let $f(t)$ be a real non-zero polynomial such that*

(A1) $f(t) \leq 0$, *for* $-1 \leq t \leq \sigma$,
   (resp. **(B1)** $f(t) \geq 0$, *for* $-1 \leq t \leq 1$*),*
(A2) *the coefficients in the ZSF expansion* $f(t) = \sum_{i=0}^{k} f_i U_i(t)$
   satisfy $f_0 > 0$, $f_i \geq 0$ *for* $i = 1, \ldots, k$.
   (resp. **(B2)** *the coefficients in the ZSF expansion* $f(t) = \sum_{i=0}^{k} f_i U_i(t)$
   satisfy $f_0 > 0$, $f_i \leq 0$ *for* $i = \tau + 1, \ldots, k$.)*

Then, $|C| \leq \Omega(f) = f(1)/f_0$ (resp. $|C| \geq \Omega(f)$).

We denote by $A_{\mathcal{M},\sigma}$ (resp. $B_{\mathcal{M},\tau}$) the set of real polynomials which satisfy the conditions **(A1)** and **(A2)** (resp. **(B1)** and **(B2)**). We will consider the quantities $A_U(\mathcal{M}, \sigma) = \min\{\Omega(f) : f(t) \in A_{\mathcal{M},\sigma}\}$ and $B_U(\mathcal{M}, \tau) = \max\{\Omega(f) : f(t) \in B_{\mathcal{M},\tau}\}$.

The universal upper (resp. lower) bounds $L(\mathcal{M}, \sigma)$ (resp. $D(\mathcal{M}, \tau)$) for the cardinality of an $(\mathcal{M}, |C|, \sigma)$-code (resp. a τ-design) can be presented in the following form [10,5]:

$$|C| \leq L(\mathcal{M}, \sigma) = \left(1 - \frac{U_{k-1+\varepsilon}^{1,0}(\sigma)}{U_k^{0,\varepsilon}(\sigma)}\right) \sum_{i=0}^{k-1+\varepsilon} r_i, \tag{6}$$

where $\varepsilon = 0$ if $t_{k-1}^{1,1} \leq \sigma < t_k^{1,0}$ and $\varepsilon = 1$ if $t_k^{1,0} \leq \sigma < t_k^{1,1}$, resp.

$$|C| \geq D(\mathcal{M}, \tau) = 2^\theta c^{0,\theta} \sum_{i=0}^{l} r_i^{0,\theta}, \tag{7}$$

where $\theta \in \{0, 1\}$ and $\tau = 2l + \theta$.

The bound (6) was obtained by Levenshtein [9] using the polynomial

$$f^{(\sigma)}(t) = (t - \sigma)(t + 1)^\varepsilon (T_{k-1}^{1,\varepsilon}(t, \sigma))^2$$

of degree $h(\sigma) = 2k - 1 + \varepsilon(\sigma)$. The polynomial $f^{(\sigma)}(t)$ has been discovered by Levenshtein [9]; the optimality proof has been given by Sidel'nikov [22]. Later new proof was proposed by Levenshtein [10].

The problem of finding lower bounds on the minimum possible size of designs in PMS was considered by many authors. Delsarte in his seminal paper [3] derived general method for obtaining the bound and found it in the case $\tau = 2l$ for finite PMS. Dunkl [6] obtained lower bounds when $\tau = 2l - 1$ for finite PMS, Rao [19] for the Hamming space, and Ray-Chaudhuri/Wilson [20] for the Johnson space. For infinite PMS, they were proved by Delsarte, Goethals and Seidel [5] for the Euclidean sphere and by Hoggar [8] for the projective spaces. All this bounds become classical lower bounds in different PMS. A classical result by Schoeneberg and Szegö [23] shows that the polynomial

$$f^{(\tau)}(t) = (t + 1)^\theta ((Q_l^{1,\theta}(t))^2$$

of degree τ is optimal. The bound (6) is called Levenshtein bound. The bound (7) we will call Delsarte bound for τ-designs and it will be the main object of our investigations. Both of these bounds were derived by using the Linear Programming Theorem.

Definition 2. *A polynomial $f(t) \in B_{\mathcal{M},\tau}$ is called $B_{\mathcal{M},\tau}$-extremal if*

$$\Omega(f) = \max\{\Omega(g) : \ g(t) \in B_{\mathcal{M},\tau}, \ \deg(g) \leq \deg(f)\}.$$

A polynomial $f(t) \in A_{\mathcal{M},\sigma}$ is called $A_{\mathcal{M},\sigma}$-extremal if

$$\Omega(f) = \min\{\Omega(g) : \ g(t) \in A_{\mathcal{M},\sigma}, \ \deg(g) \leq \deg(f)\}.$$

It is known that $f^{(\sigma)}(t)$ and $f^{(\tau)}(t)$ are $A_{\mathcal{M},\sigma}$- and $B_{\mathcal{M},\tau}$-extremal of degree $h(\sigma)$ and τ, respectively.

PMS are finite metric spaces represented by P- and Q- polynomial association schemes as well as infinite metric spaces. The most famous examples of the finite PMS are the Hamming, Johnson, Grassmann space. For finite PMS the system of orthogonal polynomials U is either Q or P, and we use $\bar{U}$ for the other one.

A stronger version of Theorem 1 is valid for the finite spaces [3].

Theorem 2. *Let $C \subset \mathcal{M}$ be an σ-code (reps.τ-design) and let $f(t) = \sum_{i=0}^{k} f_i U_i(t)$ be a real non-zero polynomial such that*

(A3) $f(1) > 0$, $f(\sigma(i)) \leq 0$, *for* $i = d, \dots, D(\mathcal{M})$,
    (resp. **(B3)** $f(1) > 0$, $f(\sigma(i)) \geq 0$, *for* $i = 1, 2, \dots, D(\mathcal{M})$*),*
(A2) $f_0 > 0$, $f_i \geq 0$ *for* $i = 1, \dots, k$.
    (resp. **(B2)** $f_0 > 0$, $f_i \leq 0$ *for* $i = \tau + 1, \dots, k$.*)*

Then, $|C| \leq \Omega(f)$ (resp. $|C| \geq \Omega(f)$).

Definition 3. *Let $\mathcal{M}$ be a PMS. We define the quantities*

$$B_U^\star(\mathcal{M}, \tau) = max\{\Omega(f) : f(t) \quad satisfy \quad (\mathbf{B3}), (\mathbf{B2})\}$$

$$A_U^\star(\mathcal{M}, \sigma) = min\{\Omega(f) : f(t) \quad satisfy \quad (\mathbf{A3}), (\mathbf{A2})\}.$$

Obviously $B_U^\star(\mathcal{M}, \tau) > B_U(\mathcal{M}, \tau)$ and $A_U^\star(\mathcal{M}, \sigma) < A_U(\mathcal{M}, \sigma)$.
As it is proved in [12,15,1] in finite PMS we have.

$$A_U^\star(\mathcal{M}, \sigma(d)) B_U^\star(\mathcal{M}, d-1) = |\mathcal{M}|. \tag{8}$$

So, optimal bounds for codes and designs in finite PMS can be obtained as a solution of either problems.

As a consequence the following bounds for the maximum cardinality of a σ-code C and the minimum cardinality of τ-design D is true:

$$|C| \leq A_Q^\star(\mathcal{M}, \sigma(d)) = \frac{|\mathcal{M}|}{B_P^\star(\mathcal{M}, d-1)}, \tag{9}$$

$$|D| \geq B_Q^\star(\mathcal{M}, \tau) = \frac{|\mathcal{M}|}{A_P^\star(\mathcal{M}, \sigma(\tau+1))}.$$

In this paper we focus on Hamming and Johnson spaces, presented by Q-polynomial association schemes.

In [18] we improve the Delsarte bound in infinite PMS and we present analytical form of the new bound for non-antipodal spaces. It turns out that these results are valid for all PMS, in particular for Hamming and Johnson spaces. In the section 2 we present these results without proofs, which can be found in [18]. In Sections 3 and 4 we apply the method from [18] in Hamming and Q-polynomial Johnson space, respectively. The asymptotic behavior of the bounds is presented as well. In this paper we will compare our results only with the Delsarte bound, nevertheless it is well known that the Delsarte bound is not the best possible (even asymptotically) in the finite PMS.

2   Preliminary Results

We consider the following linear functional, which we will call *test* functions

$$G_\tau(\mathcal{M}, f) = \frac{f(1)}{D(\mathcal{M}, \tau)} + \sum_{i=1}^{k+\theta} \rho_i^{(\tau)} f(\alpha_i)$$

where $\alpha_i, \rho_i^{(\tau)}$ are defined in [18].

This linear functional maps the set of real polynomials to the set of real numbers. We have $-1 \leq G_\tau(\mathcal{M}, Q_j) \leq 1$ and $G_\tau(\mathcal{M}, f) = f_0$ for any polynomial $f(t)$ of degree at most τ. Also $G_\tau(\mathcal{M}, f) = f(1)/D(\mathcal{M}, \tau)$ if $f(t)$ vanishes at the zeros of $f^{(\tau)}(t)$.

Now we will give necessary and sufficient conditions for improvement of the Delsarte bound. Later on we will investigate some properties of the test functions, which turn out to be very useful.

Theorem 3. *The bound $D(\mathcal{M}, \tau)$ can be improved by a polynomial $f(t) \in B_{\mathcal{M}, \tau}$ of degree at least $\tau + 1$, if and only if $G_\tau(\mathcal{M}, Q_j) < 0$ for some $j \geq \tau + 1$. Moreover, if $G_\tau(\mathcal{M}, Q_j) < 0$ for some $j \geq \tau + 1$, then $D(\mathcal{M}, \tau)$ can be improved by a polynomial in $B_{\mathcal{M}, \tau}$ of degree j.*

Lemma 1. *Let $\mathcal{M}$ be antipodal. If τ and j are odd, then $G_\tau(\mathcal{M}, Q_j) = 0$.*

Corollary 1. *Let $\mathcal{M}$ be antipodal PMS. Then*

$$G_\tau(\mathcal{M}, Q_{\tau+2}) \begin{cases} > 0, \text{ for } \tau = 2k \\ = 0, \text{ for } \tau = 2k + 1. \end{cases}$$

The investigations of the test functions for designs, Corollary 1) and the obtained necessary and sufficient conditions for improving the Delsarte bound show that the smallest possible degree of an improving polynomial is $\tau + 2$ for non-antipodal and $\tau + 3$ for antipodal PMS.

Theorem 4. *Let $\mathcal{M}$ be non-antipodal PMS. Then, any $B_{\mathcal{M}, \tau}$-extremal polynomial of degree $\tau + 2$ ($\tau = 2k + \theta$) has the form*

$$f^{(\tau)}(t; \tau + 2) = (1 + t)^{1-\theta}[q(t + 1) + (1 - t)][\eta Q_{k-1+\theta}^{1,1-\theta}(t) + Q_{k+\theta}^{1,1-\theta}(t)]^2,$$

where q, η are suitable constants.

Now we obtain the following analytical form of the new bound.

Theorem 5. *Let $\mathcal{M}$ be a non-antipodal PMS. Then*

$$B(\mathcal{M}, \tau) \geq S(\mathcal{M}, \tau; \tau + 2) = D(\mathcal{M}, \tau - 2) + R(\tau) = \Omega(f^{(\tau)}(t; \tau + 2)),$$

where $R(\tau)$ is a suitable constant.

Corollary 2. *Let $\mathcal{M}$ be a non-antipodal PMS and let τ be an integer. Then $S(\mathcal{M}, \tau; \tau + 2) > D(\mathcal{M}, \tau)$ if and only if $G_\tau(\mathcal{M}, Q_{\tau+2}) < 0$.*

Theorem 6. *Let $\mathcal{M}$ be antipodal PMS. Then, any $B_{\mathcal{M}, \tau}$-extremal polynomial of degree $\tau + 3$ ($\tau = 2k + \theta$) has the form*

$$f^{(\tau)}(t; \tau + 3) = (1 + t)^\theta[q(t + 1) + (1 - t)][\eta_1 Q_{k-1}^{1,\theta}(t) + \eta_2 Q_k^{1,\theta}(t) + Q_{k+1}^{1,\theta}(t)]^2$$

where q, η_1, η_2 are suitable constants.

Theorem 7. *Let $\mathcal{M}$ be an antipodal PMS. Then we derive a new bound*

$$B(\mathcal{M}, \tau) \geq S(\mathcal{M}, \tau; \tau + 3) = D(\mathcal{M}, \tau - 3) + R_2(\tau) = \Omega(f^{(\tau)}(t; \tau + 3)),$$

where $R_2(\tau)$ is certain constant

3   Hamming Space

The Hamming space $\mathcal{M} = H_v^n$ $(n, v = 2, 3, \ldots)$ consists of vectors $\mathbf{x} = (x_1, \ldots, x_n)$ where $x_j \in \{0, 1, \ldots, v-1\}$ with the distance $d(\mathbf{x}, \mathbf{y})$, which is equal to the number of different coordinates of the vectors $\mathbf{x}$ and $\mathbf{y}$. The Hamming space is distance invariant with the measure $w_k = v^{-n} \binom{n}{k} (v-1)^k$, $k = 0, 1, \ldots, n$. This space is self-dual (i.e. the systems Q and P coincide) polynomial graph [3] with respect to the linear standard substitution $\sigma(d) = 1 - \frac{2d}{n}$ and we have $s(\mathcal{M}) = n$, $D(\mathcal{M}) = n$, $d_k = k$. For $v = 2$ the Hamming space is antipodal and for $v > 2$ is non-antipodal. The system of ZSF $\{Q_k(t)\}_{k=0}^n$ can be defined by

$$Q_k(t) = Q_k(\sigma(d)) = \frac{K_k^{n,v}(d)}{r_k}, \quad \text{where} \quad K_k^{n,v}(z) = \sum_{j=0}^k (-1)^j (v-1)^{k-j} \binom{z}{j} \binom{n-z}{k-j}$$

is the Krawtchouk polynomial of degree k and the corresponding constants are $r_k = w_k v^n = \binom{n}{k}(v-1)^k$. Considering (4) one can find for this case $m_k = \frac{2(v-1)(n-k)}{vn}$, $c_k = \frac{2k}{vn}$

The corresponding adjacent systems of polynomials and constants (2),(3),(5) as calculated by Levenshtein in [14] are

$$c^{0,1} = \frac{v}{2}, \quad r_k^{0,1} = \binom{n-1}{k}(v-1)^k,$$

$$Q_k^{0,1}(\sigma(d)) = \frac{K_k^{n-1,v}(d)}{r_k^{0,1}} \quad \text{for} \quad k = 0, \ldots, n-1$$

$$c^{1,0} = \frac{v}{2(v-1)}, \quad r_k^{1,0} = \frac{\left(\sum_{j=0}^k \binom{n}{j}(v-1)^j\right)^2}{\binom{n-1}{k}(v-1)^k},$$

$$Q_k^{1,0}(\sigma(d)) = \frac{K_k^{n-1,v}(d-1)}{\sum_{j=0}^k \binom{n}{j}(v-1)^j} \quad \text{for} \quad k = 0, \ldots, n-1$$

$$c^{1,1} = \frac{nv^2}{4(n-1)(v-1)}, \quad r_k^{1,1} = \frac{\left(\sum_{j=0}^k \binom{n-1}{j}(v-1)^j\right)^2}{\binom{n-2}{k}(v-1)^k},$$

$$Q_k^{1,1}(\sigma(d)) = \frac{K_k^{n-2,v}(d-1)}{\sum_{j=0}^k \binom{n-1}{j}(v-1)^j} \quad \text{for} \quad k = 0, \ldots, n-2$$

The designs in the Hamming space are called *orthogonal arrays*, commonly denoted by $OA_\lambda(\tau, n, v)$. Their cardinality satisfy $|C| = \lambda v^\tau$.

Since $|\mathcal{M}| = v^n$ we will present well known pairs of universal bounds, i.e. inequalities which are valid for all codes $C \subseteq H_v^n$, which follows from Theorem 2.

The first pair is the *Singleton bound* [21] for a code $C \subseteq H_v^n$

$$v^\tau \leq |C| \leq v^{n-d+1}, \tag{10}$$

where any of the bounds is attained if and only if $d + \tau = n + 1$.

The second pair of bounds is formed by *Rao* [19] and *Hamming* [7] bounds for a code $C \subseteq H_v^n$.

$$D(H_v^n, \tau) \leq |C| \leq \frac{v^n}{D(H_v^n, d - 1)} \tag{11}$$

Codes, which cardinality is equal to the left-hand side or the right-hand side of (11) are called *tight designs* and *perfect* codes, respectively. Notice that for $\tau = 2l + \theta$ we have $D(H_v^n, \tau) = v^\theta \sum_{i=0}^{l} \binom{n-\theta}{i}(v - 1)^i$.

The third pair universal bounds for any code $C \subseteq H_v^n$ is the *Levenshtein* bound [12].

$$\frac{v^n}{L(H_v^n, \sigma(\tau + 1))} \leq |C| \leq L(H_v^n, \sigma(d)) \tag{12}$$

The first two pairs of bounds are obtained by means of combinatorial methods, but all of them can be obtained using Theorem 1 or Theorem 2. In fact the Delsarte bound for τ-designs in Hamming space coincide with the Rao bound.

Now we give the exact intervals for the dimension n, when the test functions are negative. Further on we will omit the proofs, which are too technical.

Theorem 8. *Let* $\mathcal{M} = H_v^n$, $v \geq 3$. *Then* $G_\tau(H_v^n, Q_{\tau+2}) < 0$ *for*

$$\tau + 2 \leq n \leq \tau + (k^2 + k)\frac{v - 2}{v - 1} \quad if \tau = 2k$$
$$\tau + 2 \leq n \leq \tau + k^2 + k \quad if \tau = 2k + 1.$$

Theorem 9. *Let* $\mathcal{M} = H_v^n$, $v = 2$. *Then* $G_\tau(H_2^n, Q_{\tau+3}) < 0$ *for*

$$\tau + 3 \leq n \leq \lfloor k^2/2 + 3k + 2 + 1/2\sqrt{k^4 + 4k^3 + 4k^2 + 24k + 20} \rfloor \quad if \tau = 2k$$
$$\tau + 3 \leq n \leq \tau + (k + 1)^2 \quad if \tau = 2k + 1.$$

Applying (8), (9) for our bound we arrive at the following theorem.

Theorem 10. *For a code* $C \subseteq H_v^n$ *the following bounds are valid*

$$S(H_v^n, \tau; \tau + \epsilon) \leq |C| \leq \frac{v^n}{S(H_v^n, d - 1; d - 1 + \epsilon)},$$

where $\epsilon = 2$ *and* $\epsilon = 3$ *for antipodal and non-antipodal spaces, respectively.*

We have investigated the following asymptotic behavior of the bound when τ grows with n.

Theorem 11. *Let* $v > 2$ *and* $\tau = 2k + \theta$. *If* $\lim\limits_{n \to \infty} \dfrac{k}{n} = \delta$, *where* $\frac{1}{2} > \delta > 0$, *then*

$$S(H_v^n, \tau; \tau + 2) \sim D(H_v^n, \tau) const(v, \delta).$$

The constant is bigger than 1 and it is equal to
$$const(v, \delta) = 1 + \frac{(v-1)^2(1-2\delta)+\delta^2(v^2-v+1)}{\delta v(v-1)(1-\delta)}, \text{ when } \tau \text{ is odd and}$$
$$const(v, \delta) = 1 + \frac{v^3(v-2)(1-\delta)^3+v^2(1-\delta)(1-\delta+\delta^2)+2v(1-\delta)\delta^2+\delta^3}{(v-1)(1-\delta)(v^2(1-\delta)+2v\delta(1-\delta)+\delta^2)}, \text{ when } \tau \text{ is even.}$$

Note that if $\delta \sim 0$ when τ is even, then $const \sim v$, but when τ is odd, then $const \sim 1 + \frac{v-1}{v\delta}$ which increases, when δ decreases.

Let us consider Rao-Hamming bounds (11). Till now we proposed a way to improve the Delsarte (Rao) bound. In fact we can use the same test functions to check when it is possible to improve the Hamming bound. So, the test functions can be useful in order to investigate the existence of perfect codes.

One can define $R(n, d) = \dfrac{log_v(M(H_v^n, d))}{n}$ to be the rate of the best code, where $M(H_v^n, d)$ is the largest possible cardinality of the $(H_v^n, |C|, \sigma(d))$-code. For each real number $0 \le \delta \le 1$ is defined

$$R(\delta) = \lim sup_{n \to \infty} R(n, d)$$

to be the *rate of a code*, where $d/n \to \delta$. It is known that $R(0) = 1$, and $R(\delta) = 0$ for $1/2 \le \delta \le 1$. Obviously any bound for the cardinality of the codes (designs) gives a bound for $R(\delta)$. The best known upper bound for the rate of code is found by McElliece, Rodemich, Rumsey and Welch (MRRW) [17]. Note that for fixed d and δ Levenshtein bound is stronger then MRRW-bound however, the asymptotic forms of both bounds are the same. The best known lower bound is Gilbert-Varshamov (GV) bound [16]. Unfortunately the rate R of the new bounds are the same as the classical ones (Rao or Hamming).

4   Johnson Space

We consider the Johnson space $\mathcal{M} = J_w^n$ ($n = 2, 3, \dots ; w = 1, \dots, \lfloor n/2 \rfloor$) as set $\mathcal{M}$ of all w-subsets of the n-set $\{1, \dots, n\}$ where the distance between two elements $x, y \in J_w^n$ is defined to be $w - |x \cap y|$. The Johnson space can be also considered as a subset of the binary Hamming space H_2^n consisting of all vectors which have exactly w non-zero coordinates, with distance being equal to half of the Hamming distance. That is why codes in J_w^n are usually called *constant weight* codes. For $n = 2w$ the Johnson space is antipodal and for $n > 2w$ it is non-antipodal.

It was shown by Delsarte [3] that the Johnson space is a P- and Q-polynomial graph. Here we consider only the properties of Q-polynomial Johnson space. Of course, the same calculation can be done in P-polynomial Johnson space.

With the standard substitution $\sigma(d) = 1 - 2\frac{d}{w}$ we obtain Q-polynomial Johnson space. We have $s(\mathcal{M}) = w$, $D(\mathcal{M}) = w$, $d_k = k$, measure constants $w_k = \binom{w}{k}\binom{n-w}{k}/\binom{n}{w}$, $k = 0, 1, \ldots, w$, and constants $r_k = \binom{n}{k} - \binom{n}{k-1}$, $k = 1, \ldots, w$, $(r_0 = 1)$.

The system of ZSF $\{Q_k(t)\}_{k=0}^n$ in this case can be defined by

$$Q_k(t) = Q_k(\sigma(d)) = J_k^{n,w}(d), \quad \text{where } J_k^{n,w}(z) = \sum_{j=0}^{k} (-1)^j \frac{\binom{k}{j}\binom{n+1-k}{j}}{\binom{w}{j}\binom{n-w}{j}} \binom{z}{j}$$

are the Hahn polynomials of degree k.

Considering recurrence (4) one can find for this case

$$m_k = \frac{2(w-k)(n-w-k)(n-k+1)}{w(n-2k)(n-2k+1)}, \quad c_k = \frac{2(w-k+1)(n-w-k+1)}{w(n-2k+1)(n-2k+2)}$$

The corresponding adjacent systems of polynomials (2),(3) as calculated by Levenshtein in [14] are

$$c^{0,1} = \frac{n}{2w}, \quad r_k^{0,1} = \binom{n-1}{k} - \binom{n-1}{k-1},$$

$$Q_k^{0,1}(\sigma(d)) = J_k^{n-1,w-1}(d) \quad \text{for } k = 0, \ldots, w-1$$

$$c^{1,0} = \frac{n}{2(n-w)}, \quad r_k^{1,0} = \binom{n}{k} \frac{w(n-2k)(w-n)}{n(w-k)(w+k-n)},$$

$$Q_k^{1,0}(\sigma(d)) = \frac{(w-k)(n-w-k)}{(n-2k)} \cdot \frac{J_k^{n,w}(d) - J_{k+1}^{n,w}(d)}{d} \quad \text{for } k = 0, \ldots, w-1$$

$$c^{1,1} = \frac{n(n-1)}{4(w-1)(n-w)}, \quad r_k^{1,1} = \binom{n-1}{k} \frac{(n-w)(w-1)(n-2k-1)}{(n-1)(n-w-k)(w-k-1)},$$

$$Q_k^{1,1}(\sigma(d)) = \frac{(w-k-1)(n-w-k)}{(n-2k-1)} \cdot \frac{J_k^{n-1,w-1}(d) - J_{k+1}^{n-1,w-1}(d)}{d}, \quad k = 0, \ldots, w-2.$$

Notice that the values $Q_k^{1,0}(1) = Q_k^{1,1}(1) = 1$ coincide with the corresponding limits of the right-hand sides when d tends to 0.

The designs in the Johnson space are called *block designs*, commonly denoted by $S_\lambda(\tau, n, w)$. Their cardinality satisfies $|C| = \lambda \frac{\binom{n}{\tau}}{\binom{w}{\tau}}$.

Since $|\mathcal{M}| = \binom{n}{w}$, we will present well known pairs of universal bounds, i.e. inequalities which are valid for all codes $C \subseteq J_w^n$, which follows from Theorem 2.

The first pair is for a code $C \subseteq J_w^n$

$$\frac{\binom{n}{\tau}}{\binom{w}{\tau}} \leq |C| \leq \frac{\binom{n}{w+1-d}}{\binom{w}{w+1-d}},$$

where any of the bounds is attained if and only if $d + \tau = w + 1$.

The following universal pair of bounds is formed by *Ray-Chaudhuri* and *Wilson* [20] bounds (for τ even) and *Levenshtein* [14] (for τ odd) for a code $C \subseteq J_w^n$.

$$D_Q(J_w^n, \tau) \leq |C| \leq \frac{\binom{n}{w}}{D_P(J_w^n, d-1)},$$

where $D_Q(J_w^n, \tau) = \binom{n}{w}^\theta \binom{n-\theta}{l}$ and $D_P(J_w^n, \tau) = \sum_{i=0}^{l} \binom{w-\theta}{i} \binom{n-w+\theta}{i+\theta}$ are the Delsarte bound for $\tau = 2l + \theta$-design in $Q-$ polynomial Johnson space and in $P-$ polynomial Johnson space respectively.

The following bound was proved by Levenshtein [14].

$$\frac{\binom{n}{w}}{L_P(J_w^n, \sigma(\tau+1))} \leq |C| \leq L_Q(J_w^n, \sigma(d)).$$

Now we are ready to investigate Delsarte bound in this case. First we give the exact intervals for w and fixed n, where the test functions are negative.

Theorem 12. *Let* $\mathcal{M} = J_w^n$, $n > 2w$ *(i.e.* $\mathcal{M}$ *is non-antipodal space). Then* $G_\tau(J_w^n, Q_{\tau+2}) < 0$ *for*

$$\tau + 2 \leq w \leq \frac{n}{2} - \frac{1}{2}\sqrt{-\frac{16k^3 + (8-20n)k^2 + (8n^2 - 12n - 8)k + 3n^2 - n^3 - 4}{4k^2 + 4k + n + 1}}$$
$$\text{if } \tau = 2k.$$

$$\tau + 2 \leq w \leq \frac{n}{2} \frac{k^2 + 3k + 2}{k^2 + k + n/2} \quad \text{if } \tau = 2k+1.$$

Let us define the following functions of w

$$S_{odd}(w) = 4w^3 + (-2k^2 - 20k - 18)w^2 + (37k^2 + 6k^3 + 50k + 14)w$$
$$+ 12 - 4k^4 - 21k^3 - 2k - 29k^2$$
$$S_{even}(w) = 32w^7 + (-16k^2 - 288k - 208)w^6 + (1440k + 112k^3 + 1160k^2 + 368)w^5$$
$$+ (-4156k^2 - 56 - 320k^4 - 2632k^3 - 1840k)w^4$$
$$+ (6388k^3 + 3622k^2 + 480k^5 - 206 + 96k + 3568k^4)w^3$$
$$+ (136k^2 + 606k - 5508k^4 - 3466k^3 - 2864k^5 - 400k^6 + 43)w^2$$
$$+ (-594k^2 + 1256k^6 - 338k^3 + 2524k^5 + 1586k^4 + 12 - 98k + 176k^7)w$$
$$+ 162k^4 - 270k^5 - 24k - 32k^8 - 12 + 182k^3 + 22k^2 - 480k^6 - 232k^7.$$

Theorem 13. *Let* $\mathcal{M} = J_w^n$, $n = 2w$ *(i.e. antipodal space). Then* $G_\tau(J_w^2, Q_{\tau+3}) < 0$ *for*

$$\tau + 3 \le n \le \lfloor \alpha_{even} \rfloor \ \text{if} \ \tau = 2k.$$
$$\tau + 3 \le n \le \lfloor \alpha_{odd} \rfloor \ \text{if} \ \tau = 2k + 1,$$

where α_{odd} *and* α_{even} *are the greatest zeros of the polynomials* $S_{odd}(w)$ *and* $S_{even}(w)$ *respectively.*

Applying Theorem 5 and Theorem 7 we improve the Delsarte bound in $Q-$ polynomial Johnson space.

Theorem 14. *For a code* $C \subseteq J_w^n$ *we have the following bound*

$$S(J_w^n, \tau; \tau + \epsilon) \le |C|,$$

where $\epsilon = 2$ *if* $n > 2w$ *(non-antipodal Johnson space) and* $\epsilon = 3$ *if* $n = 2w$ *(antipodal Johnson space).*

Finally for non-antipodal Johnson space we present the asymptotic behavior of the bound when τ grows with n and w.

Theorem 15. *Let* $\mathcal{M}$ *be non-antipodal Johnson space* $(n > 2w)$. *If* $\lim\limits_{n \to \infty} \dfrac{\tau}{n} = \dfrac{\delta_1}{2}$, *and* $\lim\limits_{n \to \infty} \dfrac{\tau}{w} = \delta_2$ *where* $0 < \delta_1 < \delta_2 < 1$, *then*

$$S(\mathcal{M}, \tau) \sim D(\mathcal{M}, \tau) const(\delta_1, \delta_2).$$

The constant is bigger than 1 and is equal to
$const(\delta_1, \delta_2) = \frac{2}{\delta_2}$, when τ is odd and
$const(\delta_1, \delta_2) = \frac{(4 - \delta_1)\delta_2}{2\delta_1}$, when τ is even.

Acknowledgements. The authors would like to thank to Philipe Delsarte for the encouragement and for the helpful discussions and comments.

References

1. J.Bierbrauer, K.Gopalakrishnan, D.R.Stinson, A note on the duality of linear programming bounds for orthogonal arrays and codes, *Bulletin of the ICA*, 22, (1998), 17-24.
2. P.G.Boyvalenkov, D.P.Danev, On Linear Programming Bounds for the Codes in Polynomial Metric Spaces, *Problems of Information Transmission 2*, (1998), 108-120. (English translation from *Problemy Peredachi Informatsii*).
3. P.Delsarte, An Algebraic Approach to Association Schemes in Coding Theory, Philips Research Reports Suppl. 10, 1973.
4. P.Delsarte, Application and Generalization of the MacWilliams transform in Coding Theory, *Proc. 15th Symp. on Inform. Theory in Benelux*, Louvain-la-Neuve, Belgium, (1994), 9-44.

5. P.Delsarte, J.M.Goethals, J.J.Seidel, Spherical codes and designs, *Geometricae Dedicata* 6, (1977), 363-388.

6. C.F.Dunkl, Discrete quadrature and bounds on *t*-designs, *Mich. Math. J.* 26, (1979), 81-102.

7. R.W.Hamming, Error detecting and error correcting codes, *Bell Syst. Tech. J.*, 29, (1950), 147-160.

8. S.G.Hoggar, *t*-designs in projective spaces, *Europ. J. Combin.* 3, (1982), 233-254.

9. V.I.Levenshtein, On choosing polynomials to obtain bounds in packing problems, In *Proc. Seventh All-Union Conf. on Coding Theory and Information Transmission* Part II, Moscow, Vilnus, (1978), 103-108 (in Russian).

10. V.I.Levenshtein, Designs as Maximum Codes in Polynomial Metric Spaces, *Acta Applicandae Mathematicae* 29, (1992), 1-82.

11. V.I.Levenshtein, Packing and Decomposition Problems for Polynomial Association Schemes, *Europ. J. Combinatorics* 14, (1993), 461-477.

12. V.I.Levenshtein, Krawtchouk Polynomials and Universal Bounds for Codes and Designs in Hamming Spaces, *IEEE Transactions on Information Theory* 41 (5), (1995), 1303-1321.

13. V.I.Levenshtein, On designs in compact metric spaces and a universal bound on their size, *Discrete Mathematics* 192, (1998), 251-271.

14. V.I.Levenshtein, Universal bounds for codes and designs, Chapter 6 in *Handbook of Coding Theory*, ed. V.Pless and W.C.Huffman, 1998, Elsevier Science B.V., 449-648.

15. V.I.Levenshtein, Equivalence of Delsarte's bounds for codes and designs in symmetric association schemes, and some applications, *Discrete Mathematics* 197/198, (1999), 515-536.

16. F.J.MacWilliams and N.J.A.Sloane, *The Theory of Error-Correcting Codes*, North Holland, Amsterdam, 1977.

17. R.J.McElliece, E.R.Rodemich, H.Rumsey and L.R.Welch, New Upper Bounds on the rate of a code via the Delsarte-MacWilliams Inequalities, *IEEE Transactions on Information Theory* 23 (2), (1977), 157-166.

18. S.I.Nikova, V.S.Nikov, Improvement of the Delsarte bound for τ-designs when it is not the best bound possible, submitted in *Designs Codes and Cryptography*.

19. C.R. Rao, Factorial experiments derivable from combinatorial arrangements of arrays, *J.Roy. Stat. Soc.* 89, (1947), 128-139.

20. D.K.Ray-Chaudhuri, R.M.Wilson, On *t*-designs, *Osaka J. Math.* 12, (1975), 737-744.

21. R.C.Singleton, Maximum distance *q*-ary codes, *IEEE Trans. Inform. Theory* 10, (1964), 116-118.

22. V.M.Sidel'nikov, Extremal polynomials used in bounds of code volume, *Problemy Peredachi Informatsii* 16 (3), (1980), 17-30 (in Russian). English translation in *Problems on Information Transmission* 16 (3), (1980), 174-186.

23. I.Schoeneberg, G.Szegö, An extremum problem for polynomials, *Composito Math.* 14, (1960), 260-268.

24. G.Szegö, *Orthogonal polynomials*, AMS Col. Publ., vol.23, Providence, RI, 1939.

Statistical Properties of Digital Piecewise Linear Chaotic Maps and Their Roles in Cryptography and Pseudo-Random Coding

Shujun Li[1a], Qi Li[2], Wenmin Li[3], Xuanqin Mou[1b], and Yuanlong Cai[1c]

[1] Institute of Image Processing, School of Electronics and Information Engineering, Xi'an Jiaotong University, Xi'an, Shaanxi 710049, P. R. China
hooklee@mail.com[1a], {xqmou[1b],ylcai[1c]}@mail.xjtu.edu.cn
[2] Department of Electrical Engineering and Electronics, The University of Liverpool, Brownlow Hill, Liverpool L69 3GJ, UK
andyli@liv.ac.uk
[3] Department of Electircal and Electronic Engeering, Imperial College, Exhibition Road, London SW7 2BT, UK
wen.li@ic.ac.uk

Abstract. The applications of digital chaotic maps in discrete-time chaotic cryptography and pseudo-random coding are widely studied recently. However, the statistical properties of digital chaotic maps are rather different from the continuous ones, which impedes the theoretical analyses of the digital chaotic ciphers and pseudo-random coding. This paper detailedly investigates the statistical properties of a class of digital piecewise linear chaotic map (PLCM), and rigorously proves some useful results. Based on the proved results, we further discuss some notable problems in chaotic cryptography and pseudo-random coding employing digital PLCM-s. Since the analytic methods proposed in this paper can essentially extended to a large number of PLCM-s, they will be valuable for the research on the performance of such maps in chaotic cryptography and pseudo-random coding.

1 Introduction

Chaotic systems have many interesting properties, such as the sensitive dependence on initial conditions and control parameters, ergodicity, mixing and exactness properties, etc. [1]. Most properties can be connected with some requirements in cryptography and pseudo-random coding [2, 3, 4]. From 1990s, more and more researchers devote their contributions to a new field – chaotic cryptography; many analog and digital chaotic encryption systems have been proposed [2, 5, 6, 7, 3, 8, 9] and analysed [10, 11, 12]. As a general method to design chaotic stream ciphers, chaotic pseudo-random coding techniques are commonly used to construct PRBG-s (Pseudo-Random Bits Generators) [5, 6, 9]. At the same time, chaotic pseudo-random coding techniques have also developed separately in other areas, such as electronics, communications [13, 14, 15] and computer physics [16].

B. Honary (Ed.): Cryptography and Coding 2001, LNCS 2260, pp. 205–221, 2001.
© Springer-Verlag Berlin Heidelberg 2001

As we know, piecewise linear chaotic maps (PLCM) are the simplest kind of chaotic maps from the viewpoint of realization. What's more, they have uniform invariant density and good correlation functions [17], which is very useful for cryptography and pseudo-random coding [18]. In fact, many researchers have used them to realize chaotic ciphers and PRBG-s [14, 9, 8, 6, 7].

It seems that chaotic systems are perfect as a new rich source of cryptography and pseudo-random coding. Unfortunately, when chaotic systems are realized in finite computing precision, their digital dynamical properties will be far different from the continuous ones. Some severe problems will arise, such as short cycle length, non-ideal distribution and correlation functions, etc. Assume the finite precision is L (bits) and fixed-point arithmetic is adopted, it is the following reasons to cause such degradation: 1) All values represented with finite precision are binary rational decimals formulated as $a/2^L (a = 0 \sim 2^L - 1)$. Since the Lebesgue measure of all the decimals is zero, they cannot represent the right dynamical behaviors of the chaotic systems defined on a real interval with positive measure; 2) There are only 2^L digital values to represent the chaotic orbits, so the cycle length of the orbits will not be larger than 2^L, generally it will be much smaller than 2^L; 3) The quantization errors, which are introduced into the iterations of chaotic systems, will make the chaotic orbits depart from the theoretical ones with uncontrolled manners (it is impossible to know the exact errors).

Some researchers have noticed the degradation of digital chaotic systems [10, 11, 9, 19, 20, 13], and several remedies have been suggested: using higher finite precision [11, 19], the perturbation-based algorithm [9, 20], and cascading multiple chaotic systems [13]. Because it is difficult to measure the statistical properties of digital chaotic maps theoretically, experiments are generally used as the analytic tools to estimate the performance of the above remedies. However, sometimes experiments are not enough to tell us the right things about digital chaotic systems. The theoretical tools for digital chaotic systems are needed.

2   Outline of Our Works

In this paper, we strictly prove some interesting statistical properties about a class of digital PLCM with finite computing precision. Based on our proved results, we can explain some statistical degradation of digital PLCM-s theoretically. Such degradation will cause the chaotic ciphers insecure, and cause chaotic pseudo-random sequences unbalanced. Furthermore, we discuss the performance of the three proposed remedies, and point out none of them can essentially improve such degradation. But the perturbation-based algorithm is still useful in practice, since it can be carefully used to enhance the performance of digital chaotic ciphers and pseudo-random coding.

For other digital chaotic maps, we have not yet obtained exact corresponding results. But our proof techniques may probably be extended to many other digital chaotic maps conceptually. If one chaotic map contains a control parameter that is proportional to uniformly distributed final output, the digital chaotic map may be weak from the viewpoint of this control parameter. In the future, we will try to find more delicate results.

This paper is organised as follows. In Sect. 3, we firstly introduce some preliminary knowledge. In the following Sect. 4, we focus on the mathematically rigorous proofs of the interesting properties of digital PLCM-s. Since the whole proof is rather lengthy, it is divided into several parts. Based on the proved properties, we explain what they mean in chaotic cryptography and pseudo-random coding in Sect. 5. A brief conclusion is given in the last section.

3 Preliminary Knowledge

3.1 Piecewise Linear Chaotic Map (PLCM)

Generally, given a real interval $X = [\alpha, \beta] \subset \mathbb{R}$, a piecewise linear chaotic map $F : X \to X$ is a multi-segmental map: $i = 1 \sim m, F(x)|C_i = F_i(x) = a_i x + b_i$, where $\{C_i\}_{i=1}^m$ is a partition of X, which satisfies $\bigcup_{i=1}^m C_i = X$ and $C_i \cap C_j = \varnothing, \forall i \neq j$. Each element of the partition is mapped to X by F_i: $\forall i = 1 \sim m, F_i : C_i \to X$. Such a map has the following statistical properties on its definition interval X: 1) it is chaotic, its Lyapunov exponent λ satisfies $0 < \lambda < \ln m$; 2) it is exact, mixing and ergodic, and has uniform invariant density function $f(x) = 1/(\beta - \alpha)$; 3) the correlation $\tau(n) = \frac{1}{\sigma^2} \lim_{N \to \infty} \frac{1}{N} \sum_{i=0}^{N-1} (x_i - \bar{x})(x_{i+n} - \bar{x})$ will go to zero as $n \to \infty$, where $\bar{x}, \sigma$ are the mean value and the variance of x respectively; especially, if some conditions are satisfied, $\tau(n) = \delta(n)$ [17,1].

As we know [1], the uniform invariant density function means that uniform input will generate uniform output, and that the chaotic orbit from almost every initial condition will lead to the same uniform distribution $f(x) = 1/(\beta - \alpha)$. But such a fact is not true for a digital chaotic map, this paper will point out that uniform digital input cannot generate uniform digital output for all control parameters. Such a fact will subsequently cause serious dynamical degradation when the maps are iterated again and again. Because it is inconvenient to analyse chaotic maps with uncertain formulas, in this paper, we focus our attention on the following specific PLCM used in [8]:

$$F(x, p) = \begin{cases} x/p, & x \in [0, p) \\ (x - p)/(1/2 - p), & x \in [p, 1/2] \\ F(1 - x, p), & x \in [1/2, 1] \end{cases} \tag{1}$$

where p is the control parameter, which satisfies $0 < p < 1/2$.

In order to facilitate the descriptions and proofs of the statistical properties in Sect. 4, we give some definitions in Sect. 3.2 and related results in Sect. 3.3.

3.2 Preliminary Definitions

Definition 1. *A discrete set* $S_n = \{a \mid a = \sum_{i=1}^n a_i \cdot 2^{-i}, a_i \in \{0, 1\}\}$ *is called a* ***digital set*** *with* ***resolution*** *n;* $\forall i < j$, S_i *is called the* ***digital subset*** *with* ***resolution*** *i of* S_j. *Specially, define* $S_0 = \{0\}, S_\infty = [0, 1)$, *then we have* $\{0\} = S_0 \subset S_1 \subset \ldots \subset S_i \subset \ldots \subset S_\infty = [0, 1)$.

Definition 2. *Define* $V_i = S_i - S_{i-1}(i \geq 1)$ *and* $V_0 = S_0$. $V_i(0 \leq i \leq n)$ *is called the **digital layer** with **resolution** i; $\forall p \in V_i$, i is called the **resolution** of p. The partition of S_n, $\{V_i\}_{i=0}^n$, is called the **complete multi-resolution decomposition** of S_n; $\{V_i\}_{i=0}^{\infty}$ is called the **complete multi-resolution decomposition** of $S_{\infty} = [0, 1)$. For S_n, its resolution n is also called **decomposition level**, $\bigcup_{i=0}^n V_i = S_n$, and $\forall i \neq j, V_i \cap V_j = \varnothing$.*

Definition 3. *$\forall n > m$, $D_{n,m} = S_n - S_m$ is called the **digital difference set** of the two digital sets with parameters n and m. $\{V_i\}_{i=m}^n = \{S_i - S_{i-1}\}_{i=m}^n$ is called the **complete multi-resolution decomposition** of $D_{n,m}$, $n - m + 1$ is called the **decomposition level**.*

Definition 4. *A function $G : \mathbb{R} \to \mathbb{Z}$ is called an **approximate transformation function (ATF)**, if $\forall x \in \mathbb{R}$, $|G(x) - x| < 1$. Three basic ATF-s are: 1) $\lfloor x \rfloor$ – the maximal integer not greater than x; 2) $\lceil x \rceil$ – the minimal integer not less than x; 3) round(x) – the rounded integer of x. $\forall x \in \mathbb{R}$, define its **decimal part** $x - \lfloor x \rfloor$ as function **dec(x)**. The above **three** ATF-s have the following useful properties (please note **not all** ATF-s):*

$$\text{ATF Property 1} : \forall m \in \mathbb{Z}, G(x + m) = G(x) + m; \tag{2}$$
$$\text{ATF Property 2} : a < x < b \Rightarrow \lfloor x \rfloor \leq G(x) \leq \lceil x \rceil. \tag{3}$$

The proofs of the two properties are rather simple, we omit them here.

Definition 5. *A function $G_n : S_{\infty} \to S_n$ is called a **digital approximate transformation function (DATF)** with **resolution** n, if $\forall x \in S_{\infty} = [0, 1)$, $|G_n(x) - x| < 1/2^n$. The following three DATF-s are concerned in this paper (they are also the most frequently adopted DATF-s in digital computing algorithms): 1) $\text{floor}_n(x) = \lfloor x \cdot 2^n \rfloor / 2^n$; 2) $\text{ceil}_n(x) = \lceil x \cdot 2^n \rceil / 2^n$; 3) $\text{round}_n(x) = \text{round}(x \cdot 2^n)/2^n$.[1] The above **three** DATF-s have the following useful properties (please note **not all** DATF-s):*

$$\text{DATF Property 1} : \forall m \in \mathbb{Z}, G_n(x + m/2^n) = G_n(x) + m/2^n; \tag{4}$$
$$\text{DATF Property 2} : a < x < b \Rightarrow \text{floor}_n(a) \leq G_n(x) \leq \text{ceil}_n(b). \tag{5}$$

The two properties are easily derived from the ATF Property 1–2.

3.3   Preliminary Lemmas about the Three Basic ATF-s

For the three basic ATF-s – $\lfloor \cdot \rfloor$, $\lceil \cdot \rceil$ and round$(\cdot)$, we have two fundamental lemmas and one corollary, which will be useful in the proofs of the theorems in the next section.

---

[1] Consider $1 \notin S_{\infty}$, without loss of generality, define $\text{ceil}_n(x) = 0$ if $\lceil x \cdot 2^n \rceil = 2^n$, and define $\text{round}_n(x) = 0$ if $\text{round}(x \cdot 2^n) = 2^n$. Such redefinitions will not essentially influence the following results since $\text{dec}(1) = 0$.

Lemma 1. $\forall n \in \mathbb{Z}^+, a \geq 0$, the following three facts are true:

1. $n \cdot \lfloor a \rfloor \leq \lfloor n \cdot a \rfloor \leq n \cdot \lfloor a \rfloor + (n-1)$, and $n \cdot \lfloor a \rfloor = \lfloor n \cdot a \rfloor$ when and only when $\mathrm{dec}(a) \in \left[0, \frac{1}{n}\right)$;

2. $n \cdot \lceil a \rceil - (n-1) \leq \lceil n \cdot a \rceil \leq n \cdot \lceil a \rceil$, and $n \cdot \lceil a \rceil - (n-1) = \lceil n \cdot a \rceil$ when and only when $\mathrm{dec}(a) \in \left(1 - \frac{1}{n}, 1\right) \bigcup \{0\}$;

3. $n \cdot \mathrm{round}(a) - \lfloor n/2 \rfloor \leq \mathrm{round}(n \cdot a) \leq n \cdot \mathrm{round}(a) + \lfloor n/2 \rfloor$, and $n \cdot \mathrm{round}(a) - \lfloor n/2 \rfloor = \mathrm{round}(n \cdot a)$ when and only when $\mathrm{dec}(a) \in \left[0, \frac{1}{2n}\right) \bigcup \left[1 - \frac{1}{2n}, 1\right)$.

The proof of this lemma is given in Appendix A.

Corollary 1. $\forall n \in \mathbb{Z}^+, a \geq 0$, we have the following results:

1. $\lfloor n \cdot a \rfloor \equiv 0 \ (\mathrm{mod} \ n)$ when and only when $\mathrm{dec}(a) \in \left[0, \frac{1}{n}\right)$;

2. $\lceil n \cdot a \rceil \equiv 0 \ (\mathrm{mod} \ n)$ when and only when $\mathrm{dec}(a) \in \left(1 - \frac{1}{n}, 1\right) \bigcup \{0\}$;

3. $\mathrm{round}(n \cdot a) \equiv 0 \ (\mathrm{mod} \ n)$ when and only when $\mathrm{dec}(a) \in \left[0, \frac{1}{2n}\right) \bigcup \left[1 - \frac{1}{2n}, 1\right)$.

Proof. This corollary can be derived directly from the above lemma.

Lemma 2. $\forall j, N, N' \in \mathbb{Z}^+$, and N, N' are odd integers satisfying $2^j | (N + N')$, we have $\lfloor N/2^j \rfloor + \lfloor N'/2^j \rfloor = (N + N')/2^j - 1$.

The proof of this lemma is given in Appendix B.

4   Statistical Properties of Digital PLCM

Give a one-dimensional chaotic map $F(x, p) : I \to I$, where $I = S_\infty = [0, 1)$. When the finite precision is n, its digital version can be expressed by $F_n(x, p) = G_n \circ F(x, p) : S_n \to S_n$, where $G_n(\cdot)$ is a DATF, $\mathrm{floor}_n(\cdot)$, $\mathrm{ceil}_n(\cdot)$ or $\mathrm{round}_n(\cdot)$. Denote the corresponding ATF of $G_n(\cdot)$ as $G_0(\cdot)$.

Assume P_j denotes the probability of the lowest j bits of $F_n(x, p)$ are all zeros, i.e., the probability of $F_n(x, p)$ belongs to S_{n-j}: $P_j = P\{F_n(x, p) \in S_{n-j}\}$. For the map denoted by $(1)^2$, $\forall p \in V_i \subset S_i \subseteq S_n (2 \leq i \leq n)$, we can deduce some interesting results about $P_j (1 \leq j \leq n)$, which are rather different from the expected ones based on the perfect continuous statistical properties of the map. Moreover, the results can be essentially extended to all digital PLCM-s described in Sect. 3.1.

Because the whole proof is rather lengthy, we divide it into several parts: firstly a fundamental lemma, then the results about $P_j (i \leq j \leq n)$ and the ones about $P_j (1 \leq j < i)$, finally two comprehensive theorems.

---

[2] Because $1 \notin S_\infty$, redefine $F_n(1/2, p) = 0$. Consider $F^2(1/2, p) = 0$ and $\mathrm{dec}(1) = 0$, such redefinition will not essentially influence the following results.

4.1   A Fundamental Lemma

Firstly, we introduce Lemma 3, which gives some useful results about the highest $n - i$ bits and the lowest i bits of $F_n(x, p)$. This lemma is the fundamental of the following proofs. At the same time, this lemma reflects some facts about the local linearity of the PLCM-s, which makes the obtained results in this paper conceptually available for other PLCM-s.

Lemma 3. $\forall p \in D_{i,0} = S_i - \{0\}(1 \leq i \leq n), x \in S_n$. Assume $p = N_p/2^i, x = N_x/2^n$, where N_p, N_x are integers satisfying $1 \leq N_p \leq 2^i - 1$ and $0 \leq N_x \leq 2^n - 1$. we have the following three results:

$$1. \ G_n(x/p) \in S_{n-i} \Leftrightarrow N_x \equiv 0 \ (\mathrm{mod} \ N_p), \tag{6}$$

$$2. \ \mathrm{floor}_{n-i}(G_n(x/p)) = \frac{\lfloor N_x/N_p \rfloor}{2^{n-i}}, \tag{7}$$

$$3. \ G_n(x/p) \bmod \frac{1}{2^{n-i}} = \frac{G_0(2^i \cdot (N_x \bmod N_p)/N_p)}{2^n}. \tag{8}$$

Proof. Because $x/p = \dfrac{N_x/2^n}{N_p/2^i} = \dfrac{N_x/N_p}{2^{n-i}} = \dfrac{\lfloor N_x/N_p \rfloor + (N_x \bmod N_p)/N_p}{2^{n-i}}$, we

have $G_n(x/p) = \dfrac{G_0(2^i \cdot \lfloor N_x/N_p \rfloor + 2^i \cdot (N_x \bmod N_p)/N_p)}{2^n}$. From *ATF Property 1*, we can rewrite $G_n(x/p)$ as follows

$$G_n(x/p) = \frac{\lfloor N_x/N_p \rfloor}{2^{n-i}} + \frac{G_0(2^i \cdot (N_x \bmod N_p)/N_p)}{2^n}. \tag{9}$$

Let us discuss the above equation under the following two conditions:

a) When $N_x \bmod N_p = 0$: $G_n(x/p) = \dfrac{\lfloor N_x/N_p \rfloor}{2^{n-i}} + 0 \in S_{n-i}$;

b) When $N_x \bmod N_p = k \neq 0$: Obviously $1 \leq k \leq N_p - 1$. Considering $p < 1$, we have $2^i/N_p > 1$, then $1 < 2^i \cdot (N_x \bmod N_p)/N_p < 2^i - 1$. Thus, from *ATF Property 2*, $1 \leq G_0(2^i \cdot (N_x \bmod N_p)/N_p) \leq 2^i - 1$. Therefore,

$$\frac{\lfloor N_x/N_p \rfloor}{2^{n-i}} + \frac{1}{2^n} \leq G_n(x, p) \leq \frac{\lfloor N_x/N_p \rfloor}{2^{n-i}} + \frac{2^i - 1}{2^n} \Rightarrow G_n(x, p) \notin S_{n-i}. \tag{10}$$

From a) and b), we can deduce $G_n(x/p) \in S_{n-i} \Leftrightarrow N_x \equiv 0 \ (\mathrm{mod} \ N_p)$.

At the same time, when $N_x \bmod N_p = 0$, $\mathrm{floor}_{n-i}(G_n(x/p)) = \dfrac{\lfloor N_x/N_p \rfloor}{2^{n-i}}$;

when $N_x \bmod N_p = k \neq 0$, $\mathrm{floor}_{n-i}(G_n(x/p)) \geq \dfrac{\lfloor \lfloor N_x/N_p \rfloor + 1/2^i \rfloor}{2^{n-i}} = \dfrac{\lfloor N_x/N_p \rfloor}{2^{n-i}}$

and $\mathrm{floor}_{n-i}(G_n(x/p)) \leq \dfrac{\lfloor \lfloor N_x/N_p \rfloor + (2^i - 1)/2^i \rfloor}{2^{n-i}} = \dfrac{\lfloor N_x/N_p \rfloor}{2^{n-i}}$, so finally we

can get $\mathrm{floor}_{n-i}(G_n(x/p)) = \dfrac{\lfloor N_x/N_p \rfloor}{2^{n-i}}$.

From the above result and (9), the following result is true:

$$G_n(x/p) \bmod \frac{1}{2^{n-i}} = \frac{G_0(2^i \cdot (N_x \bmod N_p)/N_p)}{2^n}.$$

The proof is complete.

4.2   Results about $P_j (i \le j \le n)$

Theorem 1. *Assume random variable x distributes uniformly in S_n, for the digital PLCM (1), $\forall p \in D_{i,1}(2 \le i \le n)$, we have: $P_i = 4/2^i$.*

Proof. Assume $p = N_p/2^i, x = N_x/2^n$, where N_p, N_x are integers that satisfy $1 \le N_p \le 2^i - 1$ and $0 \le N_x \le 2^n - 1$. Because x distributes uniformly in S_n, N_x will distribute uniformly in integer set $[0, 2^n - 1]$. Since the chaotic map is defined piecewisely, we consider it on different segments:

a) $x \in [0, p) \Rightarrow N_x \in [0, 2^{n-i} \cdot N_p - 1]$: $F_n(x, p) = G_n(x/p)$, from Lemma 3, we know $F_n(x, p) \in S_{n-i}$ when and only when $N_x \equiv 0 \pmod{N_p}$. Because N_x distributes uniformly in $[0, 2^n - 1]$, the probability of $F_n(x, p) \in S_{n-i}$ will be $2^{n-i}/(2^{n-i} \cdot N_p) = 1/N_p$. That is to say, $P_i | x \in [0, p) = 1/N_p$.

b) $x \in [p, 1/2)$: Assume $x' = x - p, p' = 1/2 - p$, we have $F_n(x, p) = x'/p'$, where $x' \in [0, p')$. Similarly to a), define $p' = N'_p/2^i, x' = N'_x/2^n$, we will get $P_i | x \in [p, 1/2) = P_i | x' \in [0, p') = 1/N'_p$.

c) $x \in [1/2, 1)$: Consider the map is even symmetric to $x = 1/2$, we can easily get the following two results: $P_i | x \in (1/2, 1 - p] = 1/N'_p$ and $P_i | x \in ((1 - p, 1) \cup \{1/2\}) = 1/N_p$. Here please note that $1 \notin S_n$ and $1/2$ takes its position that is symmetrical to 0, which will not make any difference to P_i.

From a) – c) and the total probability rule, we can deduce:

$$
\begin{aligned}
P_i &= P(x \in [0, p)) \cdot P_i | x \in [0, p) + P(x \in [p, 1/2)) \cdot P_i | x \in [p, 1/2) \\
&\quad + P(x \in (1/2, 1 - p]) \cdot P_i | x \in (1/2, 1 - p] \\
&\quad + P(x \in ((1 - p, 1) \cup \{1/2\})) \cdot P_i | x \in ((1 - p, 1) \cup \{1/2\}) \\
&= p \cdot \frac{1}{N_p} + p' \cdot \frac{1}{N'_p} + p' \cdot \frac{1}{N'_p} + p \cdot \frac{1}{N_p} = \frac{1}{2^i} + \frac{1}{2^i} + \frac{1}{2^i} + \frac{1}{2^i} = \frac{4}{2^i} \ .
\end{aligned}
$$

The proof is complete. $\quad\blacksquare$

Theorem 2. *Assume random variable x distributes uniformly in S_n, for the digital PLCM (1), $\forall p \in D_{i,1}(2 \le i \le n)$, floor$_{n-i}(F_n(x, p))$[3] distributes uniformly in S_{n-i}.*

Proof. Similarly to the proof of Theorem 1, assume $p = N_p/2^i, x = N_x/2^n$, we separately consider the map on different segments:

a) $x \in [0, p) \Rightarrow N_x \in [0, 2^{n-i} \cdot N_p - 1]$: $F_n(x, p) = G_n(x/p)$, from Lemma 3, we have floor$_{n-i}(F_n(x, p)) = \lfloor N_x/N_p \rfloor / 2^{n-i}$. Because x distributes uniformly in S_n, N_x distributes uniformly in $[0, 2^{n-i} \cdot N_p - 1]$. Thus $\lfloor N_x/N_p \rfloor$ distributes uniformly in $[0, 2^{n-i} - 1]$, i.e., floor$_{n-i}(F_n(x, p))$ distributes uniformly in S_{n-i} when $x \in [0, p)$.

b) $x \in [p, 1/2)$: Assume $x' = x - p, p' = 1/2 - p$, we have $F_n(x, p) = x'/p'$, where $x' \in [0, p')$. Similarly to a), we can prove floor$_{n-i}(F_n(x, p))$ distributes uniformly in S_{n-i} when $x \in [p, 1/2)$.

---

[3] The highest $n - i$ bits of $F_n(x, p)$.

c) $x \in [1/2, 1)$: Because the map is even symmetrical to $x = 1/2$, it can be easily deduced that $\text{floor}_{n-i}(F_n(x, p))$ distributes uniformly in S_{n-i} when $x \in [1/2, 1)$.

From a) – c), we know it is true that $\text{floor}_{n-i}(F_n(x, p))$ distributes uniformly in S_{n-i}. The proof is complete.

Theorem 3. *Assume random variable x distributes uniformly in S_n, for the digital PLCM (1), $\forall p \in D_{i,1}(2 \le i \le n)$ and $i \le j \le n$, $P_j = 4/2^j$ holds.*

Proof. Let us discuss the different conditions when $j = i$ and $j > i$.

a) $j = i$: From Theorem 1, $P_j = 4/2^i = 4/2^j$;

b) $i < j \le n$: Assume $b_m(m = 1 \sim n)$ represents the mth bit (from the lowest bit to the highest one) of $F_n(x, p)$, $P_j = P\left\{ F_n(x, p) \in S_{n-i} \wedge b_j \overbrace{\cdots b_{i+1} = 0 \cdots 0}^{j-i} \right\}$.

Recall the proof of Theorem 3, when $F_n(x, p) \in S_{n-i}$ (i.e., $N_x \mod N_p = 0$), $\lfloor N_x / N_p \rfloor$ (the highest $n - i$ bits of $F_n(x, p)$) still distributes uniformly in $[0, 2^{n-i} - 1]$. So we can get $P_j = P\{F_n(x, p) \in S_i\} \cdot \frac{1}{2^{j-i}} = \frac{4}{2^i} \cdot \frac{1}{2^{j-i}} = \frac{4}{2^j}$.

From a) and b), we have: $i \le j \le n \Rightarrow P_j = 4/2^j$. The proof is complete.

4.3   Results about $P_j (1 \le j < i)$

Firstly, we introduce Lemma 4 and Corollary 2, which will be used to facilitate the proof of Theorem 4.

Lemma 4. *Assume n is an odd integer, random integer variable K distributes uniformly in $\mathbb{Z}_n = [0, n - 1]$, the following fact is true: $K' = f(K) = (2^i \cdot K) \mod n$ distributes uniformly in $\mathbb{Z}_n$, i.e., $\forall k \in [0, n - 1], P\{K' = k\} = 1/n$.*

Proof. As we know, $(\mathbb{Z}_n, +)$ is a finite cyclic group of degree n, and a is its generator when and only when $\gcd(a, n) = 1$, where "+" is defined as "$(a + b) \mod n$" (see Theorem 2 on page 60 of [21]). Therefore, $a = 2^i \mod n$ is one generator of $\mathbb{Z}_n$ since $\gcd(a, n) = \gcd(2^i, n) = 1$. Consider $K' = (2^i \cdot K) \mod n = (a \cdot K) \mod n$, we can see $f : \mathbb{Z}_n \to \mathbb{Z}_n$ is a bijection. Then we will immediately deduce: $K' = f(K)$ distributes uniformly in $\mathbb{Z}_n$ because K distributes uniformly in $\mathbb{Z}_n$. That is to say, $\forall k \in [0, n - 1], P\{K' = k\} = 1/n$. The proof is complete.

Corollary 2. *Assume n is an odd integer, random integer variable K distributes uniformly in $\mathbb{Z}_n = [0, n - 1]$. Then $\dec(2^i \cdot K/n)$ distributes uniformly in $S = \{x | x = k/n, k \in \mathbb{Z}_n\}$.*

Proof. This corollary is the straightforward result of the above lemma.

Theorem 4. *Assume random variable x distributes uniformly in S_n, for the digital PLCM (1), $\forall p \in V_i(2 \le i \le n)^4$ and $1 \le j \le i - 1$, we have:*

---

[4] Please note the condition $p \in V_i$, **NOT** $p \in D_{i,1}$ in Theorem 1–3.

$$P_j = \begin{cases} 1/2^j + 2/2^i & , \quad G_n(\cdot) = \text{floor}_n(\cdot) \text{ or } \text{ceil}_n(\cdot) \\ 1/2^j, \ 1 \le j \le i-2 \\ 4/2^i, \quad j = i-1 \end{cases}, \quad G_n(\cdot) = \text{round}_n(\cdot)$$

Proof. $p = N_p/2^i, x = N_x/2^n$, where N_p, N_x are integers that satisfy $1 \le N_p \le 2^i - 1$ and $0 \le N_x \le 2^n - 1$. Because x distributes uniformly in S_n, N_x will distribute uniformly in integer set $[0, 2^n - 1]$. Let us consider the digital map on different segments:

a) $x \in [0, p) \Rightarrow N_x \in [0, 2^{n-i} \cdot N_p - 1]$: $F_n(x, p) = G_n(x/p)$, from Lemma 3, we know the lowest i bits of $F_n(x, p)$ are determined by $G_0(2^i \cdot (N_x \bmod N_p)/N_p)$. Then we can deduce $F_n(x, p) \in S_{n-j} \Leftrightarrow G_0(2^i \cdot (N_x \bmod N_p)/N_p) \equiv 0 \bmod 2^j$. Define $\hat{N} = N_x \bmod N_p$, which distributes uniformly in $[0, N_p - 1]$ because of the uniform distribution of N_x. Define $a = (2^{i-j} \cdot \hat{N})/N_p$, we can re-write $G_0(2^i \cdot (N_x \bmod N_p)/N_p)$ as $G_0(2^j \cdot a)$. From Corollary 1, we can get:

$$G_0(2^j \cdot a) \equiv 0 \ (\bmod \ 2^j) \Leftrightarrow \text{dec}(a) \in \begin{cases} \left[0, \frac{1}{2^j}\right) & , G_0(\cdot) = \lfloor \cdot \rfloor \\ \left(1 - \frac{1}{2^j}, 1\right) \cup \{0\} & , G_0(\cdot) = \lceil \cdot \rceil \\ \left[0, \frac{1}{2^{j+1}}\right) \cup \left[1 - \frac{1}{2^{j+1}}, 1\right), G_0(\cdot) = \text{round}(\cdot) \end{cases}$$

(11)

From Corollary 2 (please note $p \in V_i$ ensures N_p is an odd integer), we know

$$\text{dec}(a) = k/N_p(k = 0 \sim N_p - 1) \text{with uniform probability.} \quad (12)$$

Based on (11) and (12), we can deduce:

$$k \in \begin{cases} \left[0, \frac{N_p}{2^j}\right) & , G_0(\cdot) = \lfloor \cdot \rfloor \\ \left(N_p - \frac{N_p}{2^j}, N_p\right) \cup \{0\} & , G_0(\cdot) = \lceil \cdot \rceil \\ \left[0, \frac{N_p}{2^{j+1}}\right) \cup \left[N_p - \frac{N_p}{2^{j+1}}, N_p\right), G_0(\cdot) = \text{round}(\cdot) \end{cases}$$

(13)

Consider k is an integer, we can get the probability

$$P\{G_0(2^j \cdot a) \equiv 0 \ (\bmod \ 2^j)\} = \begin{cases} \dfrac{\lfloor N_p/2^j \rfloor + 1}{N_p} & , \quad G_0(\cdot) = \lfloor \cdot \rfloor \text{ or } \lceil \cdot \rceil \\ \dfrac{2 \cdot \lfloor N_p/2^{j+1} \rfloor + 1}{N_p}, & G_0(\cdot) = \text{round}(\cdot) \end{cases}$$

(14)

b) $x \in [p, 1/2)$: Assume $x' = x - p, p' = 1/2 - p$, we have $F_n(x, p) = x'/p'$, where $x' \in [0, p')$. Similarly to a), define $p' = N_p'/2^i, x' = N_x'/2^n$, we will get

$$P\{G_0(2^j \cdot a') \equiv 0 \ (\bmod \ 2^j)\} = \begin{cases} \dfrac{\lfloor N_p'/2^j \rfloor + 1}{N_p'} & , \quad G_0(\cdot) = \lfloor \cdot \rfloor \text{ or } \lceil \cdot \rceil \\ \dfrac{2 \cdot \lfloor N_p'/2^{j+1} \rfloor + 1}{N_p'}, & G_0(\cdot) = \text{round}(\cdot) \end{cases}$$

(15)

where $a' = (2^{i-j} \cdot \hat{N}')/N_p', \hat{N}' = N_x' \bmod N_p'$.

From (14) and (15), we can get the conditional probability $P_j | x \in [0, 1/2)$. Consider the map is even symmetrical to $x = 1/2$, the final probability will be $P_j = 2 \cdot (P_j | x \in [0, 1/2))$. In the following, we separately consider the condition of $G_n(\cdot) = \text{floor}_n(\cdot)$ or $\text{ceil}_n(\cdot)$ and $G_n(\cdot) = \text{round}_n(\cdot)$:

i) $G_n(\cdot) = \text{floor}_n(\cdot)$ or $\text{ceil}_n(\cdot)$, i.e., $G_0(\cdot) = \lfloor \cdot \rfloor$ or $\lceil \cdot \rceil$: $p + p' = 1/2 \Rightarrow N_p + N_p' = 2^{i-1} \Rightarrow 2^j | (N_p + N_p')$, from Lemma 2, we can deduce:

$$P_j = 2 \left(p \cdot \frac{\lfloor N_p/2^j \rfloor + 1}{N_p} + p' \cdot \frac{\lfloor N_p'/2^j \rfloor + 1}{N_p'} \right)$$
$$= 2 \left(\frac{\lfloor N_p/2^j \rfloor + \lfloor N_p'/2^j \rfloor + 2}{2^i} \right) = \frac{2^{i-j-1} - 1 + 2}{2^{i-1}} = \frac{1}{2^j} + \frac{2}{2^i} . \tag{16}$$

ii) $G_n(\cdot) = \text{round}_n(\cdot)$, i.e., $G_0(\cdot) = \text{round}(\cdot)$: When $j < i - 1$, $N_p + N_p' = 2^{i-1} \Rightarrow 2^{j+1} | (N_p + N_p')$, from Lemma 2, we can get:

$$P_j = 2 \left(p \cdot \frac{2 \cdot \lfloor N_p/2^{j+1} \rfloor + 1}{N_p} + p' \cdot \frac{2 \cdot \lfloor N_p'/2^{j+1} \rfloor + 1}{N_p'} \right)$$
$$= 2 \left(\frac{2 (\lfloor N_p/2^{j+1} \rfloor + \lfloor N_p'/2^{j+1} \rfloor) + 2}{2^i} \right) = \frac{2 (2^{i-j-2} - 1) + 2}{2^{i-1}} = \frac{1}{2^j} . \tag{17}$$

When $j = i - 1$, $N_p + N_p' = 2^{i-1} \Rightarrow 2^{j+1} \nmid (N_p + N_p')(j+1 = i > i-1)$, Lemma 2 cannot be used, but we can calculate the probability P_j by directly observing (14) and (15): $N_p < 2^i$, $N_p' < 2^i$, so $N_p/2^{j+1} < 1 \Rightarrow \lfloor N_p/2^{j+1} \rfloor = 0$, $N_p'/2^{j+1} < 1 \Rightarrow \lfloor N_p'/2^{j+1} \rfloor = 0$, then we have

$$P_j = 2 \left(p \cdot \frac{2 \cdot 0 + 1}{N_p} + p' \cdot \frac{2 \cdot 0 + 1}{N_p'} \right) = 2 \cdot \frac{2}{2^i} = \frac{4}{2^i} . \tag{18}$$

From (16) – (18), we can directly get the final result. The proof is complete.

4.4 Comprehensive Results about $P_j(1 \leq j \leq n)$

In the above subsections, we have separately proved the results about $P_j(i \leq j \leq n)$ and $P_j(1 \leq j < i)$ for any $p \in V_i \subset S_i \subseteq S_n(2 \leq i \leq n)$. To make the above "rough-and-tumble" results tidier, we rearrange them into two new theorems, which are easier to be understood and to be used in practice.

Theorem 5. *Assume random variable x distributes uniformly in S_n, $\forall p \in V_i(2 \leq i \leq n)$, the following results are true for the digital PLCM (1):*

1. When $G_n(\cdot) = \text{round}_n(\cdot)$, $P_j = \begin{cases} 4/2^j, & i \leq j \leq n \\ 4/2^i, & j = i - 1 \\ 1/2^j, & 1 \leq j \leq i - 2 \end{cases}$;

2. When $G_n(\cdot) = \text{floor}_n(\cdot)$ or $\text{ceil}_n(\cdot)$, $P_j = \begin{cases} 4/2^j, & i \leq j \leq n \\ 1/2^j + 2/2^i, & 1 \leq j \leq i - 1 \end{cases}$;

3. $\forall k \in [0, 2^{n-i} - 1]$, $P \{ \text{floor}_{n-i}(F_n(x, p)) = k/2^{n-i} \} = 1/2^{n-i}$.

Proof. the first two parts are the combinations of Theorem 3 and 4, the last part is just equivalent to Theorem 3.

Remark 1. If x distributes uniformly in the digital set S_n, $F_n(x,p)$ does not distribute uniformly in S_n (but its highest $n-i$ bits does in S_{n-i}, $\forall p \in S_i$), since $P_j = 1/2^j$ if $F_n(x,p)$ distributes uniformly in S_n. To understand what Theorem 5 really means, see Fig. 1 for more visual details.

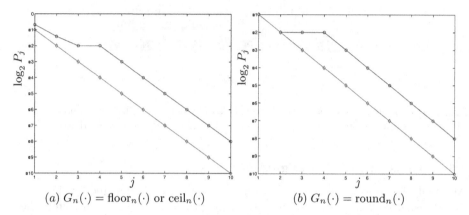

(a) $G_n(\cdot) = \text{floor}_n(\cdot)$ or $\text{ceil}_n(\cdot)$ (b) $G_n(\cdot) = \text{round}_n(\cdot)$

Fig. 1. $P_j(1 \leq j \leq n)$ when $p = 3/16 \in V_4 \subset S_4$, where the finite precision $n = 10$ (The line marked with diamond signs denotes the probability under digital uniform distribution $1/2^j$, and the other line denotes the probability P_j)

Remark 2. Note there is an absolutely weak control parameter $p = 1/4 \in V_2 \subset S_2$, which satisfies $P_2 = 4/2^2 = 1$. That is to say, the lowest 2 bits of $F_n(x,p)$ will always be zeros. In addition, $\forall x_0 \in V_i (2 \leq i \leq n)$, after at most $\lceil i/2 \rceil$ iterations, the chaotic orbit will converge at zero: $\forall m \geq \lceil i/2 \rceil, F^m(x_0) = 0$.

Theorem 6. *Assume random variable x distributes uniformly in S_n, and $P_i = P\{F_n(x,p) \in S_{n-i}\}$. The following results are true for the digital PLCM (1):*

1. $\forall p \in D_{i,1} = S_i - S_1 = \bigcup_{k=2}^i V_i, P_i = 4/2^i$;
2. $\forall p \in V_{i+1}, P_i = 4/2^{i+1}$;
3. $\forall p \in V_j (j \geq i + 2), P_i = \begin{cases} 1/2^i & , G_n(\cdot) = \text{round}_n(\cdot) \\ 1/2^i + 2/2^j & , G_n(\cdot) = \text{floor}_n(\cdot) \text{ or } \text{ceil}_n(\cdot) \end{cases}$.

Proof. This theorem is an equivalent form of Theorem 5.

Remark 3. Theorem 6 tells us: for the control parameters p with different resolution (i.e., in different digital layers of $D_{n,1}$), rather large difference exists in the generated chaotic orbits. Hence, from the observation of $P_1 \sim P_n$, one can get the resolution of the control parameter p. In Fig. 2, we give the experimental result of P_5 with respect to p when $n = 10$ and $G_n(\cdot) = \text{floor}_n(\cdot)$, which entirely coincides with Theorem 6.

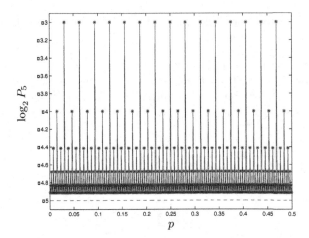

Fig. 2. $P_5 = P\{F_n(x, p) \in S_{n-5}\}$ with respect to p, where $n = 10, G_n(\cdot) = \text{floor}_n(\cdot)$ (The dashed line denotes 2^{-5}, the ideal probability under digital uniform distribution)

4.5   Extension to Other Digital PLCM-s

Although the above results are based on the specific PLCM denoted by (1), they can be essentially extended to all PLCM-s described in Sect. 3.1, of course the exact results will be different for different maps. From the proofs of theorems in above sub-sections, we can see that the statistical degradation occurs because of the piecewise linearity (Lemma 3 and 4) and the essential properties of the three ATF-s (Lemma 1 and 2). Employing Lemma 1–4 and Corollary 1–2 on other PLCM-s[5], we can easily obtain results corresponding to Theorem 5 and 6. For example, we can get the results about the following chaotic map:

$$F(x, p) = \begin{cases} x/p & , x \in [0, p) \\ (1 - x)/(1 - p), & x \in [p, 1] \end{cases}, \tag{19}$$

where p satisfies $0 < p < 1$. This map is one of the simplest PLCM-s, and generally called **tent map**.

Theorem 5'. *Assume random variable x distributes uniformly in S_n, $\forall p \in V_i (1 \leq i \leq n)$, the following results are true for digital tent map:*

1. When $G_n(\cdot) = \text{round}_n(\cdot)$, $P_j = \begin{cases} 2/2^j & , i \leq j \leq n \\ 2/2^i & , j = i - 1 \\ 1/2^{j-1}, & 1 \leq j \leq i - 2 \end{cases}$;

2. When $G_n(\cdot) = \text{floor}_n(\cdot)$ or $\text{ceil}_n(\cdot)$, $P_j = \begin{cases} 2/2^j & , i \leq j \leq n \\ 1/2^j + 1/2^i, & 1 \leq j \leq i - 1 \end{cases}$;

3. $\forall k \in [0, 2^{n-i} - 1], P\{\text{floor}_{n-i}(F_n(x, p)) = k/2^{n-i}\} = 1/2^{n-i}$.

Experiments show the results absolutely right. Of course there is the corresponding Theorem 6', we omit it here since it is just another form of Theorem 5'.

---

[5] Any PLCM defined on interval $[\alpha, \beta]$ can be re-scaled to its topologically conjugated PLCM defined on $[0, 1]$ with a linear function $h(x) = (x - \alpha)/(\beta - \alpha)$.

5   The Roles of Digital PLCM-s in Cryptography and Pseudo-Random Coding

From *remark 1*, we can know that a uniformly distributed digital signal will lead to non-uniform distribution after iterations of a digital PLCM. Such non-uniformity will become more and more severe as the iterations go, see Fig. 3 for some intuitional view (compare it with Fig. 2, the probability at most control parameters increases, and the probability at $p = 1/16$ even reaches to 1). We can use the probability P_i to denote the degree of such non-uniformity: for a fixed control parameter, the larger P_i is, the larger the degradation will be. In *remark 2*, $p = 1/4 \in V_2$ corresponds to the most serious degradation, so it is the weakest control parameter. The less weak control parameters are ones in V_3; then those in V_4, V_5, $\cdots$.

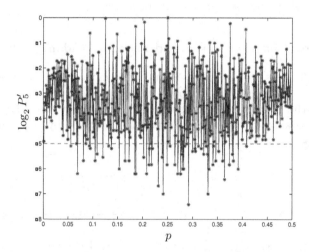

Fig. 3. $P'_5 = P\{F_n^{32}(x, p) \in S_{n-5}\}$ with respect to p, where $n = 10$, $G_n(\cdot) = \mathrm{floor}_n(\cdot)$ (P'_5 is the probability after 32 chaotic iterations of the digital PLCM (1), the dashed line denotes 2^{-5}, the ideal probability under digital uniform distribution)

5.1   Performance of the Three Remedies to Digital PLCM-s

In Sect. 1, we have mentioned three remedies proposed by other researchers. In this subsection, we discuss whether they will work well to improve the degradation of digital PLCM-s.

Apparently, cascading multiple digital chaotic maps cannot essentially improve the weaknesses, since multiple cascading PLCM-s are just equivalent to a new PLCM with more segments.

Using higher precision cannot change the weaknesses of any fixed control parameter either. For example, for the map (1), $p = 1/4$ will always be absolutely weak for any finite precision, and $\forall p \in V_i$ will always be same weak for any finite precision $n \geq i$. But higher precision will introduce more stronger digital layers[6] and then improve the overall weakness, which makes the condition better.

Now assume the perturbation-based algorithm is used to improve the degradation of digital PLCM-s. We find there exists a "strange" paradox: assume the chaotic orbit $\{x(m)\}_{m=1}^{\infty}$ is improved to obey nearly uniform by perturbation, according to Theorem 5 and 6, the chaotic sub-orbit $\{x(m)\}_{m=2}^{\infty}$ will not obey uniform distribution because $\{x(m)\}_{m=2}^{\infty} = \{F_n(x(m), p)\}_{m=1}^{\infty}$; thus $\{x(m)\}_{m=1}^{\infty}$ will not either. What does such a fact mean? It implies the non-uniformity revealed by the above theorems is the lower bound of the degradation of digital chaotic orbits. In other words, the perturbation-based algorithm cannot **essentially** improve the degradation to a better condition than the one depicted in Theorem 5 and 6. However, as we will point out in the next subsection, the perturbation-based algorithm is still useful to enhance the digital chaotic ciphers and pseudo-random coding with careful considerations.

5.2   Notes on Chaotic Ciphers and Pseudo-Random Coding

If the digital PLCM-s are directly used in chaotic ciphers and the control parameter are used as the secret key (as most chaotic ciphers do), the cryptographic properties of the ciphers will not be perfect, and many weak keys will arise (see Fig. 3), because of the severe degradation induced by the digital chaotic iterations.

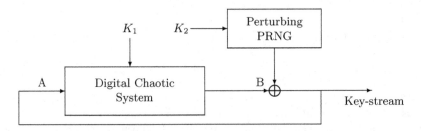

Fig. 4. Digital chaotic cipher with secretly exerted perturbation (The perturbation should be secretly exerted at position B not A)

To escape from such a bad condition and enhance the security, we suggest using the perturbation-based algorithm as follows: the perturbation is **secretly** exerted and the chaotic orbit is output **after** perturbation (See Fig. 4). It is based on the following fact: if $\{x(m)\}_{m=1}^{\infty}$ can be observed by one intruder, he will probably judge the resolution i of the right key through the probabilities $P_j(j = 1 \sim n)$ (see Theorem 6 and *Remark 3*), and then search the key only in

---

[6] When finite precision increases from n to n', $n' - n$ stronger digital layers $V_{n'-n+1} \sim V_{n'}$ will be added, although n old digital layers $V_1 \sim V_n$ remain.

the digital layer V_i that is smaller than the whole key space (the smaller i is, the faster the search will be and the weaker the key). If the perturbation is exerted secretly at point B, one intruder can only observe perturbed $\{x(m)\}_{m=1}^{\infty}$ not $\{x(m)\}_{m=1}^{\infty}$ itself, then it is relatively more difficult for him to get information about K_1 without knowing K_2. But it is obvious that K_1 will still be weak if K_2 is broken, and vice versa. It means the final key entropy will be smaller than the sum of the two sub ones: $H(K) = H((K_1, K_2)) < H(K_1) + H(K_2)$.

If the digital PLCM-s are used to generate pseudo-random bits, the generated binary sequences may be unbalanced since the chaotic orbits are not uniform. For example, if the map denoted by (1) with $p = 1/4$ is selected and the lowest 2 bits of chaotic orbit are used to generate pseudo-random bits, we can see they will be $000\cdots$. Fortunately, from Theorem 3, we can use the highest $n-i$ bits to construct desired pseudo-random bits. Here please note (approximately) uniform distribution of chaotic input is required. The perturbation-based algorithm will be useful for such a task.

6   Conclusion

We have rigorously proved some statistical properties of digital piecewise linear chaotic maps (PLCM) and explained their roles in chaotic cryptography and pseudo-random coding. Our works will be useful for the design and performance analyses of chaotic ciphers with theoretical security and PRBG-s with really good statistical properties.

For other chaotic maps, our results cannot straightforward be extended. But the proofs made in this paper depend on some essentially properties of ATF-s (Lemma 1 and 2) and the following fact: on every monotonic segment of digital chaotic maps, one control parameter is proportional to the uniformly distributed final output (Lemma 3 and 4). Consider the uniform final output is always desired for cryptography and pseudo-random coding, the proofs may be available for other digital chaotic maps that can be used in the two areas. In the future, we will try to find results concerning more generic digital chaotic maps.

Acknowledgement. The authors wish to thank Dr. Di Shuang-liang at Xi'an Jiaotong University for his valuable suggestions, and Miss Han Lu at Xi'an Foreign Language University for her help in the preparation of the final paper.

Appendix A: The Proof of Lemma 1

Proof. We prove the three sub-lemmas separately:

1. Because $a = \lfloor a \rfloor + \text{dec}(a)$, $n \cdot a = n \cdot \lfloor a \rfloor + n \cdot \text{dec}(a)$. Considering $0 \leq \text{dec}(a) < 1$, $0 \leq n \cdot \text{dec}(a) < n \Rightarrow 0 \leq \lfloor n \cdot \text{dec}(a) \rfloor \leq n - 1$. From the definition of $\lfloor \cdot \rfloor$, we can get $\lfloor n \cdot a \rfloor = \lfloor n \cdot (\lfloor a \rfloor + \text{dec}(a)) \rfloor = n \cdot \lfloor a \rfloor + \lfloor n \cdot \text{dec}(a) \rfloor \Rightarrow n \cdot \lfloor a \rfloor \leq \lfloor n \cdot a \rfloor \leq n \cdot \lfloor a \rfloor + (n - 1)$, where $n \cdot \lfloor a \rfloor = \lfloor n \cdot a \rfloor \Leftrightarrow \lfloor n \cdot \text{dec}(a) \rfloor = 0$, that is to say, $0 \leq n \cdot \text{dec}(a) < 1 \Leftrightarrow \text{dec}(a) \in \left[0, \frac{1}{n}\right)$.

2. i) When $\dec(a) = 0$: $\lceil n \cdot a \rceil = n \cdot a = n \cdot \lceil a \rceil$; ii) When $\dec(a) \in (0, 1)$: Assume $\dec'(a) = 1 - \dec(a) \in (0, 1)$, then $a = \lceil a \rceil - \dec'(a)$, then $n \cdot a = n \cdot \lceil a \rceil - n \cdot \dec'(a)$. Considering $0 < n \cdot \dec'(a) < n$, $n \cdot \lceil a \rceil - n < n \cdot a = n \cdot \lceil a \rceil - n \cdot \dec'(a) < n \cdot \lceil a \rceil$. From the definition of $\lceil \cdot \rceil$, we can get $n \cdot \lceil a \rceil - (n - 1) \le \lceil n \cdot a \rceil \le n \cdot \lceil a \rceil$, where $n \cdot \lceil a \rceil = \lceil n \cdot a \rceil \Leftrightarrow n \cdot \dec'(a) \in (0, 1)$, then $\dec(a) \in (1 - \frac{1}{n}, 1)$. As a whole, we have $n \cdot \lceil a \rceil - (n - 1) \le \lceil n \cdot a \rceil \le n \cdot \lceil a \rceil$, and $n \cdot \lceil a \rceil = \lceil n \cdot a \rceil$ when and only when $\dec(a) \in \left(1 - \frac{1}{n}, 1\right) \bigcup \{0\}$.

3. From the definition of $\round(\cdot)$, we have $\round(a) - 1/2 \le a \le \round(a) + 1/2$. Thus $n \cdot \round(a) - n/2 \le n \cdot a < n \cdot \round(a) + n/2$. i) When n is an even integer, it is obvious that $n \cdot \round(a) - n/2 \le \round(n \cdot a) < n \cdot \round(a) + n/2$. ii) When n is an odd integer, $n \cdot \round(a) - n/2 + 1/2 \le \round(n \cdot a) < n \cdot \round(a) + n/2 - 1/2$, that is to say, $n \cdot \round(a) - (n - 1)/2 \le \round(n \cdot a) < n \cdot \round(a) + (n - 1)/2$. As a whole, we can deduce: $n \cdot \round(a) - \lfloor n/2 \rfloor \le \round(n \cdot a) \le n \cdot \round(a) + \lfloor n/2 \rfloor$, where $n \cdot \round(a) = \round(n \cdot a) \Leftrightarrow n \cdot \round(a) - 1/2 \le n \cdot a < n \cdot \round(a) + 1/2$, that is to say, $\dec(a) \in \left[0, \frac{1}{2n}\right) \bigcup \left[1 - \frac{1}{2n}, 1\right)$.

The proof is complete.

Appendix B: The Proof of Lemma 2

Proof. Because $a = \lfloor a \rfloor + \dec(a)$, $\lfloor N/2^j \rfloor + \lfloor N'/2^j \rfloor = \left(N/2^j - \dec(N/2^j)\right) + \left(N'/2^j - \dec(N'/2^j)\right)$. Assume $N = n_1 \cdot 2^j + n_2, N' = n_1' \cdot 2^j + n_2'$ and $N + N' = 2^k (k \ge j)$, we have $\dec(N/2^j) = (N \bmod n)/2^j = n_2/2^j, \dec(N'/2^j) = (N' \bmod n)/2^j = n_2'/2^j$. Since N, N' are odd integers, we can get $n_2 > 0, n_2' > 0$. From $2^j | (N + N')$, it is obvious that $n_2 + n_2' = 2^j \Rightarrow \dec(N/2^j) + \dec(N'/2^j) = 1$, thus $\lfloor N/2^j \rfloor + \lfloor N'/2^j \rfloor = (N + N')/2^j - 1$. The proof is complete.

References

1. Andrzej Lasota and Michael C. Mackey. *Chaos, Fractals, and Noise - Stochastic Aspects of Dynamics.* Springer-Verlag, New York, second edition, 1997.
2. Ljupčo Kocarev, Goce Jakimoski, Toni Stojanovski, and Ulrich Parlitz. From chaotic maps to encryption schemes. In *Proc. IEEE Int. Symposium Circuits and Systems 1998*, volume 4, pages 514–517. IEEE, 1998.
3. Jiri Fridrich. Symmetric ciphers based on two-dimensional chaotic maps. *Int. J. Bifurcation and Chaos*, 8(6):1259–1284, 1998.
4. R. Brown and L. O. Chua. Clarifying chaos: Examples and counterexamples. *Int. J. Bifurcation and Chaos*, 6(2):219–249, 1996.
5. R. Matthews. On the derivation of a 'chaotic' encryption algorithm. *Cryptologia*, XIII(1):29–42, 1989.
6. Zhou Hong and Ling Xieting. Generating chaotic secure sequences with desired statistical properties and high security. *Int. J. Bifurcation and Chaos*, 7(1):205–213, 1997.
7. T. Habutsu, Y. Nishio, I. Sasase, and S. Mori. A secret key cryptosystem by iterating a chaotic map. In *Advances in Cryptology - EuroCrypt'91*, Lecture Notes in Computer Science 0547, pages 127–140, Berlin, 1991. Spinger-Verlag.

8. Hong Zhou and Xie-Ting Ling. Problems with the chaotic inverse system encryption approach. *IEEE Trans. Circuits and Systems I*, 44(3):268–271, 1997.

9. Sang Tao, Wang Ruili, and Yan Yixun. Perturbance-based algorithm to expand cycle length of chaotic key stream. *Electronics Letters*, 34(9):873–874, 1998.

10. D. D. Wheeler. Problems with chaotic cryptosystems. *Cryptologia*, XIII(3):243–250, 1989.

11. D. D. Wheeler and R. Matthews. Supercomputer investigations of a chaotic encryption algorithm. *Cryptologia*, XV(2):140–151, 1991.

12. E. Biham. Cryptoanalysis of the chaotic-map cryptosystem suggested at Euro-Crypt'91. In *Advances in Cryptology - EuroCrypt'91*, Lecture Notes in Computer Science 0547, pages 532–534, Berlin, 1991. Spinger-Verlag.

13. Ghobad Heidari-Bateni and Clare D. McGillem. A chaotic direct-sequence spread-spectrum communication system. *IEEE Trans. Communications*, 42(2/3/4):1524–1527, 1994.

14. Shin'ichi Oishi and Hajime Inoue. Pseudo-random number generators and chaos. *Trans. IECE Japan*, E 65(9):534–541, 1982.

15. Tohru Kohda and Akio Tsuneda. Statistics of chaotic binary sequences. *IEEE Trans. Information Theory*, 43(1):104–112, 1997.

16. Jorge A. González and Ramiro Pino. A random number generator based on unpredictable chaotic functions. *Computer Physics Communications*, 120:109–114, 1999.

17. A. Baranovsky and D. Daems. Design of one-dimensional chaotic maps with prescribed statistical properties. *Int. J. Bifurcation and Chaos*, 5(6):1585–1598, 1995.

18. Bruce Schneier. *Applied Cryptography – Protocols, algorithms, and souce code in C*. John Wiley & Sons, Inc., New York, second edition, 1996.

19. Julian Palmore and Charles Herring. Computer arithmetic, chaos and fractals. *Physica D*, D 42:99–110, 1990.

20. Zhou Hong and Ling Xieting. Realizing finite precision chaotic systems via perturbation of m-sequences. *Acta Eletronica Sinica*(In Chinese), 25(7):95–97, 1997.

21. Hu Guanhua. *Applied Modern Algebra*. Tsinghua University Press, Beijing, China, second edition, 1999.

22. Pan Chengdong and Pan Chengbiao. *Concise Number Theory*. Beijing University Press, Beijing, China, 1998.

23. The Committee of *Modern Applied Mathematics Handbook*. *Modern Applied Mathematics Handbook – vol. Probability Theory and Stochastic Process*. Tsinghua University Press, Beijing, China, 2000.

The Wide Trail Design Strategy

Joan Daemen[1] and Vincent Rijmen[2]

[1] ProtonWorld, Zweefvliegtuigstraat 10, B-1130 Brussel, Belgium
Joan.Daemen@protonworld.com
[2] CRYPTOMAThIC, Lei 8A, B-3001 Leuven, Belgium
Vincent.Rijmen@cryptomathic.com

Abstract. We explain the theoretical background of the wide trail design strategy, which was used to design Rijndael, the Advanced Encryption Standard (AES). In order to facilitate the discussion, we introduce our own notation to describe differential and linear cryptanalysis. We present a block cipher structure and prove bounds on the resistance against differential and linear cryptanalysis.

1 Introduction

The development of differential [2] and linear cryptanalysis [7] has led to several design theories for block ciphers. The most important requirement for a new cipher is that it resists state-of-the-art cryptanalytic attacks. Preferably, this can be demonstrated in a rigorous, mathematical way. The second requirement is a good performance and an acceptable 'cost', in terms of CPU requirements, memory requirements, ...

The Wide trail strategy is an approach to design the round transformations of block ciphers that combine efficiency and resistance against differential and linear cryptanalysis. The strategy has been used in the design of Rijndael, the block cipher which has been selected to become the Advanced Encryption Standard (AES). In this article we describe the application of the strategy to the design of a certain type of block ciphers only, but the strategy can easily be extended to more general block cipher structures. Moreover, the wide trail strategy can also be applied to the design of synchronous stream ciphers and hash functions.

In order to explain the wide trail strategy, we introduce our own notation for differential and linear cryptanalysis. We are convinced that a good notation helps to understand the reasonings, and our notation is suited very well to understand the wide trail strategy. We introduce a general block cipher model and explain how linear correlations and difference propagation probabilities are built up in block ciphers designed according to this model. Subsequently, we explain the basic principles of the wide trail strategy and introduce our new diffusion measure, the *branch number*. We explain its relevance in providing bounds for the probability of differential trails and the correlation of linear trails over two rounds. We then introduce a cipher structure that combines efficiency with high resistance against linear and differential cryptanalysis. The resistance against linear and differential cryptanalysis is based on a theorem that lower bounds the

B. Honary (Ed.): Cryptography and Coding 2001, LNCS 2260, pp. 222–238, 2001.

diffusion after four rounds of the cipher structure. In this paper, we emphasize the theoretical foundations of the wide trail design strategy. More explanation about the practical constructions can be found in [4].

In the following, the symbols $+$ and $\sum$ are used to denote bit-wise addition (XOR). The results can be generalized to other definitions for addition.

2   A General Block Cipher Model

We introduce a model for block ciphers that can be analyzed easily for their resistance against linear and differential cryptanalysis.

2.1   Key-Alternating Block Ciphers

A *block cipher* transforms *plaintext blocks* of a fixed length n_b to *ciphertext blocks* of the same length under the influence of a key k. An *iterative block cipher* is defined as the application of a number of key-dependent Boolean permutations. The Boolean transformations are called the *round transformations*. Every application of a round transformation is called a *round*. We denote the number of rounds by r. We have:

$$\beta[k] = \rho^{(r)}[k^{(r)}] \circ \cdots \circ \rho^{(2)}[k^{(2)}] \circ \rho^{(1)}[k^{(1)}] \ . \tag{1}$$

In this expression, $\rho^{(i)}$ is called the i-th *round* of the block cipher and $k^{(i)}$ is called the i-th round key. For instance, the DES has 16 rounds. Every round uses the same round transformation, so we say there is only one round transformation. The round keys are computed from the cipher key. Usually, this is specified with an algorithm, called the *key schedule*.

A *key-alternating block cipher* is an iterative block cipher with the following properties:

- Alternation: the cipher is defined as the alternated application of key-independent round transformations and the application of a round key. The first round key is applied before the first round and the last round key is applied after the last round.
- Binary Key Addition: the round keys are applied by means of a simple XOR: to each bit of the intermediate state a round key bit is XORed.

We have:

$$\beta[k] = \sigma[k^{(r)}] \circ \rho^{(r)} \circ \sigma[k^{(r-1)}] \circ \cdots \circ \ \sigma[k^{(1)}] \circ \rho^{(1)} \circ \sigma[k^{(0)}] \ . \tag{2}$$

As, hopefully, will become clear soon, key-alternating block ciphers lend themselves very well to analysis with respect to the resistance against cryptanalysis.

2.2   The $\gamma\lambda$ Round Structure

In the wide trail strategy, the round transformations are composed of two invertible steps:

- γ: a local non-linear transformation. By local, we mean that any output bit depends on only a limited number of input bits and that neighboring output bits depend on neighboring input bits.
- λ: a linear mixing transformation providing high diffusion. What is meant by high diffusion will be explained in the following sections.

Hence we have a round transformation ρ:

$$\rho = \lambda \circ \gamma. \tag{3}$$

and refer to this as a $\gamma\lambda$ round transformation.

A typical construction for γ is the so-called *bricklayer mapping* consisting of a number of invertible S-boxes. In this construction, the bits of input vector a are partitioned into n_t m-bit *bundles* $a_i \in Z_2^m$ with $i \in \mathcal{I}$ by the so-called *bundle partition*. $\mathcal{I}$ is called the *index space*. Clearly, the inverse of γ consists of applying the inverse substitution boxes to the bundles. The block size of the cipher is given by $n_b = mn_t$. In the case of the AES, the bundle size m is 8, hence bundles are bytes. This is by no means a necessity. For instance, Serpent [1] and Noekeon [5] also can be described in this framework, but have a bundle size of 4 bits. 3-WAY [3] uses 3-bit bundles.

For the purpose of this analysis, the S-boxes of γ need not to be specified. Since the use of different S-boxes for different bundles does not result in a plausible improvement of the resistance against known attacks, we propose to use the same S-box for all bundles. This allows to reduce the code size in software, and the required chip area in hardware implementations.

The transformation λ combines the bundles linearly: each bundle at the output is a linear function of bundles at the input. λ can be specified at the bit level by a simple $n_b \times n_b$ binary matrix M. We have

$$\lambda : b = \lambda(a) \Leftrightarrow b = Ma \tag{4}$$

λ can also be specified at the bundle level. For example, the bundles can be considered as elements in $\mathrm{GF}(2^m)$ with respect to some basis. In its most general form, we have:

$$\lambda : b = \lambda(a) \Leftrightarrow b_i = \sum_j \sum_{0 \le \ell < m} C_{i,j,\ell} a_j^{2^\ell} \tag{5}$$

In most instances a more simple linear function is chosen that is a special case of (5):

$$\lambda : b = \lambda(a) \Leftrightarrow b_i = \sum_j C_{i,j} a_j \tag{6}$$

Figure 1 gives a schematic representation of a $\gamma\lambda$ round transformation, followed by a key addition.

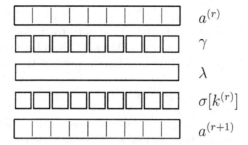

Fig. 1. Schematic representation of a $\gamma\lambda$ round, followed by a key addition.

3   Propagation in Key-Alternating Block Ciphers

In the following subsections we describe the anatomy of correlations and difference propagations in key-alternating block ciphers. This is used to determine the number of rounds required to provide resistance against linear and differential cryptanalysis. We assume that the round transformations do not exhibit correlations with amplitude 1 or difference propagation with probability 1. The limitation to the key-alternating structure allows us to reason more easily about linear and differential trails as the effect of the key addition on the propagation is quite simple.

3.1   Differential Cryptanalysis

We assume that the reader has a basic understanding of the principles of differential cryptanalysis as explained in [2]. We give a very short overview and introduce our notation.

Consider a couple of n-bit vectors a and a^* with bitwise difference $a + a^* = a'$. Let $b = h(a), b^* = h(a^*)$ and $b' = b + b^*$. The difference a' propagates to the difference b' through h. In general, b' is not fully determined by a' but depends on the value of a (or a^*).

Definition 1. *A difference propagation probability* $\mathrm{P}^h(a', b')$ *is defined as*

$$\mathrm{P}^h(a', b') = 2^{-n} \sum_a \delta(b' + h(a + a') + h(a)) \ . \tag{7}$$

Here $\delta(a)$ denotes the Kronecker delta function, which outputs zero, except when the input equals zero: $\delta(0) = 1$. If a pair is chosen uniformly from the set of all pairs (a, a^*) with $a + a^* = a'$, $\mathrm{P}^h(a', b')$ is the probability that $h(a) + h(a^*) = b'$.

Let β be a Boolean mapping operating on n-bit vectors that is composed of r mappings: $\beta = \rho^{(r)} \circ \rho^{(r-1)} \circ \ldots \circ \rho^{(2)} \circ \rho^{(1)}$. A *differential trail* A over a composed mapping consist of a sequence of $r + 1$ difference patterns:

$$Q = (q^{(0)}, q^{(1)}, q^{(2)}, \ldots, q^{(r-1)}, q^{(r)}) \ . \tag{8}$$

A differential trail has a *probability*. The probability of a differential trail is the number of values a_0 for which the difference patterns follow the differential trail divided by the number of possible values for a_0. This differential trail is composed of r differential steps $(q^{(i-1)}, q^{(i)})$, that have a propagation probability:

$$\mathrm{P}^{\rho^{(i)}}(q^{(i-1)}, q^{(i)}) \ . \tag{9}$$

Differential cryptanalysis exploits difference propagations $(q^{(0)}, q^{(r)})$ with large probabilities. The probability of difference propagation (a', b') is the sum of the probabilities of all r-round differential trails with initial difference a' and terminal difference b', i.e.,

$$\mathrm{P}(a', b') = \sum_{q^{(0)}=a', q^{(r)}=b'} \mathrm{P}(Q) \ . \tag{10}$$

3.2   Achieving Low Difference Propagation Probabilities

For a successful classical differential cryptanalysis attack, the cryptanalyst needs to know an input difference pattern that propagates to an output difference pattern over all but a few (2 or 3) rounds of the cipher, with a probability that is significantly larger than 2^{1-n_b}. To avoid this, we choose the number of rounds so that there are no differential trails with a probability above 2^{1-n_b}.

This strategy does not guarantee that there are no such difference propagations with a high probability. Equation (10) shows that in principle, many trails with each a low probability may add up to a difference propagation with high probability. As a matter of fact, for any Boolean mapping, a difference pattern at the input must propagate to some difference pattern at the output, and the sum of the difference propagation probabilities over all possible output differences is 1. Hence, there must be difference propagations with probability equal to or larger than 2^{1-n_b}. This also applies to the Boolean mapping formed by a cipher for a given value of the cipher key. Hence, the presence of difference propagations with a high probability over any number of rounds of the cipher is a mathematical fact which can't be avoided by design.

Let us analyze a difference propagation with probability y for a given key value. A difference propagation probability y means that there are exactly $y2^{n_b-1}$ pairs with the given input difference pattern and the given output difference pattern. Each of these pairs follows a particular differential trail.

Assuming that the pairs are distributed over the trails according to a Poisson distribution, the expected number of pairs that, for a given key value, follow a differential trail with propagation probability 2^{-z}, is 2^{n_b-1-z}. Consider a differential trail with a propagation probability 2^{-z} smaller than 2^{1-n_b} that is followed by at least one pair. The probability that this trail is followed by more than one pair, is approximately 2^{n_b-1-z}. It follows that if there are no differential trails with a propagation probability above 2^{1-n_b}, the $y2^{n_b-1}$ pairs that have the correct input difference pattern and output difference pattern, follow almost $y2^{n_b-1}$ different differential trails.

If there are no differential trails with a low weight, difference propagations with a large probability are the result of multiple differential trails that happen to be followed by a pair in the given circumstances, i.e. for the given key value. For another key value, each of these individual differential trails may be followed by a pair, or not. This makes predicting the input difference patterns and output difference patterns that have large difference propagation probabilities practically infeasible. This is true if the key is known, and even more so if it is unknown.

We conclude that restricting the probability of difference propagations is a sound design strategy. However, it doesn't result in a proof of security.

3.3   Linear Cryptanalysis

We assume again that the reader is familiar with the basic principles of linear cryptanalysis [7]. However, instead of using the notions *probability of a linear approximation*, and *deviation*, we prefer to use our own formalism, based on *correlation*.

Definition 2. *The correlation $C(f, g)$ between two Boolean functions $f(a)$ and $g(a)$ is defined as*

$$C(f, g) = 2 \cdot \text{Prob}(f(a) = g(a)) - 1 \ . \tag{11}$$

It follows that $C(f, g) = C(g, f)$. A *parity* of a Boolean vector is a Boolean function that consists of the XOR of a number of bits. A parity is determined by the positions of the bits of the Boolean vector that are included in the XOR. The *selection pattern* w of a parity is a Boolean vector value that has a 1 in the components that are included in the parity and a 0 in all other components. Analogous to the inner product of vectors in linear algebra, we express the parity of vector a corresponding with selection pattern w as $w^t a$. In this expression the t suffix denotes transposition of the vector w.

Note that for a vector a with n bits, there are 2^n different parities. The set of parities of a Boolean vector is in fact the set of all linear Boolean functions of that vector.

A *linear trail* U over a composed mapping consist of a sequence of $r + 1$ *selection patterns*

$$U = \left(u^{(0)}, u^{(1)}, u^{(2)}, \ldots, u^{(r-1)}, u^{(r)} \right) \ . \tag{12}$$

This linear trail is composed of r linear steps $(u^{(i-1)}, u^{(i)})$, that have a correlation:

$$C(u^{(i)^t} \rho^{(i)}(a), u^{(i-1)^t} a)$$

The *correlation contribution* C_p of a linear trail is the product of the correlation of all its steps:

$$C_p(U) = \prod_i C_{u^{(i)} u^{(i-1)}}^{\rho^{(i)}} \ . \tag{13}$$

3.4 Achieving Low Correlation Amplitudes

For a successful linear cryptanalysis attack, the cryptanalyst needs to know an input parity and an output parity after all but a few rounds of the cipher that have a correlation with an amplitude that is significantly larger than $2^{-n_b/2}$. To avoid this, we choose the number of rounds so that there are no linear trails with a correlation contribution above $n_k^{-1}2^{-n_b/2}$.

This does not guarantee that there are no high correlations over r rounds. From Parseval's equality, it follows that for any output parity, the sum of the squared amplitudes of the correlations with all input parities is 1. In the assumption that the output parity is equally correlated to all 2^{n_b} possible input parities, the correlation to each of these input parities has amplitude $2^{-n_b/2}$. In practice it is very unlikely that such a uniform distribution will be attained and correlations will exist that are orders of magnitude higher than $2^{-n_b/2}$. This also applies to the Boolean mapping formed by a cipher for a given value of the cipher key. Hence, the presence of high correlations over (all but a few rounds of) the cipher is a mathematical fact that can't be avoided by design.

However, in the absence of local clustering of linear trails, high correlations can only occur as the result of 'constructive interference' of many linear trails that share the same initial and final selection patterns. Specifically, any such correlation with an amplitude above $2^{-n_b/2}$ must be the result of at least n_k different linear trails. The condition that a linear trail in this set contributes constructively to the resulting correlation imposes a linear relation on the round key bits. From the point that more than n_k linear trails are combined, it is very unlikely that all such conditions can be satisfied by choosing the appropriate cipher key value.

The strong key-dependence of this interference makes it very unlikely that if a specific output parity has a high correlation with a specific input parity for a given key, that this will also be the case for another value of the key. In other words, although it follows from Parseval's Theorem that high correlations over the cipher will exist whatever the number of rounds, the strong round key dependence of interference makes locating the input and output selection patterns for which high correlations occur practically infeasible. This is true if the key is known, and even more so if it is unknown.

Again we conclude that restricting the amplitude of the correlation between input parities and output parities is a sound design strategy. However, it doesn't result in a proof of security.

3.5 Weight of a Trail

γ is a bricklayer mapping consisting of S-boxes. It is easy to see that the correlation over γ is the product of the correlations over the different S-box positions for the given input and output selection patterns. We define the *weight* of a correlation as the negative logarithm of its amplitude. The correlation weight for an input selection pattern and output selection pattern is the sum of the correlation weights of the different S-Box positions. If the output selection pattern is non-zero for a particular S-box position or bundle, we call this S-box or bundle *active*.

Similarly, the weight of the difference propagation over γ is defined as the negative logarithm of its probability. The weight of the difference propagation over γ is given by the sum of the weights of the difference propagations of the S-box positions for the given input difference pattern and output difference pattern. If the input difference pattern is non-zero for a particular S-box position or bundle, we call this S-box or bundle *active*.

The correlation contribution of a linear trail is the product of the correlation of all its steps. The weight of such a trail is defined as the sum of the weights of its steps. As the weight of a step is the sum of the weights of its active S-box positions, the weight of a linear trail is the sum of that of its active S-boxes. An upper limit to the correlation is a lower limit to the weight per S-box. Hence, the weight of a linear trail is equal to or larger than the number of active bundles in all its selection patterns times the minimum (correlation) weight per S-box. We call the number of active bundles in a pattern or a trail its *bundle weight*.

A differential trail is defined by a series of difference patterns. The weight of such a trail is the sum of the weights of the difference patterns of the trail. Completely analogous to linear trails, the weight of a differential trail is equal to or larger than the number of active S-boxes times the minimum (differential) weight per S-box.

3.6   Wide Trails

The reasoning above suggests two possible mechanisms to eliminate low-weight trails:

1. Choose S-boxes with high minimum differential and correlation weight.
2. Design the round transformation such a way that there are no relevant trails with low bundle weight.

The maximum correlation amplitude of an m-bit invertible S-box is above $2^{m/2}$ yielding an upper bound for the minimum (correlation) weight of $n/2$. The maximum difference propagation probability is at least 2^{m-2}, yielding an upper bound for the minimum (differential) weight of $m - 2$. This seems to suggest that one should take large S-boxes. This is not the approach we follow in the wide trail design strategy.

Instead of spending most of the resources on large S-boxes, the wide trail strategy aims at designing the round transformation(s) such that there are no trails with a low bundle weight. In ciphers designed by the wide trail strategy, a relatively large amount of resources is spent in the linear step to provide high multiple-round diffusion.

4   Diffusion

Diffusion is the term introduced by C. Shannon to denote the quantitative spreading of information [9]. Diffusion is a rather vague concept the exact meaning of which strongly depends on the context in which it is used. We will explain now what we mean by diffusion in the context of the wide trail strategy.

Inevitably, the mapping γ provides some interaction between the different bits within the bundles that may be referred to as diffusion. However, it does not provide any inter-bundle interaction: difference propagation and correlation over γ stays confined within the bundles. In the context of the wide trail strategy, it is not this kind of diffusion we are interested in. We use the term diffusion to indicate properties of a mapping that increase the minimum bundle weight of linear and differential trails. In this sense, all diffusion is realized by λ. γ does not provide any diffusion at all.

For single-round trails, obviously the bundle weight of a single round trail, differential or linear, is equal to the number of active bundles at its input. It follows that the minimum bundle weight of a single-round trail is 1, independent of λ. The situation becomes interesting as soon as we consider two-round trails.

4.1   Branch Numbers and Two-Round Trails

In two-round trails, the bundle weight is the sum of the number of active bundles in the (selection or difference) patterns at the beginning of the first and the input of the second round. We will see that the bundle weight of two-round trails can be expressed elegantly by using *branch numbers*.

Consider a partition α that divides the different bit positions of a state into n_α sets called α-sets. An example of this is the partition that divides the bits in a number of bundles. The weight of a state value with respect to a partition α is equal to the number of α-sets that have at least one non-zero bit. This is denoted by $w_\alpha(a)$. If this is applied to a difference pattern a', $w_\alpha(a')$ is the number of active α-sets in a'. Applied to a selection pattern v, $w_\alpha(v)$ is the number of active α-sets in v. If α is the partition that forms the bundles, $w_\alpha(a)$ is the number of active bundles in the pattern a and is denoted by $w_b(a)$.

We make a distinction between the differential and the linear branch number of a transformation.

Definition 3. *The differential branch number of a transformation ϕ with respect to a partition α is defined by*

$$\mathcal{B}_d(\phi, \alpha) = \min_{a,b \neq a} \{w_\alpha(a \oplus b) + w_\alpha(\phi(a) \oplus \phi(b))\} \tag{14}$$

For a linear transformation $\lambda(a) \oplus \lambda(b) = \lambda(a \oplus b)$, and (14) reduces to:

$$\mathcal{B}_d(\lambda, \alpha) = \min_{a' \neq 0} \{w_\alpha(a') + w_\alpha(\lambda(a'))\} \ . \tag{15}$$

An upper bound for the differential branch number of a Boolean transformation ϕ with respect to a partition α is given by n_α, since the output difference corresponding to an input difference with a single non-zero bundle can have at most weight n_α. Therefore, the differential branch number of ϕ with respect to α is upper bounded by

$$\mathcal{B}_d(\phi, \alpha) \leq n_\alpha + 1. \tag{16}$$

Analogous to the differential branch number, we can define the linear branch number.

Definition 4. *The* linear branch number *of a transformation ϕ with respect to a is given by*

$$\mathcal{B}_l(\phi, \alpha) = \min_{\alpha, \beta, c(\alpha^t x, \beta^t \phi(x)) \neq 0} \{w_\alpha(\alpha) + w_\alpha(\beta)\} \tag{17}$$

Many of the following discussions are valid both for differential and linear branch numbers and both $\mathcal{B}_d$ and $\mathcal{B}_l$ are denoted simply by $\mathcal{B}$. Moreover, in many cases the partition is clear from the context and $\mathcal{B}(\phi, \alpha)$ is expressed as $\mathcal{B}(\phi)$.

4.2   Some Properties

In general, the linear and differential branch number of a transformation with respect to a partition are not equal. From the symmetry of Definition 3 and 4 it follows that the branch number of a transformation and that of its inverse are the same. Moreover, we have the following properties:

- a (differential or selection) pattern a is not affected by a key addition and hence its weight $w_\alpha(a)$ is not affected. This property holds independently of the partition α.
- a bricklayer permutation compatible with α cannot turn an active α-subset into a non-active one or vice versa. Hence, it does not affect the weight $w_\alpha(a)$.

Assume we have a transformation ϕ composed of a transformation ϕ_1 and a bricklayer transformation ϕ_α operating on α-subsets, i.e., $\phi = \phi_\alpha \circ \phi_1$. As ϕ_α does not affect the number of active α-subsets in a propagation pattern, the branch number of ϕ and ϕ_1 are the same. More general, if propagation of patterns is analyzed at the level of α-subsets, bricklayer transformations compatible with α may be ignored.

If we apply this to the bundle weight of a $\gamma\lambda$ round transformation ρ, it follows immediately that the (linear or differential) bundle branch number of ρ is that of its linear part λ.

4.3   A Two-Round Propagation Theorem

The following theorem relates the value of $\mathcal{B}(\lambda)$ to a bound on the number of active bundles in a trail. The proof is valid both for linear and differential trails: in the case of linear trails $\mathcal{B}$ stands for $\mathcal{B}_l$ and in the case of differential trails $\mathcal{B}$ stands for $\mathcal{B}_d$.

Theorem 1 (Two-Round Propagation Theorem).

For a key-alternating block cipher with a $\gamma\lambda$ round structure the number of active bundles of any two-round trail is lower bounded by the (bundle) branch number of λ.

Proof. Figure 2 depicts two rounds. Since the transformations γ and $\sigma[k]$ operate on each bundle individually, they do not affect the propagation of patterns. Hence it follows that $w_b(a^{(1)}) + w_b(a^{(2)})$ is only bounded by the properties of the linear transformation λ of the first round. Definition 3 and 4 imply that the sum of the active bundles before and after λ of the first round is lower bounded by $\mathcal{B}(\lambda)$.
□

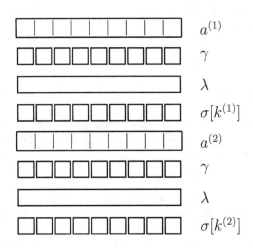

$a^{(1)}$

γ

λ

$\sigma[k^{(1)}]$

$a^{(2)}$

γ

λ

$\sigma[k^{(2)}]$

Fig. 2. Transformations relevant in the proof of Theorem 1.

5   An Efficient Key-Alternating Structure

In trails of more than two rounds, the desired diffusion properties of ρ are less trivial. It is clear than any $2n$-round trail can be decomposed in n 2-round trails and hence that its bundle weight is lower bounded by n times the branch number of ρ. The 'greedy' approach to eliminate low-weight trails is to consider Theorem 1 only and to design a round transformation with a maximum branch number. However, transformations that provide high branch numbers have a tendency to have a high implementation cost. More efficient designs can be achieved in the following way. We build a key-alternating block cipher that consists of an alternation of two different round transformations defined by:

$$\rho^a = \theta \circ \gamma \tag{18}$$
$$\rho^b = \Theta \circ \gamma \tag{19}$$

The transformation γ is defined as before and operates on n_b m-bit bundles.

5.1   The Diffusion Transformation θ

With respect to θ, the bundles of the state are grouped into a number of *columns* by a partition Ξ of the index space $\mathcal{I}$. We denote a column by ξ and the number of columns by n_Ξ. The column containing an index i is denoted by $\xi(i)$ and the number of indices in a column ξ by n_ξ. The size of the columns relates to the block length by

$$m \sum_{\xi \in \Xi} n_\xi = mn_t.$$

θ is a bricklayer mapping with component mappings that each operate on a column. Within each column, bundles are linearly combined. We have

$$\theta : b = \theta(a) \Leftrightarrow b_i = \sum_{j \in \xi(i)} C_{i,j} a_j \qquad (20)$$

The bricklayer transformation θ only needs to realize diffusion within the columns and has hence an implementation cost that is much lower.

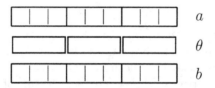

Fig. 3. The diffusion transformation θ

Similar to active bundles, we can speak of active columns. The number of active columns of a propagation pattern a is denoted by $w_s(a)$. The round transformation $\rho^a = \theta \circ \gamma$ is a bricklayer transformation operating independently on a number of columns. Taking this bricklayer structure into account we can extend the result of Section 4.1 slightly. The branch number of θ is given by the minimum branch number of its component transformations. Applying (16) to the component mappings defined by the matrices C_ξ results in the following upper bound:

$$\mathcal{B}(\theta) \le \min_\xi n_\xi + 1. \qquad (21)$$

Hence, the smallest column imposes the upper limit for the branch number. The Two-Round Propagation Theorem (Theorem 1) implies the following Lemma.

Lemma 1. *The bundle weight of any two-round trail in which the first round has a $\gamma\theta$ round transformation is lower bounded by $N\mathcal{B}(\theta)$, where N is the number of active columns at the input of the second round.*

Proof. Theorem 1 can be applied to each of the component mappings of the bricklayer mapping ρ^a separately. For each active column there are at least $\mathcal{B}(\theta)$ active bundles in the two-round trail. $\square$

5.2   The Linear Transformation Θ

Θ mixes bundles across columns.

$$\Theta : b = \Theta(a) \Leftrightarrow b_i = \sum_j C_{i,j} a_j \qquad (22)$$

The goal of Θ is to provide inter-column diffusion. The design criterion for Θ is to have a high branch number with respect to Ξ. This is denoted by $\mathcal{B}(\Theta, \Xi)$ and called its *column branch number*.

5.3   A Lower Bound on the Bundle Weight of 4-Round Trails

The combination of the bundle branch number of θ and the column branch number of Θ allows us to prove a lower bound on the bundle weight of any trail over 4 rounds starting with ρ^a.

Theorem 2 (Four-round Propagation Theorem for $\theta\Theta$ construction).
For a key-alternating block cipher with round transformations as defined in (18) and (19), the bundle weight of any trail over

$$\rho^b \circ \rho^a \circ \rho^b \circ \rho^a$$

is lower bounded by $\mathcal{B}(\theta) \times \mathcal{B}(\Theta, \Xi)$.

Proof. Figure 4 depicts four rounds. As the key additions play no role in the propagation of patterns, they have been left out. It is easy to see that the linear transformation of the fourth round plays no role. The sum of the number of active

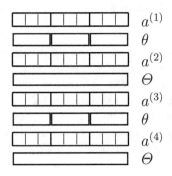

Fig. 4. Relevant transformations for the proof of Theorem 2.

columns in $a^{(2)}$ and $a^{(3)}$ is lower bounded by $\mathcal{B}(\Theta, \Xi)$. According to Lemma 1, for each active column in $a^{(2)}$ there are at least $\mathcal{B}(\theta)$ active bundles in the corresponding columns of $a^{(1)}$ and $a^{(2)}$. Similarly, for each active column in $a^{(3)}$ there are at least $\mathcal{B}(\theta)$ active bundles in the corresponding columns of $a^{(3)}$ and $a^{(4)}$. Hence the total number of active bundles is lower bounded by $\mathcal{B}(\theta) \times \mathcal{B}(\Theta, \Xi)$. $\square$

5.4   An Efficient Construction for Θ

As opposed to θ, Θ does not operate on different columns independently and hence may have a much higher implementation cost. In this we present a construction of Θ in terms of θ and bundle transpositions denoted by π. We define

$$\Theta = \pi \circ \theta \circ \pi \ . \tag{23}$$

In the following we will define π and prove that if π is well chosen the column branch number of Θ can be made equal to the bundle branch number of θ.

The bundle transposition π. The bundle transposition π is defined as

$$\pi : b = \pi(a) \Leftrightarrow b_i = a_{p(i)} \ , \tag{24}$$

with $p(i)$ a permutation of the index space $\mathcal{I}$. The inverse of π is defined by $p^{-1}(i)$. Observe that a bundle transposition π does not affect the bundle weight of a propagation pattern and hence that the branch number of a transformation is not affected if it is composed with π.

Contrary to θ, π provides *inter-column diffusion*. Intuitively, good diffusion for π would mean that it distributes the different bundles of a column to as many different columns as possible. We say π is *diffusion-optimal* if the different bundles in each column are distributed over all different columns. More formally, we have:

Definition 5. π *is* diffusion-optimal *if and only if*

$$\forall i, j \in \mathcal{I}, i \neq j : (\xi(i) = \xi(j)) \Rightarrow (\xi(p(i)) \neq \xi(p(j))). \tag{25}$$

It is easy to see that this implies the same condition for π^{-1}. A diffusion-optimal bundle transposition π implies

$$w_s(\pi(a)) \geq \max_{\xi}(w_b(a_\xi)) \ .$$

Therefore a diffusion-optimal transformation can only exist if $n_\Xi \geq \max_i(n_{\xi_i})$. In words, π can only be diffusion-optimal if there are at least as many columns as there are bundles in the largest column. If π is diffusion-optimal, we can prove that the column branch number of the mapping Θ is lower bounded by the branch number of θ.

Lemma 2. *If π is a diffusion-optimal transposition of bundles, the column branch number of $\pi \circ \phi \circ \pi$ is lower bounded by the bundle branch number of ϕ*

Proof. We refer to Figure 5 for the notations used in this proof. We have to demonstrate that

$$w_s(a) + w_s(d) \geq \mathcal{B}(\phi) \ .$$

For any active column in b, the number of active bundles in that column and the corresponding column of c is at least $\mathcal{B}(\phi)$. π moves all active bundles in an active column of c to different columns in d and π^{-1} moves all active bundles in an active column of b to different columns in a. It follows that the sum of the number of active columns in a and in d is lower bounded by the bundle branch number of ϕ. $\square$

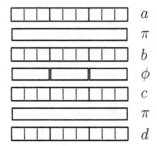

Fig. 5. Transformations relevant in the proof of Lemma 2.

6   Using Identical Round Transformations

The efficient structure described in Section 5 uses two different round transformations. It is possible to define a block cipher structure with only one round transformation, that achieves the same bound. This is the round structure used in the AES and related ciphers. The advantage of having a single round transformation is a reduction in software code in software implementations and chip area in dedicated hardware implementations. For this purpose, λ is composed of two types of the mappings:

- θ: the linear bricklayer mapping that provides high local diffusion, as defined in Section 5.1, and
- π: the transposition mapping that provides high *dispersion*, as defined in Section 5.4.

Hence we have for the round transformation:

$$\rho^c = \theta \circ \pi \circ \gamma \tag{26}$$

Figure 6 gives a schematic representation of the different transformations of a round. These component transformation are defined in such a way that they impose strict lower bounds on the number of active S-boxes in four-round trails. For two-round trails it can be seen that the number of active bundles is lower bounded by $\mathcal{B}(\rho^c) = \mathcal{B}(\lambda) = \mathcal{B}(\theta)$. For four rounds, we can prove the following important theorem:

Theorem 3 (Four-Round Propagation Theorem).
For a key-iterated block cipher with a $\gamma\pi\theta$ round transformation and diffusion-optimal π, the number of active S-boxes in a four-round trail is lower bounded by $(\mathcal{B}(\theta))^2$.

Proof. Firstly, we show that the transformation formed by 4 applications of the round transformation ρ^c as defined in (26) is equivalent to four rounds of the construction with ρ^a and ρ^b as defined in (18) and (19). For simplicity, we leave

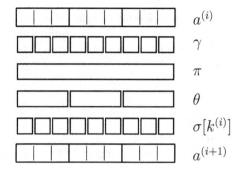

Fig. 6. Schematic representation of the different steps of a $\gamma\pi\theta$ round transformation, followed by a key addition.

out the applications of the key additions, but the proof works in the same way if the key additions are present. Let $\mathcal{A}$ be defined as:

$$\mathcal{A} = \rho^c \circ \rho^c \circ \rho^c \circ \rho^c$$
$$= (\theta \circ \pi \circ \gamma) \circ (\theta \circ \pi \circ \gamma) \circ (\theta \circ \pi \circ \gamma) \circ (\theta \circ \pi \circ \gamma) \ .$$

γ is a bricklayer mapping, operating on every bundle separately and operating independently of the bundle's position. Therefore γ commutes with π, which only moves the bundles to different positions. We get:

$$\mathcal{A} = (\theta \circ \gamma) \circ (\pi \circ \theta \circ \pi \circ \gamma) \circ (\theta \circ \gamma) \circ (\pi \circ \theta \circ \pi \circ \gamma)$$
$$= \rho^a \circ \rho^b \circ \rho^a \circ \rho^b \ ,$$

with Θ of ρ^b defined exactly as in (23). Now we can apply Lemma 2 and Theorem 2 to finish the proof.   □

7   Conclusion and Open Problems

We have shown how the application of the wide trail design strategy leads to the definition of a round transformation as the one used in Rijndael. The proposed round transformation allows us to give provable bounds on the correlation of linear trails and the weight of differential trails while at the same time allowing efficient implementations.

An interesting open problem is the effect of *trail clustering*. Theorems 1, 2 and 3 give lower bounds on the weight of trails. As mentioned in Section 3, the probability of input-output difference propagations as well as the correlation between input parities and output parities are a sum over many trails. If the trails follow indeed a Poisson distribution, then the results can be applied straightforwardly. However, it has already been observed that in some cases, the trails *don't* follow a Poisson distribution. Instead, they tend to cluster and as a result the probability of a difference propagation can be significantly higher [6]. A similar effect for correlations has been studied in [8]. It remains an open problem whether trail clustering occurs and impacts the security for the cipher structure described here.

References

1. R. Anderson, E. Biham, and L. R. Knudsen. Serpent. In *Proceedings of the first AES candidate conference*, Ventura, August 1998.
2. E. Biham and A. Shamir. Differential cryptanalysis of DES-like cryptosystems. *Journal of Cryptology*, 4(1):3–72, 1991.
3. J. Daemen, R. Govaerts, and J. Vandewalle. A new approach to block cipher design. In Vaudenay [10], pages 18–32.
4. J. Daemen, L. R. Knudsen, and V. Rijmen. Linear frameworks for block ciphers. *Designs, Codes and Cryptography*, 22(1):65–87, January 2001.
5. J. Daemen, M. Peeters, G. V. Assche, and V. Rijmen. Noekeon. In *First open NESSIE Workshop*, Leuven, November 2000.
6. L. R. Knudsen. Truncated and higher order differentials. In B. Preneel, editor, *Fast Software Encryption '94*, volume 1008 of *Lecture Notes in Computer Science*, pages 196–211. Springer-Verlag, 1995.
7. M. Matsui. Linear cryptanalysis method for DES cipher. In T. Helleseth, editor, *Advances in Cryptology, Proceedings of Eurocrypt '93*, volume 765 of *Lecture Notes in Computer Science*, pages 386–397. Springer-Verlag, 1994.
8. K. Nyberg. Linear approximation of block ciphers. In A. D. Santis, editor, *Advances in Cryptology, Proceedings of Eurocrypt '94*, volume 950 of *Lecture Notes in Computer Science*, pages 439–444. Springer-Verlag, 1995.
9. C. E. Shannon. Communication theory of secrecy systems. *Bell Syst. Tech. Journal*, 28:656–715, 1949.
10. S. Vaudenay, editor. *Fast Software Encryption '98*, volume 1372 of *Lecture Notes in Computer Science*. Springer-Verlag, 1998.

Undetachable Threshold Signatures*

Niklas Borselius, Chris J. Mitchell, and Aaron Wilson

Mobile VCE Research Group, Information Security Group,
Royal Holloway, University of London, Egham, Surrey TW20 0EX, UK

Abstract. A major problem of mobile agents is their inability to authenticate transactions in a hostile environment. Users will not wish to equip agents with their private signature keys when the agents may execute on untrusted platforms. Undetachable signatures were introduced to solve this problem by allowing users to equip agents with the means to sign signatures for tightly constrained transactions, using information especially derived from the user private signature key. However, the problem remains that a platform can force an agent to commit to a sub-optimal transaction. In parallel with the work on undetachable signatures, much work has been performed on threshold signature schemes, which allow signing power to be distributed across multiple agents, thereby reducing the trust in a single entity. We combine these notions and introduce the concept of an undetachable threshold signature scheme, which enables constrained signing power to be distributed across multiple agents, thus reducing the necessary trust in single agent platforms. We also provide an RSA-based example of such a scheme based on a combination of Shoup's threshold signature scheme, [1] and Kotzanikolaou et al's undetachable signature scheme, [2].

1   Introduction

A digital signature is the electronic counterpart to a written signature. Thus one way to commit to an electronic transaction is by the use of a digital signature. Recently, the use of mobile agents to commit to transactions for a user has become a topic of interest. Mobile agents, however, face the problem of having to execute in a hostile environment where the host executing the agent has access to all the data that an agent has stored (for instance the private signature key). Consequently, the problem of allowing an agent to sign a transaction on behalf of a user is one of interest.

Undetachable signatures were first proposed by Sander and Tschudin [3] to solve this problem, and are based on the idea of computing with encrypted functions. The host executes a function $s \circ f$, where f is an encrypting function, without having access to the user's private signature function s. The security of the

---

* The work reported in this paper has formed part of the Software Based Systems area of the Core 2 Research Programme of the Virtual Centre of Excellence in Mobile & Personal Communications, Mobile VCE, www.mobilevce.com, whose funding support, including that of the EPSRC, is gratefully acknowledged. More detailed technical reports on this research are available to Industrial Members of Mobile VCE.

B. Honary (Ed.): Cryptography and Coding 2001, LNCS 2260, pp. 239–244, 2001.

method lies in the encrypting function f. Whilst Sander and Tschudin were unable to propose a satisfactory scheme, more recently Kotzanikolaou, Burmester and Chrissikopoulos [2] have presented an RSA-based scheme which appears to be secure.

The idea of an undetachable signature is as follows. Suppose a user wishes to purchase a product from an electronic shop. The agent can commit to the transaction only if the agent can use the signature function s of the user. However as the server where the agent executes may be hostile, the signature is protected by a function f to obtain $g = s \circ f$. The user then gives the agent the pair (f, g) of functions as part of its code. The server then executes the pair (f, g) on an input x (where x encodes transaction details) to obtain the undetachable signature pair

$$f(x) = m \text{ and } g(x) = s(m).$$

The pair of functions allows the agent to create signatures for the user whilst executing on the server without revealing s to the server. The parameters of the function f are such that the output of f includes the user's constraints. Thus m links the constraints of the customer to the bid of the server. This is then certified by the signature on this message. The main point is that the server cannot sign arbitrary messages, because the function f is linked to the user's constraints.

However, one problem with this approach is that the agent is still given the power to sign any transaction it likes, subject to the requirement that the transaction must be consistent with the constraints used to construct f. Thus, for example, whilst the constraints may limit the nature and/or value of a transaction, a malicious host may force an agent to commit to a transaction much less favourable than could be achieved.

Thus, to protect further against malicious hosts, a user may wish to use more than one agent and have the agents agree on a bid before committing to it. Hence, a user may send out n agents with the criteria that k of them must agree before committing to a purchase. The obvious solution to such a requirement is to employ a *threshold signature scheme*, meaning that agents can all sign the bid they think 'best' given the user's requirements, and then, on receipt of a sufficient number of these bids, the user's signature can be reconstructed.

However, such a scheme does not possess the means to constrain the power given to a quorum of agents. This motivates the introduction of the concept of an *undetachable threshold signature* which both distributes signature authority across multiple agents and simultaneously constrains the signatures that may be constructed.

The rest of the paper is as follows. In Section 2 we outline the undetachable signature scheme of [2], and in Section 3 we briefly review threshold signatures and give a method of Shoup [1] to construct such a scheme. Finally, in Section 4 we define the concept of an undetachable threshold signature, and show how an example of such a scheme may be obtained by combining the schemes of [2] and [1].

2   RSA Undetachable Signatures

We briefly present the RSA undetachable signature scheme given in [2]. The user sets up an RSA signature pair in the usual manner, that is the user selects an RSA modulus n which is the product of two primes p and q, and a number e such that $1 \leq e \leq \phi(n) = (p-1)(q-1)$ and $\gcd(e, \phi(n)) = 1$. Let d be such that $1 \leq d \leq \phi(n)$ and $ed = 1 \bmod \phi(n)$. The user then publishes the verification key (n, e) and keeps d as the private signing key.

Let I be an identifier for the user and R the encoded requirements of the user for a purchase (we assume that R is encoded in a manner which is understood by all parties). Let h be an appropriate hash-function (*i.e.* one giving a value in $\mathbb{Z}_n$). The user then forms $H = h(I, R)$.

The user then gives an agent the user identifier, the requirements, and the pair (H, G) as its undetachable signature, where $G = H^d \bmod n$. To sign a bid B (which we assume is in the same format as R), the executing host calculates $x = h(B)$. The undetachable signature is then the pair (H^x, G^x). We note that,

$$G^x = (H^d)^x = H^{dx} = H^{xd} = (H^x)^d$$

so that the server has signed the value H^x with the user's private key.

We briefly note that this scheme appears secure, and a proof of this fact is given in [2]. To forge a signature on a different set of requirements R' a malicious host would need to forge $H' = h(I, R')$, $G' = (H')^d$ and $(G')^x$. Clearly the only work needed here is to forge G', and this would require knowledge of a user's private key. Having said this, there is nothing in this scheme to prevent a host from signing more than one bid, or presenting a bid that just meets the requirements of the user (as opposed to a possibly better standard offer).

3   Threshold Signatures

The idea of a threshold scheme is to take a secret, and divide it into pieces called shares which are distributed among a group of entities. Then any subset of these entities of a given size can reconstruct this secret, but a smaller group can learn no information about the secret. An example of such a scheme is given in [4].

Threshold cryptography was first proposed by Desmedt [5]. One important type of threshold cryptosystem is known as a *threshold signature*. In such a scheme, any set of k parties from a total of l parties can sign any document, and any coalition of less than k parties cannot sign a document. Such schemes tend to rely on a combiner which is not necessarily trusted. Several schemes have been proposed based on both El Gamal and RSA cryptography (see, for example, [1] for a short survey). Recently Shoup [1] proposed an RSA scheme which is as efficient as possible; the scheme uses only one level of secret sharing, each server sends a single part signature to a combiner, and must do work that is equivalent, up to a constant factor, to computing a single RSA signature.

Although in some sense not perfect as a threshold signature scheme (as it relies on a trusted party to form the shares) this scheme is ideal in our setting, where the user dispatching the agent will always (one would hope) trust them-selves. (Note that an alternative scheme without a trusted dealer is given in [6]. This scheme also improves on [1] by not relying on an RSA modulus made up of 'safe primes'). An example of an El Gamal based scheme is given in [7].

We next briefly outline the threshold signature scheme of [1].

The user (dealer) forms the following:

- An RSA modulus $n = pq$ where $p = 2p' + 1$ and $q = 2q' + 1$ are safe primes, i.e. p', q' are prime.
- A public exponent e where e is prime and a private key d, where $de \equiv 1$ (mod $p'q'$).
- A polynomial $f(x) = \sum_{i=0}^{k-1} a_i x^i$ where $a_0 = d$ and $a_i \in \{0, \ldots, p'q' - 1\}$ (selected at random) for $1 \leq i \leq k$.
- $L(n)$, the bit length of n, and L_1, a secondary security parameter — Shoup [1] suggests $L_1 = 128$.
- The l signature key shares of the scheme s_i, where each s_i is selected at random from the set $\{s | 0 \leq s \leq 2^{L(n)+L_1}, s \equiv f(i) \bmod (p'q')\}$.
- The verification keys $\text{VK} = v$ and $\text{VK}_i = v^{s_i}$ where $v \in Q_n$, the subgroup of squares of $\mathbb{Z}_n^*$.
- A global hash function h mapping into $\mathbb{Z}_n^*$.
- A second hash function g whose output is an L_1-bit integer.

In this scheme a shareholder signs a message m in the following manner. Firstly the shareholder calculates the hash of the message, i.e. $x = h(m)$. The signature share of a shareholder i then consists of

$$x_i = x^{2\Delta s_i}$$

and a 'proof of correctness' (note that $\Delta = l!$). The proof of correctness is basically just a proof that the discrete logarithm of x_i^2 to the base $x^{4\Delta}$ is the same as the discrete logarithm of v_i to the base v. Let $L(n)$ be the bit length of n. The shareholder then chooses a random number $r \in \{0, \ldots, 2^{L(n)+3L_1} - 1\}$ and computes

$$v' = v^r, \ x' = x^{4\Delta}r, \ c = g(v, x^{4\Delta}, v_i, v', x'), \ z = s_i c + r.$$

The proof of correctness is then (z, c) which can be verified by calculating

$$c = g(v, x^{4\Delta}, v_i, x_i^2, v^z v_i^{-c}, x^{4\Delta z} x_i^{-2c}).$$

To combine the shares the combiner acts as follows. Assume we have valid shares from a set $S = \{i_1, i_2, \ldots, i_k\}$ of shareholders. The combiner computes

$$\lambda_{0,j}^S = \Delta \prod_{i \in S \setminus \{j\}} \frac{i}{(i - j)}.$$

These values are derived from the standard Lagrange interpolation formula. These values are integers and it is clear that they are easy to compute. We also have, from the Lagrange interpolation formula that,

$$\Delta \cdot f(0) = \sum_{j \in S} \lambda_{0,j}^S f(j) \bmod (p'q').$$

In other words we have,

$$d \cdot \Delta = \sum_{j \in S} \lambda_{0,j}^S s_j$$

The combiner then computes,

$$
\begin{aligned}
w &= x_{i_1}^{2\lambda_{0,i_1}^S} \cdots x_{i_k}^{2\lambda_{0,i_k}^S} \\
&= x^{4\Delta^2 \sum_{j \in S}(s_j \lambda_{0,j}^S)} \\
&= x^{4\Delta^5 d}.
\end{aligned}
$$

To check this signature we note that $w^e = x^{4\Delta^5}$ where $\gcd(e, 4\Delta^5) = 1$. As e is coprime to $4\Delta^5$ we can find a, b such that $a(4\Delta^5) + be = 1$ so that we finally have the signature

$$y^e = (w^a x^b)^e = x.$$

4   Undetachable Threshold Signatures

We now introduce the notion of an undetachable threshold signature. Suppose a user has a private signature key s and a public verification key v. Suppose also that the user has a 'constraint string' R, which will define what types of signature can be created. Then an undetachable threshold signature scheme will enable the user to provide n entities with 'shares' of the private signature key (where the shares will be a function of R), where the following properties must be satisfied:

- each entity can use their share to sign a message M of their choice to obtain a 'signature share';
- the 'correctness' of a signature share can be verified independently of any other signature shares;
- any entity, when equipped with k different signature shares for the same message M, can construct a signature on the message M which will be verifiable by any party with a trusted copy of the public key of the user, and which will also enable the string R to be verified;
- knowledge of less than k different signature shares for the same message M cannot be used to construct a valid signature on the message M;
- knowledge of any number of different signature shares for messages other than M will not enable the construction of a valid signature on message M;
- knowledge of any number of different signature shares for constraints strings other than R will not enable the construction of a valid signature with associated constraint string R.

As discussed above, the motivation for introducing this concept is that the use of a threshold signature scheme or a detachable signature scheme on its own would not protect against all possible attacks in a mobile agent scenario. We now describe an example of such a scheme. For brevity, we only give the necessary changes to the threshold scheme in section 3 to form the undetachable threshold signature scheme.

Recall that the secret share for shareholder i consists of a number s_i. Let h be an appropriate hash function. The signature share of this shareholder for a message m is then

$$x_i = x^{2 \cdot \Delta \cdot s_i}.$$

where l is the total number of shares, $\Delta = l!$ and $x = h(m)$ is a hash of the message.

As in Section 2 let I be the identifier of a user and let R be the user requirements. Let $H = h(I, R)$ be a hash of the requirements. We replace the share s_i with a pair $(H, t_i = H^{2 \cdot \Delta \cdot s_i})$. To sign a bid B the shareholder calculates $C = h(B)$ and

$$t_i^C = (H^{2 \cdot \Delta \cdot s_i})^C = H^{2 \cdot \Delta \cdot s_i C} = (H^C)^{2 \cdot \Delta \cdot s_i}.$$

Thus, when all the shares are combined the combiner will have a signed copy of H^C, thus achieving a signed undetachable signature.

We observe that a proof of security is given for the scheme in Section 3 provided that k is one greater than the number of corrupt servers (in the case where k exceeds the number of corrupt servers by a greater number a slightly adapted scheme is used). With this information to hand we note that this scheme is secure as long as the undetachable scheme given in [2] is secure, and that this scheme appears to be sound.

References

1. Shoup, V.: Practical threshold signatures. In Preneel, B., ed.: Advances in Cryptology — EUROCRYPT 2000. Number 1807 in LNCS, Springer-Verlag, Berlin (2000) 207–220
2. Kotzanikolaou, P., Burmester, M., Chrissikopoulos, V.: Secure transactions with mobile agents in hostile environments. In Dawson, E., Clark, A., Boyd, C., eds.: Information Security and Privacy, Proceedings of the 5th Australasian Conference ACISP 2000. Number 1841 in LNCS, Springer-Verlag (2000) 289–297
3. Sander, T., Tschudin, C.: Protecting mobile agents against malicious hosts. In Vigna, G., ed.: Mobile Agents and Security. Number 1419 in LNCS, Springer-Verlag, Berlin (1998) 44–60
4. Shamir, A.: How to share a secret. Communications of the ACM **22** (1979) 612–613
5. Desmedt, Y.: Society and group oriented cryptography. In Pomerance, C., ed.: Advances in Cryptology — Crypto '87. Number 293 in LNCS, Springer-Verlag, Berlin (1988) 120–127
6. Damgård, I., Koprowski, M.: Practical threshold RSA signatures without a trusted dealer. In Pfitzmann, B., ed.: Advances in Cryptology — EUROCRYPT 2001. Number 2045 in LNCS, Springer-Verlag, Berlin (2001) 152–165
7. Langford, S.K.: Threshold DSS signatures without a trusted party. In Coppersmith, D., ed.: Advances in Cryptology — Crypto '95. Number 963 in LNCS, Springer-Verlag, Berlin (1995) 397–409

Improving Divide and Conquer Attacks against Cryptosystems by Better Error Detection / Correction Strategies

Werner Schindler[1], François Koeune[2], and Jean-Jacques Quisquater[2]

[1] Bundesamt für Sicherheit in der Informationstechnik (BSI)
Godesberger Allee 185–189
53175 Bonn, Germany
Werner.Schindler@bsi.bund.de
[2] Université catholique de Louvain
Place du Levant 3
1348 Louvain-la-Neuve, Belgium
{fkoeune,jjq}@dice.ucl.ac.be

Abstract. Divide and conquer attacks try to recover small portions of cryptographic keys one by one. Usually, a wrong guess makes subsequent ones useless. Hence possible errors should be detected and corrected as soon as possible. In this paper we introduce a new (generic) error detection and correction strategy. Its efficiency is demonstrated at various examples, namely at a power attack, two timing attacks against RSA implementations with and without Chinese Remainder Theorem, and a timing attack against the future AES (Rijndael). As the design of efficient countermeasures requires a good understanding of an attack's actual power, the possible improvement induced by sophisticated error detection and correction should not be neglected. Although divide and conquer attacks are typical for side-channel attacks, we would like to stress that they are not restricted to that field, as will be illustrated by Siegenthaler's attack.

Keywords: Error detection, error correction, timing attack, power attack.

1 Introduction

Cryptographic algorithms (encryption schemes, random generators, ...) often gather their security on one (or a few) secret parameter(s), whereas the rest of the design is left public.

To seize this secret parameter (usually denoted as the *key*), a frequent attack scenario assumes that the pirate is able to observe the output of the algorithm, possibly with access – or even control – of this algorithm's input during a limited period of time. The attacker will use these observations to deduce information on the key.

B. Honary (Ed.): Cryptography and Coding 2001, LNCS 2260, pp. 245–267, 2001.

In particular, divide and conquer attacks basically consist in dividing the key into smaller pieces, whose size makes exhaustive search possible, and then handling these pieces separately. Such attacks are very efficient, but will of course only be feasible provided that it is possible to guess parts of the key separately. In other words, it must be possible, with reasonable probability, to confirm or invalidate a partial key guess without knowing the other parts of the key.

In many cases, a wrong partial key guess makes subsequent ones worthless. Consequently, it is desirable to be able to validate the guesses made so far. This suggests the following iterated process:

1. guess a part of the key
2. check whether guess is correct so far (error detection);
3. if yes, repeat step 1 with next part of the key;
4. otherwise, identify probable error position(s) among previously guessed parts (error location), and go back to step 1 trying another guess for that part (error correction).

Many different attacks can be put in the divide and conquer class, and the relative importance of the different parts (guess, error detection, ...) may vary greatly from one to another. However, these parts are usually more or less independent, and can therefore be subject to independent improvements, any of which would result in a global performance improvement.

This paper focuses on the error management [1] part. Error management faces several efficiency constraints. First, it must keep sample size small: a simple error check, for example, consists in repeating Step 1 with a new observation; however, this reduces the efficiency of the attack by at least a factor 2, which is not acceptable for many realistic scenarios. Second, it must be time-efficient, in the sense that a wrong guess must be identified as quickly as possible; but this must be counterbalanced with the risk of hindering the attack's success: a too restrictive strategy involves the risk of definitively rejecting a guess that was correct, with the consequence that the attack will fail completely, whereas a too permissive strategy will make the attack longer, possibly up to impractical running time.

In this paper we do not concentrate on the attacks themselves, i.e. on strategies to guess parts of the key (Step 1) and on countermeasures which prevent these attacks but on error detection and error correction strategies (Steps 2 and 4). In particular, we mainly suggest a new type of error detection strategy which may be characterized as a *three-option decision strategy*. Everytime the error detection is applied, it can either:

2.a. Validate the correctness of previous partial key guesses up to a certain point (which need not necessarily be the current one). This yields a "confirmed index" (i.e. we definitively assume that that part of the key is correct) which will facilitate later error locations/corrections, as only the partial

---

[1] The idea that errors could be identified and corrected in a timing attack was first mentioned by Kocher [7].

key guesses which were derived *after* the actual confirmed index will be later considered.

2.b. Conclude an error has occurred between this point and the last confirmed index. In this case, we enter the error-correction phase.

2.c. If there are no convincing indicators for any of these two cases, do not conclude. In this case, we simply continue the attack, and postpone decision to a future error detection step.

This concept is rather generic and can be adapted to various types of attacks. We will consider five examples which may seem to be very different at first sight.

Section 2 briefly introduces the subject by presenting an extreme example of divide and conquer attack, in which the error correction and detection strategy is so efficient that it constitutes the most important part of the attack. Section 3, which is the main part of this paper, then develops the error management strategy in detail, on a timing attack example. In this example, the sophisticated error detection and correction strategy, together with an optimized bit estimation strategy, allowed to improve the efficiency of a timing attack on RSA ([6]) by about factor 50. Section 4 shows how a similar technique can be applied to a timing attack against another RSA implementation (using the Chinese Remainder Theorem (CRT) and thus being resistant against the attacks considered in [6] and [13]). Section 5 treats of a timing attack on a careless AES (Rijndael) implementation (cf. [8]) which is atypical as the key parts principally can be guessed independently and errors do not influence the later guesses. However, also in this case an efficient error location strategy is evaluated. We would like to insist on the fact that the proposed error management strategy is in fact much more general, and can be applied to many other divide and conquer attacks. Although the examples developed here correspond to physical attacks (timing attack, power analysis), divide and conquer attacks may also appear in algorithmically based attacks. This will be sketched in Sect. 6.

We point out that this paper was not written as a "manual" for potential attackers but shall help designers to assess the realism and the true threat of certain divide and conquer attacks.

2   Messerges et al.'s Power Analysis

Messerges et al.'s Multiple Exponent, Single Data (MESD) attack ([10]) corresponds to some extreme form of divide and conquer attack. Although this attack's efficiency leaves very little room for improvement through a new error-detection policy, we believe that, due to its simplicity, it constitutes a good example to start with. The reader may consider it as a witness of how much an adequate error-detection policy can improve an attack's performances.

The context is that of a smart card accepting to exponentiate a constant value with user-supplied exponents[2]. The attacker is able to measure the power

---

[2] We will not discuss this attack's realism here; see [10] for a brief discussion on the subject.

consumption of that card, and tries to guess the secret exponent used by that card. The attack could be described as follows. Assume the attacker already knows the most significant t bits $k_{w-1}, \ldots, k_{w-t}$ of the key:

Guessing phase attacker builds an exponent $e = k_{w-1} \cdots k_{w-t} r_{w-t-1} \cdots r_0$, where bits r_i are chosen at random, and submits this exponent to the card;

Error detection/location attacker compares the power consumption of exponentiation with e to the power consumption of exponentiation with the secret key k, and notes the position s of the first bit after which the two curves differ (due to measurement errors, this may require averaging on several exponentiations with same input);

Error correction attacker builds a new exponent $e' = k_{w-1} \cdots k_s \overline{k}_{s-1} r'_{s-2} \cdots r'_0$, where $\overline{k}_i$ denotes $1 - k_i$ and the r'_i are chosen at random.

In this case, the error location method is so efficient, allowing to point the error position almost exactly, that the guessing process can be limited to its simplest form (random guess). Similarly, this precision in error location allows a very simple error correction (bit inversion).

3   Timing Attack against RSA without CRT

In this section we describe a timing attack against RSA signature. The attack ([6]) was initially developed against a preliminary version of the Cascade ([2]) smart card, although it would work equally well against many other modular exponentiation algorithms without CRT (see [6,13]). Our approach – the optimal decision strategy derived in [13] combined with a very efficient error detection and correction strategy – increases the efficiency of the attack by about factor 50. As this attack constitutes a systematic approach to exploit side-channel information in an optimal way, we will describe the attack and the development of our error management policy in detail.

Remark 1. (i) The attack presented here is a pure timing attack, in the sense that the only information we dispose of is a set of messages and, for each of them, the total time required for signature;
(ii) in view of [6], the final version of the Cascade cryptographic library was later modified to resist against timing attacks ([5]).

3.1   Definitions and Mathematical Background

To compute $y^d (\mathrm{mod}\, M)$ the Cascade chip uses the simple square and multiply algorithm. Modular multiplications are carried out with Montgomery's algorithm ([11]). In its simplest variant $R := 2^\omega > M$ where ω fits to the device's hardware architecture. Let $R^{-1} \in Z_M := \{0, \ldots, M-1\}$ denote the multiplicative inverse of R in Z_M, i.e. $RR^{-1} \equiv 1 \ (\mathrm{mod}\, M)$. The integer $M' \in Z_R$ satisfies the integer equation $RR^{-1} - MM' = 1$. To simplify our notation we introduce the functions $\Psi, \Psi_*: Z \rightarrow Z_M$ defined by $\Psi(x) := xR(\mathrm{mod}\, M)$ and $\Psi_*(x) := xR^{-1}(\mathrm{mod}\, M)$. For $a' = \Psi(a)$ and $b' = \Psi(b)$ Montgomery's algorithm returns $s := \Psi_*(\Psi(a)\Psi(b)) = \Psi(ab)$.

Montgomery's algorithm

$z := a'b';$
$r := (z(\mathrm{mod}\, R))M' \;(\mathrm{mod}\, R)$
$s := \frac{z+rM}{R}$
if $(s \geq M)$ then $s := s - M$
return $s \quad (= \Psi_*(a'b') = a'b'R^{-1})$

Algorithm 1 (square and multiply using Montgomery's algorithm)

temp $:= \Psi(y);$
for i=w-2 downto 0 do {
    temp $:= \Psi_*(\mathrm{temp}^2);$
    if $(d_i = 1)$ temp $:= \Psi_*(\mathrm{temp} * \Psi(y));$
}
return $\Psi_*(\mathrm{temp});$

The secret exponent d has binary representation $(d_{w-1}d_{w-2}\ldots d_0)_2$ where d_{w-1} denotes its most significant bit. Further, $\mathrm{ham}(d)$ denotes the Hamming weight of d. The subtraction $s := s - M$ in Montgomery's algorithm is called *extra reduction* while $y \mapsto \Psi(y)$ and temp $\mapsto \Psi_*(\mathrm{temp})$ are the *pre-multiplication* and the *post-multiplication*. For any $a', b' \in Z_M$ we have $Time(\Psi_*(a', b')) = c$ if no extra reduction is necessary and $= c + c_{\mathrm{ER}}$ else. The sources of our timing attack are time differences which are caused by different numbers of extra reductions within the for-loop of Algorithm 1.

Remark 2. Many implementations (among which Cascade) use a more efficient multiprecision variant of Montgomery's algorithm (see e.g. [9], Algorithm 14.36) than the one listed above. This influences the absolute value of the constants c and c_{ER} but *not* the fact whether an extra reduction is necessary ([14], Remark 1). We hence clearly analyze the simplest variant of Montgomery's algorithm described above.

Let $t := Time(y^d(\mathrm{mod}\, M))$. For a sample $y_{(1)}, \ldots, y_{(N)} \in Z_M$ the attacker measures $\tilde{t}_{(1)} := t_{(1)} + t_{\mathrm{Err}(1)}, \ldots, \tilde{t}_{(N)} := t_{(N)} + t_{\mathrm{Err}(N)}$ where $t_{\mathrm{Err}(j)}$ denotes the measurement error. More precisely,

$$\tilde{t}_{(j)} = t_{\mathrm{Err}(j)} + t_{S(j)} + (w + \mathrm{ham}(d) - 2)c + r_{(j)}\, c_{\mathrm{ER}} \qquad (1)$$

where $t_{S(j)}$ denotes the time needed for set-up operations such as input, output, increasing the loop variable, evaluating the if-statements and, above all, the pre- and post-multiplication. Finally, $r_{(j)}$ denotes the number of extra reductions needed within the for-loop of Algorithm 1, i.e. $r_{(j)} = w_{1(j)} + w_{2(j)} + \cdots$ where $w_{i(j)} = 1$ if the ith Montgomery multiplication in Algorithm 1 requires an extra reduction for basis $y_{(j)}$ while $w_{i(j)} = 0$ else. From (1) we derive the "discretized running time"

$$\tilde{t}_{d(j)} := \frac{\tilde{t}_{(j)} - t_{S(j)} - (w + \mathrm{ham}(d) - 2)c}{c_{\mathrm{ER}}}. \qquad (2)$$

If $t_{\mathrm{Err}(j)} = 0$ (i.e. for exact time measurement), $\tilde{t}_{d(j)}$ equals $r_{(j)}$, the total number of extra reductions needed within the for-loop of Algorithm 1. The values $w_{1(j)}, w_{2(j)}, \ldots$ can be interpreted as realizations (i.e. values assumed by) of a particular non-stationary sequence of random variables $W_{1(j)}, W_{2(j)}, \ldots$ which are closely related with the Montgomery multiplications within Algorithm 1 (cf. [13], Sect. 6). (Numerous empirical experiments confirmed perfectly the suitability of this mathematical model.) In particular, the definition of $W_{i(j)}$ explicitly depends on the basis $y_{(j)}$ and whether the ith Montgomery multiplication within the for-loop in Algorithm 1 is a squaring (shortly: $type(i) ='\, Q'$) or a multiplication with $\Psi(y_{(j)})$ (shortly: $type(i) ='\, M'$), resp. To derive an optimal decision strategy, the sequence $W_{1(j)}, W_{2(j)}, \ldots$ has to be studied first. We briefly give the main results. For a proof the interested reader is referred to [13] (Lemma 6.3). In particular, the expectation of $W_{i(j)}$ (which equals the probability for an extra reduction in the ith Montgomery multiplication for basis $y_{(j)}$) is given by

$$E(W_{i(j)}) = \begin{cases} p_* := \frac{1}{3}\frac{M}{R} & \text{if } type(i) ='\, Q' \\ p_j := \frac{\Psi(y_{(j)})}{2M}\frac{M}{R} & \text{if } type(i) ='\, M'. \end{cases} \tag{3}$$

The covariance $\mathrm{Cov}(W_{i(j)}, W_{i+1(j)})$ equals

$$\begin{cases} \mathrm{cov}_{MQ(j)} := 2p_j^3 p_* - p_j p_* & \text{if } (type(i), type(i+1)) = ('M','Q') \\ \mathrm{cov}_{QM(j)} := \frac{9}{5}p_j p_*^2 - p_j p_* & \text{if } (type(i), type(i+1)) = ('Q','M') \\ \mathrm{cov}_{QQ} := \frac{27}{7}p_*^4 - p_*^2 & \text{if } (type(i), type(i+1)) = ('Q','Q') \end{cases} \tag{4}$$

whereas $W_{i(j)}$ and $W_{h(j)}$ are independent if $|i - h| > 1$. $\tag{5}$

3.2   Guessing d_k

Our attack estimates the exponent bits $d_{w-1}, \ldots, d_0$ successively. We assume that the attacker has already estimated the exponent bits $d_{w-1}, \ldots, d_{k+1}$ and that his estimators $\widetilde{d}_{w-1}, \ldots, \widetilde{d}_{k+1}$ have been correct. From these estimators he determines the respective temp values before the if-statement in the for-loop for $i = k$ for all bases $y_{(1)}, \ldots, y_{(N)}$ in his sample. Finishing the exponentiation $y_{(j)}^d (\mathrm{mod}\, M)$ still requires k squarings. The number m of remaining multiplications with $\Psi(y_{(j)})$ results from w, $ham(d)$ and $\widetilde{d}_{w-1}, \ldots, \widetilde{d}_{k+1}$. Subtracting the extra reductions already carried out from the discretized running time $t_{d(j)}$ yields the *remaining discretized running time* $\widetilde{t}_{\mathrm{drem}(j)}$ which is interpreted as realization of random variable

$$\widetilde{T}_{\mathrm{drem}(j)} = T_{\mathrm{dErr}(j)} + W_{w+ham(d)-k-m-1(j)} + \cdots + W_{w+ham(d)-2(j)}. \tag{6}$$

It is reasonable to assume that the (random) measurement error is independent of the running time and hence that $T_{\mathrm{dErr}(j)}$ is independent of $W_{w+ham(d)-k-m-1(j)} + \cdots + W_{w+ham(d)-2(j)}$. Further, we assume that it is normally distributed with expectation 0 and variance $\alpha^2 := \sigma_{\mathrm{Err}}^2 / c_{\mathrm{ER}}^2$.

From the 4-tuples $(\widetilde{t}_{\mathrm{drem}(1)}, u_{M(1)}, u_{Q(1)}, t_{Q(1)}), \ldots, (\widetilde{t}_{\mathrm{drem}(N)}, u_{M(N)}, u_{Q(N)}, t_{Q(N)})$ the attacker derives an estimator $\widetilde{d}_k$ for the unknown exponent bit d_k. Here $t_{Q(j)}$ (resp., $u_{M(j)}$, resp. $u_{Q(j)}$) $\in \{0,1\}$ equals 1 iff the next Montgomery multiplication (i.e., squaring, resp. multiplication by $\Psi(y_{(j)})$, resp. squaring after this multiplication) requires an extra reduction. Formally, this can be interpreted a statistical decision problem where the attacker has to decide between the two hypotheses $\theta = 0$ (corresponding to the case $d_k = 0$) and $\theta = 1$ (corresponding to $d_k = 1$) (cf. [13]). Theorem 1 gives the optimal decision strategy for the next bit value[3], i.e. a strategy which minimizes the probability that $\widetilde{d}_k \neq d_k$. As before, the letter N stands for the sample size.

Notations. Within theorem 1 we use the abbreviations
$$hn(0,j) := (k-1)p_*(1-p_*) + mp_j(1-p_j) + 2(m-1)\mathrm{cov}_{\mathrm{MQ}(j)} + 2(m-1)\mathrm{cov}_{\mathrm{QM}(j)} +$$
$$2(k-m-1)\mathrm{cov}_{\mathrm{QQ}} + 2\tfrac{k-m}{k-1}\mathrm{cov}_{\mathrm{QM}(j)} + 2\tfrac{m-1}{k-1}\mathrm{cov}_{\mathrm{QQ}} + \alpha^2,$$
$$hn(1,j) := (k-1)p_*(1-p_*) + (m-1)p_j(1-p_j) + 2(m-2)\mathrm{cov}_{\mathrm{MQ}(j)} +$$
$$2(m-2)\mathrm{cov}_{\mathrm{QM}(j)} + 2(k-m)\mathrm{cov}_{\mathrm{QQ}} + 2\tfrac{k-m+1}{k-1}\mathrm{cov}_{\mathrm{QM}(j)} + 2\tfrac{m-2}{k-1}\mathrm{cov}_{\mathrm{QQ}}$$
$$+\alpha^2,$$
$$ew(0,j \mid b) := (k-1)p_* + mp_j + \tfrac{k-m}{k-1}(p_{*Q(b)} - p_*) + \tfrac{m-1}{k-1}(p_{jQ(b)} - p_j)$$
$$ew(1,j \mid b) := (k-1)p_* + (m-1)p_j + \tfrac{k-m+1}{k-1}(p_{*Q(b)} - p_*) + \tfrac{m-2}{k-1}(p_{jQ(b)} - p_j)$$
with
$$p_{*Q(1)} := \tfrac{27}{7}p_*^3, \; p_{*Q(0)} := \tfrac{p_* - p_* p_{*Q(1)}}{1-p_*}, \; p_{jQ(1)} := \tfrac{9}{5}p_* p_j \text{ and } p_{jQ(0)} := \tfrac{p_j - p_* p_{jQ(1)}}{1-p_*}.$$

Theorem 1. *(i) Assume that the estimators $\widetilde{d}_{w-1}, \ldots, \widetilde{d}_{k+1}$ have been correct and that $d_k + \cdots + d_0 = m$, i.e. for each exponentiation m Montgomery multiplications of type 'M' still have to be carried out. Let*

$$\psi_{N,d} : (\,\mathbb{R} \times \{0,1\}^3)^N \to \mathbb{R}, \qquad \psi_{N,d}((\widetilde{t}_{\mathrm{drem}(1)}, u_{M(1)}, \ldots, u_{Q(N)}, t_{Q(N)})) :=$$
$$-\frac{1}{2}\sum_{j=1}^{N}\left(\frac{(\widetilde{t}_{\mathrm{drem}(j)} - t_{Q(j)} - ew(0,j \mid t_{Q(j)}))^2}{hn(0,j)} - \right.$$
$$\left.\frac{(\widetilde{t}_{\mathrm{drem}(j)} - u_{M(j)} - u_{Q(j)} - ew(1,j \mid u_{Q(j)}))^2}{hn(1,j)}\right).$$

Then the deterministic decision strategy $\tau_d \colon (\mathbb{R} \times \{0,1\}^3)^N \to \{0,1\}$ defined by

$$\tau_d = 1_{\psi_{N,d} < \log(\frac{m-1}{k-m+1}) + \frac{1}{2}\sum_{j=1}^{N}\log(1+c_j)} \qquad \text{with } c_j := \frac{hn(0,j) - hn(1,j)}{hn(1,j)} \quad (7)$$

is optimal.

A nice property of the optimal decision strategy described above is that it allows to detect errors. It can be shown that, after an error has occurred (i.e. bit $\widetilde{d}_{k*} \neq d_{k*}$), the probability $\mathrm{Prob}(\widetilde{d}_k = 1)$ to guess 1 for subsequent bits is about 0.20 (although the exact value is parameter-dependent and thus changes during the attack). The proof of this fact, as well as precise probabilities, can be found in [13] (Theorem 6.5); for the sake of simplicity, however, we will skip these – rather complex – expressions here, and focus on the way we can exploit this error witness.

---

[3] For a proof of this theorem, the interested reader is referred to [13] (proof of Theorem 6.5(i)).

Remark 3. It can also be shown that the error probability $\text{Prob}(\widetilde{d}_k \neq d_k)$ decreases as the attack proceeds. To quantify this probability, the distribution of $Z := \psi_{N,d}(\widetilde{T}_{\text{drem}(1)}, U_{M(1)}, \ldots, U_{Q(N)}, T_{Q(N)})$ has to be determined for both alternatives $\theta = 0$ and $\theta = 1$. (We interpret $\widetilde{t}_{\text{drem}(1)}, \ldots, t_{Q(N)}$ as realizations of specific random variables $\widetilde{T}_{\text{drem}(1)}, \ldots, T_{Q(N)}$.) The minimal sample size which guarantees a particular error probability (e.g., 0.01) is essentially linear in k. Once again, we refer the interested reader to [13] (Theorem 6.5) or to [15] (theorem without proof) for precise expressions.

Let us illustrate these two facts (low probability to guess a 1 after an error has occurred and decreasing error probability as the attack proceeds), by an example (cf. [13]):

Example 1. In (a) and (b) below we assume that the estimators $\widetilde{d}_{w-1}, \ldots, \widetilde{d}_{k+1}$ have been correct. For randomly chosen bases $y_{(1)}, \ldots, y_{(N)}$ the following inequalities hold in good approximation.
Let $M/R = 0.7$, $\alpha^2 = 0$, $N \geq 5620$, and further ....
   (a) ... $(k, m) = (510, 255)$. Then $\text{Prob}(\widetilde{d}_k \neq d_k) \leq 0.01$.
   (b) ... $(k, m) = (440, 220)$. Then $\text{Prob}(\widetilde{d}_k \neq d_k) \leq 0.0064$.
   (c) ... $(k^*, m^*, k, m) = (505, 250, 470, 235)$.
         Then $\text{Prob}(\widetilde{d}_k = 1) \leq 0.1879$ and $\text{Prob}(\psi_{N,d} < E_{\theta=1}(Z) \mid \widetilde{d}_k = 1) \leq$
0.0170.

The error detection and correction strategy described below is more efficient than its pre-variant described in [13]. Roughly speaking, to estimate the exponent bits the attacker derives a sequence of $\psi_{N,d}$-values on which his decisions are based on. These values themselves can be interpreted as realizations of random variables (cf. Fig. 1 below) whose distribution changes noticeably after the first wrong guessing.

3.3   Error Detection

The following diagram illustrates the facts we will base our error detection on. The curves g_0, g_1 and g_f are the density functions of $Z := \psi_{N,d}(\cdot)$ defined in Theorem 1 if $d_k = 0$, $d_k = 1$, or if $\widetilde{d}_{k'}$ was false for any $k' > k$, resp. (In particular, g_0, g_1 and g_f are normal densities; cf. the proof of Theorem 6.5 in [13].) As we have seen, in the latter case, it is unlikely to derive $\widetilde{d}_k = 1$. If this yet happens the $\psi_{N,d}(\cdot)$-value is $\geq E_{\theta=1}(Z)$ (the expectation of Z if $\theta = 1$) with high probability. Both observations will be the basis of our error detection and correction strategy. The arrow in Fig. 1 points to the area corresponding to the probability that $\psi_{N,d} < E_{\theta=1}(Z)$.

The task of an error detection strategy is to check whether an error has occurred. The error probability decreases when k decreases so that estimation errors usually occur at the beginning of the attack, at least if $\alpha^2 = 0$ and if the attacker knows the values $w, ham(d), c, c_{\text{ER}}$ and t_S or has guessed them correctly (cf. Subsect. 3.5). Note that there is also a possibility for errors to

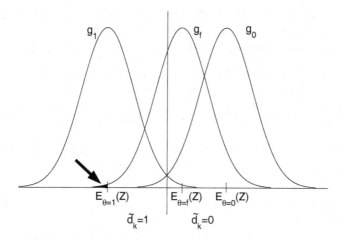

Fig. 1. Density functions of $\psi_{N,d}$ depending on d_k

occur at the very end of the attack if the parameter guesses are not absolutely correct; this, however, is no serious harm as the few final exponent bits may be checked exhaustively: we will thus leave this case aside.

The basic idea of our error-detection strategy is the following: before guessing a new bit d_k, we will look at a "window" of the f preceding bits, and test the two (non-complementary!) hypotheses:

(A) The estimators $\tilde{d}_{w-2}, \ldots, \tilde{d}_{k+1}$ are correct.

(B) There is an index $k' > k + f$ for which $\tilde{d}_{k'}$ is wrong.

Roughly speaking, a false hypothesis (A) will be witnessed by an unusually large number of 0s in the window, whereas a large number of 1s tends to infirm hypothesis (B). If we can neither reject (A) nor (B), then we simply increase the window's length and test the hypotheses again; if the window reaches its maximum size with yet no decision being possible, the attack advances one position, deciding the next bit d_k as usual.

If hypothesis (B) is rejected, then we can set up a *confirmed index* at position $k + f + 1$, meaning that no error has occurred before this point. Therefore we will not try to modify any bit located before it, i.e. with greater or equal index than a confirmed index.

If hypothesis (A) is rejected ("alarm"), we must choose one index $k'' \in \{con - 1, \ldots, k + 1\}$, revert that decision, and start the attack from that point over again. The attack will continue until either:

− a new confirmed index is established, in which case the algorithm "forgets" an alarm had occurred and continues the attack (having corrected the error at k''), or

– a further alarm occurs, in which case we conclude the index k'' was not the first error position; we then choose a new index in the same set $\{con - 1, \ldots, k+1\}$ and restart from that point.

To determine the threshold values that will conduct to the rejection of hypotheses (A) and (B), we interpret the sum $nones := \widetilde{d}_{k+f} + \cdots + \widetilde{d}_{k+1}$ as a realization of a binomial random variable X_1. If (A) is true, then this variable's distribution is $X_1 \sim B(f, 0.5)$; we therefore reject (A) if $\text{Prob}(X_1 \leq nones) < p_{al} := 0.0001$.

The condition to reject (B) is a bit more complicated: in a setup-step, we used Theorem 6.5(iii) ([13]) to obtain rough approximators p_{err} and p_{len}, resp., for the average probabilities $\text{Prob}(\widetilde{d}_k = 1)$ or $\text{Prob}(\psi_{N,d} < E_{\theta=1}(Z) \mid \widetilde{d}_k = 1)$, resp., within the initial stage of the attack under the condition that $\widetilde{d}_{k'}$ was false for a $k' > k$. We reject (B) if $len := |\{k+1 \leq i \leq k+f \mid \widetilde{d}_i = 1, \psi_{N,d} < E_{\theta=1}(Z)\}| \geq 2$ and $\text{Prob}(X_2 \geq len) < p_{con} := 0.0005$ where X_2 is $B(nones, p_{len})$-distributed. The choice of the probabilities p_{al} and p_{con} was somewhat arbitrary. The values 0.0001 and 0.0005 turned out to be suitable.

3.4   Error Correction

We have not detailed yet in which order the successive k'' are chosen during the successive alarms: after a "new" alarm (i.e. an alarm when any preceding one has been "forgotten") the indices $\{con - 1, \ldots, k+1\}$ are ordered with respect to the rank function defined below. The k'' will then be chosen in that order, the index with the smallest rank first, until the next "new" alarm.

The rank function is determined on the basis of two criteria: first, for reasonable sample size N the error probability for a single decision is small and thus the (wrong) decision for the first false estimator should be "close". To quantify this idea, let $\text{absdiff}(i)$ denote the number of indices s in $\{i+8, \ldots, i+1\}$ for which the absolute value $|\psi_{N,d} - decbound|$ is smaller than the respective term for index i. Here $decbound$ corresponds to the decision boundary, marking the limit between a decision towards 0 and towards 1 in the decision strategy, and is defined as $\log(\frac{m-1}{k-m+1}) + \frac{1}{2}\sum_{j=1}^{N} \log(1 + c_j)$ (cf. Theorem 1). Intuitively, $\text{absdiff}(i)$ denotes the number of recent (relative to i) positions for which the decision was more difficult to make than for index i.

Second, from the $\psi_{N,d}(\cdot)$-values from which the estimators $\widetilde{d}_{con-1}, \ldots, \widetilde{d}_{k+1}$ were derived we guess the "region" where the first false estimator is located. For this, we define the intervals $I_1 := (-\infty, E_{\theta=1}(Z)]$, $I_2 := (E_{\theta=1}(Z), decbound]$, $I_3 := (decbound, E_{\theta=0}(Z)]$, and $I_4 := (E_{\theta=0}(Z), \infty)$ (these regions can be easily visualized on Fig. 1). For $s \in \{con - 1, \ldots, k+1\}$ we set $iv(s) := j$ if the respective $\psi_{N,d}(\cdot)$-value was contained in I_j. Let us now consider the distribution of successive estimators among these regions, and compare these distributions before and after an error. Denote by k^* the first error position:

– before that error, the estimators should be equidistributed among the 4 regions; that is, $\text{Prob}(iv(s) = j) \approx 0.25$ for $s > k^*$;

- at the error position, the probability is high for the estimator to be in I_2 or I_3; in other words, $\mathrm{Prob}(iv(k^*) = 2), \mathrm{Prob}(iv(k^*) = 3) \approx 0.5$;
- finally, we have seen that, after the error has occurred, decisions will be biased towards 0; the precise distribution is given by Theorem 6.5(iii) in [13], that we omit here for simplicity.

To determine the region where the first error was located, we define $L(t) :=$ $0.25^{\mathrm{con}-t-1} \prod_{s=k+1}^{t} \mathrm{Prob}(iv(s)))$ and retain as k'' the value of t that maximized this term. (In fact, the terms $\mathrm{Prob}(iv(s))$ equal the respective probabilities under the assumption that the first error occurred at index t. The term $L(t)$ is an approximation for the probability for the observed $iv(\cdot)$-values to have occurred if the first error position was t (cf. also [15]).)

Finally, we define posmax: $\{\mathrm{con} - 1, \ldots, k + 1\}$, $\mathrm{posmax}(i) := |k'' - i|$ and use the rank function

$$\mathrm{rank}(i) := 7.0 * \mathrm{absdiff}(i) + 3.0 * \mathrm{posmax}(i) . \qquad (8)$$

In the same way as for the probabilities p_{err} and p_{len} we used Theorem 6.5(iii) in [13] to pre-compute average values for the probabilities $\mathrm{Prob}(iv(s) = j)$ to save computation time. The rank function turned out to be very efficient (cf. Subsect. 3.5).

3.5   Practical Experiments / Efficiency

Although the original attack also used an error correction strategy, about $200000 - 300000$ time measurements were necessary to recover a 512-bit key. In this section we present empirical results where we applied the optimal decision strategy stated in Subsect. 3.2 and the error detection and correction strategy from Subsects. 3.3 and 3.4. For our experiments we distinguished two cases. In the *ideal case* we assumed that the time measurements are exact, the attacker knows the constants and parameters c, c_{ER}, w and $\mathrm{ham}(d)$ and he is able to determine the setup time $t_{(S)}$ exactly. (This corresponds to a computer emulation of Algorithm 1 with the number of extra reductions as output.) However, if the attacker does not have exact knowledge of the smart card implementation these assumptions may be not realistic. Therefore, we also conducted a suggestively called *real-life attack* where the attacker's knowledge and his abilities were assumed to be lower, using actual timing measurements from a smart card running a cryptographic library ([2]). In this case, we assumed that the attacker does not have precise implementation knowledge and exploited various relations to estimate these values. We refer the interested reader to [15] (Sect. 6) for further details.

Remark 4. As no physical smart card is available yet, timing measurements were in fact performed using an emulator [1] which predicts the running time (in clock cycles) of a program. The code we used was the ready-for-transfer version of the Cascade library, i.e. with critical routines directly written in the card's native assemble language. Since the emulator is designed to allow implementors to optimize their code before "burning" the actual smart cards, its predictions should match almost perfectly. Consequently, physical attacks on the smart card should not induce many measurement errors more.

Tables 1 and 2 contain empirical results where we used the optimal decision strategy without (Table 1) and combined with error detection and correction (Table 2). The last two columns respectively correspond to an ideal case, in which all $t_{d(j)}$ are measured exactly, and to the real-life attack of the Cascade chip.

Table 1. Optimal decision strategy without error correction

Key size (bits)	number of measurements	Success rate ideal case	real-life attack
512	5 000	12%	15%
512	6 000	35%	32%
512	7 000	55%	40%
512	8 000	65%	46%
512	9 000	95%	72%
512	10 000	98%	92%

Table 2. Optimal decision strategy with error correction

Key size (bits)	number of measurements	Success rate ideal case	real-life attack
512	5 000	85%	74%
512	6 000	95%	85%
512	7 000	98%	89%
512	8 000	100%	91%
512	9 000	100%	94%
512	10 000	100%	100%

Compared with the original attack presented in [6], our approach – optimal decision strategy combined with the error detection and correction strategy described below – has improved the efficiency by a factor of about 50 for 512-bit keys, at no cost in prerequisites, generality, or complexity. This improvement factor is expected to grow even further with larger keys. The main portion of the improvement is obviously due to the optimal bit estimation strategy but a brief comparison between both tables shows that the applied error handling itself reduces the sample size by about a 40 per cent. In the ideal case in more than 70 per cent of the trials the index of the first false decision was ranked on position one or two. In the real-life attack the efficiency of the rank function is somewhat lower.

To conclude this section, an important point to note is that, in this case, the detection and correction of errors does not cost any additional time measurement.

4   Timing Attack against RSA with CRT

4.1   Description of the Attack

Let $n = p_1 p_2$ denote an RSA modulus where p_1 and p_2 are large primes. If the CRT is used the computation $y \mapsto y^d (\mathrm{mod}\, n)$ decomposes into three steps:

Step 1: $y_1 := y(\mathrm{mod}\, p_1)$. Compute $x_1 := (y_1)^{d'} (\mathrm{mod}\, p_1)$
Step 2: $y_2 := y(\mathrm{mod}\, p_2)$. Compute $x_2 := (y_2)^{d''} (\mathrm{mod}\, p_2)$
Step 3: Return $(b_1 x_1 + b_2 x_2)(\mathrm{mod}\, n)$

The parameters d', d'', b_1 and b_2 are precomputed once. In particular, $d' := d(\mathrm{mod}\, (p_1 - 1))$, $d'' := d(\mathrm{mod}\, (p_2 - 1))$ while $b_1 \equiv 1 \pmod{p_1}$ and $b_1 \equiv 0 \pmod{p_2}$, and similarly, $b_2 \equiv 0 \pmod{p_1}$ and $b_2 \equiv 1 \pmod{p_2}$. Unlike as in Sect. 3 the attacker knows neither the bases y_i nor the moduli p_i in Steps 1 and 2. In particular, the "classical" timing attacks ([7], [6], [13]) do not work. In [14] however, a new timing attack was introduced which enables the factorization of n if the modular multiplications $(\mathrm{mod}\, p_i)$ are carried out with Montgomery's algorithm. We briefly describe the attack. For details, the interested reader is referred to [14].

As the primes p_1 and p_2 are of similar size it is reasonable to assume the Montgomery constant R (cf. Sect. 3) is equal for both multiplications (mod p_1) and (mod p_2). For simplicity, for the moment we assume that the modular exponentiations in Steps 1 and 2 are carried out with the square & multiply algorithm. In accordance to Sect. 3 we define the mappings $\Psi_i: Z \to Z_{p_i}$ by $\Psi_i(z) := zR (\mathrm{mod}\, p_i)$. Let $0 < u_1 < u_2 < n$ with $u_2 - u_1 < p_1, p_2$. Three cases are possible:

Case A: $\{u_1 + 1, \dots, u_2\}$ does not contain a multiple of p_1 or p_2.
Case B: $\{u_1 + 1, \dots, u_2\}$ contains a multiple of one of p_1 or p_2 but not of both.
Case C: $\{u_1 + 1, \dots, u_2\}$ contains a multiple of both p_1 or p_2.

Let R^{-1} denote the multiplicative inverse of R modulo n. For input $uR^{-1}(\mathrm{mod}\, n)$ clearly $\Psi_i(uR^{-1} (\mathrm{mod}\, n)) = u (\mathrm{mod}\, p_i)$ for $i = 1, 2$. In Step i the probability for an extra reduction in a Montgomery multiplication with $u (\mathrm{mod}\, p_i)$ equals $(u (\mathrm{mod}\, p_i)/2R)$ (cf. (3) or Theorem 1 in [14]) while the probability for an extra reduction in a squaring is $p_i/3R$, independent of the base. The running time for the input $uR^{-1} (\mathrm{mod}\, n)$, denoted with $T(u)$, is interpreted as a realization of a random variable X_u (cf. [14]). The expectation of the difference $X_{u_2} - X_{u_1}$ depends essentially on the fact whether Case A, Case B or Case C is true:

$$E(X_{u_2} - X_{u_1}) \approx \begin{cases} 0 & \text{in Case A} \\ -\frac{c_{ER}}{8} \frac{\sqrt{n}}{R} & \text{in Case B} \\ -\frac{c_{ER}}{8} \frac{2\sqrt{n}}{R} & \text{in Case C.} \end{cases} \tag{9}$$

This observation is essential for the attack which falls in three phases: In Phase 1 an "interval set" $\{u_1 + 1, \ldots, u_2\}$ has to be found which contains an integer multiple of p_1 or p_2. Starting from this set, in Phase 2 a sequence of decreasing interval subsets has to be determined, each of which containing an integer multiple of p_1 or p_2. More precisely, in each step of Phase 2 it is checked whether the upper subset $\{u_3 + 1, \ldots, u_2\}$ with $u_3 := \lceil (u_1 + u_2)/2 \rceil$ contains such a multiple or not. The decisions in Phase 1 and 2 are based on the time differences $T(u_2) - T(u_1)$ or $T(u_2) - T(u_3)$, resp., where the attacker decides for "Case A" iff $T(u_2) - T(u_1) > -c_{\mathrm{ER}} \sqrt{n}/16R$ or $T(u_2) - T(u_3) > -c_{\mathrm{ER}} \sqrt{n}/16R$, resp. (Note that there is no need to distinguish between Cases B and C.) When the actual subset $\{u_1 + 1, \ldots, u_2\}$ is sufficiently small Phase 3 begins where $\gcd(u, n)$ is calculated for all u contained in this subset. If all decisions within Phase 1 and 2 were correct then the final subset indeed contains a multiple of p_1 or p_2. Then Phase 3 delivers the factorization of n.

4.2   Detecting Errors

However, at any instant within Phase 2 the attacker can verify with high probability whether his decisions were correct so far, i.e. whether the actual interval $\{u_1 + 1, \ldots, u_2\}$ really contains a multiple of p_1 or p_2. He just has to apply the decision rule to a time difference for neighbouring values of u_1 and u_2, resp., e.g. to $T(u_2 - 1) - T(u_1 + 1)$. If this leads to the same decision it is confirmed with overwhelming probability that the interval $\{u_1 + 1, \ldots, u_2\}$ truly contains a multiple of p_1 or p_2. (We then call $\{u_1 + 1, \ldots, u_2\}$ a *confirmed interval*.)

Otherwise, the attacker evaluates a further time difference (e.g. $T(u_2 - 2) - T(u_1 + 2)$). Depending on this difference he either finally confirms the interval $\{u_1 + 1, \ldots, u_2\}$ or restarts the attack at the preceding confirmed interval $\{u_{1;c} + 1, \ldots, u_{2;c}\}$ using values u_1' and u_2' close to $u_{1;c}$ and $u_{2;c}$, resp.

As opposed to the attack of section 3, this error detection method requires additional time measurements to be carried out, and has therefore significant impact on the attack's efficiency. It would thus be useful to be able to reduce the number of detection steps applied.

In [14] a static error detection and correction is applied: After a pre-assigned number of steps (adapted to the parameters n and R) a new confirmed interval is tried to be established. If this fails (due to a preceding error) the attack is restarted at the preceding confirmed interval. For $n \approx 0.7 \cdot 2^{1024}$ and $R = 2^{512}$ practical attacks required 570 time measurements in average where confirmed intervals were tried to establish after each 42 steps.

Remark 5. (i) This attack is somewhat atypical as it does not directly reconstruct the secret exponent d itself. Instead, the interval sequence delivers the bit representation of a multiple of p_i, beginning with the most significant bits. This multiple, however, may be viewed as a "key" as its knowledge enables the factorization of the modulus.

(ii) If the the initial value u_2 in Phase 1 is chosen sufficiently small the attacker will find a prime factor p_i itself rather than just a multiple of it. Using an algorithm of Coppersmith (cf. [3]), it then suffices to reconstruct the upper half of the bit representation of p_i which almost halves the number of time measurements in Phase 2. For $n \approx 0.7 \cdot 2^{1024}$ and $R = 2^{512}$, for example, about 300 time measurements are sufficient for the whole attack (cf. [14] (Remark 6)).

(iii) Although its efficiency decreases the attack also works if table methods are used in Steps 1 and 2 of the CRT. For a 4-ary table (storing $2^4 - 1 = 15$ values), for example, about 17700 time measurements are required in average ([14], Sect. 7).

4.3   Dynamic Error Detection and Correction

However, this static error detection and correction strategy can still be improved, using a similar argument to the one of section 3.3. In fact, after a wrong decision in Phase 2 the following decisions should always be "no multiple of p_1 or p_2 in the upper subset $\{u_3 + 1, \ldots, u_2\}$". If all decisions were correct so far, however, within Phase 2 this event should occur with probability $1/2$.

This suggests the following error detection and correction strategy: If in none of the preceding v (let's say $v = 13$) steps a multiple of p_1 or p_2 was assumed in the respective upper subinterval then try to confirm the actual interval $\{u_1 + 1, \ldots, u_2\}$ by evaluating $T(u_2 + 1) - T(u_1 + 1)$. If this attempt fails, restart the attack at the last but one interval where a multiple of p_1 or p_2 was assumed in the upper subset (restart with neighbored values).

For $n \approx 0.7 \cdot 2^{1024}$ and $R = 2^{512}$, for example, the static strategy described above requires about 45 of 570 time measurements in average for error detection or correction reasons. However, the probability for a single decision in Phase 1 or 2 to be wrong is about 0.001. Hence about 0.5 errors are expected within the attack whereas the probability for a "false alarm" is about $2^{8-13} = 1/16$. Thus the proposed new error detection and correction strategy costs about $0.5(2+15+2) + 2/32 \approx 10$ time measurements which reduces the average number of time measurements to about 535 which is a reduction by 6 per cent. If the modular exponentiations in Steps 1 and 2 of the CRT use tables, then the portion of time measurements carried out due to error detection and correction reasons is considerably larger than for the square & multiply algorithm. Consequently, the gain of efficiency caused by a dynamic error detection and correction strategy (with an adapted value v) also increases.

5   Timing Attack against AES (Rijndael)

5.1   Brief Description of Rijndael and the Vulnerable Model

A complete description of Rijndael can be found in [4]. We will focus here on the parts of interest for the attack.

A Rijndael encryption consists in an initial round key addition, followed by M round transformations, the last round being slightly different from the others. The different transformations applied during each round operate on an array of bytes, named the *state*, composed of 4 lines and M_b columns (where M_b is the block size, in 32-bit words). Basically, each round, except the last one, consists of the following steps: ByteSub (byte-by-byte substitution S), ShiftRow (fixed permutation of bytes), MixColumn (described below) and AddRoundKey (round key $\oplus$ state) where "$\oplus$" denotes the bitwise XOR-addition. One AddRoundKey operation is performed before the first round. In the final round there is no MixColumn operation.

The MixColumn transformation operates on the columns of the state and applies them the following matrix multiplication:

$$
\begin{bmatrix} b_0 \\ b_1 \\ b_2 \\ b_3 \end{bmatrix} = \begin{bmatrix} 02 & 03 & 01 & 01 \\ 01 & 02 & 03 & 01 \\ 01 & 01 & 02 & 03 \\ 03 & 01 & 01 & 02 \end{bmatrix} \begin{bmatrix} a_0 \\ a_1 \\ a_2 \\ a_3 \end{bmatrix} \tag{10}
$$

where the multiplication is defined in $GF(2^8)$ as multiplication of binary polynomials modulo the irreducible polynomial $x^8 + x^4 + x^3 + x + 1$.

Due to the choice of the matrix and irreducible polynomial, MixColumn can be implemented very efficiently: first, it is easy to see that, as '03'='02'+'01', the only multiplications that will actually have to be performed are by '02'; second, it can be showed (cf. [4]) that multiplication by '02' in GF(256) can be implemented as follows:

- shift byte one position left,
- if a carry occurs, XOR the result with hexadecimal '1B'.

Assumptions. If the implementation is careless, the multiplications with '02' and '03' will not take constant time, but will take longer in the case where a carry occurs. Throughout the rest of this section, we assume such a behaviour. Moreover, we assume that the time needed for the remaining operations, i.e. for substitutions, permutations and XOR additions, does not depend on the plaintext and thus is constant for all encryptions. (However, slight deviations could be interpreted as a part of the measurement error; cf. Subsection 5.3.) For simplicity, we further assume that the length of the key ($=F$ bytes) is not larger than that of a plaintext block ($=4M_b$ bytes). Finally, we assume that the attacker knows at least one pair $(p, c := Enc(p; k))$ where $Enc(\cdot; k)$ stands for a Rijndael encryption with the secret (unknown) key k. This pair will be used to check whether a key candidate is correct or not.

5.2   Basic Idea

Let us focus on what will happen to a given byte (say, the first) of a known plaintext during the first few encryption sub-steps:

- before entering the first round, that byte will be XOR-ed with the first byte of the first round key (equals the first byte of k) – call it R_1 – whose value is unknown, but constant independent of the plain text;

- then a substitution S, according to a known S-box, will take place;
- the byte will then be moved around (to a known place) by ShiftRow, without being modified;
- finally, MixColumn will be applied; during this operation, the byte will be multiplied at least once by '02' (the result may be stored and reused for the multiplication with '03'). By assumption the multiplications will take longer if the first bit of the byte is set.

Could we observe the time taken by that peculiar multiplications, we would then be able to deduce the value of R_1 with about 8 encryptions. Of course, this is not the case, as we have no access to partial timings, but only to the total encryption time. However, if we encrypt a large amount of random messages, we can expect all other operations to behave as random noise and therefore hope to be able to "filter out" the information we are interested in. Naturally, the same method can be applied to guess the other bytes $R_2, \ldots, R_F$ of the searched secret key k.

To make the attack robust against errors induced by random noise, we will not build a single answer, but rather a set of possible keys, small enough to make exhaustive search easy. More precisely, for each index $j \leq F$ the attacker determines a (small) set of "candidates", denoted with Ca_j. After the "guessing phase" has been finished the attacker computes $Enc(p; ca_1\|ca_2\| \cdots \|ca_F)$ for all possible combinations of key byte candidates (i.e. $ca_j \in Ca_j$ for each $j \leq F$). If $Enc(p; ca_{1;0}\|ca_{2;0}\| \cdots \|ca_{F;0}) = c$ then $k = ca_{1;0}\|ca_{2;0}\| \cdots \|ca_{F;0}$ with overwhelming probability.

5.3   The Pure Attack

We begin with a definition.

Definition 1. *The Greek letter Δ denotes the extra time needed for the multiplications of a byte with '02' and '03' if a shift occurs. The term $N(\mu, \sigma^2)$ denotes a normal distribution with mean μ and variance σ^2. In particular, $f(t) := e^{-(t-\mu)^2/2\sigma^2}/\sqrt{2\pi}\sigma$ denotes its Lebesgue density.*

In the first step the attacker randomly chooses plaintexts $p_1, \ldots, p_N$ and measures the running times $\widetilde{t}_1, \ldots, \widetilde{t}_N$ needed for their encryption with the secret key k. In particular, $\widetilde{t}_j = t_j + t_{Err}$ with $t_j = Time(Enc(p_j; k))$ while t_{Err} denotes the measurement error. Then he initializes tables $U_1, \ldots, U_F$ where $U_s := (u_s[i][j])_{0 \leq i \leq 255; 1 \leq j \leq N}$. The component $u_s[i][j]$ equals the most significant bit of $S(i \oplus s^{th}$ byte of $p_j)$. In other words, $u_s[i][j]$ tells – assuming the sth byte of the key, R_s, is equal to i – whether the first multiplication by '02' applied to the corresponding byte involves a shift or not.

If $R_s \neq i$ the measured running time $\widetilde{t}_j$ can be interpreted as a realization of a normally distributed random variable $\widetilde{T}_j \sim N(\mu, \sigma^2)$ where $\sigma^2 = \sigma_\Delta^2 + \sigma_{Err}^2$ with $\sigma_\Delta^2 := 4M_b(M-1)\Delta^2/4$ while σ_{Err}^2 denotes the variance of the measurement error. However, if $R_s = i$ the table entry $u_s[i][j]$ provides an additional piece of information (concerning particular multiplications by '02' and '03').

Consequently, $\widetilde{T}_j \sim N(\mu + \Delta(u_s[i][j] - 0.5), \sigma^2 - \Delta^2/4)$. The original question, namely whether $R_s = i$, can be viewed as a more general problem which is independent of any cryptographic context: namely, whether $\widetilde{T}_j \sim N(\mu, \sigma)$ or $\widetilde{T}_j \sim N(\mu + \Delta(u_s[i][j] - 0.5), \sigma^2 - \Delta^2/4)$ for $j \leq N$. The measured running times $\widetilde{t}_1, \ldots, \widetilde{t}_N$ is the only information available. For this, statistical decision theory can be applied.

Remark 6. (parameter estimation) For simplicity, first assume that the attacker is able to use a device identical to the one he wants to attack with a known key. He then randomly generates plaintexts $p'_1, \ldots, p'_{N1}$ and $p''_1, \ldots, p''_{N1}$ with constant first byte in each subset such that *for the known key* for the respective multiplications by '02' and '03' a shift occurs (subset 1), resp. does not occur (subset 2). The difference in the mean values delivers an estimator for Δ (and thus for σ^2_Δ) and their arithmetical mean an estimator for μ. If the attacker cannot use a known key the estimation strategy is similar; he just has to identify two subsets with the properties from above, which can easily be done as follows: he starts by building two subsets of plaintexts (A, B), in which the first byte of every element of A and the first byte of every element of B are fixed to (two different) constants. If the average time for processing subset A significantly differs from that for subset B, then the attacker has identified the two desired sets; otherwise he simply repeats the operation with a third subset, comparing its average processing time with that of the second etc.

Roughly speaking, in a statistical decision problem the statistician (here: the attacker) estimates the unknown distribution of random variable(s), p_θ, on the basis of an observation $\omega \in \Omega$. The set Θ describes all possible alternatives. We will not consider statistical decision problems in full generality (cf., e.g. [17]) but apply the mechanisms to our specific problem. Here the parameter set Θ equals $\{0, 1\}$ where $\theta = 0$ denotes the case $i = R_s$ while $\theta = 1$ stands for $i \neq R_s$. The probability that for byte i the hypothesis $\theta = 0$ (resp., $\theta = 1$) is true equals $\eta_0 = 1/256$ (resp., $\eta_1 = 255/256$). If the attacker decides for $\theta = 0$ although $\theta = 1$ is true he needlessly adds one candidate i to Ca_s which increases the time for final exhaustive search. Erroneously deciding for $\theta = 1$, i.e. cancelling the true byte value R_s, is much worse as the attack must fail in the end. To quantify these considerations, we introduce the loss function $s(\theta, \theta') \geq 0$ where the first component denotes the correct parameter and the second the estimated one. Of course, $s(0, 0) = s(1, 1) = 0$ (correct decisions). We further set[4] $s(1, 0) = 1$ and $s(0, 1) = 100$. Of course, the attacker uses the decision strategy d_{opt} which minimizes the expected loss, that is, $d_{opt}(\widetilde{t}_1, \ldots, \widetilde{t}_N) := u \in \{0, 1\}$ if u minimizes the term

$$\sum_{\theta=0}^{1} \eta_\theta s(\theta, u) \prod_{j=1}^{N} f_{\theta;j}(\widetilde{t}_j). \tag{11}$$

---

[4] The ratio $s(0, 1)/s(1, 0)$ is somewhat arbitrary. Increasing $s(0, 1)$ reduces the probability that the correct value is cancelled but simultaneously increases the candidate set Ca_s.

For the moment, $f_{1;j}$ (resp., $f_{0;j}$) denotes a normal density with mean μ and variance σ^2 (resp., with mean $\mu + \Delta(u_s[i][j] - 0.5)$ and variance $\sigma^2 - \Delta^2/4$). As the variances for $\theta = 0$ and $\theta = 1$ are almost equal the attacker may use the simplified decision strategy $d'_{opt}(\widetilde{t}_1, \ldots, \widetilde{t}_N) = 0$ instead iff

$$\sum_{j=1}^{N} \left((\mu + \Delta(u_s[i][j] - 0.5) - \widetilde{t}_j)^2 - (\mu - \widetilde{t}_j)^2 \right) < 2\sigma^2(\log(s(0,1) - \log(255))$$

$$(12)$$

with only minimal loss of efficiency. The attacker applies this decision rule to each $i \in \{0, \ldots, 255\}$ which yields the candidate set Ca_s.

5.4   Error Location and New Guesses

This timing attack on Rijndael is an atypical divide and conquer attack as a wrong key estimator does not influence the following partial key guesses. In principle, the attacker could thus apply the decision strategy from the preceding section independently to all key bytes and completely resign on any intermediate error detection strategy. On the other hand, the candidate sets must be kept small enough for the verification phase to remain practically feasible. We propose the following strategy.

We assume that the attacker has already determined the candidate sets $Ca_1, \ldots, Ca_{s-1}$. To derive Ca_s we initialize a matrix $(c_m, i)_{1 \leq m \leq |Ca_{s-1}|; 0 \leq i \leq 255}$ where $c_m \in Ca_{s-1}$. Our aim is to "tick" all components for which at least one component is a correct candidate (for R_{s-1} or R_s, resp.). If this can be managed with reasonable probability, many ticks should occur in the row indexed by the correct candidate $c_x = R_{s-1}$ and in the column indexed by $i = R_s$. This suggests the following strategy: cancel all elements of Ca_{s-1} besides y (let's say $y \leq 4$) candidates whose rows had the most ticks; similarly, build a set Ca_s containing z (let's say $z \leq 20$) bytes for which in the corresponding columns the most ticks occur.

An efficient "ticking strategy" remains to be derived. For each component (c_m, i), four cases are possible: Case A: $(c_m = R_{s-1}, i = R_s)$, Case B: $(c_m = R_{s-1}, i \neq R_s)$, Case C: $(c_m \neq R_{s-1}, i = R_s)$ and Case D: $(c_m \neq R_{s-1}, i \neq R_s)$. Then $\widetilde{T}_j$ is normally distributed with mean $\mu + \Delta(u_{s-1}[c_m][j] + u_s[i][j] - 1.0)$, $\mu + \Delta(u_{s-1}[c_m][j] - 0.5)$, $\mu + \Delta(u_s[i][j] - 0.5)$ or μ, resp., and variance $\sigma^2 - \Delta^2/2$, $\sigma^2 - \Delta^2/4$, $\sigma^2 - \Delta^2/4$, or σ^2, resp. For the moment the respective densities are suggestively denoted with $f_{A;j}$, $f_{B;j}$, $f_{C;j}$, $f_{D;j}$. The cases A, B, C and D occur with probabilities $\eta_A = 1/256|Ca_{s-1}|$, $\eta_B = 255/256|Ca_{s-1}|$, $\eta_C = (|Ca_{s-1}| - 1)/256|Ca_{s-1}|$ and $\eta_D = 1 - \eta_A - \eta_B - \eta_C$. If Case D is true, the loss for yet ticking the component (c_m, i) is set $s_D := 1$. If Case B or C is true, but the component is not ticked, the loss equals $s_{BC} := 30$ whereas $s_A := 2s_{BC}$ for Case A. (Recall that we are only interested in the total number of ticks in a row or column but not in their positions.) This leads to the following decision rule:

Tick the component (c_m, i) iff

$$\eta_D s_D \prod_{j=1}^{N} f_{D;j}(\tilde{t}_j) < \eta_A s_A \prod_{j=1}^{N} f_{A;j}(\tilde{t}_j) + s_{BC} \left(\eta_B \prod_{j=1}^{N} f_{B;j}(\tilde{t}_j) + \eta_C \prod_{j=1}^{N} f_{C;j}(\tilde{t}_j) \right).$$

(13)

As the attack itself, its error handling is also atypical. There is no error correction but some candidates (which are assumed to be false) are cancelled. In the original meaning of the word there is no error detection either, as errors must trivially have occured if the previous candidate set contains more than one element. However, some false estimators are located and cancelled. The location of previous errors and the estimation of new key byte candidates are done simultaneously. Compared with the pure attack, the number of operations in the guessing phase increases linearly with the average size of the candidate sets before their reduction. On the other hand, the number of operations in the verifying phase decreases by an (exponential) factor of about β^F where β denotes the average ratio between the size of a candidate set *after* cancelling wrong candidates and the size *before* the cancelling. (Note that Ca_F can finally be reduced by applying error location to $Ca_F \times \{0, \dots, 255\}$.) Consequently, before the cancelling, the candidate sets may be much larger than for the pure attack, which means that the number of time measurements can be reduced significantly.

5.5   Practical Results

In [8] an attack was experimented against the 128-bit block, 128-bit key version of Rijndael with $M = 10$ rounds. It turned out that, with 3000 samples per key byte, the complete key was recovered with very high probability at negligible cost. The attack described above reduced the total number of time measurements from $16 \cdot 3000 = 48000$ to 4000, with a success rate of more than 90%. The set Ca_1 was determined with the pure attack, $Ca_2, \dots, Ca_F$ with the strategy described in the preceding subsection. The same time measurements were used for all key bytes. The decisions within the pure attack and the error location strategy for $s = 2$ were strongly correlated. Hence for $s = 2$ we used $Ca_1' := \{0, \dots, 255\}$ instead of Ca_1. A more time-efficient variant which yet requires more time measurements is to use another sample of size 2000 for the pure attack.

Note that the sets Ca_i are in fact ordered, with the most probable candidates first. Therefore the final "exhaustive search" phase generally succeeds after exploring a very small portion of the search space.

Although the new attack is much more efficient than the one in [8] both attacks are based on the same basic idea (cf. Subsect. 5.2) and exploit the same implementation weakness.

6   Algorithmically Based Divide and Conquer Attacks

An important design criterion for cryptographic algorithms is that it must not be possible to confirm or validate partial key guesses, with reasonable probability, using algorithmically based attacks, i.e. attacks which do not consider implementation details. Divide and conquer attacks thus typically occur in side channel attacks. However, they are not restricted to that field. Siegenthaler's attack (cf. [16]), for example, has been well-known for almost 20 years.

In [16] Siegenthaler analyzes a divide and conquer attack on a particular class of stream ciphers for which the key stream $k_1, k_2, \ldots$ (single bits) is generated by m linear feedback shift registers $LFSR_1, \ldots, LFSR_m$ of length $r_1, \ldots, r_m$ with primitive feedback polynomials $q_1, \ldots, q_m$ whose output values at time t, denoted with $x_{1(t)}, \ldots, x_{m(t)}$, are memoryless combined with $f : \{0,1\}^m \to \{0,1\}$. More precisely, $k_t := f(x_{(1)t}, \ldots, x_{(m)t})$ and $c_t := p_t \oplus k_t$ where p_t and c_t denote the tth plaintext and ciphertext bit, resp. The searched key is the initial values of $LFSR_1, \ldots, LFSR_m$ and, if the feedback polynomials are unknown, also of q_i. Let the random variables $X_1, \ldots, X_m$ be independent and equidistributed on $\{0,1\}^{r_1}, \ldots, \{0,1\}^{r_m}$, resp. If $\epsilon_i := \text{Prob}(X_i = f(X_1, \ldots, X_m)) - 0.5 \neq 0$, the ith part of the key (the initial value of $LFSR_i$ and eventually q_i) may be guessed independently from the remainder. More precisely, if the candidate for the ith key part is correct then $\#\{1 \leq t \leq N \mid x_{i(t)} = f(x_{1(t)}, \ldots, x_{m(t)})\} \approx (0.5 + \epsilon_i)N$ can be expected whereas $\approx 0.5N$ else. If $\epsilon_i \neq 0$ for all i, this enables a straight-forward divide and conquer attack with known plaintext. In fact, even a ciphertext-only-attack is feasible (cf. [16]). Vice versa, the use of correlation immune combiners of high order (eventually with memory) prevents such attacks (cf., e.g., [16], [12]) as the attacker then had to guess many key parts simultaneously which is practically infeasible.

As in the timing attack on Rijndael but unlike in the other attacks treated in this paper the key parts can be guessed independently. If the available sample size N is small (in relation to the ϵ_i), for each key part so many candidates may remain that the verifying phase may be very costly. Depending on the concrete combiner f, however, there may exist a very obvious and near at hand error location strategy. Assume, for example, that the absolute value $|\epsilon_{j_1, \ldots, j_b}| := |\text{Prob}(X_{j_1} \oplus \cdots \oplus X_{j_b} = f(X_1, \ldots, X_m)) - 0.5|$ is fairly large. Then we can pre-check the cartesian product $Ca_{j_1} \times \cdots \times Ca_{j_b}$ of candidate sets. However, unlike in the timing attack against Rijndael, there is no tendency for "ticks" in particular regions of $Ca_{j_1} \times \cdots \times Ca_{j_b}$. More precisely, we can expect correlations only for the b-tuple for which all components are correct; other pre-confirmed b-tuples result from statistical deviations. We hence do not pre-confirm subsets of $Ca_{j_1}, \ldots, Ca_{j_b}$ but a subset of $Ca_{j_1} \times \cdots \times Ca_{j_b}$ whose elements are the remaining candidates for $LFSR_{j_1}, \ldots, LFSR_{j_b}$.

7   Conclusion

This paper proposed a method to improve the error detection/correction step in divide and conquer attacks, independently of the attack itself. Although the method is not a "ready-to-use" tool, in the sense that some work and analysis is necessary to instantiate it to a particular divide and conquer problem, we believe the above examples have highlighted its basic principles, allowing a motivated user to apply it with reasonable effort.

Moreover, the efficiency of an adequate error management policy is often big enough (sometimes it reduces the necessary sample size up to factor 2) to make it worth this effort, especially in a real-world attack.

A good understanding of the potential power of an attack is necessary to be able to design adequate countermeasures. This paper, focusing on an in-depth analysis of the available data in order to exploit them in a nearly-optimal way, attempted to give an insight of that potential power.

References

1. *ARM Software Development Toolkit version 2.11.* Advanced RISC Machines Ltd, 1997. User's guide document number: ARM DUI 0040C.
2. Cascade (Chip Architecture for Smart CArds and portable intelligent DEvices). Project funded by the European Community, see http://www.dice.ucl.ac.be/crypto/cascade.
3. D. Coppersmith: Small Solutions to Polynomial Equations, and Low Exponent RSA Vulnerabilities. J. Cryptology **10** (no. 4) (1997) 233–260.
4. J. Daemen, V. Rijmen: AES proposal: Rijndael. In: Proc. first AES conference, August 1998. Available on-line from the official AES page: `http://csrc.nist.gov/encryption/aes/aes_home.htm`.
5. J.F. Dhem.: Design of an Efficient Public-Key Cryptographic Library for RISC-Based Smart Cards. PhD thesis, Université catholique de Louvain - UCL Crypto Group - Laboratoire de microélectronique (DICE), May 1998.
6. J.-F. Dhem, F. Koeune, P.-A. Leroux, P.-A. Mestré, J.-J. Quisquater, J.-L. Willems: A Practical Implementation of the Timing Attack. In: J.-J. Quisquater and B. Schneier (eds.): Smart Card – Research and Applications, Springer, Lecture Notes in Computer Science, Vol. **1820**, Berlin (2000), 175–191.
7. P. Kocher: Timing Attacks on Implementations of Diffie-Hellman, RSA, DSS and Other Systems. In: N. Koblitz (ed.): Advances in Cryptology – Crypto '96, Springer, Lecture Notes in Computer Science **1109**, Berlin (1996), 104–113.
8. F. Koeune, J.-J. Quisquater: A Timing Attack against Rijndael. Université catholique de Louvain, Crypto Group, Technical report CG-1999/1, 1999.
9. A.J. Menezes, P.C. van Oorschot, and S.C. Vanstone: Handbook of Applied Cryptography, Boca Raton, CRC Press (1997).
10. T.S. Messerges, E.A. Dabbish, R.H. Sloan: Power Analysis Attacks of Modular Exponentiation in Smartcards. In: Ç.K. Koç, C. Paar (eds.): Cryptographic Hardware and Embedded Systems — CHES 1999, Springer, Lecture Notes in Computer Science, Vol. **1717**, Berlin (1999), 144–157.
11. P.L. Montgomery: Modular Multiplication without Trial Division, Math. Comp. **44**, no. 170, 519–521 (April 1985).
12. R.A. Rueppel: Analysis and Design of Stream Ciphers, Springer, Berlin (1986).
13. W. Schindler: Optimized Timing Attacks against Public Key Cryptosystems. To appear in Statistics & Decisions.

14. W. Schindler: A Timing Attack against RSA with the Chinese Remainder Theo-
    rem. In: Ç.K. Koç, C. Paar (eds.): Cryptographic Hardware and Embedded Sys-
    tems — CHES 2000, Springer, Lecture Notes in Computer Science **1965**, Berlin
    (2000), 110–125.
15. W. Schindler, F. Koeune, J.-J. Quisquater: Unleashing the Full Power of Timing
    Attacks. Université catholique de Louvain, Crypto Group, Technical report CG-
    2001/3, 2001.
16. T. Siegenthaler: Decrypting a Class of Stream Ciphers Using Ciphertext Only.
    IEEE Transactions on Computers. C-34 (1985), 81-85.
17. H. Witting: Mathematische Statistik I, Stuttgart, Teubner (1985).

Key Recovery Scheme Interoperability – A Protocol for Mechanism Negotiation

Konstantinos Rantos and Chris J. Mitchell

Information Security Group,
Royal Holloway, University of London,
Egham, Surrey TW20 0EX, UK.
{K.Rantos, C.Mitchell}@rhul.ac.uk

Abstract. This paper investigates interoperability problems arising from the use of dissimilar key recovery mechanisms in encrypted communications. The components that can cause interoperability problems are identified and a protocol is proposed where two communicating entities can negotiate the key recovery mechanism(s) to be used. The ultimate goal is to provide the entities a means to agree either on a mutually acceptable KRM or on different, yet interoperable, mechanisms of their choice.

Keywords: Key recovery, interoperability, negotiation protocol

1 Introduction

As business increases its use of encryption for data confidentiality, the threats arising from the lack of access to decryption keys grow [2]. Although transient keys typically should not be retained, there is a potential need to access these keys during their lifetime, i.e. during the communication session (or afterwards if the company logs any communications). Corporations might want to access encrypted communications to check for malicious software or to track leakage of sensitive information.

Key recovery mechanisms (KRMs) address this problem [3,10] by providing the means to recover decryption keys. They can be divided into two types: *key escrow* and *key encapsulation* mechanisms. A *key escrow* mechanism, [8], is a method of *key recovery* (KR) where the secret or private keys, key parts or key-related information are stored by one or more key escrow agents. In a *key encapsulation* mechanism, keys, key parts, or key-related information are enclosed in a KR block, which is typically attached to the data and encrypted specifically for the *key recovery agent* (KRA). Here the terms *key escrow agent* and *key recovery agent* are considered synonymous, and refer to the trusted entity which responds to key recovery requests and potentially holds users' key-related material.

The variety of KRMs so far proposed, in conjunction with the lack of a standard for KRMs, means that interoperability problems are likely to arise from the use of dissimilar KRMs in encrypted communications [11]. Interoperability,

B. Honary (Ed.): Cryptography and Coding 2001, LNCS 2260, pp. 268–276, 2001.

as in [5], means the ability of entity A, using KRM KRM_A, to establish a KR-enabled cryptographic association with entity B, using KRM KRM_B. Entities deploying dissimilar KRMs may not know whether the remote party can deal with their KRM's demands. This may force them to avoid key recovery, with associated increased risks. Note that, even if KRMs were standardised, interoperability problems may still arise; standards tend to provide a variety of sound but not necessarily interoperable mechanisms.

This paper addresses these interoperability problems. The components that can cause interoperability problems are identified and a protocol is proposed that, to a great extent, overcomes the interoperability problems. The protocol offers communicating parties the ability to agree either on a mutually acceptable KRM or on different, yet interoperable, mechanisms for their encrypted communications.

2   Key Recovery Enabled Communications

In the context of communications between entities A and B, we give two scenarios where KRM use can affect the establishment of a cryptographic association.

1. Entities A and B make use of KRMs KRM_A and KRM_B respectively, which might be identical, compatible or dissimilar mechanisms. In the case of identical or compatible mechanisms the two entities are not expected to face any problems. Problems, however, might arise if the two entities make use of dissimilar mechanisms. They are unlikely to be able to establish a secure communication while using their respective KRMs, as this would typically demand each entity to fulfil the requirements of the peer's KRM.
2. Entity A uses KRM_A while B does not use KR. The issues that arise here are whether B will be able to cope with KRM_A's needs, and whether A will be able to generate valid KR information. For the two entities to be able to communicate, assuming that A manages to generate valid KR information, B should at least be aware that A makes use of a KRM. This is important as B should not discard incoming traffic because of unrecognised KR fields that B cannot interpret. Another potential problem is whether A's policy will permit the acceptance of incoming traffic that does not make use of KR. If A operates within a corporate environment this requirement is likely to be crucial, as the company might want to check incoming data for malicious software before they reach their destination.

Communicating entities wanting KR functionality for encrypted communications without the above problems must use interoperable KRMs. Also, any deployed cryptographic mechanisms with embedded KR functionality should be compatible with cryptographic products not using KR. These requirements will ensure that neither of the above scenarios will prevent the establishment of a secure session.

3 Factors That Can Affect Interoperability

Many KRMs, especially key escrow mechanisms, demand the use of a specific mechanism for session key generation, and as such they can be considered as part of the key establishment protocol. This restriction is a cause of KRM interoperability problems. A KRM with this property demands compatibility of the underlying key establishment protocols, a requirement that is not always fulfilled. Key escrow mechanisms suffer more from this problem, as most require the use of a specific key establishment protocol. By contrast, key encapsulation schemes appear to be more adaptable in this respect, since they simply wrap the generated data encryption key under the KRA's public encryption key (and hence potentially work with any key establishment protocol).

Flexibility of key encapsulation mechanisms with respect to the underlying key establishment protocols does not always rule out interoperability problems. Interoperability very much depends on what additional requirements exist. For example, problems arise if the recipient of encrypted data needs to validate KR information, or the receiver relies on the sender to generate KR information. These needs will typically demand interaction between the two parties during KR information generation/verification. If either parties' mechanism cannot meet the peer's demands, interoperability problems are likely to arise. This problem is not restricted to key encapsulation schemes. In the case of key escrow mechanisms, a requirement for participation of both entities in generating KR information will have the same effect. We can therefore divide KRMs into two classes, depending on their communications requirements during KR information generation/verification.

1. KRMs where each entity generates KR information for its own use only, without peer assistance. If neither party requires verification of the peer's KR information prior to decryption, interoperability issues become of minor importance and the parties will be able to use their respective KRMs.
2. KRMs that require interaction between the two entities for the generation/verification of KR information. Interaction might be needed in the following cases.
   - Exchange of data is required for KR information generation.
   - The sender generates KR information both for his own and the peer's needs. This is particularly relevant to single-message communications.
   - Either party wishes, e.g. for policy reasons, to verify the KR information generated by the peer.
   In situations like these interoperability is an issue that must be dealt with; otherwise, it is likely to lead to a failure to establish secure communications.

In summary, the two factors that are likely to affect the interoperability of KRMs in encrypted communication sessions are:

1. the KRMs' dependence on the underlying key establishment protocol, and
2. the interaction requirements between the communicating parties for the generation and/or verification of KR information.

4   Interoperable Mechanisms

Based on the above analysis, a mechanism which is neither dependent on the underlying key establishment protocol nor needs any interaction with the peer for generation or verification of KR information, will always be interoperable with a KRM with the same requirements. The two mechanisms can work independently regardless of the underlying key establishment protocol. A mechanism with these requirements can also inter-operate with one that is dependent on an underlying key establishment protocol which both entities can deal with, as long as it does not require interaction with the peer for KR information generation/verification.

Interoperability problems are likely to arise in the following cases (we assume that the communicating parties can deal with all possible underlying key establishment protocols):

1. Both KRMs use specific key establishment mechanisms regardless of interaction requirements. In this case the interoperability of the KRMs depends on the compatibility of the underlying cryptographic mechanisms.
2. At least one KRM demands peer participation in KR information generation/verification. For example, if the policy demands that the KR information for the receiver should be generated by the sender, then the sender must be able to handle the receiver's mechanism. Otherwise, it is likely that establishment of secure communications will fail.

The chances of interoperability problems are therefore considerable, and a solution is needed. However, due to the significant differences in the characteristics of existing KRMs it is unlikely that a single model, such as the one proposed by the Key Recovery Alliance in [5] (see also [11]), can apply to all of them. This latter model mainly addresses problems arising from the transmission of KR information in proprietary formats, and suggests as a solution the use of a wrapper, namely the "Common Key Recovery Block". However, as described in [9], it fails to achieve one of its main objectives, which is to offer the ability for validation of KR information by the peer, and does not deal with the situation where KR information cannot be generated because of the use of dissimilar mechanisms.

A different approach is described in this paper, which requires the entities to be able to deal with more than one KRM. This enables some of the difficulties described above to be avoided.

5   A KRM Negotiation Protocol

We now describe a protocol designed to enable two communicating parties to negotiate the KRM to be used in an encrypted communication session. Its main objective is to deal with situations where the parties wish to make use of different, non-interoperable, KRMs.

A similar protocol specifically designed to allow the negotiation of KRMs using the Internet Security Association and Key Management Protocol (ISAKMP)

[7] is described in [1]. This model, however, adopts the mechanism described in [5] for the transmission of KR information, which, as mentioned in the previous section, has been shown to have problems. A more generalised model is described here that considers the different requirements of various mechanisms and the additional requirements that might arise regarding the exchange of cryptographic certificates. Moreover, the proposed protocol can be used to provide key recovery functionality in the application layer, in contrast to the mechanism proposed by the Key Recovery Alliance which targets the IP layer.

Note that in the protocol description we refer to the two parties as 'Client' and 'Server'; this is so as to follow the client-server model terminology as closely as possible.

5.1   The Proposed Scheme

The protocol consists of the following steps (messages in SMALL CAPS are optional; more detailed descriptions of the exchanged messages are given in the next section).

Client		Server
ClientHello	⟶	
		ServerHello
		CERTIFICATE
		CERTIFICATEREQUEST
	⟵	ServerHelloDone
CERTIFICATE		
KRPARAMETERS		
Finished	⟶	
	⟵	Finished

The client first sends the *ClientHello* message, to which the server responds with the *ServerHello*. With these two messages the two entities exchange the parameters necessary for KRM negotiation. After the *ServerHello* (if required by the selected KRM(s)) the server sends the *Certificate* message containing the appropriate certificates, requests client certificates with the *CertificateRequest*, and, finally, sends the *ServerHelloDone* message. The client responds with the optional *Certificate* message, containing the certificates specified in the *CertificateRequest*, the optional *KRParameters* message, containing any additional information required by the selected KRMs, and, finally, the *Finished* message. The server verifies the received *Finished* message and, if successful, responds with a similar *Finished* message. On receipt, the client verifies it and, if successful, the negotiation terminates successfully.

5.2   Exchanged Messages

In the following sections the exchanged messages are described in detail.

Client Hello. The client, as previously mentioned, initiates the protocol by sending the first message of the negotiation protocol (*ClientHello*). The *ClientHello* contains a list of KRMs (from a complete list of mechanisms the protocol supports) that the client is willing to use, in decreasing order of preference. A default mechanism that all parties are assumed to be able to use can be included in the list.

With the *ClientHello* the client must also inform the server whether he wants to resume a previous session by including the identifier in the appropriate field. If this field is empty a new session id should be assigned by the server.

Server Hello. If the client does not request the resumption of a previous session, or if the server wants to initiate a new one, the server must assign a new session id, which will be sent with the *ServerHello* message. Otherwise, the server will respond with the session id included in the *ClientHello* and proceed with the *Finished* message.

If the server initiates a new session, he indicates the mechanism that he wants the client to use, and the mechanism that the server will use. The two mechanisms, if not identical, must be interoperable. For this purpose, a list of all possible matches of interoperable mechanisms has to be kept by both entities. If an acceptable match is not found in the list, the server can either terminate the negotiation protocol unsuccessfully, or choose the default mechanism if this is in the list sent by the client. Otherwise the server drops the session.

If the selection of the mechanism for both entities is a KRM that can itself handle the exchange of certificates and related KR parameters, the two parties can terminate the negotiation protocol and leave this KRM to take charge. To achieve this the server will send a *Finish* message (after the *ServerHello*) to indicate that control is now to be passed to the negotiated KRM(s).

Finally, within the *ServerHello* the server also includes the *KRParameters* field, which carries any additional information that the client has to possess to be able to deal with the server's KRM.

Certificate and KRM related information exchange. Depending on the selection of the KRM, and if the server has not sent a *Finish* message, the server proceeds with the *Certificate* message. This message is optional and contains the required certificates (for the chosen mechanisms) for the generation and/or verification of KR information. Following that, and depending on the requirements of the chosen KRM(s), the server can also send a request for the corresponding client's certificates using the *CertificateRequest*. The purpose of the *CertificateRequest* is to give the client a list of specific types of certificates needed by the server, and a list of certification authorities trusted by the server. After the *CertificateRequest* the server sends the *ServerHelloDone* message, which indicates that the server has completed his *Hello* messages.

On receipt of the *ServerHelloDone*, if the client has received a *CertificateRequest* he responds with his *Certificate* message, which contains the requested certificates, assuming that he is in possession of the appropriate ones.

Further, the client sends in the optional *KRParameters* message any additional information required by the selected for the client KRM. Note that the corresponding *KRParameters* for the server's KRM is sent as part of the *ServerHello* message.

Finish messages. If the client is satisfied with the current selection of mechanisms he sends the *Finish* message, which indicates that the client is willing to proceed with the current selection of mechanisms. Subsequently, the client waits for the corresponding server's *Finish* message, whose receipt indicates successful execution of the protocol.

5.3   Protecting the Integrity of the *Hello* Messages

After the execution of the above protocol the two entities are not sure whether any of the exchanged messages have been altered during transmission by an adversary, as the specified protocol includes no proper integrity checks. Moreover, neither of the communicating parties authenticates the other. Assuming, however, that the KRMs that can be negotiated are sound, the protocol does not introduce any vulnerabilities to the secrecy of the session key. The only attack that an adversary can mount against the protocol is to alter the *Hello* messages exchanged between the two entities in an attempt to downgrade the negotiated KRM(s) to one(s) that the attacker regards as weaker. Such an attack will only force the two entities to make use of less favourable mechanisms.

To avoid such problems we propose enhancing the previously proposed protocol. These enhancements provide the following security services.

- Integrity of the exchanged messages.
- Assurance that the *Hello* messages exchanged are not a replay from a previous session.

Additionally, the mechanism provides mutual authentication of the communicating parties. Note, however, that mutual authentication is not a requirement for the negotiation protocol. It is a property derived from the use of digital signatures. The modifications proposed are as follows.

- The client generates a random value $randC$, which he sends to the server with the *ClientHello* message.
- The server generates a random value $randS$, which he sends to the client with the *ServerHello* message.
- The client's *Finish* message becomes

$$S_C(ClientHello \parallel ServerHello \parallel randC \parallel randS)$$

and the server's *Finish* message becomes

$$S_S(ServerHello \parallel ClientHello \parallel randS \parallel randC)$$

where $S_U(M)$ is U's signature on data M and "$\parallel$" denotes concatenation.

The rest of the messages remain as previously defined. On receipt of the respective *Finish* messages the two entities check the signatures and if either of the two verification checks fails the protocol terminates unsuccessfully (this indicates that at least one of the *ClientHello, ServerHello* might have been altered during transmission). The modified protocol deals with the threat of modification of the exchanged *Hello* messages by an adversary. The generated random values prevent against replay attacks, i.e. where an adversary uses old exchanged messages to subvert the protocol. The cost of this countermeasure, however, is the introduction of signatures which have to be supported by an appropriate public key infrastructure. In practice the two variants could co-exist, and an extra field in the *Hello* messages could be used to indicate which variants of the protocol are supported.

6   Properties and Discussion

The proposed protocol offers the communicating parties a means of negotiating the KRMs to be used for their encrypted communications. Given that this negotiation will affect the selection of the key establishment protocol, execution of the proposed protocol must take place before the establishment of any session keys. It might also be the case that the two entities are obliged to use a specific key establishment protocol. This will simply restrict the number of mechanisms that the two entities will be able to negotiate.

The negotiating entities will be able to choose different KRMs as long as there are no conflicts between the underlying key establishment mechanisms. Therefore, the choice of the key recovery mechanism(s) will only be affected by the compatibility of the underlying key establishment protocols. If these are compatible, the two parties will be able to use the negotiated KRMs, overcoming efficiently any interoperability problems that the two parties would have otherwise faced.

Finally, note that in order to achieve the degree of agreement needed, the KRM negotiation process and the KRMs to be negotiated need to be subject of a standardisation process of some type (e.g. via the IETF). This standard will need to include agreed identifiers for a large set of KRMs.

7   Conclusions

The introduction of a large number of KRMs and their use in encrypted communications is likely to lead to interoperability problems between KR-enabled encryption products. In this paper the factors that can cause interoperability problems have been identified. Following a different approach to the single model solution proposed by the Key Recovery Alliance, a protocol has been proposed that gives communicating entities the means to negotiate the KRMs to be used. The parties can make use of different, yet interoperable, KRMs matching their needs for the specific communication session.

References

1. Balenson, D., Markham, T.: ISAKMP key recovery extensions. Computers & Security, **19(1)** (2000) 91–99.
2. Denning, D.E.: Information Warfare and Security. Addison Wesley, (1998).
3. Denning, D.E., Branstad, D.K.: A taxonomy of key escrow encryption systems. Communications of the ACM, **39(3)** (1996) 34–40.
4. Dierks, T., Allen, C.: The TLS protocol, Version 1.0. RFC 2246 (1999).
5. Gupta, S.: A common key recovery block format: Promoting interoperability between dissimilar key recovery mechanisms. Computers & Security, **19(1)** (2000) 41–47.
6. Kennedy, J., Matyas Jr., S.M., Zunic, N.: Key recovery functional model. Computers & Security, **19(1)** (2000) 31–36.
7. Maughan, D., Schertler, M., Turner, J.: Internet security association and key management protocol (ISAKMP). RFC 2408.
8. National Institute of Standards and Technology: Requirements for key recovery products. Available at http://csrc.nist.gov/keyrecovery/ (1998).
9. Rantos, K., Mitchell, C.: Remarks on KRA's key recovery block format. Electronics Letters, **35** (1999) 632–634.
10. Smith, M., van Oorschot, P., Willett, M.: Cryptographic information recovery using key recovery. Computers & Security, **19(1)** (2000) 21–27.
11. Williams, C., Zunic, N.: Global interoperability for key recovery. Computers & Security, **19(1)** (2000) 48–55.

Unconditionally Secure Key Agreement Protocol

Cyril Prissette

Signal - Information - Systèmes
Université de Toulon et du Var
83130 La Garde, France `prissette@univ-tln.fr`

Abstract. The key agreement protocol are either based on some computational infeasability, such as the calculus of the discrete logarithm in [1], or on theoretical impossibility under the assumption that Alice and Bob own specific devices such as quantum channel [2]. In this article, we propose a new key agreement protocol called CHIMERA which requires no specific device. This protocol is based on a generalization we propose of the reconciliation algorithm. This protocol is proved unconditionally secure.

1 Introduction

The security of cryptographic systems is based either on a computational infeasability or an a theoretical impossibility. However, some cryptographic problems have no known unconditionally secure solution. For example, the key agreement problem has computational secure solutions, as the Diffie-Hellman protocol [1], but no unconditional secure solution under the assumption that Alice and Bob has no specific equipment such as quantum channel, deep-space radio source, or satellite.

Our work is inspired by these protocols and uses a generalized version of an interactive error-correcting algorithm proposed by C.H. Bennett and G. Brassard in [2]. This algorithm, called reconciliation, fits the parameter of the quantum channel, but is insecure for our protocol because of some properties of the sequences we use. The first part of this paper is a presentation of the generalization of the reconciliation algorithm.

The next part is a presentation of CHIMERA, which is a key agreement protocol with unconditional security. It uses information-theoretic algorithms such as generalized reconciliation and extended Huffman coding.

In [3], U. Maurer gives a general description of key agreement protocols and the conditions a key agreement protocol must satisfy to be secure [4],[5]. We recall these conditions and prove that CHIMERA satisfy all this conditions if the value of a parameter of the protocol is in a given range. Next, we propose a particular value of this parameter in the given range to optimize the length of the key created by CHIMERA.

B. Honary (Ed.): Cryptography and Coding 2001, LNCS 2260, pp. 277–293, 2001.

2   Generalized Reconciliation

2.1   Bennett and Brassard's Reconciliation

The reconciliation process is, as describe in [2], an iterative algorithm which destroy errors between two binary sequences A and B owned by Alice and Bob. The destruction of the errors is secure even if Eve listen the insecure channel used by Alice and Bob to perform reconciliation. The algorithm does not destroy all errors between the two sequences in one round, but it can be repeated several times to destroy statistically all the errors. The price to pay to obtain to identical sequence is the sacrifice of bits of the sequences and thus, the reduction of the length of the sequences.

Here is the algorithmic description of one round of reconciliation :

Alice and Bob cut their sequences A and B into subsequences of length k. For each sub-sequence $(A_i, \ldots, A_{i+k-1})$ from A and $(B_i, \ldots, B_{i+k-1})$ from B, they send each other (on the public insecure channel) the parity of their sub-sequence.

- If the parity of the sub-sequence $(A_i, \ldots, A_{i+k-1})$ differs from the parity of the sub-sequence $(B_i, \ldots, B_{i+k-1})$, Alice and Bob destroy their respective sub-sequences.
- Else Alice and Bob destroy respectively A_{i+k-1} and B_{i+k-1}, and keep $(A_i, \ldots, A_{i+k-2})$ and $(B_i, \ldots, B_{i+k-2})$.

The principle is simple : if the parities differ, then the sub-sequences differ. if Alice and bob destroy these sub-sequences, they destroy (at least) one error between the two sequences.

On the other hand, if the parities are equal. This does not mean that the two sequences are equal. However Eve knows one bit of information about the subsequence : so, Alice and Bob destroy one bit from their subsequence.

Obviously, the reconciliation works only if the sequences A abd b are close enough, and is secure only if Eve has no information about A and B before the reconciliation. For example, if she knows with certainty the value of one bit from A and B and if Alice and Bob use sub-sequences of length two, she learns from the parities of the sequences the whole sequences and so the bit kept if the parity are equals.

2.2   Generalized Reconciliation

Sometimes, in particular in CHIMERA, the parity of a sub-sequence reveals more information than the entropy of one bit of the subsequence. This happens, for example, when $p(A_i = 0) < p(A_i = 1)$.

The generalized reconciliation algorithm REC(k,n), which is as follows, let Alice and Bob sacrifice n symbols (instead of only one) of their sub-sequences of length k when the parities are equals.

Alice and Bob cut their sequences A and B into subsequences of length k. For each sub-sequence $(A_i, \ldots, A_{i+k-1})$ from A and $(B_i, \ldots, B_{i+k-1})$ from B, they send each other (on the public insecure channel) the parity of their sub-sequence.

- If the parity of the sub-sequence $(A_i, \ldots, A_{i+k-1})$ differs from the parity of the sub-sequence $(B_i, \ldots, B_{i+k-1})$, Alice and Bob destroy their respective sub-sequences.
- Else Alice and Bob destroy respectively A_{i+k-n} and B_{i+k-n}, and keep $(A_i, \ldots, A_{i+k-n-1})$ and $(B_i, \ldots, B_{i+k-n-1})$.

The principle is the same than in Bennett and Brassard reconciliation R(k,1) : if the parities differs,then the sub-sequence contain errors, so Alice and Bob destroy the sub-sequences. Otherwise, Alice and Bob destroy more information than the information revealed by the parities.

The generalization of the reconciliation algorithm is very useful in our protocol, called CHIMERA, which uses REC(3,2). Actually, in this protocol the sequences are biased but the entropy of two bits is always greater than the entropy of the parity of three bits. This property is proved in the section (7).

3   Presentation of CHIMERA

The CHIMERA is a key agreement protocol. we present it with some parameters which are optimal and insure its security. The choice of the values used in CHIMERA is explain in the study of the protocol which follows this presentation.

The following protocol allows Alice and Bob to build a secret common quantity of length 128 bits.

- Alice builds a binary sequence $A^{[0]}$ with the following properties :
  - $|A^{[0]}| = 2000000$
  - $\forall_i \ p(A_i^{[0]} = 1) = p_b = \frac{3}{16}$
- Bob builds a binary sequence $B^{[0]}$ with the following properties :
  - $|B^{[0]}| = 2000000$
  - $\forall_i \ p(B_i^{[0]} = 1) = p_b = \frac{3}{16}$
- Alice and Bob repeat 6 times the following reconciliation algorithm REC(3,2) on their respective sequences (We note $A^{[k]}$ and $B^{[k]}$ Alice and Bob's sequences after k rounds of reconciliation).

```
l=0
forall i such as (i < |A[k]| − 2 and i mod 3 = 0)
        if (⊕²ⱼ₌₀ A[k]ᵢ₊ⱼ = ⊕²ⱼ₌₀ B[k]ᵢ₊ⱼ)) then
            A[k+1]ₗ ← A[k]ᵢ
            B[k+1]ₗ ← B[k]ᵢ
            l ← l + 1
        end if
    end forall
```

- Alice compresses the sequence $A^{[6]}$ with the extended Huffman code $\mathcal{H}$ using 11-tuples as symbols of the language. The resulting sequence is the key S.

– Bob compresses the sequence $B^{[6]}$ with the extended Huffman code $\mathcal{H}$ using 11-tuples as symbols of the language. The resulting sequence is the key S'.

Alice and Bob have the same quantity $S = S'$ of length 128.

4   Properties of Key Agreement Protocols

In [3], U. Maurer gives the properties a key agreement have to satisfy. These properties come from [4] and [5]. They are conditions of soundness and security.

Considering that Eve is passive, a key agreement protocol which creates binary sequences S and S' by exchanging between Alice and Bob t messages $C_1, \ldots, C_t$ must satisfy the three conditions

– $P[S \neq S'] \approx 0$: Alice and bob must obtain with a very high probability the same sequence.
– $H(S) \approx |S|$: the key must be very close to uniformly distributed.
– $I(S; C^t Z) \approx 0$: Eve has no information about S, considering her initial knowledge Z and her eavesdropping of the insecure channel.

Moreover, the goal of the key-agreement is to make the length of the key S as long as possible.

The CHIMERA satisfied each of these properties. The proof that each property is satisfied is given in the three following sections of this paper. For each proof, we assume that the bias p_b of the initial sequences $A^{[0]}$ and $B^{[0]}$ is in the range $[0 : \frac{1}{2})$, and we search the conditions on this parameter the CHIMERA have to respect to work and be sure. We also assume the reconciliation needs r round to create identical sequences and the extended Huffman code uses n-tuples.

Then, under the conditions on p_b obtained in each proof we explain the choice of the values $p_b = \frac{3}{16}$, $r = 6$ and $n = 11$.

5   Proof of the Property $P[S \neq S'] \approx 0$

The proof of the property $P[S \neq S'] \approx 0$ is based on the study of the distance evolution between Alice's sequence $A^{[i]}$ and Bob's sequence $B^{[i]}$ after i rounds of reconciliation.

5.1   Definition : Normalized Distance

The normalized distance $d_N(A, B)$ between to sequences of bits A and B is defined as the ration between Hamming distance $d_H(A, B)$ and the length $|A|$ of the sequences.

$$d_N(A, B) = \frac{d_H(A, B)}{|A|}. \tag{1}$$

5.2   Initial Normalized Distance $d_N(A^{[0]}, B^{[0]})$

Let p_b be the biased probability of the random generators. The initial normalized distance is a function of p_b. The following table presents the four possible values of the couple $(A_i^{[0]}, B_i^{[0]})$ with their occurrence probability.

Table 1. Possible values of $(A_i^{[0]}, B_i^{[0]})$ with occurence probability.

$A_i^{[0]}$	$B_i^{[0]}$	$p(A_i^{[0]}, B_i^{[0]})$
0	0	$(1 - p_b)^2$
1	1	p_b^2
0	1	$p_b(1 - p_b)$
1	0	$p_b(1 - p_b)$

In the two last cases $A_i^{[0]}$ and $B_i^{[0]}$ differs, so $p(A_i^{[0]} \neq B_i^{[0]}) = 2p_b(1 - p_b)$. This result can be extended to the whole sequences to obtain the average Hamming distance $d_H(A^{[0]}, B^{[0]}) = |A^{[0]}|2p_b(1 - p_b)$. So the initial normalized distance between $A^{[0]}$ and $B^{[0]}$ is :

$$d_N(A^{[0]}, B^{[0]}) = \frac{d_H(A^{[0]}, B^{[0]})}{|A^{[0]}|} = 2p_b(1 - p_b). \tag{2}$$

In CHIMERA, we set $p_b \in [0 : \frac{1}{2})$. So we have the following range for the initial normalized distance between S and S' which is a function of the bias of the random generators used to build $A^{[0]}$ and $B^{[0]}$:

$$d_N(A^{[0]}, B^{[0]}) \in [0 : \frac{1}{2}). \tag{3}$$

5.3   Evolution of the Normalized Distance $d_N(A^{[k]}, B^{[k]})$

Let $d_N(A^{[k]}, B^{[k]})$ be the normalized distance between $A^{[k]}$ and $B^{[k]}$ after k rounds of reconciliation with the algorithm $REC(3, 2)$. The following table presents the 32 possible values of the two 3-tuples $(A_i^{[k]}, A_{i+1}^{[k]}, A_{i+2}^{[k]})$ and $(B_i^{[k]}, B_{i+1}^{[k]}, B_{i+2}^{[k]})$ with their occurrence probability when the bits $A_i^{[k]}$ and $B_i^{[k]}$ are kept (i is a multiple of 3).

The 16 first cases give $A_i^{[k]} = B_i^{[k]}$, which means that the reconciliation $REC3$ works and the distance reduces. At the opposite, the 16 last cases gives $A_i^{[k]} \neq B_i^{[k]}$, the reconciliation REC(3,2) fails and the distance increases.

The normalized distance $d_N(A^{[k+1]}, B^{[k+1]})$ after one more round of reconciliation REC(3,2) is a function of $d_N(A^{[k]}, B^{[k]})$. It is given by the ratio between the probability of the 16 last cases and the probability of the 32 cases (we set $d_N = d_N(A^{[k]}, B^{[k]})$) :

$$d_N(A^{[k+1]}, B^{[k+1]}) = \frac{2(1 - d_N)d_N^2}{3(1 - d_N)d_N^2 + (1 - d_N)^3}. \tag{4}$$

Table 2. Possibles values of $(A_i^{[k]}, A_{i+1}^{[k]}, A_{i+2}^{[k]}, B_i^{[k]}, B_{i+1}^{[k]}, B_{i+2}^{[k]})$ with occurrence probability

$A_i^{[k]}$	$A_{i+1}^{[k]}$	$A_{i+2}^{[k]}$	$B_i^{[k]}$	$B_{i+1}^{[k]}$	$B_{i+2}^{[k]}$	$p(A_i^{[k]}, A_{i+1}^{[k]}, A_{i+2}^{[k]}, B_i^{[k]}, B_{i+1}^{[k]}, B_{i+2}^{[k]})$
0	0	0	0	0	0	$(1 - d_N(A^{[k]}, B^{[k]}))^3$
0	0	0	0	1	1	$(1 - d_N(A^{[k]}, B^{[k]}))(d_N(A^{[k]}, B^{[k]}))^2$
0	0	1	0	0	1	$(1 - d_N(A^{[k]}, B^{[k]}))^3$
0	0	1	0	1	0	$(1 - d_N(A^{[k]}, B^{[k]}))(d_N(A^{[k]}, B^{[k]}))^2$
0	1	0	0	0	1	$(1 - d_N(A^{[k]}, B^{[k]}))(d_N(A^{[k]}, B^{[k]}))^2$
0	1	0	0	1	0	$(1 - d_N(A^{[k]}, B^{[k]}))^3$
0	1	1	0	0	0	$(1 - d_N(A^{[k]}, B^{[k]}))(d_N(A^{[k]}, B^{[k]}))^2$
0	1	1	0	1	1	$(1 - d_N(A^{[k]}, B^{[k]}))^3$
1	0	0	1	0	0	$(1 - d_N(A^{[k]}, B^{[k]}))^3$
1	0	0	1	1	1	$(1 - d_N(A^{[k]}, B^{[k]}))(d_N(A^{[k]}, B^{[k]}))^2$
1	0	1	1	0	1	$(1 - d_N(A^{[k]}, B^{[k]}))^3$
1	0	1	1	1	0	$(1 - d_N(A^{[k]}, B^{[k]}))(d_N(A^{[k]}, B^{[k]}))^2$
1	1	0	1	0	1	$(1 - d_N(A^{[k]}, B^{[k]}))(d_N(A^{[k]}, B^{[k]}))^2$
1	1	0	1	1	0	$(1 - d_N(A^{[k]}, B^{[k]}))^3$
1	1	1	1	0	0	$(1 - d_N(A^{[k]}, B^{[k]}))(d_N(A^{[k]}, B^{[k]}))^2$
1	1	1	1	1	1	$(1 - d_N(A^{[k]}, B^{[k]}))^3$
0	0	0	1	0	1	$(1 - d_N(A^{[k]}, B^{[k]}))(d_N(A^{[k]}, B^{[k]}))^2$
0	0	0	1	1	0	$(1 - d_N(A^{[k]}, B^{[k]}))(d_N(A^{[k]}, B^{[k]}))^2$
0	0	1	1	0	0	$(1 - d_N(A^{[k]}, B^{[k]}))(d_N(A^{[k]}, B^{[k]}))^2$
0	0	1	1	1	1	$(1 - d_N(A^{[k]}, B^{[k]}))(d_N(A^{[k]}, B^{[k]}))^2$
0	1	0	1	0	0	$(1 - d_N(A^{[k]}, B^{[k]}))(d_N(A^{[k]}, B^{[k]}))^2$
0	1	0	1	1	1	$(1 - d_N(A^{[k]}, B^{[k]}))(d_N(A^{[k]}, B^{[k]}))^2$
0	1	1	1	0	1	$(1 - d_N(A^{[k]}, B^{[k]}))(d_N(A^{[k]}, B^{[k]}))^2$
0	1	1	1	1	0	$(1 - d_N(A^{[k]}, B^{[k]}))(d_N(A^{[k]}, B^{[k]}))^2$
1	0	0	0	0	1	$(1 - d_N(A^{[k]}, B^{[k]}))(d_N(A^{[k]}, B^{[k]}))^2$
1	0	0	0	1	0	$(1 - d_N(A^{[k]}, B^{[k]}))(d_N(A^{[k]}, B^{[k]}))^2$
1	0	1	0	0	0	$(1 - d_N(A^{[k]}, B^{[k]}))(d_N(A^{[k]}, B^{[k]}))^2$
1	0	1	0	1	1	$(1 - d_N(A^{[k]}, B^{[k]}))(d_N(A^{[k]}, B^{[k]}))^2$
1	1	0	0	0	0	$(1 - d_N(A^{[k]}, B^{[k]}))(d_N(A^{[k]}, B^{[k]}))^2$
1	1	0	0	1	1	$(1 - d_N(A^{[k]}, B^{[k]}))(d_N(A^{[k]}, B^{[k]}))^2$
1	1	1	0	0	1	$(1 - d_N(A^{[k]}, B^{[k]}))(d_N(A^{[k]}, B^{[k]}))^2$
1	1	1	0	1	0	$(1 - d_N(A^{[k]}, B^{[k]}))(d_N(A^{[k]}, B^{[k]}))^2$

5.4 Limit of the Normalized Distance $d_N(A^{[k]}, B^{[k]})$

Proving that $\forall d_N(A^{[0]}, B^{[0]}) \in [0 : \frac{1}{2}), \lim_{r \to +\infty} d_N(A^{[k]}, B^{[k]}) = 0$ is equivalent to prove $P[S \neq S'] \approx 0$. We do not consider the last computation of the protocol (the Huffman coding of the sequences S and S') because Alice and Bob obtain the same sequence after this compression if they have the same sequence before this compression. So we only have to prove the normalized distance between $A^{[r]}$

and $B^{[r]}$ to be equal to zero before the Huffman coding, i.e after the reconciliation rounds.

The limits of $d_N(A^{[k]}, B^{[k]})$ are the roots of the equation

$$d = \frac{2d^2}{3(1-d)^2 + (1-d)^3}. \tag{5}$$

This equation can be re-write as :

$$d(1-d)(d - \frac{1}{2})^2 = 0. \tag{6}$$

Obviously, the roots of this equation, and so the possible limits of the normalize distance between $A^{[k]}$ and $B^{[k]}$ after k rounds of reconciliation, are $\{0, \frac{1}{2}, 1\}$.

$$\lim_{k \to +\infty} d_N^{[k]}(S, S') \in \{0, \frac{1}{2}, 1\}. \tag{7}$$

Let us consider now the case $d_N(A^{[0]}, B^{[0]}) \in [0 : \frac{1}{2})$ seen in (3) which is encountered in CHIMERA and study the limit of the normalized distance between $A^{[k]}$ and $B^{[k]}$ for this initial range of value. In this range, the next inequality is true :

$$\forall d_N^{[0]}(S, S') \in [0 : \frac{1}{2}), \frac{2d^2}{3(1-d)^2 + (1-d)^3} < d. \tag{8}$$

So, re-writing the equation with the normalized distance evolution function (4), we have:

$$\forall d_N^{[0]}(S, S') \in [0 : \frac{1}{2}), d_N(A^{[k+1]}, B^{[k+1]}) < d_N(A^{[k]}, B^{[k]}). \tag{9}$$

For $d_N(A^{[0]}, B^{[0]}) \in [0 : \frac{1}{2})$, the sequence $\{d_N(A^{[k+1]}, B^{[k+1]})\}_{k \geq 0}$ is decreasing and bounded. So it is convergent and its limit is 0.

$$\forall d_N(A^{[0]}, B^{[0]}) \in [0 : \frac{1}{2}), \lim_{k \to +\infty} d_N(A^{[k]}, B^{[k]}) = 0. \tag{10}$$

So after enough rounds, noted r, of reconciliation the normalized distance between $A^{[k]}$ and $B^{[k]}$ becomes as close to zero as wanted. This means that the sequences are equal, with a very high probability.

$$\forall d_N(A^{[0]}, B^{[0]}) \in [0 : \frac{1}{2}), \forall \epsilon > 0, \exists r, d_N(A^{[r]}, B^{[r]}) < \epsilon. \tag{11}$$

Choosing ϵ very close to 0, we can write :

$$P[A^{[r]} = B^{[r]}] \approx 0. \tag{12}$$

Obviously, the Huffman coding $\mathcal{H}$ does not change this result. We note $\mathcal{H}(A^{[r]})$ and $\mathcal{H}(B^{[r]})$, the Huffman coding of $A^{[r]}$ and $B^{[r]}$ respectively. So,

$$P[\mathcal{H}(A^{[r]}) = \mathcal{H}(B^{[r]})] \approx 0. \tag{13}$$

As defined in CHIMERA, the sequences $\mathcal{H}(A^{[r]})$ and $\mathcal{H}(B^{[r]})$ are the keys and can be noted, in accordance with [3], S and S'. So, we have :

$$P[S \neq S'] \approx 0. \tag{14}$$

6   Proof of the Property $|S| \approx H(S)$

The proof of the property $|S| \approx H(S)$ is based on the evaluation of the normalized weight of the sequences $A^{[r]}$ and $B^{[r]}$ and on a property of the Huffman code.

6.1   Definition : Normalized Weight

The normalized weight $\omega_N(A)$ of the binary sequence A is defined as the ratio between Hamming weight $\omega_H(A)$ and the length $|A|$.

$$\omega_N(A) = \frac{\omega_H(A)}{|A|}. \tag{15}$$

Of course, the initial normalized weight of the sequences $A^{[0]}$ and $B^{[0]}$ is equal to p_b.

6.2   Residual Normalized Weight

We consider the residual normalized weight of the sequences $A^{[r]}$ and $B^{[r]}$, i.e. when the condition $(P[S \neq S'] = 0)$ is satisfied. We note p_k the probability of keeping a bit after r rounds of reconciliation. This probability, we will not evaluate now, is function of the number of reconciliation rounds (each round divide by three, at least, the length of the sequences) and of the normalized distance of the sequences for each round of reconciliation (the closest the sequences are, the highest is the probability to keep a given bit).

As we keep only identical bits and sacrifice a certain amount of bits for security, the following table presents the two values the ith bit of $A^{[r]}$ and $B^{[r]}$ can have, with the probability associated to each case.

Table 3. Possibles values of $A_i^{[r]}$ and $B_i^{[r]}$ with occurrence probability

$A_i^{[r]}$	$B_i^{[r]}$	$p(A_i^{[r]}, B_i^{[r]})$
0	0	$(1 - p_b)^2 p_k$
1	1	$p_b^2 p_k$

Obviously, the normalized weight of $A^{[r]}$ (and $B^{[r]}$) at the end of the reconciliation is :

$$\omega_N(A^{[r]}) = \frac{p_b^2 p_k}{(1 - p_b)^2 p_k + p_b^2 p_k} = \frac{p_b^2}{(1 - p_b)^2 + p_b^2}. \tag{16}$$

This result is validated by simulations as one can see in the following graph representing $\omega_N(A^{[r]})$ as a function of p_b :

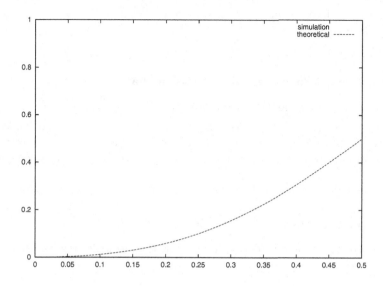

Fig. 1. This graph shows $\omega_N(A^{[r]}$ as a function of p_b. The curve is given by the theory. The dots are simulation results

Note that for $p_b > \frac{1}{4}$, the simulation results are noisy because the residual length of the sequence becomes too small. So, we will avoid this range of value for the bias of the random generators used to build Alice and Bob's sequences.

6.3   Entropy of $\mathcal{H}(A^{[r]})$

As $\omega_N(A^{[r]}) < \frac{1}{2}$, the entropy of $A^{[r]}$ is not maximal [6]. However, the last stage of the protocol is the compression of the sequences with an extended Huffman code. It is well known that using big t-tuples as the symbols of the language improves the compression ratio. With big enough t-tuples, the compression ratio is near of the entropy of the sequence. Noting $\mathcal{H}$, the extended Huffman code, we have :

$$|\mathcal{H}(A^{[r]})| \approx H(\mathcal{H}(A^{[r]})). \tag{17}$$

As $\mathcal{H}(A^{[r]})$ is the sequence S, we can rewrite the preceding equation :

$$H(S) \approx |S|. \tag{18}$$

7   Proof of the Property $I(S; C^t Z) \approx 0$

The proof of the property $I(S; C^t Z) \approx 0$ is based on the comparison of the amount of information revealed and sacrificed by the reconciliation algorithm. We will only study the cases in which bits are kept : when the bits are destroyed because they are different, the information that Eve can gather is useless.

Moreover, as Eve has no information about $A^{[0]}$ and $B^{[0]}$, we can forget Z and just prove that

$$I(S; C^t) \approx 0. \tag{19}$$

7.1 Information Sacrificed by the Reconciliation

Let us consider the reconciliation of the 3-tuples $(A_i^{[k]}, A_{i+1}^{[k]}, A_{i+2}^{[k]})$ from Alice's sequence and $(B_i^{[k]}, B_{i+1}^{[k]}, B_{i+2}^{[k]})$ from Bob's sequence (i is a multiple of 3). When for a given 3-tuples one bit is kept, then 2 bits are destroyed. Moreover, the sacrificed bits are independent from each other. So, the amount of information sacrificed is

$$H_s = 2H(\omega_N(A^{[k]})). \tag{20}$$

7.2 Information Revealed by the Reconciliation

Now, let us consider the information revealed by the reconciliation, i.e. the parity of the 3-tuple $(A_i^{[k]}, A_{i+1}^{[k]}, A_{i+2}^{[k]})$:

$$H(C_i^{2k+1}) = H(\bigoplus_{j=0}^{2} A_{i+j}^{[k]}). \tag{21}$$

The following table gives the probability of incidence of each case :

Table 4. Possible values of $(A_i^{[k]}, A_{i+1}^{[k]}, A_{i+2}^{[k]})$ with occurence probability.

$A_i^{[k]}$	$A_{i+1}^{[k]}$	$A_{i+2}^{[k]}$	$\bigoplus_{j=0}^{2} A_{i+j}^{[k]}$	$p(A_i^{[k]}, A_{i+1}^{[k]}, A_{i+2}^{[k]})$
0	0	0	0	$(1 - \omega_N(A^{[k]}))^3$
0	1	1	0	$(1 - \omega_N(A^{[k]}))\omega_N(A^{[k]})^2$
1	0	1	0	$(1 - \omega_N(A^{[k]}))\omega_N(A^{[k]})^2$
1	1	0	0	$(1 - \omega_N(A^{[k]}))\omega_N(A^{[k]})^2$
1	0	0	1	$(1 - \omega_N(A^{[k]}))^2 \omega_N(A^{[k]})$
0	1	0	1	$(1 - \omega_N(A^{[k]}))^2 \omega_N(A^{[k]})$
0	0	1	1	$(1 - \omega_N(A^{[k]}))^2 \omega_N(A^{[k]})$
1	1	1	1	$\omega_N(A^{[k]})^3$

From the four last cases, we have :

$$\omega_N(\bigoplus_{j=0}^{j \leq 2} A_{i+j}^{[k]}) = 3(1 - \omega_N(A^{[k]}))^2 \omega_N(A^{[k]}) + (\omega_N(A^{[k]}))^3. \tag{22}$$

Which give us, the entropy of the parity :

$$H(\bigoplus_{j=0}^{j \leq 2} A_{i+j}^{[k]}) = H(3(1 - \omega_N(A^{[k]}))^2 \omega_N(A^{[k]}) + (\omega_N(A^{[k]}))^3). \tag{23}$$

So,

$$H(C_i^{2k+1}) = H(3(1 - \omega_N(A^{[k]}))^2 \omega_N(A^{[k]}) + (\omega_N(A^{[k]}))^3). \qquad (24)$$

7.3   Comparison between Hs and $H(C_i^{2k+1})$

Obviously, we want th amount of information sacrificed to be greater than the amount of information revealed :

$$H_s \geq H(C_i^{2k+1}). \qquad (25)$$

With (20) and (24), it becomes

$$2H(\omega_N(A^{[k]})) \geq H(3(1 - \omega_N(A^{[k]}))^2 \omega_N(A^{[k]}) + (\omega_N(A^{[k]}))^3). \qquad (26)$$

The following graph shows $H(C^{2k+1})$ and Hs as functions of p_b.

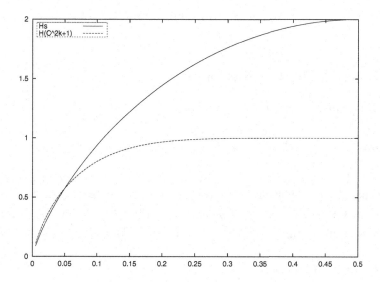

Fig. 2. This graph shows $H(C^{2k+1})$ and Hs as functions of p_b. For $p_b > \frac{1}{20}$, the amount of information revealed is lesser than the amount of information sacrificed

This inequality is true for $\omega_N(A^{[k]}) \in [\frac{1}{20} : \frac{1}{2}]$. To insure the security of the protocol, the inequality must be true for each round of the reconciliation :

$$\forall k \leq r\omega_N(A^{[k]}) \in [\frac{1}{20} : \frac{1}{2}]. \qquad (27)$$

As $\{\omega_N(A^{[k]})\}_{0 \leq k \leq r}$ is decreasing and $\omega_N(A^{[0]}) \leq \frac{1}{2}$, we just have to prove that the normalized weight of the residual sequence $A^{[r]}$ after the reconciliation is greater than $\frac{1}{20}$

$$\omega_N(A^{[r]}) \geq \frac{1}{20}. \tag{28}$$

Using (16), this inequality becomes

$$\frac{p_b^2}{(1 - p_b)^2 + p_b^2} \geq \frac{1}{20}. \tag{29}$$

So, the reconciliation algorithm REC(3,2) is secure if

$$p_b \geq \frac{\sqrt{19} - 1}{18}. \tag{30}$$

It means that eve gather no information from the communications C^t between Alice and Bob if the initial normalized weight of the sequences is in the range $[\frac{\sqrt{19}-1}{18} : \frac{1}{2}]$. Under this condition, we have :

$$I(S; C^t) \approx 0. \tag{31}$$

Moreover, as Eve has no initial sequence Z, we can write :

$$I(S; C^t Z) \approx 0. \tag{32}$$

8   Choice of the Parameter p_b

8.1   Constraints on the Choice of p_b

The bias of the random generators used to build $A^{[0]}$ and $B^{[0]}$ is the most important parameter of CHIMERA, as the security and the efficiency of the protocol depend on the value of p_b.

As seen in the proof of the property $|S| \approx H(S)$, the bias p_b should not be greater than $\frac{1}{4}$ to be efficient. Moreover, the proof of the property $I(S; C^t Z) \approx 0$ stands that CHIMERA is safe if p_b is greater than $\frac{\sqrt{19}-1}{18}$. So, the bias of the random generators must be choose in the range $[\frac{\sqrt{19}-1}{18} : \frac{1}{4}]$.

8.2   Simulation Results

We have made simulations with sequences $A^{[0]}$ and $B^{[0]}$ of length $2 \cdot 10^8$ bits. The bias of the random generators is set in the range $[0 : \frac{1}{2})$ (although only the range $[\frac{\sqrt{19}-1}{18} : \frac{1}{4}]$ is really useful in CHIMERA) and the reconciliation round is repeated while Alice and Bob's sequences are different.

Then, we have consider the residual length of the sequences weighted by the entropy of the normalized weight of the sequences, i.e. the length of the sequences compressed with an optimal compression code (like the extended Huffman code). The results of these simulations are presented in the following graph. The x-axis is the bias p_b and the y-axis is the residual length $|S|$.

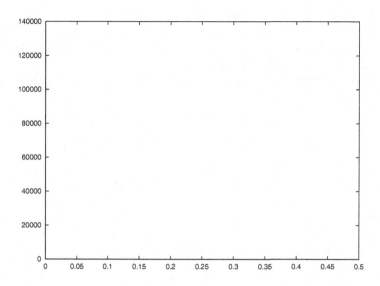

Fig. 3. This graph shows the residual length $|S|$ as a function of p_b. The value p_b we propose is 0.1875

As stands in [3], the goal is to make $|S|$ as large as possible. In the range $[\frac{\sqrt{19}-1}{18} : \frac{1}{4}]$, we have two clouds of points ; the first one, located in $[\frac{\sqrt{19}-1}{18} :\approx 0.22]$, re-groups the results of the simulations with six rounds of reconciliation. The other cloud of points re-groups the results of the simulations with seven rounds of reconciliation.

As one can see, in the range $[\frac{\sqrt{19}-1}{18} :\approx 0.22]$ the residual length $|S|$ is greater than the length $|S|$ in the range $[\approx 0.22 : \frac{1}{4}]$. Moreover, in the first range six rounds of reconciliation, instead of seven rounds, are needed. So we have to chose $p_b \in [\frac{\sqrt{19}-1}{18} :\approx 0.22]$.

Moreover, as the first cloud decreases with p_b, the bias of the random generator should be close to $\frac{\sqrt{19}-1}{18}$. For implementation convenience, we propose to use :

$$p_b = \frac{3}{16}. \tag{33}$$

8.3   Creation of a Biased Random Generator for $p_b = \frac{3}{16}$

The bias p_b can be easily obtain with a combination of non-biased random generators. For example, considering the outputs a, b, c and d of four non-biased random generators, the logical combination

$$p = a \cdot b \cdot c + a \cdot b \cdot d^1. \tag{34}$$

---

[1] $\cdot$ denotes the logical operator AND, $+$ denotes the logical operator OR

is a biased random generator of bias $p_b = \frac{3}{16}$.

A such simple construction can be implement in any environment and let Alice and Bob build their initial sequences with very light calculus. As the other parts of the protocol need very light calculations (XOR and Huffman coding with pre-calculated trees), our intend is to make the creation of the sequence as easy as the rest of the protocol.

9   Parameter of the Extended Huffman Code

The efficiency of the Huffman code depends on the number of symbols of the language on which is based the Huffman tree. For example if only two symbols appears, whatever their frequencies, the Huffman tree will be a simple root. But, if you consider n-tuples of symbols as the symbols of a language, the Huffman code become more and more efficient as n increases. The compression ratio is, of course, bounded by the entropy of the language.

For the last stage of CHIMERA, we have to find a size of the n-tuples such as a 128 bit key created with CHIMERA as at least 127 bits of entropy. The method to find n is simple : we calculate the minimum-redundancy code for an increasing n, with the algorithm presented in [7] until we found a compression ration $\mathcal{R}_n$ such as :

$$128 \cdot \frac{H(\omega_N(A^{[6]}))}{\mathcal{R}_n} \geq 127. \tag{35}$$

The following table present, for a given n, the compression ratio of the minimal-redundancy code obtained with n-tuples as symbols, and the entropy of a 128 bits key created with this minimal-redundancy code:

Table 5. Compression ratio and entropy of the key for a given length of the extended Huffman code.

n	$\mathcal{R}_n$	H(S)
1	1	36.9
2	0.5745	64.3
3	0.4347	85.1
4	0.3685	100.2
5	0.3378	109.4
6	0.3179	116.3
7	0.3056	121.0
8	0.3007	122.9
9	0.2971	124.4
10	0.2936	125.8
11	0.2905	127.2

The compression ratio for $n = 11$ is close enough to entropy of $H(A^{[6]}) \approx 0.28878$ to obtain a key with an entropy greater than 127.

Considering bigger n-tuples, one has a better approximation of the entropy. Nevertheless, the Huffman tree need more memory. with 11-tuples, the compression table (i.e. the Huffman tree) needs

10   Average Length of the Keys

The length of the keys can be easily calculated knowing the length $|A^{[r]}|$. As we need empirically 6 rounds of reconciliations to have $P(S \neq S') \approx 0$, we set $r = 6$ for $p_b = \frac{3}{16}$.

10.1   Residual Length $|A^{[6]}|$

The amount of bits kept after a reconciliation round is a function of the normalized distance between the sequences : the closer the sequences are, the fewer 3-tuples are destroyed.

As one bit is kept when the 3-tuples have the same parity, and none if the parities differ, noting $R(d_N(A^{[k]}, B^{[k]})) = \frac{A^{[k+1]}}{A^{[k]}}$ the reduction factor of the sequence, we have (with i multiple of 3):

$$R(d_N(A^{[k]}, B^{[k]})) = \frac{P((\bigoplus_{j=0}^2 A_{i+j}^{[k]}) = (\bigoplus_{j=0}^2 B_{i+j}^{[k]}))}{3}. \tag{36}$$

Considering the 3-tuples with the same parity, the table in the section (5.3) gives, setting $d_N = d_N(A^{[k]}, B^{[k]})$:

$$R(d_N) = \frac{(1 - d_N)^3 + 3(1 - d_N)d_N^2}{3}. \tag{37}$$

As the reconciliation is an iterative process, the length $|A^{[6]}|$ is reduced six times, with a ratio depending on the normalized distance between Alice and Bob's before each round of reconciliation REC(3,2). so, the length $|A^{[6]}|$ is :

$$|A^{[6]}| = |A^{[0]}| \prod_{i=0}^5 R(d_N(A^{[k]}, B^{[k]})). \tag{38}$$

Of course, $d_N(A^{[k]}, B^{[k]})$ is given for each iteration by (4).

10.2   Length of the Key S

At the end of the reconciliation, Alice and Bob own respectively the sequences $A^{[6]}$ and $B^{[6]}$, of length $k = |A^{[6]}|$ and of normalized weight $\omega_N(A^{[6]})$. The normalized weight is given by (16) :

$$\omega_N(A^{[6]}) = \frac{(\omega_N(A^{[0]}))^2}{(1 - \omega_N(A^{[0]}))^2 + (\omega_N(A^{[0]}))^2}. \tag{39}$$

These sequences equal with a very high probability are compressed at the end of the protocol with an extended Huffman code which compression ratio is very close to the entropy of the sequence. Thus, the length of the key is :

$$|S| = H(\omega_N(A^{[6]})) \cdot |A^{[6]}|. \tag{40}$$

From (38) and (39), we have :

$$|S| = H(\frac{(\omega_N(A^{[0]}))^2}{(1 - \omega_N(A^{[0]}))^2 + (\omega_N(A^{[0]}))^2})|A^{[0]}| \prod_{i=0}^{5} R(d_N(A^{[k]}, B^{[k]})). \tag{41}$$

With the extended Huffman code of length $n = 11$, the practical length of the keys is :

$$|S| = \mathcal{R}_{11}|A^{[0]}| \prod_{i=0}^{5} R(d_N(A^{[k]}, B^{[k]})). \tag{42}$$

The evaluation of this formula gives:

$$|S| \approx 6.37 \cdot 10^{-5}|A^{[0]}|. \tag{43}$$

So Alice and Bob can create a common key of 128 bits with initial sequences of length 2000000 bits.

11   Conclusion

The main points addressed in this paper are :

- A generalized definition of reconciliation has been proposed to let the users destroy more than one symbol of their sequences. The generalization is useful when the entropy of the reconciled sequences is not maximal.
- A unconditionally secure key agreement protocol, called CHIMERA, has been proposed. Its soundness and its security has been proved. The CHIMERA uses no specific devices unlike other unconditionally secure key agreement protocol.
- Convenient parameters has been given for practical implementation of the CHIMERA.

Acknowledgments. We are grateful to Alistair Moffar for the source code of his in-place calculation of minimum redundancy codes which was helpful in the determination of the characteristics of the extended Huffman code. We are also grateful to Sami Harari for his advice in the writing of this paper.

References

1. W. Diffie and M. Hellman. New directions in cryptography, 1976.
2. Charles Bennett, H., François Bessette, Gilles Brassard, and Louis Salvail. Experimental quantum cryptography. *Journal of Cryptology: the journal of the International Association for Cryptologic Research*, 5(1):3–28, ???? 1992.

3. Ueli Maurer. Information-theoretic cryptography. In Michael Wiener, editor, *Advances in Cryptology - CRYPTO '99*, volume 1666 of *Lecture Notes in Computer Science*, pages 47–64. Springer-Verlag, 1999.
4. Ueli M. Maurer. Information-theoretically secure secret-key agreement by NOT authenticated public discussion. In *Theory and Application of Cryptographic Techniques*, pages 209–225, 1997.
5. Stefan Wolf. Strong security against active attacks in information-theoretic secret-key agreement. In *Advances in Cryptology – ASIACRYPT 98: International Conference on the Theory and Application of Cryptology*, volume 1514 of *Lecture Notes in Computer Science*, pages 405–419. Springer-Verlag, 1998.
6. D. A. Huffman. A method for the construction of minimum redundancy codes. *Proceedings of the Institute of Electronics and Radio Engineers*, 40:1098–1101, 1952.
7. A. Moffat and J. Katajainen. In-place calculation of minimum-redundancy codes. In S.G. Akl, F. Dehne, and J.-R. Sack, editors, *Proc. Workshop on Algorithms and Data Structures*, pages 393–402, Queen's University, Kingston, Ontario, August 1995. LNCS 955, Springer-Verlag.

An Efficient Stream Cipher Alpha1 for Mobile and Wireless Devices

N. Komninos, Bahram Honary, and Michael Darnell

HW Communications Limited White Cross Industrial Estate
mdarnell@hwcomms.com   http://www.hwcomms.com

Abstract. Encryption in wireless networks is essential to protect sensitive information and to prevent fraud. Furthermore, wireless devices, such as palmtops, bluetooth devices, and mobiles phones require an efficient encryption algorithm which is secure and small, fast and simple to implement. There are several stream ciphers available with the best known being A5/1 used in GSM and RC4 used in 802.11 standards. However, these stream ciphers are weak against cryptographic attacks [1] [2] [3] [4] [5] [6]. In this paper, a new synchronous stream cipher (Alpha1) is proposed. Alpha1 is robust against these attacks, small, fast and very simple to implement in small wireless devices with low processing power and size (i.e. bluetooth, mobile phones, access points etc.).

1   Introduction

Stream ciphers can be designed to be exceptionally fast, much faster in fact than any block cipher. While block ciphers operate on large blocks of data, stream ciphers typically operate on smaller units of plaintext, usually bits. A stream cipher generates a *keystream* and encryption is provided by combining the keystream with the plaintext, usually with the bitwise XOR operation. The generation of the keystream can be independent of the plaintext and ciphertext (yielding what is termed a *synchronous stream cipher*) or it can depend on the data and its encryption (in which case the stream cipher is said to be *self-synchronizing*). The majority of the encryption algorithms that protect the air interface in mobile and wireless networks (i.e. GSM, bluetooth etc.) use synchronous stream ciphers build on linear feedback shift registers.

A *Linear Feedback Shift Register* (LFSR) is a mechanism for generating a sequence of binary bits. The register consists of a series of cells that are set by an initialisation vector that is, most often, the secret key. The behaviour of the register is regulated by a clock and at each clocking instant, the contents of the cells of the register are shifted right/left by one position, and the XOR of a subset of the cell contents is placed in the leftmost/rightmost cell. One bit of output is usually derived during this update procedure. LFSRs are fast and easy to implement in both hardware and software. With a judicious choice of *feedback taps* the sequences that are

B. Honary (Ed.): Cryptography and Coding 2001, LNCS 2260, pp. 294–300, 2001.
© Springer-Verlag Berlin Heidelberg 2001

generated can have a good statistical appearance. However, LFSRs are useful as building blocks in more secure systems (i.e. A5/1 in GSM) but still suffer from two resent cryptographic attacks as described by Biryukov and Shamir, the biased birthday attack and the random subgraph attack.

The main idea of the biased birthday attack [1] is to consider sets A and B from the LFSRs which are not chosen with uniform probability distribution among all the possible states. The observation which makes this attack efficient is that in LFSR-based stream ciphers there is a huge variance in the weights of various states that begin with a specific a-bit pattern. The register bits that affect clock control and the register bits that affect the output are unrelated for about a clock cycles because of the single clock control tap. This decreases the states to be sampled to $2^n \cdot 2^{-a} = 2^{n-a}$. It was also found in A5/1 that the weight of a large percentage of the states was zero because their trees died out before reaching depth 100. As a result, efficient determination of initial states can be obtained since the exact location of a and the depth of the initial state is known. This is made possible by the frequent initialisation and poor choice of the clocking taps.

The main idea of the random subgraph attack is to make most of the special states accessible by simple computations from the subset of special states which are actually stored in the hard disk [1]. The attack is based on a function f that maps special states in an easy computable way and Hellman's time-memory tradeoff, $M\sqrt{T} = |U|$, for block ciphers described in [4] can be used to invert it easily. In A5/1 the attack is applied to the subspace of 2^{48} by the fact that it can be efficiently sampled. The time trade-off formula of the random subgraph attack results to $M = 2^{36}$ and just $T = 2^{24}$. This number of steps can be carried out in several minutes on a fast PC. Therefore, it is essential to create a stream cipher which is robust against these attacks, small in size and very simple to implement in hardware.

2   Description

In general, Alpha1 is a synchronous clock controlled cipher and operates by expanding a short key into an infinite pseudo-random key stream. The sender uses a boolean exclusive-OR (XOR) gate, to XOR the key stream with the plaintext to produce ciphertext. The receiver has a copy of the same key, and uses it to generate identical key stream. XORing the key stream with the ciphertext yields the original plaintext. The key stream can be generated based on a combination of the system specifications that is used.

The cipher uses four linear feedback shift registers (LFSRs) of length 29, 31, 33, and 35 denoted by R1, R2, R3, and R4 respectively. The four registers are maximum length of LFSRs with maximum periods $2^{29} - 1$, $2^{31} - 1$, $2^{33} - 1$, and $2^{35} - 1$ respectively. When R2 and R3 registers are clocked their output are combined using a boolean AND gate; the output of this AND gate is then an input to a boolean exclusive-OR (XOR) gate, together with the outputs of R1 to R4, to produce the final output stream (Figure 1). The AND gate was added to the cipher to increase the linear complexity and avoid attacks based on linear statistical weakness [3]. Moreover, several complicated combinations of AND gates has been designed and tested in Alpha1 to increase mathematically the linear complexity. However, it is not recommended in real systems because the stream cipher becomes slow and impractical in mobile and wireless devices.

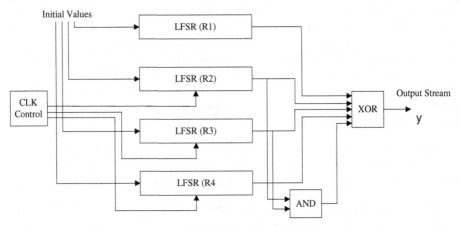

Fig. 1. Alpha1 Stream Cipher

The total length of the registers is 128 and the feedback polynomials used are all primitive. The primitive polynomials have been generated by computer software based on [7]. The Hamming weight of all the feedback polynomials is chosen to be five for R1, R2, and R3, and six for R4 (see Table 1); this represents a reasonable trade-off between reducing the number of required XOR gates in the hardware realization and obtaining good statistical properties for the generated sequences.

Table 1. Primitive Polynomials and Hamming Weight

	Polynomials $f_i(x)$	Weight
R1	$x^{29} + x^{27} + x^{24} + x^8 + 1$	5
R2	$x^{31} + x^{28} + x^{23} + x^{18} + 1$	5
R3	$x^{33} + x^{28} + x^{24} + x^4 + 1$	5
R4	$x^{35} + x^{30} + x^{22} + x^{11} + x^6 + 1$	6

Let x_1, x_2, x_3, and x_4 denote the element of the polynomial with highest degree in R1, R2, R3, and R4 respectively. The output y of Alpha1 is given by the following equation:

$$y = x_1 \oplus x_2 \oplus x_3 \oplus x_4 \oplus x_2 \cdot x_3 \tag{1}$$

If L_1, L_2, L_3, and L_4 denote the highest degree of the primitive polynomials in R1, R2, R3, and R4 respectively then, according to (1), the linear complexity of Alpha1 is given by [7]:

$$L = L_1 + L_2 + L_3 + L_4 + L_2 L_3 = 29 + 31 + 33 + 35 + 1023 = 1151 \tag{2}$$

The linear complexity is in practice greater than 1151 because of the clock control mechanism used in Alpha1. In the clock control mechanism of each register, apart from R1, there are two clocking taps: bits 10, 21 for R2, bits 10, 22 for R3 and bits 11, 24 for R4. The clocking taps, which control the operation of Alpha1, divide each of the three registers into three, almost equal, parts. Every clocking tap per register is checked in a triangular scheme; the bits 10, 22, and 11 (blue colour) are checked with bits 21, 10, and 24 (green colour) of R2, R3, and R4 respectively, as shown in Figure 2.

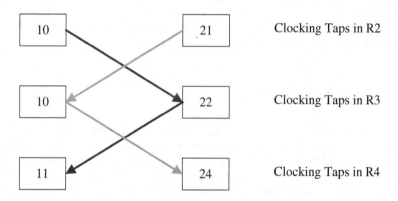

Fig. 2. Clock Control Mechanism of Alpha1

For each clock cycle, the registers whose clocking taps agree with the majority bit in the triangle are shifted. Note here that R1 is shifted in every clock cycle. For example, if B, C, and D represent the clocking taps of R2, R3 and R4 respectively, then Table 2 shows all the combinations for shifting.

Table 2. Shift per Clock Register

One Clock Cycle	Majority Bit Of 10, 22, 11 = B, C, D	Majority Bit Of 21, 10, 24 = B, C, D	Candidates For Shift 2
Case 1	B=C	B=C	R2, R3
Case 2	B=D	B=C	R2
Case 3	C=D	B=C	R3
Case 4	B=C	B=D	R2
Case 5	B=D	B=D	R2, R4
Case 6	C=D	B=D	R4
Case 7	B=C	C=D	R3
Case 8	B=D	C=D	R4
Case 9	C=D	C=D	R3, R4
Case 10	B=C	B=C=D	R2, R3
Case 11	B=D	B=C=D	R2, R4
Case 12	C=D	B=C=D	R3, R4
Case 13	B=C=D	B=C=D	R2, R3, R4

For instance in case one, if the clocking taps of R2 in positions 10 and 21 agree with the clocking taps of R3 in positions 22 and 10 respectively then only R2 and R3 are shifted. Likewise, in case 2 if the clocking tap of R2 in positions 10 agrees with the clocking tap of R4 in position 11 and the clocking tap of R2 in position 21 agrees with the clocking tap of R3 in position 10 then only R2 is shifted. Finally, note that at each clock cycle R1 and/or R2 or/and R3 or/and R4 registers are shifted. Registers R2, R3 and R4 move with probability $\cong \frac{7}{13}$.

3   Cryptanalysis of Alpha1

As mentioned in the introduction section, stream ciphers, such as A5/1, suffer from two recent cryptographic attacks, the biased birthday attack and the random subgraph attack, as described by Biryukov, Shamir and Wagner in the Fast Software Encryption Workshop 2000. The poor initialisation and choice of the clocking taps in LFSR-based stream ciphers make both attacks applicable by identifying special states in the registers.

The clock control mechanism in Alpha1 was designed in such way that the register bits that affect the clock control and the register bits that affect the output are unrelated for only 10 clock cycles. Therefore, in the biased birthday attack there is no variance in the weights of various states that begin with a 10-bit pattern. As a result the 10-bit pattern reduces the number of states that required to be sampled from 2^{128} to $2^{128} * 2^{-10} = 2^{118}$. In this case the biased birthday attack becomes very inefficient which results to an impractical random subgraph attack.

The random subgraph attack is applied to the subspace of 2^{118} regardless if it can be efficiently sampled or not. The time trade-off formula of the random subgraph attack results to $M = 2^{70}$ and just $T = 2^{48}$. This number of steps makes random subgraph attack impossible.

4   Implementation

Alpha1 has been implemented on a complex programmable logic device (CPLD) using VHSIC Hardware Description Language (VHDL). The XC95216 Xilinx chip was used to implement Alpha1 stream cipher. XC95216 was the smallest CPLD chip found in Xilinx. It has 216 macrocells and 216 registers (about 4,800 gates) out of which 131 macrocells (60%) and 128 registers (59%) are used for Alpha1. Thus, Aphal1 requires about 2,911 gates. The maximum clock speed of XC95216 is about 33.3MHz.

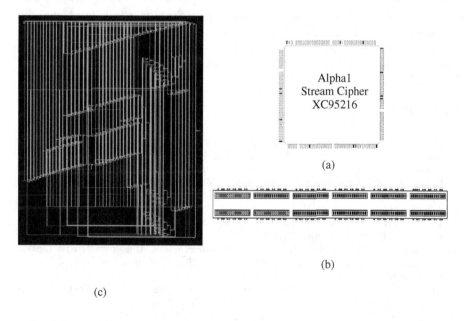

(a)

(b)

(c)

Fig. 3. External / Internal Structure of XC95216 (CPLD) Chip for Alpha1 Stream Cipher

In Figure 3, the external and internal structure of XC95216 chip is shown. In particularly, the chip is illustrated in Figure 3a. The registers are shown in Figure 3b and finally the logical gates are illustrated in Figure 3c.

5   Conclusion

There are a vast number of stream ciphers that have been proposed in cryptographic literature as well as an equally vast number that appear in implementations and products worldwide. Many are based on the use of LFSRs since such ciphers tend to be more amenable to analysis and it is easier to assess the security that they offer. Alpha1, which is similar to A5/1, has been designed to increase the security in LFSR-based stream ciphers used in mobile and wireless devices. High linear complexity, ease of implement and small in size are the main advantages of Alpha1. Moreover, the operation of Alpha1 has been designed to prevent special states being identified by their output sequences [1]; this is achieved by the clock control mechanism described which ensures that the register bits that affect the clock control and the register bits that affect the output are unrelated for only 6 clock cycles. This makes the biased birthday attack and the random subgraph attack [1] inefficient. Moreover, Anderson and Roe [6] have proposed an attack on A5/1 that uses only 3 LFSRs, based on guessing the bits of the shorter registers, and deriving the other bits of the longer register from the output. However, they have to guess the clocking sequence and the complexity of the attack is about 2^{40} in A5/1. Such an attack on Alpha1 has complexity greater than $\geq 2^{99}$ because of the number of states $n = 2^{128}$. Moreover, the high linear complexity of Alpha1 makes attack in [2] impractical because it is based on a solution of a system with linear equations.

References

1. A. Biryukov, A.Shamir, D.Wagner, *Real Time Cryptanalysis of A5/1 on a PC*, Fast Software Encryption Workshop 2000, New York, USA, April 10-12, 2000
2. J. Golic, *Cryptanalysis of Alleged A5 Stream Cipher*, proceedings of EUROCRYPT'97, Lecture Notes in Computer Science, pages 239-255, Konstanz, Germany, May 1997, Springer-Verlag.
3. J. Golic, "*Linear statistical weakness of alleged RC4 keystream generator*", proceeding of EUROCRYPT' 97, Lecture Notes in Computer Science, pages 226-238, Konstanz, Germany, May 1997, Springer-Verlag.
4. S. Babbage, *A Space / Time Tradeoff in Exhaustive Search Attacks on Stream Ciphers*, European Convention on Security and Detection, IEE Conference publication, No 408, May, 1995
5. M. E. Hellman, *A Cryptanalytic Time – Memory trade-Off*, IEEE Transactions on Information Theory, Vol. IT – 26, N 4, pp.401-406, July 1980.
6. R. Anderson, M. Roe, A5, http://jya.com/crack-a5.htm, 1994.
7. A. J. Menezes, P.C. van Oorschot, S. A. Vanstone, *Handbook of Applied Cryptography*, CRC Press LLC, 1997

Investigation of Linear Codes Possessing Some Extra Properties

Viktoriya Masol

Kyiv National Taras Shevchenko University,
vul. Volodymyrska, 64
01033 Kyiv, Ukraine
vicamasol@pochtamt.ru

Abstract. The paper considers additive representation of the number of linear codes. The summands of the representation are numbers of linear codes of weight 1, 2, ... Formulae for calculating the summands with respect to weight 1 and weight 2 are stated. The contribution of the summands mentioned to the whole sum is estimated. The property of strong logarithmic concavity of the number of linear codes of minimal weight is established.

1 Introduction

Let GF(q) be a finite field consisted of q elements. An arbitrary k-dimantional subspace of an n-dimentional vector space over the field GF(q) will be called [n, k] linear code over the field GF(q) or [n, k]-code. Denote by G(n, k) the number of all [n, k]-codes different in pairs. To know the number of [n, k]-codes, which possess some extra properties, is of interest for the problem of information coding [3]. The weight of a linear code is considered to be such a property. (Recall that by the weight of a linear code we mean the minimal weight of a nonzero vector that belongs to the linear code. In its turn, by the weight of an n-dimentional vector we mean the number of its nonzero components.) Let us denote by G(n, k, ϖ) the number of all different in pairs [n, k]-codes each of which has weight ϖ, $\varpi = 1, 2, ..., n$. Then the number G(n, k) can be obviously represented in the following form

$$G(n, k) = G(n, k, 1) + G(n, k, 2) + \cdots . \tag{1}$$

The paper is devoted to the statement of the properties of the summands G(n, k, 1) and G(n, k, 2), and to the estimation of their contribution to the right-hand side of (1).

2 Formulae for G(n, k, 1) and G(n, k, 2) and Their Applications

Theorem 1. For integers $1 \le k \le n$ we have

$$G(n, k, 1) = G(n-1, k, 1) + G(n-1, k-1, 1) (B(m) - 1) + G(n-1, k-1) , \tag{2}$$

where $B(a) = \exp\{a \ln q\}$, $a \in \mathbf{R}$, $m = n-k$, $G(d, p, 1) = 0$ if either $p = 0$ or $p > d$.

B. Honary (Ed.): Cryptography and Coding 2001, LNCS 2260, pp. 301-306, 2001.

Theorem 2. For integers $1 \leq k \leq n$ we have

$$G(n, k, 2) = G(n-1, k, 2) + G(n-1, k-1, 2) (B(m) - 1) + \tag{3}$$

$$+ \{G(n-1, k-1) - G(n-1, k-1, 1) - G(n-1, k-1, 2)\} (n-1) (q-1),$$

where $G(d, p, 2) = 0$ if either $p = 0$ or $p > d$.

Theorem 3. Condition $B(m) - (q-1) m - 1 < (q-1) k$, where $1 \leq k \leq n$, is necessary and sufficient for the following relation to be valid:

$$G(n, k) = G(n, k, 1) + G(n, k, 2),$$

where $G(n, k, 1)$ and $G(n, k, 2)$ are found by means of (2) and (3) respectively.

Denote by $P(n, k) = G(n, k, 1)/G(n, k)$, $1 \leq k \leq n$.

Theorem 4. If $m = o(\ln k)$, $n \to \infty$, then $P(n, k) \to 1$, $n \to \infty$; if $k = x B(m + o(1))$ as $n \to \infty$, where x is a chosen positive number, then $P(n, k) \to 1 - \exp\{-x\}$, $n \to \infty$; if, finally, $k = o(B(m))$, $n \to \infty$, then $P(n, k) \to 0$, $n \to \infty$.

Theorem 5. If

$$k = y (B(m) - (q-1) m),$$

where y is a chosen number, and $1 \leq (q-1) y < \infty$, then $G(n, k, 2)/G(n, k) \to \exp\{-y\}$ as $n \to \infty$.

The proofs of theorems 1, 3–5 are given in [4]. Similary by (2) we can verify relation (3).

The main problem of the theory of linear codes is to construct a code possessing the greatest weight ϖ at chosen parameters n and k. From the point of view of the advancement in solving this problem Theorem 4 can be interpreted as follows. If n and k differ insignificantly, for instance by a quantity not exceeding $o(\ln k)$, i.e., if $n-k \ll \ln k$, then, asymptotically as $n \to \infty$, all [n, k]-codes have weight 1. If n differs from k by a quantity of order of ln k, then we have exponential behavior of the number $G(n, k, 1)/G(n, k,)$ as $n \to \infty$. Finally, if $n \gg k + \ln k$, then asymptotically as $n \to \infty$, all [n, k]-codes have weight at least 2. Theorem 5 admits analogous interpretation.

3 Property of Strong Logarithmic Concavity

Theorem 6 states that the numbers $G(n+1, k, 1)$, $1 \leq k \leq n+1$, $n = 0, 1, 2,...$, satisfy the property of strong logarithmic concavity. Note that Gaussian coefficients $G(n, k)$, Stirling numbers of the first and second kind, etc. satisfy this property as well (see, for example, [1], [2], etc.).

Theorem 6. For integers $1 \leq k \leq n+1$, $n = 0, 1, 2, ...$

$$2 \ln G(n+1, k, 1) > \ln G(n+1, k-1, 1) G(n+1, k+1, 1), \tag{4}$$

where $G(d, p, 1) = 0$ if either $p = 0$ or $p > d$.

We preface the proof of Theorem 6 by the following lemmas.

3.1 Auxiliary Assertions

Lemma 1. For integers $1 \leq k \leq n$, $n = 1, 2, \ldots$

$$G(n, k+1) \, G(n, k, 1) \leq G(n, k) \, G(n, k+1, 1) . \tag{5}$$

Proof. Using the well-known equality

$$G(n, k) = (B(n) - 1) \cdots (B(m+1) - 1)/(B(k) - 1) \cdots (B(1) - 1) \tag{6}$$

we can rewrite (5) as

$$(B(m) - 1) \, G(n, k, 1) \leq (B(k+1) - 1) \, G(n, k+1, 1) . \tag{7}$$

Relation (7) is obvious at $n = 1$. The induction step from $n-1$ to n, $n \geq 2$, can be proved as follows. Lets consider $1 \leq k \leq n-1$ (if $k = n$ then (7) is obvious). Applying Theorem 1 to relation (6) and using the induction assumption we obtain

$$(B(m-1) - 1) \, (B(k) + \{G(n-1, k-1)/G(n-1, k, 1)\}) \leq \tag{8}$$

$$\leq (B(m-1) - 1) \, B(k+1) + (B(k+1) - 1) \, G(n-1, k)/G(n-1, k, 1)) .$$

It follows from relation (6) and estimate $G(n-1, k, 1) \leq G(n-1, k)$ that inequality (8) holds true. This completes the proof of Lemma 1.

The following Lemmas 2–5 can be established similary, however it requires more computations to set their proofs in.

Lemma 2. For integers $1 \leq k \leq n$, $n = 1, 2, \ldots$

$$G(n, k-1) \, G(n, k+2, 1) \leq G(n, k) \, G(n, k+1, 1) .$$

Lemma 3. For integers $1 \leq k \leq n$, $n = 1, 2, \ldots$

$$(B(m+1) - 1) \, G(n, k+1) \, G(n, k-1, 1) \leq (B(m) - 1) \, G(n, k) \, G(n, k, 1) .$$

Lemma 4. For integers $1 \leq k \leq n$, $n = 1, 2, \ldots$

$$(B(m-1) - 1) \, G(n, k-1) \, G(n, k+1, 1) \leq (B(m) - 1) \, G(n, k) \, G(n, k, 1) .$$

Lemma 5. For integers $1 \leq k \leq n$, $n = 1, 2, \ldots$
$$(B(m+2) - 1) \, G(n, k-2, 1) \, G(n, k+1, 1) \leq$$

$$\leq (B(m+1) - 1) \, G(n, k-1, 1) \, G(n, k, 1) .$$

3.2 Proof of Theorem 6

If $n = 0$ then relation (4) is obvious. Let us prove the induction step from $n-1$ to n, $n \geq 1$. Using Theorem 1 we can represent left-hand and right-hand sides of relation (4) in the form

$$G^2(n+1, k, 1) = D_1 + \ldots + D_6 , \tag{9}$$

where

$$D_1 = G^2(n, k, 1) ,$$

$$D_2 = (G(n, k-1, 1) (B(m+1) - 1))^2 ,$$

$$D_3 = G^2(n, k-1) ,$$

$$D_4 = 2 G(n, k, 1) G(n, k-1, 1) (B(m+1) - 1) ,$$

$$D_5 = 2 G(n, k, 1) G(n, k-1) ,$$

$$D_6 = 2 G(n, k-1, 1) G(n, k-1) (B(m+1) - 1) ,$$

and

$$G(n+1, k-1, 1) G(n+1, k+1, 1) = F_1 + \ldots + F_9 , \qquad (10)$$

where

$$F_1 = G(n, k-1, 1) G(n, k+1, 1) ,$$

$$F_2 = G(n, k-1, 1) G(n, k, 1) (B(m) - 1) ,$$

$$F_3 = G(n, k-1, 1) G(n, k) ,$$

$$F_4 = G(n, k-2, 1) G(n, k+1, 1) (B(m+2) - 1) ,$$

$$F_5 = G(n, k-2, 1) G(n, k, 1) (B(m+2) - 1) (B(m) - 1) ,$$

$$F_6 = G(n, k-2, 1) G(n, k) (B(m+2) - 1) ,$$

$$F_7 = G(n, k-2) G(n, k+1, 1) ,$$

$$F_8 = G(n, k-2) G(n, k, 1) (B(m) - 1) ,$$

$$F_9 = G(n, k-2) G(n, k) .$$

By virtue of the induction assumption we have

$$D_1 > F_1 , \qquad (11)$$

and, taking into consideration the inequality

$$(B(m+1) - 1)^2 > (B(m+2) - 1) (B(m) - 1)) ,$$

$$D_2 > F_5 . \qquad (12)$$

Using the property of strong logarithmic concavity of Gaussian coefficients $G(n, k)$ we get

$$D_3 > F_9 . \qquad (13)$$

It follows from Lemma 5 that

$$D_4 \geq F_2 + F_4 . \qquad (14)$$

Lemmas 1 and 2 imply

$$D_5 \geq F_3 + F_7 .$$ (15)

And, finally, Lemmas 3 and 4 yield

$$D_6 \geq F_6 + F_8 .$$ (16)

Relations (9)–(16) prove (4). This completes the proof of Theorem 6.

4 Problems and Examples

For $\varpi = q = 2$ and $k \leq n \leq 10$ the numbers $G(n, k, 2)$ are given in Table 1 and Table 2. (The numbers in the tables are found by means of relations (2), (3) and (6).)

Table 1. Numbers $G(n, k, 2)$ for $2 \leq n \leq 6$

k	n 2	3	4	5	6
1	1	3	6	10	15
2	0	1	13	75	305
3	0	0	1	40	640
4	0	0	0	1	121
5	0	0	0	0	1
6	0	0	0	0	0
7	0	0	0	0	0
8	0	0	0	0	0
9	0	0	0	0	0
10	0	0	0	0	0

Table 2. Numbers $G(n, k, 2)$ for $7 \leq n \leq 10$

k	n 7	8	9	10
1	21	28	36	45
2	1022	3038	8346	21720
3	6265	46522	292068	1647480
4	4781	109991	1800213	23830197
5	364	34041	1786866	64127334
6	1	1093	239380	27853180
7	0	1	3280	1678940
8	0	0	1	9841
9	0	0	0	1
10	0	0	0	0

Making use of Table 1and Table 2 it is easy to show that for $\varpi = q = 2$ and $k \leq n \leq 10$ numbers $G(n, k, 2)$ satisfy the property of strong logarithmic concavity.

The question about whether the numbers $G(n, k, \varpi)$, $\varpi \geq 2$, satisfy the property of strong logarithmic concavity is still open.

References

1. Kurts, D.C.: A Note on Concavity Properties of Triangular Arrays of Numbers. J. Comb. Theory, Vol. 13 (1972) 135–139
2. Lieb, E.H.: Concavity Properties and a Generating Function for Stirling Numbers. J. Comb. Theory, Vol. 5 (1968) 203–206
3. MacWilliams, F.J., Sloane, N.J.A.: The Theory of Error-correcting Codes (1977)
4. Masol, V.I.: Asymptotic Behaviour of the Number of Certain k-dimantional Subspaces over a Finite Field. Mathematical Notes, Vol. 59 (1996) 525–530

Statistical Physics of Low Density Parity Check Error Correcting Codes

David Saad[1], Yoshiyuki Kabashima[2], Tatsuto Murayama[2], and
Renato Vicente[3]

[1] Neural Computing Research Group, Aston University, Birmingham B4 7ET, UK.
[2] Dept. of Comp. Intel. & Syst. Sci., Tokyo Institute of Technology, Yokohama
2268502, Japan.
[3] Dep. de Física Geral, Instituto de Física, Universidade de São Paulo,
Caixa Postal 66318, 05315-970 São Paulo - SP, Brazil.

Abstract. We study the performance of Low Density Parity Check
(LDPC) error-correcting codes using the methods of statistical physics.
LDPC codes are based on the generation of codewords using Boolean
sums of the original message bits by employing two randomly-
constructed sparse matrices. These codes can be mapped onto Ising spin
models and studied using common methods of statistical physics. We examine various regular constructions and obtain insight into their theoretical and practical limitations. We also briefly report on results obtained
for irregular code constructions, for codes with non-binary alphabet, and
on how a finite system size effects the error probability.

1 Introduction

Modern telecommunication relies heavily on error correcting mechanisms to com-
pensate for corruption due to noise during transmission. The information transmission code rate, measured in the fraction of informative transmitted bits, plays
a crucial role in determining the speed of communication channels. Rigorous
bounds [1] have been derived for the maximal code rate for which codes, capable
of achieving arbitrarily small error probability, can be found. However, these
bounds are not constructive and most existing practical error-correcting codes
are far from saturating them.

Two code families currently achieve the highest information transmission
rates for a given corruption level, especially in the high code rate regime. Turbo
codes [2] have been introduced less than a decade ago, and were followed by the
rediscovery of Low Density Parity Check Codes (LPDC) [3]. The latter have been
originally introduced by Gallager [4] in 1962, and abandoned in favour of other
codes due to the limited computing facilities of the time. Both codes show excellent performance and recently discovered irregular LDPC constructions nearly
saturate Shannon's bound for infinite message size [5].

LDPC codes are generally based on the introduction of random sparse matrices for generating the transmitted codeword as well as for decoding the received

B. Honary (Ed.): Cryptography and Coding 2001, LNCS 2260, pp. 307–316, 2001.
© Springer-Verlag Berlin Heidelberg 2001

corrupted codeword. Two main types of matrices have been studied: regular constructions, where the number of non-zero row/column elements in these matrices remains fixed; and irregular constructions where it can vary from row to row or column to column. Various decoding methods have been successfully employed; we will mainly refer here to the leading decoding techniques based on Belief Propagation (BP) [6].

Most analyses of LDPC codes have been obtained via methods of information theory, backed up by numerical simulations. These rely on deriving upper and lower bounds for the performance of codes, with or without making assumptions about the code used. These bounds represent a worst case analysis, and may be tight or loose depending on the accuracy and restrictiveness of the assumptions used, and the specific difference between the worst and typical cases.

The statistical physics based analysis takes a different approach, analysing directly the typical case, making use of explicit assumptions about the code used and its macroscopic characteristics. Moreover, using methods adopted from statistical physics of Ising spin systems, one can actually carry out averages over ensembles of codes with the same macroscopic properties to obtain exact performance estimates in the limit of infinitely large systems. Two methods have been used in particular, the replica method and the Bethe approximations [7], that is also linked to the Thouless-Anderson-Palmer (TAP) approach [8] to diluted systems. In this paper we will review recent studies of LDPC codes, using a statistical physics based analysis. We focus on two specific codes, Gallager's original LDPC code [4] and the MN code [3] where messages are represented by binary vectors and are communicated through a Binary Symmetric Channel (BSC) where uncorrelated bit flips appear with probability p.

A Gallager code is defined by a binary matrix $\mathcal{A} = [A \mid B]$, concatenating two very sparse matrices known to both sender and receiver, with B (of dimensionality $(M-N) \times (M-N)$) being invertible - the matrix A is of dimensionality $(M-N) \times N$.

Encoding refers to the production of a M dimensional binary codeword $\boldsymbol{t} \in \{0,1\}^M$ ($M > N$) from the original message $\boldsymbol{\xi} \in \{0,1\}^N$ by $\boldsymbol{t} = G^T \boldsymbol{\xi}$ (mod 2), where all operations are performed in the field $\{0,1\}$ and are modulo 2. The generator matrix is $G = [I \mid B^{-1}A]$ (mod 2), where I is the $N \times N$ identity matrix, implying that $AG^T = 0$ (mod 2) and that the first N bits of $\boldsymbol{t}$ are set to the message $\boldsymbol{\xi}$. In *regular* Gallager codes the number of non-zero elements in each row of A is chosen to be exactly $\hat{K}$. The number of elements per column is then $C = (1-R)\hat{K}$, where the code rate is $R = N/M$ (for unbiased messages). The encoded vector $\boldsymbol{t}$ is then corrupted by noise represented by the vector $\boldsymbol{\zeta} \in \{0,1\}^M$ with components independently drawn with probability $P(\zeta) = (1-p)\delta(\zeta) + p\delta(\zeta - 1)$. The received vector takes the form $\boldsymbol{r} = G^T \boldsymbol{\xi} + \boldsymbol{\zeta}$ (mod 2).

Decoding is carried out by multiplying the received message by the matrix $\mathcal{A}$ to produce the *syndrome* vector $\boldsymbol{z} = \mathcal{A}\boldsymbol{r} = \mathcal{A}\boldsymbol{\zeta}$ (mod 2) from which an estimate $\hat{\boldsymbol{\tau}}$ for the noise vector can be produced. An estimate for the original message is then obtained as the first N bits of $\boldsymbol{r} + \hat{\boldsymbol{\tau}}$ (mod 2). The Bayes optimal estimator (also known as *marginal posterior maximiser*, MPM) for the noise is defined as

$\widehat{\tau}_j = \text{argmax}_{\tau_j} P(\tau_j \mid z)$, where $\tau_j \in \{0, 1\}$. The performance of this estimator can be measured by the probability of bit error $P_b = 1 - 1/M \sum_{j=1}^{M} \delta[\widehat{\tau}_j; \zeta_j]$, where $\delta[;]$ is Kronecker's delta. Knowing the matrices B and A, the syndrome vector z and the noise level p, it is possible to apply Bayes' theorem and compute the posterior probability

$$P(\tau \mid z) = \frac{1}{Z} \chi [z = A\tau (\text{mod } 2)] P(\tau), \tag{1}$$

where $\chi[X]$ is an indicator function providing 1 if X is true and 0 otherwise. To compute the MPM one has to compute the marginal posterior $P(\tau_j \mid z) = \sum_{i \neq j} P(\tau \mid z)$, which in general requires $\mathcal{O}(2^M)$ operations, thus becoming impractical for long messages. To solve this problem one can use the sparseness of A to design algorithms that require $\mathcal{O}(M)$ operations to perform the same task. One of these methods is the probability propagation algorithm, also known as belief propagation (BP) [6].

The MN code has a similar structure, except for the fact that the generator matrix is $G = B^{-1}A$. The randomly-selected sparse matrices A and B are of dimensionality $M \times N$ and $M \times M$ respectively; these are characterized by K and L non-zero unit elements per row and C and L per column respectively. Correspondingly, the code rate becomes $R = N/M = K/C$. Decoding is carried out by taking the product of the matrix B and the received message $z = G^T \xi + \zeta$ (mod 2). The equation

$$z = A\xi + B\zeta = AS + B\tau \pmod{2}, \tag{2}$$

is solved via the iterative methods of BP [3] to obtain the most probable Boolean vectors S and τ; the posterior probability (1) becomes slightly more elaborate, including two sets of free variables S and τ and two priors.

2   Statistical Physics

To facilitate the statistical physics analysis we replace the $\{0, 1\}$ representation by the conventional Ising spin $\{1, -1\}$ representation, and mod 2 sums by products [9]. For instance, in Gallager's code, the syndrome vector acquires the form of a multi-spin coupling $\mathcal{J}_\mu = \prod_{j \in \mathcal{L}(\mu)} \zeta_j$ where $j = 1, \cdots, M$ and $\mu = 1, \cdots, (M - N)$. The $\widehat{K}$ indices of nonzero elements in the row μ of a matrix A, that is not necessarily a concatenation of two matrices (therefore defining a *non-structured* Gallager code), are given by $\mathcal{L}(\mu) = \{j_1, \cdots, j_{\widehat{K}}\}$, and in a column l are the C indices given by $\mathcal{M}(l) = \{\mu_1, \cdots, \mu_C\}$.

The posterior (1) can be written as the Gibbs distribution [10]:

$$P(\tau \mid \mathcal{J}) = \frac{1}{Z} \lim_{\beta \to \infty} \exp\left[-\beta \mathcal{H}_\beta(\tau; \mathcal{J})\right] \tag{3}$$

$$\mathcal{H}_\beta(\tau; \mathcal{J}) = -\sum_{\mu=1}^{M-N} \mathcal{J}_\mu \left(\prod_{j \in \mathcal{L}(\mu)} \tau_j - 1\right) - \frac{F}{\beta} \sum_{j=1}^{M} \tau_j ,$$

where $\mathcal{H}$ the Hamiltonian of the system.

The quantity that one concentrates on, in the statistical physics based analysis, is the *free energy* which is linked to the probability of finding the system in a specific configuration. In the *thermodynamic limit* of infinite system size, which is the main case considered in this work, the state of the system is dominated by configurations with the lowest free energy; finite systems are more likely to be found in configurations with lower free energy, but may also be found in other configurations with some probability.

To investigate the typical properties of a model, we calculate the partition function $\mathcal{Z}(\mathcal{A}, \mathcal{J}) = \mathrm{Tr}_{\{\boldsymbol{\tau}\}} \exp[-\beta \mathcal{H}]$ and the free energy $\langle \ln[\mathcal{Z}(\mathcal{A}, \mathcal{J})] \rangle_{\mathcal{A}, \boldsymbol{\zeta}}$ by averaging over the randomness induced by the specific code matrix $\mathcal{A}$ and the true noise vector $\boldsymbol{\zeta}$. For carrying out these averages we use the replica method [10] or the Bethe approximation [11]; both methods provide the same results.

The replica method makes use of the identity $\langle \ln \mathcal{Z} \rangle = \langle \lim_{n \to 0} 1/n \, [\mathcal{Z}^n - 1] \rangle$, by calculating averages over a product of partition function replica. Employing assumptions about replica symmetries and analytically continuing the variable n to zero, one obtains solutions which enable one to determine the state of the system. The Bethe approximation is based on a consistent solution to a tree based expansion for calculating the free energy. Details of the techniques used and of the calculations themselves can be obtained in [7] and in the corresponding papers [10] and [11].

3   Results

Once the free energy for the possible solutions is calculated, one can identify the stable dominant solutions and their overlap m with the true noise/signal vectors. In the case of Gallager's code we monitor $m = 1/M \sum_{j=1}^{M} \delta[\hat{\tau}_j; \zeta_j]$, where $\hat{\tau}$ is the noise vector MPM estimate. In the case of MN we calculate $m = 1/N \sum_{j=1}^{N} \delta[\hat{S}_j; \xi_j]$, estimating the signal vector $\hat{\boldsymbol{S}}$.

One observes three types of solutions: perfect retrieval (ferromagnetic solution) $m = 1$; catastrophic failure (paramagnetic solution) $m = 0$; and partial failure (sub-optimal ferromagnetic solution) $0 < m < 1$.

In each case one identifies two main critical noise levels: the spinodal point p_s, the noise level below which only perfect (ferromagnetic) solutions exist; and p_t, the noise level above which the ferromagnetic solution is no longer dominant. The former marks the practical decoding limit, as current practical decoding methods fail above p_s, while the latter marks the theoretical limits of the system.

The results obtained for $R = 1/4$ Gallager code are shown in Fig.1a, where we present the theoretical mean overlap between the actual noise vector ζ and the estimate $\hat{\tau}$ as a function of the noise level p, as well as results obtained using BP decoding. In Fig.1b we show the thermodynamic transition for $\hat{K} = 6$ and $R = 1/2$ compare with the theoretical upper bound, Shannon's bound and the theoretical p_s values.

Results obtained for MN code with various K, L values are presented in Fig.2. On the left - a schematic description of the free energy surface for various K

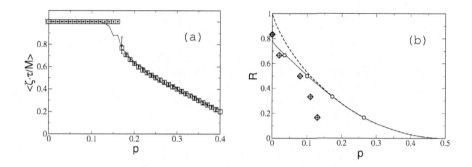

Fig. 1. (a) Mean normalized overlap between the actual noise vector ζ and decoded noise $\hat{\tau}$ for $\hat{K} = 4$ and $C = 3$ (therefore $R = 1/4$). Theoretical values (squares), experimental averages over 20 runs for code word lengths $M = 5000$ (•) and $M = 100$ (full line). (b) Transitions for $\hat{K} = 6$. Shannon's bound (dashed line), information theory based upper bound (full line) and thermodynamic transition obtained numerically (○). Theoretical (diamond) and experimental (+, $M = 5000$ averaged over 20 runs) BP decoding transitions are also shown. In both figures, symbols are chosen larger than the error bars.

values; on the right a description of the existing solutions for each noise value p and their corresponding overlap m.

For unbiased messages with $K \geq 3$ and $L > 1$. we obtain both the ferromagnetic and paramagnetic solutions either by applying the TAP approach or by solving the saddle point equations numerically. The former was carried out at the values of F_τ and $F_s = 0$) which correspond to the true noise and input bias levels (for unbiased messages $F_s = 0$) and thus to Nishimori's condition [12]. The latter is equivalent to having the correct prior within the Bayesian framework [9].

The most interesting quantity to examine is the maximal code rate, for a given corruption process, for which messages can be perfectly retrieved. This is defined in the case of $K \geq 3$ by the value of $R = K/C = N/M$ for which the free energy of the ferromagnetic solution becomes smaller than that of the paramagnetic solution, constituting a first order phase transition. The critical code rate obtained $R_c = 1 - H_2(p) = 1 + (p \log_2 p + (1 - p) \log_2(1 - p))$, coincides with *Shannon's capacity*.

The MN code for $K \geq 3$ seems to offer optimal performance. However, the main drawback is rooted in the co-existence of the stable $m = 1, 0$ solutions, which implies that from most initial conditions the system will converge to the undesired paramagnetic solution. Studying the ferromagnetic solution numerically shows a highly limited basin of attraction, which becomes smaller as K and L increase, while the paramagnetic solution at $m = 0$ *always* enjoys a wide basin of attraction.

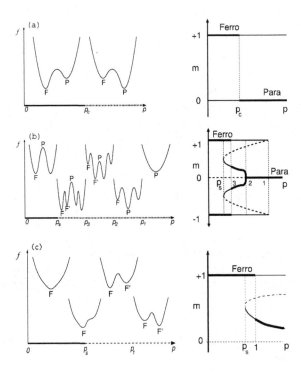

Fig. 2. Left hand figures show a schematic representation of the free energy land-scape while figures on the right show the ferromagnetic, sub-optimal ferromagnetic and paramagnetic solutions as functions of the noise rate p; thick and thin lines denote stable solutions of lower and higher free energies respectively, dashed lines correspond to unstable solutions. In all cases considered $L > 1$. (a) $K \geq 3$; the solid line in the horizontal axis represents the phase where the ferromagnetic solution (F, $m = 1$) is thermodynamically dominant, while the paramagnetic solution (P, $m = 0$) becomes dominant for the other phase (dashed line). The critical noise p_c denotes Shannon's channel capacity. (b) $K = 2$; the ferromagnetic solution and its mirror image are the only minima of the free energy over a relatively small noise level (the solid line in the horizontal). The critical point,due to dynamical considerations, is the spinodal point p_s where sub-optimal ferromagnetic solutions (F', $m < 1$) emerge. The thermodynamic transition point p_3, at which the ferromagnetic solution loses its dominance, is below the maximum noise level given by the channel capacity, which implies that these codes do not saturate Shannon's bound even if optimally decoded. (c) $K = 1$; the solid line in the horizontal axis represents the range of noise levels where the ferromagnetic state (F) is the only minimum of the free energy. The sub-optimal ferromagnetic state (F') appears in the region represented by the dashed line. The spinodal point p_s, where F' solution first appears, provides the highest noise value in which convergence to the fer-romagnetic solution is guaranteed. For higher noise levels, the system becomes bistable and an additional unstable solution for the saddle point equations necessarily appears. A thermodynamical transition occurs at the noise level p_1 where the state F' becomes dominant.

Studying the case of $K = 2$ and $L > 1$, indicates the existence of paramagnetic, ferromagnetic and sub-optimal ferromagnetic solutions depicted in Fig.2b. For corruption probabilities $p > p_s$ one obtains either a dominant paramagnetic solution or a mixture of ferromagnetic ($m = \pm 1$) and paramagnetic ($m = 0$) solutions. Reliable decoding may only be obtained for $p < p_s$, which corresponds to a spinodal point, where a unique ferromagnetic solution emerges at $m = 1$ (plus a mirror solution at $m = -1$). Initial conditions for BP decoding can be chosen randomly, with a slight bias in the initial magnetization. The results obtained point to the existence of a unique pair of global solutions to which the system converges (below p_s) from *all initial conditions*. Similarly, the case of $K = 1$, $L > 1$ presented in Fig.2c shows a dominant ferromagnetic solution below p_s and the emergence of a sub-optimal ferromagnetic solution above it, that becomes dominant at p_1.

The main differences between the results obtained for Gallager and MN codes in the case of unbiased messages are as follows. While Gallager's code allows for sub-optimal practical decoding for any $\hat{K}$ value, it saturates Shannon's bound only in the limit of $\hat{K} \to \infty$. On the other hand, MN codes can *theoretically* saturate Shannon's limit for constructions with $K \geq 3$, which are of no practical value, but they can only achieve suboptimal performance for regular configurations with $K = 1, 2$.

It should be pointed out that these results are valid only in the case of unbiased signal vectors $\boldsymbol{\xi}$. A different picture emerges in the case of biased messages; this includes the emergence of a spinodal point also in the case of $K \geq 3$ MN codes and a decrease in the noise level of the thermodynamic transition to below Shannon's limit.

It has been shown that irregular LDPC constructions can achieve better practical performance (e.g. [5,13]). In analytical studies, based on the same framework presented here [14] we investigated the position of both critical points p_s and p_t with respect to Shannon's limit and their values in regular constructions. We show that improved irregular constructions correspond to models with higher p_s values while the position of p_t changes only slightly. The possibility of employing the statistical physics based analysis for providing a principled method to optimise the code construction is still an open question.

4   Related Studies

We also studied the effect of non-binary alphabet on the performance of LDPC codes [15] as it seems to offer improved performance in many cases [16]. The alphabet used in this study is defined over Galois field $GF(q)$ [17]. Our results show that Gallager codes of this type saturate Shannon's limit as $C \to \infty$ irrespective of the value of q. For finite C, these codes exhibits two different behaviours for $C \geq 3$ and $C = 2$. For $C \geq 3$, we show that the theoretical error correcting ability of these codes is monotonically improving as q increases, i.e., the value of p_t increases with q for a given configuration. The practical decoding limit, determined by the emergence of a suboptimal solution and the value of p_s, decreases with q. On the other hand, $C = 2$ codes exhibit a continuous transition from optimal to sub-optimal solutions at a certain noise level, below which practical BP decoding converges to the (unique) optimal solution. This critical noise level monotonically increases with q and becomes even higher than that of some codes of connectivity $C \geq 3$, while the optimal decoding performance is inferior to that of $C \geq 3$ codes with the same q value.

The work described so far is limited to the case of infinite message length. In finite systems there is some probability of finding the system in a non-dominant state, what translates to an error probability which vanishes exponentially with the systems size. Significant effort has been dedicated to bounding the *reliability exponent* in the information theory literature [18]; we have also studied the reliability exponent [19] by carrying out direct averages over ensembles of Gallager codes, characterised by finite and infinite $\hat{K}$ values. In the limit of infinite connectivity our result collapses onto the best general random coding exponents reported in the IT literatures, the *random coding exponent* and the *expurgated exponent* for high and low R values respectively. The method provides one of the only tools available for examining codes of finite connectivity, and predicts the tightest estimate of the zero error noise level threshold to date for Gallager codes. It can be easily extended to investigate other linear codes of a similar type and is clearly of high practical significance.

Finally, insight gained from the analysis led us to suggest the potential use of a similar system as a public-key cryptosystem [20]. The cryptosystem is based on an MN code where the matrix G and a corruption level $p < p_s$ play the role of the public key and the matrices used to generate G play the role of the secret key and are known only to the authorised user.

In the suggested cryptosystem, a plaintext represented by an N dimensional Boolean vector $\boldsymbol{\xi} \in (0,1)^N$ is encrypted to the M dimensional Boolean ciphertext $\boldsymbol{J}$ using a predetermined Boolean matrix G, of dimensionality $M \times N$, and a corrupting M dimensional vector $\boldsymbol{\zeta}$, whose elements are 1 with probability p and 0 otherwise, in the following manner $\boldsymbol{J} = G\,\boldsymbol{\xi} + \boldsymbol{\zeta}$, where all operations are (mod 2). The corrupting vector $\boldsymbol{\zeta}$ is chosen at the transmitting end. The matrix G, which is at the heart of the encryption/decryption process is constructed by choosing two randomly-selected sparse matrices A $(M \times N)$ and B $(M \times M)$, and a dense matrix D $(N \times N)$, defining $G = B^{-1}AD$ (mod 2). The matrices A and B are similar to those used in other MN constructions; the dense invertible Boolean matrix D is arbitrary and is added for improving the system's security. Authorised decryption follows a similar procedure to decoding corrupted messages in LDPC codes (i.e., using BP), while an unauthorised user will find the decryption to be computationally hard [20].

5   Conclusions

We showed how the methods of statistical physics can be employed to investigate error-correcting codes and related areas, by studying the typical case characteristics of a given system. This approach provides a unique insight by examining macroscopic properties of stochastic systems, carrying out explicit averages over ensembles of codes that share the same macroscopic properties.

The results obtained shed light on the properties that limit the theoretical and practical performance of parity check codes, explain the differences between Gallager and MN constructions, explores the role of irregularity, finite size effects and non-binary alphabets in LDPC constructions.

We believe that methods developed over the years in the statistical physics community can make a significant contribution also in other areas of information theory. Research in some of these areas, such as CDMA and image restoration is currently underway.

Acknowledgements. Support by Grants-in-aid, MEXT (13680400) and JSPS (YK), The Royal Society and EPSRC-GR/N00562 (DS) is acknowledged. We would like to acknowledge the contribution of Kazutaka Nakamura and Naoya Sazuka to this research effort.

References

1. Shannon, C.E.: A Mathematical Theory of Communication: Bell Sys. Tech. J., **27** (1948) 379-423, 623-656.
2. Berrou, C. & A. Glavieux: Near Optimum Error Correcting Coding and Decoding - Turbo-codes: IEEE Transactions on Communications **44** (1996) 1261-1271.
3. MacKay, D.J.C.: Good Error-correcting Codes Based on Very Sparse Matrices: IEEE Transactions on Information Theory **45** (1999) 399-431.
4. Gallager, R.G.: Low-density Parity-check Codes: IRE Trans. Info. Theory **IT-8** (1962) 21-28 . Gallager, R.G.: Low-density Parity-Check Codes, MIT Press, Cambridge, MA. (1963).
5. Richardson, T., Shokrollahi, A., Urbanke, R.: Design of Provably Good Low-density Parity-check Codes: IEEE Transactions on Information Theory (1999) in press .
6. Pearl, J.: Probabilistic Reasoning in Intelligent Systems. Morgan Kaufmann, San Francisco (1988).
7. Nishimori, H.: Statistical Physics of Spin Glasses and Information Processing. Oxford University Press, Oxford UK (2001).
8. Thouless, D.J., Anderson, P.W., Palmer, R.G.: Solution of 'Solvable Model of a Spin Glass': Philos. Mag. (1977) **35**, 593-601.
9. Sourlas, N.: Spin-glass Models as Error-correcting Codes: Nature **339** (1989) 693-695.
10. Kabashima, Y., Murayama, T., Saad, D.: Typical Performance of Gallager-type Error-Correcting Codes: Phys. Rev. Lett. **84** (2000) 1355-1358.
11. Vicente, R., Saad, D., Kabashima, Y.: Error-correcting Code on a Cactus - a Solvable Model: Europhys. Lett. **51**, (2000) 698-704.
12. Nishimori, H.: Internal Energy, Specific Heat and Correlation Function of the Bond-random Ising Model: Prog.Theo.Phys. **66** (1981) 1169-1181.
13. Kanter, I., Saad, D.: Error-Correcting Codes That Nearly Saturate Shannon's Bound : Phys. Rev. Lett. **83** (1999) 2660-2663. Kanter, I., Saad, D.: Finite-size Effects and Error-free Communication in Gaussian Channels: Journal of Physics A **33** (2000) 1675-1681.

14. Vicente, R., Saad, D., Kabashima, Y.: Statistical Physics of Irregular Low-Density Parity Check Codes: Jour. Phys. A **33** (2000) 6527-6542.
15. Nakamura, K., Kabashima, Y., Saad, D.:Statistical Mechanics of Low-Density Parity Check Error-Correcting Codes Over Galois Fields: Europhys. Lett. (2001) in press.
16. Davey, M.C., MacKay, D.J.C: Low Density Parity Check Codes Over $GF(q)$: IEEE Comm. Lett., **2** (1998) 165-167.
17. Lidl, R., Niederreiter, H.: Introduction to Finite Fields and their Applications: Cambridge University Press, Cambridge, UK (1994).
18. Gallager, R.G.: Information Theory and Reliable Communication: Weily & Sons, New York (1968).
19. Kabashima, Y., Sazuka, N., Nakamura, K., Saad, D.: Tighter Decoding Reliability Bound for Gallager's Error-Correcting Code: Phys. Rev. E (2001) in press.
20. Kabashima, Y., Murayama, T., Saad, D.: Cryptographical Properties of Ising Spin Systems: Phys. Rev. Lett. **84** (2000) 2030-2033. Saad, D., Kabashima, Y., Murayama, T.: Public Key Cryptography and Error Correcting Codes as Ising Models. In: Sollich, P., Coolen, A.C.C., Hughston, L.P., Streater, R.F. (eds): Disordered and Complex Systems. American Institute of Physics Publishing, Melville, New York (2001) 89-94.

Generating Large Instances of the Gong-Harn Cryptosystem

Kenneth J. Giuliani and Guang Gong

Centre for Applied Cryptographic Research
University of Waterloo
Waterloo, Ontario, Canada, N2L 3G1
{kjgiulia, ggong}@cacr.math.uwaterloo.ca

Abstract. In 1999, Gong and Harn proposed a new cryptosystem based on third-order characteristic sequences over finite fields. This paper gives an efficient method to generate instances of this cryptosystem over large finite fields. The method first finds a "good" prime p to work with and then constructs the sequence to ensure that it has the desired period. This method has been implemented in C++ using NTL [7] and so timing results are presented.

1 Introduction

In 1998 and 1999, Gong and Harn proposed a new public-key cryptosystem (PKC) based on third-order characteristic sequences of period $q^2 + q + 1$ over $\mathbb{F}_q$ where q is a power of a prime, published in ChinaCrypt'98 [1] and in the IEEE Transactions on Information Theory [2], respectively. The security of the PKC is based on the difficulty of solving the discrete logarithm (DL) in $\mathbb{F}_{q^3}$. In 2000, Lenstra and Verheul [3] proposed the XTR public-key cryptosystem at the Crypto'2000, which is based on the third-order characteristic sequences with period $p^2 - p + 1$ by taking $q = p^2$. However, for large values of p where $q = p^r$, it seems to be difficult to check whether or not a sequence has the correct period $q^2 + q^2 + 1$ in the GH-PKC. In this paper, we give a method for constructing such instances which are assured to have the desired properties for the case $q = p^2$. The more general case $q = p^r$ will appear in the full paper.

2 Third-Order Characteristic Sequences and the Gong-Harn Cryptosystem

In this section, we present a review for 3rd-order characteristic sequences over finite fields and the GH Diffie-Hellman key agreement protocol.

2.1 3rd-Order Characteristic Sequences

Let

$$f(x) = x^3 - ax^2 + bx - 1, a, b \in \mathbb{F}_q \tag{1}$$

B. Honary (Ed.): Cryptography and Coding 2001, LNCS 2260, pp. 317–328, 2001.
© Springer-Verlag Berlin Heidelberg 2001

be an irreducible polynomial over $\mathbb{F}_q$ and α be a root of $f(x)$ in the extension field $\mathbb{F}_{q^3}$. A sequence $\{s_i\}$ is said to be a 3rd-order characteristic sequence generated by $f(x)$ if the initial state of $\{s_i\}$ is given by

$$s_0 = 3, s_1 = a, \text{ and } s_2 = a^2 - 2b$$

and

$$s_{k+3} = a s_{k+2} - b s_{k+1} + s_k, k = 0, 1, \cdots.$$

In this case, the trace representation of $\{s_i\}$ is as follows:

$$s_k = Tr(\alpha^k) = \alpha^k + \alpha^{kq} + \alpha^{kq^2}, k = 0, 1, 2, \cdots.$$

In the following, we write $s_k = s_k(a,b)$ or $s_k(f)$ to indicate the generating polynomial. Let $f^{-1}(x) = x^3 - bx^2 + ax - 1$, which is the reciprocal polynomial of $f(x)$. Let $\{s_k(b,a)\}$ be the characteristic sequence of $f^{-1}(x)$, called *the reciprocal sequence of* $\{s_k(a,b)\}_{k \geq 0}$. Then we have $s_{-k}(a,b) = s_k(b,a), k = 1, 2, \cdots.$

2.2   The GH Diffie-Hellman Key Agreement Protocol

Note that in [2], the GH-DH was presented in $\mathbb{F}_p$. However, these results are also true in $\mathbb{F}_q$ where q is a power of a prime.

GH-DH Key Agreement Protocol (Gong and Harn, 1999) [2] :

System parameters: p is a prime number, $q = p^r$ and $f(x) = x^3 - ax^2 + bx - 1$ which is an irreducible polynomial over $\mathbb{F}_q$ with period $Q = q^2 + q + 1$.

User Alice chooses $e, 0 < e < Q$, with $gcd(e, Q) = 1$ as her private key and computes (s_e, s_{-e}) as her public key. Similarly, user Bob has $r, 0 < r < Q$, with $gcd(r, Q) = 1$ as his private key and (s_r, s_{-r}) as his public key. In the key distribution phase, Alice uses Bob's public key to form a polynomial:

$$g(x) = x^3 - s_r x^2 + s_{-r} x - 1$$

and then computes the eth terms of a pair of reciprocal char-sequences generated by $g(x)$. I.e., Alice computes

$$s_e(s_r, s_{-r}) \text{ and } s_{-e}(s_r, s_{-r}).$$

Similarly, Bob computes

$$s_r(s_e, s_{-e}) \text{ and } s_{-r}(s_e, s_{-e}).$$

They share the common secret key as (s_{er}, s_{-er}).

Remark 1. The XTR [3] uses the characteristic sequences generated by an irreducible polynomial of the form $f(x) = x^3 - ax^2 + a^p x - 1$ over $\mathbb{F}_{p^2}$ with period $p^2 - p + 1$. Thus, XTR uses one characteristic sequence instead of a pair of reciprocal characteristic sequences, since in this case $s_{-k} = s_k^p$. Hence the two terms s_k and s_{-k} are dependent. For $q = p^2$, the two schemes have the same efficiency when they are applied to the DH key agreement protocol, because the GH-DH computes a pair of elements over $\mathbb{F}_{p^2}$ and shares a pair of elements over $\mathbb{F}_{p^2}$, while the XTR-DH computes one element over $\mathbb{F}_{p^2}$ and shares one element over $\mathbb{F}_{p^2}$.

3 The Approach

One approach to generating sequences of a given period is to randomly select fields and sequence polynomials and check to see if it has the desired period. However, determining the order of a sequence may be very difficult, and it may require many attempts before a good sequence is found. We give a more systematic approach to generating instances based on the following proposition (see [5] for a proof).

Proposition 1. *Let f be an irreducible polynomial of degree 3 over $GF(q)$ and t be a positive integer. The following are equivalent:*

1. *The sequence generated by f has period t.*
2. *t is the smallest integer such that f divides $x^t - 1$.*
3. *A root α of f has order t in the extension field $GF(q^3)$ of $GF(q)$ generated by f.*

Thus, if we can find an element of order $q^2 + q + 1$ in the extension field $GF(q^3)$, we can construct a polynomial whose sequence is of the desired period. Note that the multiplicative group of $GF(q^3)$ is cyclic. The following lemma helps us determine whether an element in a cyclic group has a given order.

Lemma 1. *Let G be a cyclic group of order n and let t be a number dividing n whose prime factorization is $t = p_1^{e_1} \cdots p_r^{e_r}$. Then an element $g \in G$ has order exactly t if and only if $g^t = 1_G$ and $g^{t/p_i} \neq 1_G$ for all $i = 1, \ldots, r$ where 1_G denotes the identity element of G.*

Thus, if we knew the factorization of $q^2 + q + 1$, we could easily determine whether or not an element has this order. The problem we immediately encounter is that the factorization of large numbers is a very difficult problem. One way to circumvent this problem is to choose primes p for which this factorization is much simpler.

4   Choosing "Good" Primes p

We define a prime p to be *good* if the factorization of $q^2 + q + 1$ is known where $q = p^2$. We have the following factorization

$$q^2 + q + 1 = p^4 + p^2 + 1 = (p^2 + p + 1)(p^2 - p + 1)$$

Let $p^+ = p^2 + p + 1$ and $p^- = p^2 - p + 1$. Let us now derive conditions for when a prime l divides either p^+ or p^-.

First, observe that neither 2 nor p are divisors since $q^2 + q + 1$ is odd and congruent to 1 modulo p. Next, note that no prime l divides both p^+ and p^-, for then it would also have to divide $p^+ - p^- = 2p$.

Let us now consider the case $l = 3$. If $p \equiv 1 \pmod 3$, then $p^+ \equiv 0 \pmod 3$. On the other hand, if $p \equiv 2 \pmod 3$, then $p^- \equiv 0 \pmod 3$. Thus, $q^2 + q + 1$ will always be divisible by 3.

For all other primes l, we see that l will divide p^+ (respectively p^-) if and only if p is a root of the polynomial $x^2 + x + 1$ (respectively $x^2 - x + 1$) in $\mathbb{F}_l$. Observe that any root of $x^2 + x + 1$ (respectively $x^2 - x + 1$) has order 3 (respectively order 6) in the multiplicative group $\mathbb{F}_l^*$. This will occur only if $l \equiv 1 \pmod 3$. The converse to this assertion, namely that if $l \equiv 1 \pmod 3$ then $x^2 + x + 1$ (respectively $x^2 - x + 1$) is reducible, is easily seen. Thus, the condition $l \equiv 1 \pmod 3$ is necessary for l to divide p^+ or p^-, and so we only need to consider small primes l of this type.

If $l \equiv 1 \pmod 3$, we can determine the roots of these polynomials over $\mathbb{F}_l$ using the quadratic formula. The roots of $x^2 + x + 1$ are $2^{-1}(-1 \pm \sqrt{-3})$, and the roots of $x^2 - x + 1$ are $2^{-1}(1 \pm \sqrt{-3})$, where the inverse is taken modulo l. Let $r = 2^{-1}(-1 + \sqrt{-3})$ be one of the roots of $x^2 + x + 1$. Then, $-r - 1$ is the other root. In addition, we see that $r + 1$ and $-r$ are the roots of $x^2 - x + 1$. Hence, l divides p^+ if and only if $p \equiv r, -r - 1 \pmod l$, while l divides p^- if and only if $p \equiv r + 1, -r \pmod l$. Note that $r, -r - 1, r + 1, -r$ are easily computable given l, and independent of the value of p.

We now have a strategy for finding a *good* prime p. We select a bound B and a random prime p and attempt trial division of p^+ and p^- by all primes $< B$. We then perform a primality test on the remaining large factors of p^+ and p^-. If both are prime then we have found a desirable p. Otherwise we try again with another random p.

Note that we need not attempt trial division by those primes l for which $l \equiv 2 \pmod 3$. Moreover, we can determine if l divides by p^+ and p^- by simply checking if the much smaller value p matches one of $r, -r - 1, r + 1, -r$ modulo l. Since these numbers are independent of p, they can be precomputed once for all primes p to be tested. This enables us to determine divisibility by calculating p modulo l instead of working with the much larger numbers p^+ and p^-.

If desired, one also may use some more advanced factorization methods on p^+ and p^- such as Pollard's $(p-1)$-method [6] or the elliptic curve method [4]. Regardless, the goal is to obtain a value p for which the factorizations of p^+ and p^- are known.

5   Finding Elements of Order $q^2 + q + 1$

Now that we have a prime p for which the factorization of q^2+q+1 is known, we would like to use this information to find an element of this order. Suppose we have the factorization $q^2+q+1 = p_1^{e_1} \cdots p_r^{e_r}$. We know that $\mathbb{F}_{q^3}^*$ is a cyclic group of order $q^3 - 1$. Let $G \subseteq \mathbb{F}_{q^3}^*$ be the unique cyclic subgroup of order $q^2 + q + 1$. The map

$$\psi : \mathbb{F}_{q^3}^* \to G$$
$$\alpha \mapsto \alpha^{q-1}$$

is a group homomorphism onto G. Thus, if $\alpha \in \mathbb{F}_{q^3}^*$ is selected at random, then $\beta = \alpha^{q-1}$ is a random element in G. We know that β has order dividing q^2+q+1. We can now use the lemma to determine if β is indeed an element of this order.

Now, an element $\beta \in G$ has order exactly q^2+q+1 if and only if it generates G. For a randomly chosen element, this happens with probability $\frac{\phi(q^2+q+1)}{q^2+q+1}$ where ϕ is the Euler ϕ-function. Given the factorization above, this works out to

$$\frac{\phi(q^2 + q + 1)}{q^2 + q + 1} = \frac{\phi(p_1^{e_1})}{p_1^{e_1}} \cdots \frac{\phi(p_r^{e_r})}{p_r^{e_r}} = \frac{p_1 - 1}{p_1} \cdots \frac{p_r - 1}{p_r}$$

If, for example we take $B = 2^{16} = 65536$, then in the worst case, this probability would work out to

$$\frac{2}{3} \cdot \frac{6}{7} \cdot \ldots \cdot \frac{65520}{65521} \cdot \frac{q^+ - 1}{q^+} \cdot \frac{q^- - 1}{q^-} > 0.28$$

where q^+ and q^- are the large leftover factors respectively of p^+ and p^-. Thus, one can find such an element β with very high probability after only a few tries.

6   Constructing the Sequence

Suppose that we have an element $\beta \in \mathbb{F}_{p^6}$ of order $p^4 + p^2 + 1$. Let

$$f(x) = (x - \beta)(x - \beta^{p^2})(x - \beta^{p^4}) = x^3 - ax^2 + bx - 1$$

where $a = \beta + \beta^{p^2} + \beta^{p^4}$ and $b = \beta^{-1} + \beta^{-p^2} + \beta^{-p^4}$. Then we have the following result [5].

Lemma 2. f is an irreducible polynomials of degree 3 over $\mathbb{F}_{p^2}$.

Corollary 1. f generates a sequence of period $p^4 + p^2 + 1 = q^2 + q + 1$ over $\mathbb{F}_{p^2} = \mathbb{F}_q$.

Proof. This follows immediately from the preceding lemma 2 and proposition 1.

□

7   Obtaining a Representation for the Coefficients

There is one detail remaining. The representation that we currently have for a and b are as elements in $\mathbb{F}_{p^6}$. We need to have some sort of representation for them in $\mathbb{F}_{p^2}$.

7.1   A General Method

One method to do so is as follows. Suppose $g(x) = x^2 + c_1 x + c_0$ is an irreducible polynomial over $\mathbb{F}_p$ such that the degree 6 polynomial $h(x) = x^6 + c_1 x^3 + c_0$ is also irreducible over $\mathbb{F}_p$. Consider the field extensions to $\mathbb{F}_{p^2}$ and $\mathbb{F}_{p^6}$ given by these two respective polynomials. Then we have the injective field homomorphism

$$\begin{array}{ccc} \mathbb{F}_{p^2} & \rightarrow & \mathbb{F}_{p^6} \\ u_1 x + u_0 & \mapsto & u_1 x^3 + u_0 \end{array}$$

Suppose that we had used h as our representation of $\mathbb{F}_{p^6}$. Then a and b must be of the form $a = u_1 x^3 + u_0$ and $b = v_1 x^3 + v_0$ in $\mathbb{F}_{p^6}$. Thus, we have the explicit representation of the sequence over $\mathbb{F}_{p^2}$ as $a = u_1 x + u_0$ and $b = v_1 x + v_0$ where the extension to the field $\mathbb{F}_{p^2}$ is given by g.

This method will suffice if one is not constrained with regards to the extension polynomial to be used. One can simply search for a polynomial of the form given by h which is also irreducible, from which one obtains the irreducible polynomial g. In addition, this method will work if one has a specific extension polynomial $g(x) = x^2 + c_1 x + c_0$ in mind, as long as $h(x) = x^6 + c_1 x^3 + c_0$ is also irreducible over $\mathbb{F}_p$. Note that is not always true.

7.2   A Method for a Specific Polynomial Representation

Suppose that one would like to use a specific polynomial representation given by $g(x) = x^2 + c_1 x + c_0$. If $x^6 + c_1 x^3 + c_0$ were not irreducible over $\mathbb{F}_p$, then one could not use the method given in the previous subsection. This subsection lists another method for doing so.

First, choose an irreducible degree 3 polynomial $\hat{g}(y) = y^3 + d_2 y^2 + d_1 y + d_0$. We can extend to the field $\mathbb{F}_{p^2}$ by using the representation

$$\mathbb{F}_{p^2} = \frac{\mathbb{F}_p[x]}{(g(x))}$$

We can then consider $\hat{g}$ as a polynomial over $\mathbb{F}_{p^2}$ and extend the field once more. Note that since $\hat{g}$ is irreducible over $\mathbb{F}_p$ and its degree is coprime to that of g, it is also irreducible over $\mathbb{F}_{p^2}$. Thus we have the extension

$$\mathbb{F}_{p^6} = \frac{\mathbb{F}_{p^2}[y]}{(\hat{g}(y))}$$

Note that the subfield $\mathbb{F}_{p_2}$ corresponds to those elements with no y-term associated to it. Thus, it would be easy to obtain a and b using the representation of g.

The problem we encounter with this method is that it is somewhat tedious to do a double field extension. In fact, certain software packages such as NTL [7] allow single extensions but do not support multiple field extensions. It would be much simpler if we could use only a single field extension. To this end, we now show a method to interpolate the the polynomials g and $\hat{g}$ into an irreducible degree 6 polynomial h.

Let γ and η be roots of g and $\hat{g}_2$ respectively. Thus, we have that

$$g(x) = x^2 + c_1 x + c_2 = (x - \gamma)(x - \gamma^p)$$

and

$$\hat{g}(y) = y^3 + d_2 y^2 + d_1 y + d_0 = (y - \eta)(y - \eta^p)(y - \eta^{p^2})$$

We construct the polynomial h to be the minimal polynomial of the element $\zeta = \gamma\eta$. Since $\gamma \in \mathbb{F}_{p^2} \setminus \mathbb{F}_p$ and $\eta \in \mathbb{F}_{p^3} \setminus \mathbb{F}_p$, we see that this polynomial must be of degree 6 and is explicitly given as

$$
\begin{aligned}
h(z) &= z^6 + e_5 z^5 + e_4 z^4 + e_3 z^3 + e_2 z^2 + e_1 z + e_0 \\
&= (z - \zeta)(z - \zeta^p)(z - \zeta^{p^2})(z - \zeta^{p^3})(z - \zeta^{p^4})(z - \zeta^{p^5}) \\
&= (z - \gamma\eta)(z - \gamma^p\eta^p)(z - \gamma\eta^{p^2})(z - \gamma^p\eta)(z - \gamma\eta^p)(z - \gamma^p\eta^{p^2})
\end{aligned}
$$

where we have used the relations $\gamma^{p^2} = \gamma$ and $\eta^{p^3} = \eta$.

It turns out that we can derive the coefficients of h directly from the coefficients of g and $\hat{g}$. With some work, it is easily seen that we have the following relations

$$
\begin{aligned}
e_5 &= -c_1 d_2 \\
e_4 &= c_1^2 d_1 + c_0 d_2^2 - 2 c_0 d_1 \\
e_3 &= 3 c_0 c_1 d_0 - c_1^3 d_0 - c_0 c_1 d_1 d_2 \\
e_2 &= c_0 c_1^2 d_0 d_2 - 2 c_0^2 d_0 d_2 + c_0^2 d_1^2 \\
e_1 &= -c_0^2 c_1 d_0 d_1 \\
e_0 &= c_0^3 d_0^2
\end{aligned}
$$

Elements in $\mathbb{F}_{p^6}$ in this polynomial representation have the form (replacing z with ζ)

$$
\begin{aligned}
&u_0 + u_1 \zeta + u_2 \zeta^2 + u_3 \zeta^3 + u_4 \zeta^4 + u_5 \zeta^5 \\
&= u_0 + u_1(\gamma\eta) + u_2(\gamma^2\eta^2) + u_3(\gamma^3\eta^3) + u_4(\gamma^4\eta^4) + u_5(\gamma^5\eta^5)
\end{aligned}
$$

Using the relations $\gamma^2 = -c_1\gamma - c_2$ and $\eta^3 = -d_2\eta^2 - d_1\eta - d_0$, we can reduce to a representation of the form

$$v_0 + v_1\gamma + v_2\eta + v_3\gamma\eta + v_4\eta^2 + \gamma\eta^2$$

Note that the subfield $\mathbb{F}_{p^2}$ consists of those elements with no η-term. We can use this to determine a and b in the representation using g.

Example 1. Suppose $p \equiv 3 \pmod 4$. Then $g(x) = x^2 + 1$ is irreducible over $\mathbb{F}_p$. Let $\hat{g}(y) = y^3 - y - k$ be irreducible for some $k \in \mathbb{F}_p$. Then $h(z) = z^6 + 2z^4 + z^2 + k^2$. An element can be represented as

$$u_0 + u_1\zeta + u_2\zeta^2 + u_3\zeta^3 + u_4\zeta^4 + u_5\zeta^5$$
$$= u_0 + u_1(\gamma\eta) + u_2(\gamma^2\eta^2) + u_3(\gamma^3\eta^3) + u_4(\gamma^4\eta^4) + u_5(\gamma^5\eta^5)$$
$$= u_0 + k(u_5 - u_3)\gamma + ku_4\eta + (u_1 - u_3 - ku_5)\gamma\eta + (u_4 - u_2)\eta^2 + ka_5\gamma\eta^2$$

If this element were in the subfield $\mathbb{F}_{p^2}$, then the coefficients of the terms with an η must be 0. This yields the relations $u_2 = u_4 = u_5 = 0$ and $u_1 = u_3$. Thus, such an element can be represented as $u_0 - ku_1x$ in the field $\mathbb{F}_{p^2}$ where $g(x) = x^2 + 1$ was used as the extension polynomial.

Example 2. Suppose $p \equiv 2 \pmod 3$. Then $g(x) = x^2 + x + 1$ is irreducible over $\mathbb{F}_p$. Let $\hat{g}(y) = y^3 - y - k$ be irreducible for some $k \in \mathbb{F}_p$. Then $h(z) = z^6 + z^4 - 2kz^3 + z^2 - kz + k^2$. An element can be represented as

$$u_0 + u_1\zeta + u_2\zeta^2 + u_3\zeta^3 + u_4\zeta^4 + u_5\zeta^5$$
$$= u_0 + u_1(\gamma\eta) + u_2(\gamma^2\eta^2) + u_3(\gamma^3\eta^3) + u_4(\gamma^4\eta^4) + u_5(\gamma^5\eta^5)$$
$$= u_0 + k(u_3 - u_5) - ku_5\gamma + (u_3 - u_5)\eta + (u_1 - u_5 + ku_4)\gamma\eta$$
$$+ (-u_2 - ku_5)\eta^2 + (u_4 - u_2 - ku_5)\gamma\eta^2$$

If this element were in the subfield $\mathbb{F}_{p^2}$, then the coefficients of the terms with an η must be 0. This yields the relations $u_4 = 0$ and $u_1 = u_3 = u_5 = -k^{-1}u_2$. Thus, such an element can be represented as $u_0 - ku_1x$ in the field $\mathbb{F}_{p^2}$ where $g(x) = x^2 + x + 1$ was used as the extension polynomial.

8   The Algorithm

The method is summarized here in algorithmic form.

Input (bit size b, bound $B > 3$)

Precomputation:

for each prime $l \equiv 1 \pmod 3$, $3 < l < B$
     compute $r = 2^{-1}(1 + \sqrt{-3}) \pmod l$
     store the pair (l, r) in table A
end for

Prime Generation:

loop

    generate a random prime p of b bits

    calculate p^+ and p^-

    set $q^+ = p^+$ and $q^- = p^-$

    if $p \equiv 1 \pmod 3$

        let s be the highest power of 3 dividing q^+

        divide q^+ by 3^s

        store (l, s) in table B

    else

        let s be the highest power of 3 dividing q^-

        divide q^- by 3^s

        store (l, s) in table C

    end if

    for each entry (l, r) in table A

        compute $m = p \pmod l$

        if $m = r$ or $m = l - r - 1$

            let s be the highest power of l dividing q^+

            divide q^+ by l^s

            store (l, s) in table B

        else if $m = r + 1$ or $m = l - r$

            let s be the highest power of l dividing q^-

            divide q^- by l^s

            store (l, s) in table C

        end if

    end for

    do primality tests on both q^+ and q^-

    if both q^+ and q^- are prime

        exit the loop

    else

        clear tables B and C

    end if

    return to the top of the loop

end loop

Constructing the Sequence:

select an irreducible polynomial h of degree 6 over $\mathbb{F}_p$ using one of the methods in section 7

loop

    choose a random element $\alpha \in \mathbb{F}_{p^6}$

    compute $\beta = \alpha^{p^2 - 1}$

    for $i = 3, q^+, q^-$, each prime l in tables B and C

        if $\beta^{(p^4 + p^2 + 1)/i} = 1$

            return to top of loop

>   **end if**
>   **end for**
>   exit loop
> **end loop**

compute $a' = \beta + \beta^{p^2} + \beta^{p^4}$

compute $b' = \beta^{-1} + \beta^{-p^2} + \beta^{-p^4}$

derive the irreducible degree 2 polynomial g from h using the appropriate method from section 7

derive a and b from a' and b' respectively using the appropriate method from section 7

Output (p, g, a, b)

A   Implementation and Timing Analysis

The algorithm listed in the previous section was implemented in C++ on a UNIX server using the NTL [7] number theory library. The dominant step of the algorithm was, by far, prime generation. As a result, timing analysis was performed on this portion of the algorithm with the results listed in table 1 and table 2.

The bounds $B = 2^{10}$ and $B = 2^{16}$ were both tested. 10 instances were generated for each bound and for each bit size from 160 to 320 by 8 bits. The number of primes searched before the candidate was found, the CPU time, and the bit sizes of the largest prime factors q^+ and q^- respectively of p^+ and p^- were recorded. The class of instances with bound $B = 2^{16}$ were, on average generated much faster than those with bound $B = 2^{10}$, especially for the larger bit sizes. The bit sizes of q^+ and q^- were only marginally larger with the bound $B = 2^{10}$.

In addition, attempts were made to generate instances with the bound $B = 4$. This was attempted for bit sizes 160 to 240 by 8. The average CPU time using this bound was substantially longer than with the other two bounds. Thus, for some bit sizes, fewer than 10 instances were generated. Results for this bound are listed in table 3.

B   Example Instances

This section lists some example instances.

Example 3. The prime p listed here has 160 bits.

$p = 1276593311082943972800140646397807976959837132709$

$p^+ = 1629690481901714162735730680725186288280728823955302744324479689642812431700906503271994314811391$

$p^- = (3)5432301606339047209119102269083954294269096079842498525674379338990707168027036291060248801819991$

$g(x) = x^2 + x + 1$

Table 1. Timing results for prime generation with bound $B = 2^{10}$.

bits	# searched	time (sec)	q^+	q^-
160	139.9	19	309.5	310.2
168	529.5	79.6	327.5	326.8
176	318.9	47.6	342.4	343
184	184.9	33.9	358.7	358.3
192	388.2	90	375.9	374.6
200	441.5	90.6	391.7	390.9
208	432.2	93.5	407	406.1
216	395.5	127.1	420.3	426.4
224	445	117.4	439.7	441
232	295.9	83.4	455.7	457
240	418.5	118.9	467.8	472.2
248	847.8	304	488.1	489.3
256	1098.4	396.8	499.7	503.5
264	582.2	235.2	516.4	518.6
272	796.4	350.9	535.3	533.5
280	1233.2	579	550.8	552.9
288	1299.4	645.6	567.8	563.5
296	1144.1	590.1	584.5	581.1
304	1320.8	781.4	595	596.8
312	467.6	288.8	615.4	614.2
320	885.3	585.3	634.1	634.2

Table 2. Timing results for prime generation with bound $B = 2^{16}$.

bits	# searched	time (sec)	q^+	q^-
160	106.3	19.8	307.6	306.3
168	108.2	21.8	322.2	328.1
176	115.4	23.4	335.6	334.5
184	78.9	18.9	354.6	353
192	221.4	57.9	375.2	371.7
200	162.2	43	388.4	387.7
208	236.8	66.7	397.7	398.4
216	240	76.8	421.1	399.2
224	158.9	53.6	437.9	427.2
232	224	78.3	444	448.6
240	106.5	41.1	460.8	468.3
248	433	184.2	481.6	479.1
256	327.4	141.8	493.7	498.9
264	160.7	75.3	510.7	511.9
272	331.8	186.5	533	530
280	300.9	162.4	544	547.9
288	405	230.7	560.8	560.2
296	803	476.9	584.2	575
304	604	398.7	592.5	592.6
312	291.8	202.9	611.1	608.2
320	319.4	264.8	624.3	625.7

Table 3. Timing results for prime generation with bound $B = 4$.

bits	# searched	time (sec)	q^+	q^-	# instances
160	3721.5	383.5	319.5	319.5	2
168	11727.29	1383.29	335	334.29	7
176	17520.5	2114.5	350	352	2
184	8214.8	1215.6	366.6	367.3	10
192	23773	3751	384	383	1
200	17927	2918.5	400	398	2
208	13146	2203	413	415	1
216	21012	4238	431	430	1
224	6769.5	1412	447	446.5	2
236	10056	2254	462.5	463	2
240	30651	6889	477	479	1

$a = [852223667913003644390015513849147072549735623335]x$
$\quad + [1209115664825072234387309339396575750110393685169]$
$b = [2468685823891209653406986907473626739952482400017]x$
$\quad + [8930480475689858607934582522202320598557566667683]$

Example 4. The prime p listed here has 160 bits.
$p = 1353081569040243787002953026589849378107407355807$
$p^+ = (3)6102765774921360046408448466984410943175577269572070036269426$
$\quad 142909226834429782185051963465901 9$
$p^- = 18308297324764080139225345400953232829526731808689148477428122967$
$\quad 1327089897971376679534408926544 3$
$g(x) = x^2 + 1$
$a = [96591392999283599669949832736756776816781690 4081]x$
$\quad + [1144394849915005311617080018668684639829272039 84]$
$b = [4997466869034282500770588350042075853220760775 88]x$
$\quad + [2414407143720141016530453913453586486085191663 55]$

References

1. G. Gong and L. Harn, "A new approach on public-key distribution", *ChinaCRYPT '98*, pp 50-55, May, 1998, China.
2. G. Gong and L. Harn, "Public-key cryptosystems based on cubic finite field extensions", *IEEE IT* vol 45, no 7, pp 2601-2605, Nov. 1999.
3. A. K. Lenstra and E. R. Verheul, "The XTR public key system", *Advances in Cryptology, Proceedings of Crypto'2000*, pp. 1-19, Lecture Notes in Computer Science, Vol. 1880, Springer-Verlag, 2000.
4. H. W. Lenstra Jr., "Factoring Integers with Elliptic Curves", *Ann. of Math.* vol 126, no 2, pp 649-673, 1987.
5. R. Lidl and H. Niederreiter, *Finite Fields*, Addison-Wesley Publishing Company, Reading, MA, 1983.
6. J. M. Pollard, "Theorems on Factorization and Primality Testing", *Proc. Cambridge Philos. Soc.* vol 76, pp 521-528, 1974.
7. V. Shoup, "NTL: A Library for doing Number Theory", *http://shoup.net/ntl*.

Lattice Attacks on RSA-Encrypted IP and TCP

P.A. Crouch and J.H. Davenport*

Department of Mathematical Sciences, University of Bath, Bath BA2 7AY, England
Paul@p-crouch.com   J.H.Davenport@bath.ac.uk

Abstract. We introduce a hypothetical situation in which low-exponent RSA is used to encrypt IP packets, TCP segments, or TCP segments carried in IP packets. In this scenario, we explore how the Coppersmith/Howgrave-Graham method can be used, in conjunction with the TCP and IP protocols, to decrypt specific packets when they get re-transmitted (due to a denial-of-service attack on the receiver's side). We draw conclusions on the applicability of the Coppersmith/Howgrave-Graham method, its interaction with "guessing", and the difficulties of building a secure system by combining well-known building blocks.

1 Introduction

We consider a scenario in which there are many, internally secure, TCP/IP networks. These communicate across an insecure internet. To enable secure communication, the firewalls for each secure network take the IP_{sec} packet from the secure network, encrypt it with RSA [2] and treat the encrypted packet as data for an IP_{insec} packet directed at the firewall of the destination network. This then decrypts the packet and injects it into its secure network, as shown in figure 1.

From the point of view of the IP_{sec} layers, the firewalls and the IP_{insec} communication all form a (complex) link layer over which the IP_{sec} packet travels.

The attacker is assumed to listen to the IP_{insec} packets transmitted across the insecure internet from the sender (A) to the receiver (B). By flooding B, or a switch near B, with other traffic, the attacker can (at least with high probability) cause B to miss a transmission from A. IP is inherently an unreliable protocol, so the higher levels above IP (e.g. TCP) will have mechanisms to re-transmit the lost message. What information can the attacker gain in this scenario?

The Coppersmith/Howgrave-Graham method [4,7,8,11] is encapsulated in the following theorem.

Theorem 1. *Let P be a monic polynomial of degree δ in one variable modulo an integer N (of unknown factorisation). Then one can find, in time polynomial in $(\log N, \delta)$, all integers x_0 such that $P(x_0) \equiv 0 \pmod{N}$ and $|x_0| \leq N^{1/\delta}$.*

The method forms an $h\delta \times h\delta$ lattice, where h is the control parameter. To reach $N^{1/\delta}$, one needs h to be arbitrarily large, and the actual bound on x_0 achieved varies with h, as shown in table 1 for the case of $\delta = 3$.

* The authors are grateful to Dr. D. Coppersmith, Dr. N.A. Howgrave-Graham and Mr. A.J. Holt for their contributions to this work.

B. Honary (Ed.): Cryptography and Coding 2001, LNCS 2260, pp. 329–338, 2001.
© Springer-Verlag Berlin Heidelberg 2001

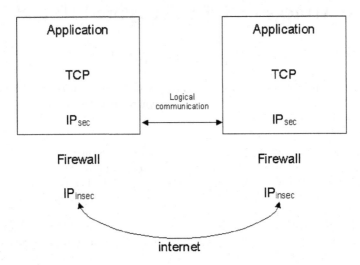

Fig. 1. Protocol layers in the scenario

Table 1. x_0 as a function of h; $\delta = 3$

$h = 2$	$h = 3$	$h = 4$	$h = 5$	$h = 6$	$h = 7$	$\ldots$	$h = 67$
$N^{0.2}$	$N^{0.25}$	$N^{0.27}$	$N^{0.286}$	$N^{0.294}$	$N^{0.3}$	$\ldots$	$N^{0.33}$

In this paper, we are concerned with recovering bit-fields from IP packets, generally 16-bit fields. Table 1 shows how many bits we can expect to recover for various values of h in various scenarios for the size of the RSA modulus (512, 1024 or 2048 bits), choices of the RSA exponent, and the presence or absence of checksum wrapping (as explained later).

Table 2. Values for ϕ for recovering $x_0 \le 2^\phi$

	$e = 3$						$e = 5$					
	No CS Wrap			CS Wrap			No CS Wrap			CS Wrap		
	512	1024	2048	512	1024	2048	512	1024	2048	512	1024	2048
$h = 2$	**34.13**	**68.27**	**136.53**	**30.12**	**60.24**	**120.47**	11.38	**22.76**	**45.51**	10.44	**20.89**	**41.8**
$h = 3$	42.66	85.33	170.67	39.38	78.77	157.54	14.63	29.26	58.51	13.84	27.68	55.35
$h = 4$	46.54	93.09	186.18	43.89	87.78	175.54	**16.17**	32.34	64.67	15.52	31.03	62.06
$h = 5$	48.76	97.52	195.04	46.55	93.09	186.18	17.07	34.14	68.26	**16.52**	33.03	66.06
$h = 6$	50.19	100.39	200.78	48.3	96.60	193.20	17.66	35.31	70.62	17.18	34.36	68.72

The complexity of the lattice reduction, which is the dominant step in the Coppersmith/Howgrave-Graham method, is

$$O\left(h^9 \delta^6 (\log N)^3\right),\tag{1}$$

assuming classical arithmetic. From this and table 1, we see the importance of minimising both h and δ.

Most of the computation for this paper was done in `Maple` but Victor Shoup's NTL[12] library for C++ was used for the time-critical lattice reduction part.

2 IP Packets

Figure 2 shows a diagram of an IP packet. To understand the attacks on IP datagrams, we need to understand the IP datagram header. A good general reference on TCP and IP is [13].

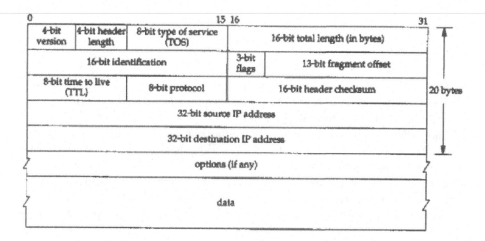

Fig. 2. IP datagram, showing the fields in the IP header

In the denial-of-service scenario mentioned in the introduction, the higher protocols above IP would cause a re-transmission. The IP layer is unaware that this is a re-transmission, and sends this as a new packet. What fields change between this re-transmission and the original transmission? The only two fields that we expect to change are the following.

- The 16-bit identification field. This normally increases by 1 every time the IP layer sends a packet.
- The one's complement sum of the header, stored in the 16-bit checksum field. Every time the 16-bit identification field is incremented, the checksum is decremented. Care must be taken if the checksum exceeds 65535, as it will then wrap around and restart from 1.

3   Attacks on IP Packets

First we need to take the denial of service attack and look at it in more detail. In the denial of service attack letting

$$c_1 = m^e$$

$$c_2 = (m + x)^e$$

gives us a basic idea of how an attack could be implemented. However we are more specifically interested in what two IP packets would look like when represented as c_1 and c_2, and the IP$_\text{sec}$ identification fields differ by α. Using the knowledge about IP packets it is possible to construct a similar general formula, which is shown and explained below.

$$c_1 = m^e$$

$$c_2 = (m + \overbrace{(2^{48}\alpha - \alpha}^{1} \underbrace{\pm 65535}_{2}) \times \overbrace{2^{64+d}}^{3})^e$$

1. This is because the id field is 48 bits from the checksum field and increases by 1 each time a packet is sent. The $-\alpha$ is because when the id field increments by 1, the checksum decrements by 1.
2. The ± 65535 only applies if the IP$_\text{sec}$ checksum wraps around, something which cannot be determined from the encrypted packets. If the checksum does not wrap around between the two packets then this value is 0.
3. This 2^{64+d} is derived from the checksum being 64-bits (32 bit source address and 32 bit destination address) from the end of the packet $+d$ which is the number of **bits** of data which are appended to the end of the packet.

Next two RSA encrypted versions of the same (apart from identification and checksum) IP$_\text{sec}$ packets are required. Define these IP$_\text{insec}$ data fields to be eip_1 and eip_2. These would correspond to the IP$_\text{sec}$ packets ip_1 and ip_2, and would have been encrypted by taking $eip_j = ip_j^e \mod N$.

Letting c_1 and c_2 be the symbolic forms of the two encrypted packets eip_1 and eip_2, it is possible to calculate resultants. This is done by taking the resultants of $c_1 - eip_1$ and $c_2 - eip_2$, i.e. of $m^e - eip_1$ and $(m + x)^e - eip_2$, with respect to m.

Taking resultants will give a polynomial in α of degree e^2 (though it may be possible to reduce this to e). This is the polynomial satisfied by the unknown, but comparatively small, α. We use the Coppersmith/Howgrave-Graham method [4, 7,8] to find α, by solving the appropriate lattice.

Once α is known, we take the greatest common divisor with respect to z of

$$\gcd(z^e - eip_1, (z + \alpha \times (2^{48} - 1) \times 2^{64+d})^e - eip_2) \mod N) = z + ip_1 + \lambda N$$

where z is an indeterminate, and λ represents the fact that we are interested in $ip_1 \pmod{N}$ (in practice $\lambda = 1$). This gives

$$z - (\gcd(z^e - eip_1, (z + (2^{48}\alpha - \alpha) \times 2^{64+d})^e - eip_2) \mod N) - \lambda N = ip_1$$

Therefore we have recovered ip_1 and broken the RSA encryption on these particular IP packets.

Symbolically, in the absence of checksum wrapping, and ignoring for simplicity the large numerical factors, the equation for α is of the following form.

$$\alpha^9 + 3(c_1 - c_2)\alpha^6 + 3(c_1^2 + 7c_1c_2 + c_2^2)\alpha^3 + (c_1 - c_2)^3 \equiv 0 \pmod{N}$$

While this is of degree 9 in α, which would imply reducing an 18×18 lattice with $h = 2$, it is in fact a polynomial of degree 3 in α^3, and solving it as such only requires reducing a 6×6 lattice, giving a $3^6 = 729$ theoretical saving[1]. Unfortunately, if the checksum does wrap, this simplification is not possible.

Shown below is a summary of the timings in seconds to perform the lattice reduction phase of this attack for various sizes of the public-key modulus..

Table 3. Times for IP reduction

NTL Timings in seconds to lattice reduce RedHat Linux 6.2 on 1Ghz Pentium III with 500Mb RAM							
Public exponent			e=3			e=5		
RSA-type		512	1024	2048	512	1024	2048
h	(control parameter)	2	2	2	4	2	2
IP	No checksum wrapping	2	9	27	8068*	177	1386
	With checksum wrapping	653	3413	3976	†	793465	§

† Not implemented due to software restrictions. This would in fact have required $h = 5$.

§ Not implemented in this report, due to the running time & resource constraints.

* Taking $\alpha \leq 2^{11}$ allowed $h = 2$, with $e = 5$ this formed a 10x10 matrix which reduced in 19 seconds. Therefore guessing all possibilities for the top five bits of a 16-bit identification field and using the Coppersmith/Howgrave-Graham method to find the bottom 11 bits would take $2^5 * 19 = 608$ seconds: over a factor of 10 faster than direct solving.

Without checksum wrapping, and with $h = 2$, the $e = 3$ examples required a 6×6 lattice, and $e = 5$ required 10×10. With checksum wrapping and $h = 2$, the

---

[1] The same happens for $e = 5$, and in fact we also save on h, being able to take $h = 4$ rather than $h = 5$, as can be seen from table 1.

$e = 3$ examples required a 18×18 lattice, and $e = 5$ required 50×50. In general, we note that doubling the length of the modulus multiplies the running time by about 8, which is what one would expect from the theoretical performance (as given in equation (1)) of the LLL algorithm [9].

4   TCP/IP Sessions

TCP (Transmission Control Protocol) is a common protocol to use above IP. It provides a reliable connection-oriented service on top of the unreliable service provided by IP. In the scenario in Figure 1, if an opening (i.e. carrying the

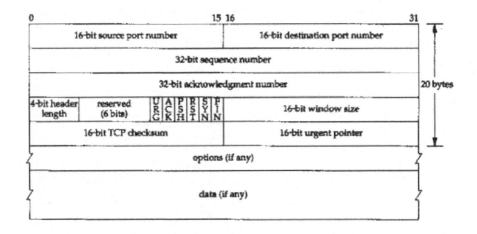

Fig. 3. TCP segment, showing the fields in the TCP header

SYN flag) TCP packet is lost, the TCP packet will simply be re-transmitted unchanged. In this case, the attack in the previous session is the only one possible. However, it is possible to deny service for the entire TCP opening sequence, in which case protocols above TCP may well start a new TCP session, which will have its own sequence number.

In a denial of service attack we are interested in what would change in a TCP packet header, shown in figure 3. This is slightly more complex because the TCP checksum is calculated (as in the case of the IP checksum) as a one's complement sum of **16-bit** words. In the case of TCP, the field that changes is the sequence number. Unfortunately this is a 32-bit field. Therefore although we know that the sequence number in many implementations (incorrectly: see [1,3, 10]) simply increments by 64000 every half second, the effect on the checksum is rather different.

If the 32-bit sequence number increments by 64000β, the high 16-bits in the 32-sequence number normally[2] increase by β and the low 16-bits decrease by 1536β so the overall change in the checksum (as it is the one's complement sum) is $+1535\beta$. This is valid only if $\beta \leq 2^{16}$ (otherwise the 32-bit nature of the field manifests itself). We assume that $\beta \not> 2^{16}$.

5   Attacks on TCP/IP Sessions

We consider two kinds of attacks: those where we assume the TCP packet alone is encrypted (i.e. the encryption takes place between TCP and IP, rather than below IP_{sec} as in figure 1) and those where the complete IP_{sec} packet is encrypted, in accordance with figure 1. These are referred to as TCP and TCP/IP in table 5. The TCP attack is similar to the IP attack: we are solving for one variable, β, and the equation to be solved is of degree e^2, which can be reduced to degree e in the event of no checksum wrapping.

Table 4. Times for TCP reduction

NTL Timings in seconds to lattice reduce RedHat Linux 6.2 on 1Ghz Pentium III with 500Mb RAM							
Public exponent			e=3			e=5		
RSA-type		512	1024	2048	512	1024	2048
h	(control parameter)	2	2	2	4	2	2
TCP	No checksum wrapping	1	9	27	9456	167	1384
	With checksum wrapping	702	4631	5129	†	§	§
TCP/IP	No checksum wrapping	Ψ	§	§	§	§	§

† Not implemented due to software restrictions. This would in fact have required $h = 5$.
Ψ Computation aborted after one month.
§ Not implemented due to the running time & resource constraints.

The TCP/IP attack is more complicated, as we are solving for two variables, α (the change in IP identification field) and β. From a polynomial point of view, this forms a bivariate modular equation. As pointed out by Coppersmith [4] and Howgrave-Graham [7,8], there is a heuristic extension of their method to bivariates. Let us assume that the IP_{sec} packets are $tcpip_1$ and $tcpip_2$, with one byte of TCP data. Taking $etcpip_1$ and $etcpip_2$ to be $tcpip_1$ and $tcpip_2$ encrypted with RSA, resultants were taken using them and c_1 and c_2 shown below, which formed a bivariate modular polynomial.

---

[2] That is to say, when the low-order part of the sequence number is greater than 1536β, which one might expect to occur $1 - \frac{1536\beta}{65536}$ of the time.

$$c_1 = m^3$$

$$c_2 = (m + (\underbrace{(2^{48} - 1)\alpha \overbrace{2^{232}}^{\Delta}}_{\text{IP packet}}) + \underbrace{((2^{80} \times (64000\beta) + (1535\beta)) \times 2^{24}))^3}_{\text{TCP packet}}$$

$$\underbrace{\qquad\qquad\qquad\qquad\qquad\qquad\qquad\qquad\qquad\qquad\qquad\qquad\qquad}_{\text{TCP/IP packet}}$$

Δ - The IP checksum field is now 232 **bits** from the end of the packet as it is 64 bits from the end of the IP header, 160 bits (20 bytes) of TCP header and 8 bits of data. 64+160+8=232

As this is an $e = 3$ case we have already attacked IP and TCP with $e = 3$ so it is safe to conclude that $h = 2$. Therefore the polynomial is never raised to a greater degree than 1.

We obtained a polynomial in α and β where the maximum degree of β is 9. This polynomial was made monic with respect to β^9. Then using the formula to calculate the dimensions of the matrix where $j = 9$, showed that a 190x190 matrix was required.

$$dim = 2j^2 + 3j + 1 = 2 \times 9^2 + 3 \times 9 + 1 = 162 + 27 + 1 = 190$$

It is possible to calculate the number of rows which must contain N and the number of rows which must contain the coefficients of $p(x)$,

$$190 - \frac{j^2 + 3j + 2}{2} = 190 - \frac{110}{2} = 135$$

So 135 rows of the matrix contain N's (scaled by the bounds for α and β) on the diagonal (and zeros elsewhere) and the remaining 55 rows contain the coefficients of the polynomial multiplied by the different combinations of the two variables α and β and scaled by the bounds for α and β, so that the leading term is on the diagonal.

The lattice reduction is extremely costly in time and resources, and at the time of finishing [5], the 190x190 lattice reduction process had been running for in excess of 750 hours.

In theory though, if the LLL reduction had been successful on this large matrix, we would have taken the first two short vectors of the LLL reduced matrix, divided them both by the numeric vector (of upper bounds), and formed two simultaneous equations. These two simultaneous equations would be solved to return α and β.

Next with α and β in hand, and $etcpip_1$ and $etcpip_2$ corresponding to the two encrypted messages, we would calculate the linear polynomial

$$\gcd(z^3 - ep_1, (z + (2^{48} - 1)\alpha 2^{232} + (2^{80} \times 64000 + 1535)\beta \times 2^{24})^3 - ep_2) \mod N$$

to recover $tcpip_1$.

6   TCP/IP Attacks Revisited

A better way to attack TCP/IP packets is by "'guessing'" β. α represents the change in the IP identification field, as this increments by 1 every time a packet is sent from the original sender (including packets internal to the sender's IP$_{\text{sec}}$ network, which cannot be detected outside), guessing this would be essentially impossible due to the number of packets sent. However β increments by 1 every half second. Remembering that we are performing a denial of service attack, if we, for example, sniffed two packets in the space of 4 seconds, there would be only eight β's to guess.

After experimentation with $e = 3$ and substituting $\beta = \{1, 2, 3, 4, 5, 6, 7, 8\}$ it was clear to see from the resultant polynomial that this was a univariate polynomial in α with maximum degree 9. Unfortunately this could not be solved as a degree 3 polynomial in α^3, but running eight 18x18 lattice reductions takes significantly less time ($8 \times 680 = 5440$ seconds) than attempting to LLL reduce the 190x190 matrix.

Taking

$$c2 := (m + ((2^{48}\alpha - \alpha) \times 2^{232}) + ((2^{80} \times (64000\beta) + (1535\beta)) \times 2^{24}))^3$$

But then for example guessing $\beta = 1$ gives:

$$c2 := (m + ((2^{48}\alpha - \alpha) \times 2^{232}) + 1329207713375312221233383113029058560)^3$$

We observe empirically that **lattice reducing a matrix with an unsuccessful guess of β takes no longer than to reduce a successful guess of β.**

A polynomial is constructed and solved using LLL reduction as in the IP case with checksum wraparound to resolve α, and now with α calculated and β guessed,

$$\gcd(\ z^3 - etcpip_1,$$
$$(z + (2^{48} - 1)\alpha 2^{232} + (2^{80} \times 64000 + 1535)\beta \times 2^{24})^3 - etcpip_2) \mod N)$$

is calculated, which is equal to $z - tcpip_1 - \lambda N$, hence RSA encryption on these TCP/IP packets is broken.

7   Conclusions

Table 1 shows that the Coppersmith/Howgrave-Graham method is eminently applicable for the decoding of low-exponent RSA-encrypted IP packets. For $e = 5$ we also have the paradoxical result that increasing the key length from 512 bits to 1024 actually reduces the time by a factor of 45, since a smaller lattice (10×10 rather than 20×20, with entries modulo N rather than N^3) is involved.

Extrapolating from table 1 shows that the key size would have to rise to 4096 bits before security is improved.

Yet again, we note that the common, but flawed, implementation of TCP initial sequence numbers is a security loophole [1,3,10]. In our case, the fact that $64000 \approx 2^{16}$ is an added weakness.

We also note the power of combining guessing some bits with the Coppersmith/Howgrave-Graham method for finding others: see note (*) in IP attacks and section 6.

More generally, we have shown that a cryptosystem built according to standard principles of protocol layering with "standard" components displays unexpected, and in some cases computationally trivial, vulnerabilities.

References

1. Bellovin, S., Defending Against Sequence Number Attacks. Internet RFC 1948, May 1996.
2. Boneh, D., Twenty Years of Attacks on the RSA Cryptosystem. *Notices of the AMS* **46** (1998) pp. 203–213.
   http://crypto.stanford.edu/~dabo/papers/RSA.ps
3. Braden, R. (Ed.), Requirements for Internet Hosts — Communication Layers. Internet RFC 1122, October 1989.
4. Coppersmith, D., Small solutions to polynomial equations, and low exponent RSA vulnerabilities. *J. Cryptology* **10** (1997) pp. 233–260.
5. Crouch, P.A., *A small public exponent RSA attack on TCP/IP packets*. Project, University of Bath Department of Mathematical Sciences, May 2001.
   http://www.p-crouch.com/rsa-tcpip.
6. Davenport, J.H., Lecture notes at LMS Durham Symposium.
   http://www.bath.ac.uk/~masjhd/Durham.{dvi,ps,pdf}
7. Howgrave-Graham, N.A., Finding Small Roots of Univariate Modular Equations Revisited. *Cryptography and Coding* (Ed. M. Darnell), Springer Lecture Notes in Computer Science 1355, 1997, pp. 131–142.
8. Howgrave-Graham, N.A., *Computational Mathematics inspired by RSA*. Ph.D. Thesis, University of Bath, 1998.
9. Lenstra, A. Lenstra, H. Lovász. Factoring Polynomials with Rational Coefficients. *Mathematische Annalen* **261** (1982) pp. 515–534. Zbl. 488.12001. MR 84a:12002.
10. Morris, R.T., A Weakness in the 4.2BSD Unix TCP/IP Software. Computing Science Technical Report 117, AT&T Bell Laboratories, Murray Hill, New Jersey, 1985.
11. Nguyen, S. and Stern, J., Lattice Reduction in Cryptography: An update. Proc. ANTS-IV (ed. W. Bosma), Springer Lecture Notes in Computer Science 1838, Springer-Verlag, 2000, pp. 85–112. Updated at http://www.di.ens.fr/~pnguyen/pub.
12. Shoup, V. NTL (Number Theory Library) for C++. http://www.shoup.net.
13. Stevens, W.R., *TCP/IP Illustrated, Volume 1*. Addison Wesley, 1994.

Spectrally Bounded Sequences, Codes, and States: Graph Constructions and Entanglement

Matthew G. Parker

Code Theory Group, Inst. for Informatikk, HIB,
University of Bergen, Norway
matthew@ii.uib.no,
http://www.ii.uib.no/~matthew/MattWeb.html

Abstract. A recursive construction is provided for sequence sets which possess good Hamming Distance and low Peak-to-Average Power Ratio (PAR) under any Local Unitary Unimodular Transform. We identify a subset of these sequences that map to binary indicators for linear and nonlinear Factor Graphs, after application of subspace Walsh-Hadamard Transforms. Finally we investigate the quantum PAR$_l$ measure of 'Linear Entanglement' (LE) under any Local Unitary Transform, where optimum LE implies optimum weight hierarchy of an associated linear code.

1 Introduction

Golay Complementary sequences of length 2^n form sequences with Peak-to-Average Power Ratio (PAR) ≤ 2 under the one-dimensional continuous Discrete Fourier Transform (DFT$_1^\infty$) [9]. The upper PAR bound of 2 follows by forming these Complementary Sequences using Rudin-Shapiro construction [25,26]. This set is the union of certain quadratic cosets of Reed-Muller (RM) $(1, n)$ [5]. Moreover the quadratic coset representatives can be viewed as 'line graphs' in Algebraic Normal Form (ANF) [21]. As these sequences are a subset of RM$(2, n)$, the Hamming Distance, D, between sequences in the set satisfies $D \geq 2^{n-2}$. The problem of finding error-correcting codes where each codeword also has low PAR has application to Orthogonal Frequency Division Multiplexing (OFDM) communications systems [11]. However the fundamental codeset identified by Davis and Jedwab [5] (DJ sequences) suffers from vanishing rate as n increases, and much higher rates are possible and desirable, where PAR $\leq O(n)$ [27,22]. A generalisation of Rudin-Shapiro construction to other starting seeds [16,17]. allows inclusion of more low PAR quadratic cosets of RM$(1, n)$ in the code, thereby improving code rate somewhat. Higher degree cosets...etc can also be added, increasing code rate at price of distance, D, which decreases. However these rate improvements are marginal. In this paper we present a construction for much larger codesets of sequences with PAR $\leq 2^t$, comprising ANFs up to degree u, where $u \leq t$ for $t > 1$, and $u = 2$ for $t = 1$ [19]. These codesets have PAR $\leq 2^t$ under **all** Linear Unimodular Unitary Transforms (LUUTs), including one and multi-dimensional continuous DFTs. As LUUTs include the Walsh-Hadamard

B. Honary (Ed.): Cryptography and Coding 2001, LNCS 2260, pp. 339–354, 2001.
© Springer-Verlag Berlin Heidelberg 2001

Transform (WHT) then our construction gives large codesets of Almost-Bent functions [3,23]. The functions are cryptographically even stronger, as the binary sequences are distant from linear sequences over <u>all</u> alphabets, not just over Z_2. We then describe a mapping of a subset of the bipolar sequences, generated using our construction, to Factor Graphs [12]. By applying tensor products of Hadamard and Identity kernels to our bipolar sequence we transform to a Factor Graph in a Normal Realisation [7] representing a linear or nonlinear error-correcting code. This transformation provides spectral characterisation for Factor Graphs (and Quantum Factor Graphs [15]). Finally we present PAR_l, which is a partial measure of quantum entanglement and measures PAR under **all** Linear Unitary Transforms (LUTs) [17,18]. We also define 'Linear Entanglement' (LE), and 'Stubborness of Entanglement' (SE), which is a series of parameters related to PAR_l over all sequence subspaces. At least in the bipartite quadratic case, a length 2^n bipolar sequence with optimal LE and SE represents a $[n, k, d]$ binary linear code with optimal weight hierarchy. We conjecture that optimally entangled subsystems represent optimal linear and nonlinear codes - and vice versa. A similar relationship between secrecy and entanglement has recently been highlighted by [4].

2   A Construction for Low PAR Error-Correcting Codes

Joint work with C.Tellambura [19]
PAR is a spectral measure. We must therefore define the transforms over which the spectrum is computed:

2.1   Definitions

Definition 1 L_n *is the infinite set of length 2^n complex linear unimodular sequences,* $l = (l_0, l_1, \ldots, l_{2^n-1})$, *where* $|l_i| = |l_j|$, $\forall i, j$, $\sum_{i=0}^{2^n-1} |l_i|^2 = 1$, *and,*

$$l = \{2^{\frac{-n}{2}} (a_0, b_0) \otimes (a_1, b_1) \otimes \ldots \otimes (a_{n-1}, b_{n-1})\}$$

where $\otimes$ means 'tensor product'.

Definition 2 *A $2^n \times 2^n$ Linear Unimodular Unitary Transform (LUUT) matrix $\mathbf{L}$ has rows taken from L_n such that $\mathbf{LL}^\dagger = \mathbf{I}_{2^n}$, where $\dagger$ means conjugate transpose, and $\mathbf{I}_{2^n}$ is the $2^n \times 2^n$ identity matrix.*

Definition 3 G_n *is the infinite set of length 2^n complex linear sequences,* $l = (l_0, l_1, \ldots, l_{2^n-1})$, *where* $\sum_{i=0}^{2^n-1} |l_i|^2 = 1$ *and,*

$$l = \{2^{\frac{-n}{2}} (a_0, b_0) \otimes (a_1, b_1) \otimes \ldots \otimes (a_{n-1}, b_{n-1})\}$$

Note that $G_n \supset L_n$.

Definition 4 *A $2^n \times 2^n$ Linear Unitary Transform (LUT) matrix $\mathbf{G}$ has rows taken from $\mathbf{G_n}$ such that $\mathbf{GG}^\dagger = \mathbf{I}_{2^n}$. LUUTs are a special case of LUT.*

Let s_i be an element of a length 2^n vector, $\mathbf{s}$. PAR($\mathbf{s}$) is computed by measuring maximum possible correlation of $\mathbf{s}$ with **any** length 2^n 'linear' unimodular sequence, $\mathbf{l} \in \mathbf{L_n}$:

Definition 5 $PAR(\mathbf{s}) = 2^n \, max_l(|\mathbf{s} \cdot \mathbf{l}|^2)$
where $\mathbf{l} \in \mathbf{L_n}$ and $\cdot$ means 'inner product' [17].

Let $\mathbf{x} = \{x_0, x_1, \ldots, x_{n-1}\}$. Then $p(\mathbf{x})$: $Z_2^n \to Z_2$ has a bipolar representation, $\mathbf{s} = (-1)^{p(\mathbf{x})} = (s_0, s_1, \ldots, s_{2^n-1})$, where $s_i = (-1)^{p(x_0=i_0, x_1=i_1, \ldots, x_{n-1}=i_{n-1})}$, and $i = \sum_{k=0}^{n-1} i_k 2^k$ is a radix-2 decomposition of i.

2.2   Construction

This paper focuses on a special case of a more general construction. Here, all x_i are two-state binary variables, and the fundamental recursion is based on Walsh-Hadamard Transform (WHT) kernels. The more general construction is presented in [19]. We now present the construction:

$$p(\mathbf{x}) = \sum_{j=0}^{L-2} \sum_{l=0}^{t-1} x_{\pi(tj+l)} f_{l,j}\big(x_{\pi(t(j+1))}, x_{\pi(t(j+1)+1)}, \ldots, x_{\pi(t(j+2)-1)}\big)$$
$$+ \sum_{j=0}^{L-1} g_j\big(x_{\pi(tj)}, x_{\pi(tj+1)}, \ldots, x_{\pi(tj+t-1)}\big) \qquad (1)$$

where $n = Lt$, π permutes Z_n, and where $f_{l,j} : Z_2^t \to Z_2$ is such that $f_{\gamma_j} = (f_{0,j}, f_{1,j}, \ldots, f_{t-1,j})$ is an invertible boolean function (permutation polynomial) from $Z_2^t \to Z_2^t$, governed by the permutation, $i' = \gamma_j(i)$, where $i' = \sum_{l=0}^{t-1} i'_l 2^l$ is a radix-2 decomposition, $i'_l = f_{l,j}(i_0, i_1, \ldots, i_{t-1})$, and each γ_j permutes Z_t. To avoid unnecessary duplications, we exclude the f_{γ_j} where one or more $f_{l,j}$ has a '+1' constant offset, and also the cases where all $f_{l,j}$ are monomials, except when f_{γ_j} is the identity function.

Theorem 1 *[19] The length $N = 2^n$ bipolar sequence $\mathbf{s} = (-1)^{\mathbf{p}}$ satisfies $PAR(\mathbf{s}) \leq 2^t$ under all LUUTs, where $\mathbf{p}$ is generated using construction (1).*

Proof. (sketch) Let m factor fully as $m = \prod_{i=0}^{F-1} p_i$, p_i not necessarily distinct. A length m vector, $\mathbf{l}$, is defined linear if it satisfies $\mathbf{l} = \bigotimes_{i=0}^{F-1} \mathbf{v_i}$ where length($\mathbf{v_i}$) = p_i, and $\sum_{j=0}^{m-1} |l_j|^2 = 1$. Let $\mathbf{E_j}$ and $\mathbf{A_j}$, $1 \leq j \leq L$, be a series of $N \times N$ and $N \times N^j$ complex matrices, respectively, where $\mathbf{A_1} = \mathbf{E_1}$ is unitary. Let the rows of $\mathbf{A_{j-1}}$, $(\mathbf{a_{0,j-1}}, \mathbf{a_{1,j-1}}, \ldots, \mathbf{a_{N-1,j-1}})$, form a complementary set of N sequences under any $N^{j-1} \times N^{j-1}$ unitary transform with linear unimodular rows. Let $\mathbf{l}$ and $\mathbf{l_j}$ be normalised linear rows of length N^{j-1} and N, respectively. Let $\mathbf{r} = \mathbf{A_{j-1}}\mathbf{l}$. Let γ permute Z_N. Construct the $N \times N^j$ matrix, $\mathbf{A_j}$, such that $\mathbf{a_{i,j}} = ((\mathbf{a_{\gamma(0),j-1}}|\mathbf{a_{\gamma(1),j-1}}| \ldots, |\mathbf{a_{\gamma(N-1),j-1}}) \odot (\mathbf{e_{i,j}} \otimes \mathbf{1}))$ where $\mathbf{x} \odot \mathbf{y} = (x_0 y_0, x_1 y_1, \ldots, x_{N^j-1} y_{N^j-1})$, $\mathbf{1}$ is the length N^{j-1} all-ones vector, $\mathbf{e_{i,j}}$ is the ith row of $\mathbf{E_j}$, and '$|$' means concatenation. The rows of $\mathbf{A_j}$ form a complementary

N-set under any unitary transform if $r' = A_j(l_j \otimes l)$ satisfies, $\sum_{k=0}^{N-1} |r_i'|^2 = 1$. This follows if $\sum_{i=0}^{N-1} |\sum_{k=0}^{N-1} (r_{\gamma(k)} e_{i,k} l_k)|^2 = 1$, for $r_k, e_{i,k}$ and l_k elements of r, $e_{i,j}$ and l_j, respectively. This is true if E_j is unitary, and if $e_{i,j} \odot l_j$ is unimodular, which follows if $e_{i,j}$ and l_j are unimodular. Construction (1) occurs when successive A_j are recursively generated, where all E_i are $2^t \times 2^t$ WHTs. The γ permutation essentially maps to f_γ, and concatenation is widened to a more general permutation, π, over all linear variables. ∎

Theorem 2 *For a fixed t, let P be the codeset of length 2^n binary sequences of degree μ or less, generated using (1). Then,*

$$\frac{|P|}{2^{n+1}} \leq \frac{(\frac{\Gamma}{t!})^{\frac{n}{t}-1} n! (2^{2^t-t-1})^{\frac{n}{t}}}{2t!}, \qquad \mu = 2$$
$$\leq \frac{((2^t-1)!)^{\frac{n}{t}-1} n! (2^{2^t-t-1})^{\frac{n}{t}}}{2t!}, \qquad \mu \geq 2 \qquad (2)$$

where $\Gamma = \prod_{i=0}^{t-1}(2^t - 2^i) = |GL(t,2)|$. (GL is the General Linear Group). (Only for $t = 1$ is the upper bound exact).

Proof. By counting arguments we can show that, for $\mu = 2$,

$$\frac{|P|}{2^{n+1}} \leq \frac{\prod_{l=1}^{t} \binom{\frac{ln}{t}}{\frac{n}{t}}}{t!} \times \frac{(\frac{n}{t})!^t}{2} \times (\frac{\Gamma}{t!})^{\frac{n}{t}-1} \times (2^{(t/2)})^{\frac{n}{t}}$$

For $\mu \geq 2$, we replace $\frac{\Gamma}{t'}$ with $\frac{(2^t)!}{2^t}$, which is the number of permutations excluding those with a constant offset, '+1'. The Theorem follows. ∎

In Section 2.4 we show how to generate all degree-one permutation polynomials, via an isomorphism to the General Linear Group, where the number of degree-one permutation polynomials is Γ.

2.3   Examples

The $2^n \times 2^n$ Walsh-Hadamard (WHT) and Negahadamard (NHT) Transform matrices are $\bigotimes_{i=0}^{n-1} H$, and $\bigotimes_{i=0}^{n-1} N$, respectively, where $H = \left(\begin{smallmatrix} 1 & 1 \\ 1 & -1 \end{smallmatrix}\right)$ and $N = \left(\begin{smallmatrix} 1 & i \\ 1 & -i \end{smallmatrix}\right)$, and $i^2 = -1$. DFT$_1^\infty$ is the set of $2^n \times 2^n$ matrices, the union of whose rows form a subset of L_n such that each row satisfies $a_i = 1$, $b_i = \omega^{ik}$ for some fixed k, and ω is a complex root of unity (see Definition 1). These three transforms are used as 'spot-checks' in the examples to validate the PAR upper-bound.

Example 1. Let γ_j be the identity permutation $\forall j$. Then, $f_{l,j}(x_{\pi(t(j+1))}, x_{\pi(t(j+1)+1)}, \ldots, x_{\pi(t(j+2)-1)}) = x_{\pi(t(j+1)+l)}$, and (1) becomes,

$$p(x) = \sum_{j=0}^{L-2} \sum_{l=0}^{t-1} x_{\pi(tj+l)} x_{\pi(t(j+1)+l)} + \sum_{j=0}^{L-1} g_j(x_{\pi(tj)}, x_{\pi(tj+1)}, \ldots, x_{\pi(tj+t-1)}) \qquad (3)$$

When $\deg(g_j) < 2$, $\forall j$, it is well-known that $s = (-1)^{p(x)}$ is Bent (PAR $= 1$ under the WHT) for L even [14] and (perhaps not known) that s has PAR $= 2^t$ under the WHT for L odd. In general, for any g_j, s has PAR $\leq 2^t$ under all LUUTs. For example, if $L = 4$ and,

$$p(x) = x_0x_3 + x_1x_4 + x_2x_5 + x_3x_6 + x_4x_7 + x_5x_8 + x_6x_9 + x_7x_{10} + x_8x_{11}$$

then $s = (-1)^{p(x)}$ has PAR $= 1.0$ under the WHT, PAR $= 1.0$ under the NHT, and PAR $= 7.09$ under DFT$_1^\infty$. Similarly, let $g_0(x_0, x_1, x_2) = x_1x_2$, $g_1(x_3, x_4, x_5) = x_3x_4x_5$, and $g_2(x_6, x_7, x_8) = 0$. Then $s' = (-1)^{p(x)+g_0+g_1+g_2}$ has PAR $= 4.0$ under the WHT, PAR $= 2.0$ under the NHT, and PAR $= 7.54$ under DFT$_1^\infty$. In all cases, PAR ≤ 8.0 under any LUUT.

Example 2, PAR ≤ 2.0. Let $t = 1$. Then we have one possible permutation polynomial, namely, $f_\gamma = x$, (we exclude $f_\gamma = x + 1$). From (1) we obtain,

$$p(x) = \sum_{j=0}^{L-2} x_{\pi(j)}x_{\pi(j+1)} + c_jx_j + d, \qquad c_j, d \in Z_2 \qquad (4)$$

This is exactly the DJ set of binary quadratic cosets of RM$(1, n)$, where $n = L$ [5]. This set has PAR ≤ 2.0 under DFT$_1^\infty$ [5]. Such sequences are Bent for n even [14,23] and, in [16,17] it was shown that such a set has PAR $= 2.0$ under the WHT for n odd, and also, under the NHT, has PAR $= 1.0$ for $n \neq 2$ mod 3 (NegaBent), and PAR $= 2.0$ for $n = 2$ mod 3. More generally the DJ set has PAR ≤ 2.0 under any LUUT [17], and this agrees with Theorem 1. For example, let $p(x) = x_0x_4 + x_4x_1 + x_1x_2 + x_2x_3 + x_1 + 1$. Then $s = (-1)^{p(x)}$ has PAR $= 2.0$ under the WHT, PAR $= 2.0$ under the NHT, and PAR $= 2.0$ under DFT$_1^\infty$. The DJ set, being cosets of $R(2, n)$, forms a codeset with Hamming Distance, $D \geq 2^{n-2}$. The rate of the DJ codeset follows $\frac{(\frac{n}{2})^{2n+1}}{2^{2n}}$ as n increases. This is their primary drawback as the code rate vanishes rapidly as n increases.

Example 3, PAR ≤ 4.0. [5,22,17,23] have all proposed techniques for the inclusion of further quadratic cosets, so as to improve rate at the price of increased PAR. We here propose an improved rate code (although still vanishing), where PAR ≤ 4.0. To achieve this we set $t = 2$ in (1). There are $\frac{(2^t)!}{2^t t!} = 3$ valid permutation polynomials, $f_\gamma = (f_0, f_1)$. These polynomials map from $Z_2^2 \rightarrow Z_2^2$, and are taken from the set,

$$f_\gamma(x_0, x_1) \in \{(x_0, x_1), (x_0 + x_1, x_1), (x_0, x_0 + x_1)\}$$

Substituting for $f_{l,j}$ and g_j in (1) gives a large set of polynomials with PAR≤ 4.0 under all LUUTs. We now list, for this construction, the $p(x)$ arising from the the 3 invertible polynomial functions, f_γ, for one 'section' of the polynomial, i.e. for $L = 2$, where we fix π to the identity permutation.

$$p(x) = x_0x_2 + x_1x_3 + c_0x_0x_1 + c_1x_2x_3 + \text{RM}(1, 4)$$
$$p(x) = x_0(x_2 + x_3) + x_1x_3 + c_0x_0x_1 + c_1x_2x_3 + \text{RM}(1, 4)$$
$$p(x) = x_0x_2 + x_1(x_2 + x_3) + c_0x_0x_1 + c_1x_2x_3 + \text{RM}(1, 4)$$

where $c_0, c_1 \in Z_2$. The quadratic part of each of these 3 functions is isomorphic to a distinct invertible boolean $t \times t$ matrix, where $t = 2$ (Section 2.4), as the permutation polynomials form a group which is isomorphic to the General Linear Group, $GL(t, 2)$, where $|GL(t, 2)| = \prod_{i=0}^{t-1}(2^t - 2^i)$ [13]. Two of the 3 quadratic functions are inequivalent under permutation of the four variable indices, e.g.,

$$p(\boldsymbol{x}) = x_0 x_2 + x_1 x_3 + c_0 x_0 x_1 + c_1 x_2 x_3 + RM(1, 4)$$
$$p(\boldsymbol{x}) = x_0(x_2 + x_3) + x_1 x_3 + c_0 x_0 x_1 + c_1 x_2 x_3 + RM(1, 4)$$

An upper bound on $|\boldsymbol{P}|$ is given by Theorem 2, (2). Substituting $t = 2$ into (2),

$$\frac{|\boldsymbol{P}|}{2^{n+1}} < n! 2^{\frac{n-4}{2}} 3^{\frac{n}{2}-1} \tag{5}$$

An exact enumeration and construction for this set remains open, due to extra 'hidden' symmetries. Computationally we are able to calculate the exact number of quadratic coset leaders for $n = 4, 6, 8, 10$, and these are compared to the upper bound of (5) in Table 1. They are also compared to the number of quadratic coset leaders, $(= \frac{n!}{2})$ in the binary DJ codeset (Example 2). By assigning $t = 2$

Table 1. The Number of Quadratic Coset Leaders for Construction (1) when $t = 2$

n	4	6	8	10
Theorem 2, (5),(2), $|\boldsymbol{P}|/2^{n+1}$	72	12960	4354560	2351462400
Exact Computation	36	9240	4086096	2317593600
$\frac{\text{DJ Code}}{2^{n+1}}$	12	360	20160	1814400
$\log_2(|\boldsymbol{P}|/2^{n+1})$	6.2	13.7	22.1	31.1
$\log_2(\text{Number of quadratics})$	6	15	28	45

we have a construction for a much larger codeset than the DJ codeset and with the same Hamming Distance, $D = 2^{n-2}$, but the price paid is that the PAR is now upper-bounded by 4.0 instead of 2.0. For example, let,
$p(\boldsymbol{x}) = x_0 x_2 + x_1 x_2 + x_1 x_6 + x_2 x_5 + x_6 x_3 + x_6 x_5 + x_5 x_4 + x_3 x_7 + x_0 x_1 + x_5 x_3 + x_7 + x_1$
Then $\boldsymbol{s} = (-1)^p$ has PAR $= 1.0$ under the WHT, PAR $= 2.0$ under the NHT, and PAR $= 3.43$ under DFT_1^∞.

Example 4, PAR $\leq$ 8.0. Set $t = 3$ in (1). There are now $\frac{(2^t)!}{2^t t!} = 840$ valid permutation polynomials, $f_\gamma = (f_0, f_1, f_2)$. These polynomials map from $Z_2^3 \to Z_2^3$. Moreover, $(2^3 - 1)(2^3 - 2)(2^3 - 2^2)/t! = \frac{168}{6} = 28$ of the polynomials are degree-one permutations leading to quadratic forms, $p(\boldsymbol{x})$, and can be represented by the following 7 permutation polynomials.

$f_\gamma(x_0, x_1, x_2) \in \{$
$(x_0, x_1, x_2), (x_0 + x_2, x_1, x_2), (x_0 + x_2, x_1 + x_2, x_2), (x_0 + x_1 + x_2, x_1, x_2),$
$(x_0 + x_1, x_1 + x_2, x_2), (x_0 + x_1 + x_2, x_1 + x_2, x_2), (x_0 + x_2, x_1 + x_0, x_2 + x_0 + x_1)\}$

Substituting for $f_{l,j}$ and g_j in (1) gives a large set of polynomials with PAR≤ 8.0 under all LUUTs. We now list, for this construction, all quadratic $p(x)$ arising from the 7 inequivalent degree-one permutation polynomials, f_γ, for one 'section' of the polynomial, i.e. for $L = 2$, where π is fixed as the identity permutation.

$$p(x) = x_0x_3 + x_1x_4 + x_2x_5 + g(x)$$
$$p(x) = x_0x_3 + x_0x_5 + x_1x_4 + x_2x_5 + g(x)$$
$$p(x) = x_0x_3 + x_0x_5 + x_1x_4 + x_1x_5 + x_2x_5 + g(x)$$
$$p(x) = x_0x_3 + x_0x_4 + x_0x_5 + x_1x_4 + x_2x_5 + g(x)$$
$$p(x) = x_0x_3 + x_0x_4 + x_1x_4 + x_1x_5 + x_2x_5 + g(x)$$
$$p(x) = x_0x_3 + x_0x_4 + x_0x_5 + x_1x_4 + x_1x_5 + x_2x_5 + g(x)$$
$$p(x) = x_0x_3 + x_0x_5 + x_1x_3 + x_1x_4 + x_2x_3 + x_2x_4 + x_2x_5 + g(x)$$

where $g(x) = c_0x_0x_1 + c_1x_0x_2 + c_2x_1x_2 + c_3x_0x_1x_2 + c_4x_3x_4 + c_5x_3x_5 + c_6x_4x_5 + c_7x_3x_4x_5 + RM(1,6)$, and $c_0, c_1, \ldots, c_7 \in Z_2$. An upper bound to $|P|$ can be computed from Theorem 2, (2), and the upper bound is compared to the total number of quadratics in n binary variables in Table 2. As with $t = 2$, an

---

Table 2. The Number of Quadratic Coset Leaders for Construction (1) when $t = 3$

n	6	9	12	15
Theorem 2, (2), $\log_2(|P|/2^{n+1})$	16.7	33.5	51.7	70.9
$\log_2$(Number of quadratics)	15	36	66	105

---

exact enumeration and construction for this set remains open, due to extra 'hidden' symmetries. By assigning $t = 3$ we have a construction for a codeset with Hamming Distance, $D \geq 2^{n-2}$ and PAR ≤ 8.0 under all LUUTs.

For $t = 3$ we can also include cubic forms in Construction (1). There are $\frac{5040 - 168}{6} = 812$ degree 2 permutation polynomials, $f_\gamma = (f_0, f_1, f_2)$, that map from $Z_2^3 \rightarrow Z_2^3$, and lead to cubic forms, $p(x)$. This set can be represented by 147 degree 2 permutation polynomials which are inequivalent under variable permutation, and these are listed at [20]. (Along with the 7 inequivalent degree 1 permutation polynomials, this makes a total of 154 inequivalent permutation polynomials for $t = 3$ [10,28]). Substituting for $f_{l,j}$ and g_j in (1) gives a large set of polynomials with PAR≤ 8.0 under all LUUTs, and Hamming Distance, $D \geq 2^{n-3}$. An upper bound to $|P|$ can be computed from Theorem 2, (2), and the upper bound is compared to the total number of quadratics and cubics in n binary variables in Table 3. Here is an example from this codeset, where ijk, uv is short for $x_ix_jx_k + x_ux_v$. Let,

$$p(x) = 034, 035, 045, 135, 145, 234, 235, 245, 367, 368, 378, 567, 568, 69A, 79A, 7AB,$$
$$89A, 345, 9AB, 03, 05, 14, 24, 25, 36, 38, 47, 58, 69, 6A, 6B, 7A, 7B, 89, 8B, 67, 78, AB$$

then $s = (-1)^{p(x)}$ has PAR $= 4.0$ under the WHT, PAR $= 6.625$ under the NHT, and PAR $= 7.66$ under DFT$_1^\infty$. In all cases, PAR ≤ 8.0.

Table 3. The Number of Cubic and Quadratic Coset Leaders for Construction (1) when $t = 3$

n	6	9	12	15
Theorem 2, (2), $\log_2(|\boldsymbol{P}|/2^{n+1})$	23.6	46.3	70.4	95.5
$\log_2$(Number of quadratics and cubics)	35	120	286	560

2.4   A Matrix Construction for All Quadratic Codes from (1)

Each degree-one permutation polynomial, f_γ from $Z_2^t \to Z_2^t$ can be viewed as a $t \times t$ binary adjacency matrix. Let $x = \{x_0, x_1, \ldots, x_{t-1}\}$. We can write,

$$M \Leftrightarrow f_\gamma(x) = (f_0(x), f_1(x), \ldots, f_{t-1}(x)), \quad M = \{m_{i,l}\}, \deg(f_l(\boldsymbol{x})) = 1, \text{ and}$$
$$m_{i,l} = 1 \quad \text{if } x_i \in f_l(x) \quad m_{i,l} = 0 \quad \text{otherwise}$$

The mapping is an isomorphism from the degree-one permutation polynomials to the General Linear Group, $G = \mathrm{GL}(t, 2)$, of all binary $t \times t$ invertible matrices [13]. To construct all quadratic sequences, $p(\boldsymbol{x})$, for a given n and t we need to construct all degree one permutation polynomials, f_γ. These can, in turn be constructed by generating all members of $G = \mathrm{GL}(t, 2)$, and this is accomplished as follows [1,2].

Definition 6 *A binary $t \times t$ 'transvection' matrix, X_{ab}, satisfies,*

$$X_{ab} = \{u_{i,j}\}, \text{ where}$$
$$u_{i,j} = 1, \quad i = j, \text{ and } i = a, j = b \quad u_{i,j} = 0, \quad \text{otherwise}$$

Definition 7 *The Borel subgroup of G over Z_2 is the $t \times t$ upper-triangular binary matrices, B.*

Definition 8 *The Weyl subgroup of G is the $t \times t$ permutation matrices, W.*

Assign a fixed ordering, O, to the $\binom{t}{2}$ matrices, X_{ab}, $a < b$. Let $w \in W$ be a permutation of Z_t and its associated $t \times t$ permutation matrix. For each w, form the matrix product, X_w, comprising all X_{ab} which satisfy $a < b = w(a) > w(b)$, where the X_{ab} in X are ordered according to O.

Theorem 3 *[1,2]*

$$G = X'_w W B \tag{6}$$

where X'_w is any sub-product of X_w that maintains the ordering of the X_{ab} matrices in X_w. This is the 'Bruhat' decomposition.

All quadratic constructions using (1) can be constructed using Theorem 3., where $|G| = \Gamma = \prod_{i=0}^{t-1}(2^t - 2^i)$.

3   Graphical Representations

Joint work with V.Rijmen [18]

We now identify a subset of the length 2^n sequence constructions of (1), where $(-1)^{p(\boldsymbol{x})}$ exhibits a bipolar $\leftrightarrow$ binary equivalence under transform by a tensor product of combinations of $\boldsymbol{H}$ and $\boldsymbol{I}$ 2×2 matrices. The resultant length 2^n binary sequences can be interpreted as indicators for binary linear or nonlinear $[n, k, d]$ error-correcting codes. In such cases, $p(\boldsymbol{x})$ is closely related to a Normal Realisation for the Factor Graph of the associated $[n, k, d]$ code [7]. Let $\boldsymbol{s} = (-1)^{p(\boldsymbol{x})}$.

Definition 9 *"H acting on i" means the action of the $2^n \times 2^n$ transform, $\boldsymbol{I} \otimes \ldots \otimes \boldsymbol{I} \otimes \boldsymbol{H} \otimes \boldsymbol{I} \otimes \ldots \otimes \boldsymbol{I}$ on $\boldsymbol{s}$, where $\boldsymbol{H}$ is preceded by i $\boldsymbol{I}$ matrices, and followed by $n - i - 1$ $\boldsymbol{I}$ matrices. We write this as $H(i)$, or $H(i)[\boldsymbol{s}]$.*

Definition 10 *Let $\mathbf{T_C}$, $\mathbf{T_{C^\perp}}$ be integer sets chosen so that $\mathbf{T_C} \cap \mathbf{T_{C^\perp}} = \emptyset$, and $\mathbf{T_C} \cup \mathbf{T_{C^\perp}} = \{0, 1, \ldots, n-1\}$. This is a bipartite splitting of $\{0, 1, \ldots, n-1\}$. Let us also partition the variable set $\mathbf{x}$ as $\mathbf{x} = \mathbf{x_C} \cup \mathbf{x_{C^\perp}}$, where $\mathbf{x_C} = \{x_i | i \in \mathbf{T_C}\}$, and $\mathbf{x_{C^\perp}} = \{x_i | i \in \mathbf{T_{C^\perp}}\}$.*

Definition 11 *$\kappa_{\mathbf{p}}$ is the set of all $s(\mathbf{x})$ of the form $s(\mathbf{x}) = (-1)^{p(\mathbf{x})}$, where $p(\mathbf{x}) = \sum_k q_k(\mathbf{x_C}) r_k(\mathbf{x_{C^\perp}})$, where $\deg(q_k(\mathbf{x_C})) = 1 \ \forall k$, and where $x_i \in p(\mathbf{x})$, $\forall i \in \{0, 1, \ldots, n-1\}$. We refer to $\kappa_{\mathbf{p}}$ as the set of 'half-linear bipartite bipolar' states. $\ell_{\mathbf{p}}$ is the subset of $\kappa_{\mathbf{p}}$ where $\deg(r_k(\mathbf{x_C})) = 1 \ \forall k$.*

Theorem 4 *[18] Let $m(\mathbf{x})$ be a binary ANF. If $s(\mathbf{x}) \in \kappa_{\mathbf{p}}$, then the action of $\prod_{i \in \mathbf{T_C}} H(i)$ on $s(\mathbf{x})$ gives $s'(\mathbf{x}) = m(\mathbf{x})$. If $s(\mathbf{x}) \in \ell_{\mathbf{p}}$, then the action of $\prod_{i \in \mathbf{T_{C^\perp}}} H(i)$ on $s(\mathbf{x})$ gives $s''(\mathbf{x}) = m(\mathbf{x})$. $s'(\mathbf{x})$ $(s''(\mathbf{x}))$ is the binary indicator for a binary linear or nonlinear $[n, n - |\mathbf{T}|, d]$ error correcting code, $\mathbf{C}$.*

Theorem 4 is particularly relevant when $p(\boldsymbol{x})$ is constructed using (1), as the 'strongest' members of $\kappa_{\mathbf{p}}$ are generated as a subclass of the construction if $\deg(g_j) < 2, \forall j$. (By considering matrices other than $\boldsymbol{H}$ it is conjectured that it is always possible to convert a bipolar sequence, $\boldsymbol{s} = (-1)^{\boldsymbol{p}}$, constructed using (1) to a binary form, even when $\deg(g_j) \geq 2$.) If $\boldsymbol{s}$ can be transformed to a binary linear indicator, $\boldsymbol{s}'$, using only tensor products of $\boldsymbol{H}$ and $\boldsymbol{I}$, then we say that $\boldsymbol{s}$ is 'HI-equivalent to' $\boldsymbol{s}'$.

Theorem 5 *[18] The set $\ell_{\mathbf{p}}$ is HI-equivalent to the set of $[n, k, d]$ binary linear codes.*

3.1   Examples

Example A. Let $t = 2$, $L = 3$. Then (1) can generate,
$$p(\boldsymbol{x}) = x_0x_2 + x_1x_3 + x_2x_4 + x_3x_5 + x_2x_5$$
Let $\mathbf{T_C} = \{0, 1, 4, 5\}$ and $\mathbf{T_{C^\perp}} = \{2, 3\}$. Applying $H(0)H(1)H(4)H(5)$ (in any order) to $\boldsymbol{s} = (-1)^{p(\boldsymbol{x})}$ gives the binary sequence, $\boldsymbol{s}' = m(x) = (x_0 + x_2 + 1)(x_1 + x_3 + 1)(x_2 + x_4 + 1)(x_2 + x_3 + x_5 + 1)$, which is the indicator for a $[6, 2, 2]$ binary linear code, $\mathbf{C}$. Graphical representations for $\boldsymbol{s}$ and $\boldsymbol{s}'$ are shown in Fig 1, where the graph for $\boldsymbol{s}'$ is a Normal Realisation of a Factor Graph [7]. If, instead, we apply $H(2)H(3)$ (in any order) to $\boldsymbol{s} = (-1)^{p(\boldsymbol{x})}$, we get the binary sequence, $\boldsymbol{s}'' = m(x) = (x_0 + x_2 + x_4 + x_5 + 1)(x_1 + x_3 + x_5 + 1)$, which is the indicator for a $[6, 4, 2]$ binary linear code, $\mathbf{C}^\perp$, the dual of $\mathbf{C}$. Applying $H(0)H(1)H(4)H(5)$ to $\boldsymbol{s}'$, followed by $H(2)H(3)$, gives $\boldsymbol{s}''$. This is the same as applying the WHT to $\boldsymbol{s}'$, and it is known that binary indicators of a linear code code, $\mathbf{C}$, and its dual, $\mathbf{C}^\perp$, are related by the WHT [14].

Example B. Let $t = 3$, $L = 3$. Then (1) can generate,
$$p(\boldsymbol{x}) = 034, 035, 045, 134, 135, 145, 234, 235, 245, 03, 05, 14, 15, 36, 47, 58$$
Let $\mathbf{T_C} = \{0, 1, 2, 6, 7, 8\}$ and $\mathbf{T_{C^\perp}} = \{3, 4, 5\}$. Applying
$H(0), H(1), H(2), H(6), H(7), H(8)$ (in any order) to $\boldsymbol{s} = (-1)^{p(\boldsymbol{x})}$ gives,

$\boldsymbol{s}' = m(x) =$
$(x_0 + x_3x_4 + x_3x_5 + x_4x_5 + x_3 + x_5 + 1)(x_1 + x_3x_4 + x_3x_5 + x_4x_5 + x_4 + x_5 + 1)$
$\times (x_2 + x_3x_4 + x_3x_5 + x_4x_5 + 1)(x_3 + x_6 + 1)(x_4 + x_7 + 1)(x_5 + x_7 + 1)$

which is the indicator for a $[9, 3, 3]$ binary nonlinear code, $\mathbf{C}$. Graphical representations for $\boldsymbol{s}$ and $\boldsymbol{s}'$ are shown in Fig 1, where the graph for $\boldsymbol{s}'$ is a Normal Realisation of a **nonlinear** Factor Graph. In this case application of $H(3)H(4)H(5)$ does not produce the dual code, $\mathbf{C}^\perp$, but the nonlinear dual could be obtained by nonlocal transform over x_3, x_4, x_5.

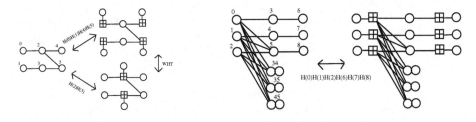

Fig. 1. Bipolar $\leftrightarrow$ Factor Graph HI-Equivalence for Examples A and B

Example C. The nonlinear $[16, 8, 6]$ Nordstrom-Robinson binary code is HI-equivalent to a half-linear bipolar bipartite sequence, $(-1)^{p(\boldsymbol{x})}$, where $p(\boldsymbol{x})$ can be constructed using (1), and has ANF comprising 96 cubic and 40 quadratic terms, and where $|T_C| = |T_{C^\perp}| = 8$. The quadratic part of $p(\boldsymbol{x})$ is HI-equivalent to a binary linear $[16, 8, 4]$ code, so we can view the 96 cubic terms of $p(\boldsymbol{x})$ as further 'doping' to increase Hamming Distance, d, from 4 to 6.

3.2   Comments

This section has identified an important subset of $\kappa_{\mathbf{p}}$ as a subset of the construction of (1), where a member of $\kappa_{\mathbf{p}}$ can be transformed to a binary sequence under selective action of $\boldsymbol{H}$. Conversely, this gives us a way of analysing a Factor Graph, by transforming it back into bipolar sequence form. A natural question to ask is which length 2^n bipolar sequences are transform-equivalent to the best $[n, k, d]$ linear and nonlinear codes? We offer offer the following conjecture,

Conjecture 1 *Optimal linear or nonlinear codes can be constructed from (1) if $L = 2$, and $(-1)^{g_j}$ is, itself, HI-equivalent to an optimal linear or nonlinear code, $\forall j$. But what f_{γ_j} should be chosen?*

In the next section we pose the related question: Which quantum n-qubit states have optimal Linear Entanglement?

4 PAR_l and Quantum 'Linear' Entanglement (LE)

Joint work with V.Rijmen [18]
In previous sections our PAR metric has been measured relative to all LUUTs. Quantum systems require that we compute our PAR metric (now called PAR_l) relative to all LUTs, of which LUUTs are a subset. It is argued in [18] that PAR_l and Linear Entanglement (LE) are good partial measures of quantum entanglement. [1] Let $\boldsymbol{s}$ be a length 2^n bipolar sequence. In the context of quantum systems we interpret (after appropriate normalisation) this sequence as a probability density function of an n-qubit quantum state. Let s_i be an element of $\boldsymbol{s}$. Then $|s_i|^2$ is the probability of measuring the quantum system in state i. We must normalise so that $\sum_{i=0}^{2^n-1} |s_i|^2 = 1$, although normalisation constants are usually omitted in this paper. An n-qubit state, $\boldsymbol{s}$, contains entanglement if $\boldsymbol{s}$ is not a member of $\boldsymbol{G_n}$. The definition of PAR_l is then identical to Definition 5 except that, now, $|l_i|$ does not have to equal $|l_j|$, i.e. $\boldsymbol{l}$ is not necessarily unimodular.

Definition 12 $PAR_l(\boldsymbol{s}) = 2^n max_l(|\boldsymbol{s} \cdot \boldsymbol{l}|^2))$
where $\boldsymbol{l}$ is any normalised linear sequence from the set, $\boldsymbol{G_n}$, and $\cdot$ means 'inner product' [17,18].

---

[1] Quantum information theorists often consider 'mixed-state' entanglement, where entanglement with the environment is unavoidable [24,8]. This is similar to the analysis of classical communications codes in the context of a corrupting channel. In this paper we only consider a closed (pure) quantum system with no environmental entanglements [6].

Linear Entanglement (LE) is then defined as,

Definition 13 $LE(s) = n - \log_2(PAR_l(s))$

Entanglement and LE are invariant under transformation of s by any LUT. Therefore PAR_l is Local Unitary (LU)-invariant, and two states, s and s', related by a transform from LUT, are LU-equivalent. Code duality under the WHT and the HI-equivalence between s and s', as discussed in Section 3, are special cases of LU-equivalence. One can also view entanglement invariance as a generalisation of code duality.

4.1 PAR_l for States from ℓ_p

Theorem 6 *[18] If $s \in \ell_p$, then s is LU equivalent to the indicator for an $[n, k, d]$ binary linear code, and,*

$$PAR_l(s) \geq 2^r, \quad where \; r = max(k, n-k)$$

Theorem 6 implies that states, s, from ℓ_p have a minimum lower bound on PAR_l (upper bound on LE) when the associated $[n, k, d]$ code, $\mathbf{C}$, satisfies $k = \lfloor \frac{n}{2} \rfloor$, with $PAR_l \geq 2^{\lceil \frac{n}{2} \rceil}$. Here is a stronger result.

Theorem 7 *[18] In (1), let $t = 1$ and $f_{\gamma j}$ be the identity permutation $\forall j$. Using (1), we can generate $s(\mathbf{x}) = (-1)^{p(\mathbf{x})}$ for $p(\mathbf{x})$ constructed using (4). Then $PAR_l(s) = 2^{\lceil \frac{n}{2} \rceil}$.*

Definition 14 $PA(s) = 2^n max_i(|s_i|^2)$

We now compute PA for any *HI* transform of a member of ℓ_p. Let $s \in \ell_p$. Recalling Definition 10, let $k = |\mathbf{T}_{\mathbf{C}^\perp}|$, $k^\perp = |\mathbf{T}_{\mathbf{C}}|$, and $k + k^\perp = n$. Without loss of generality we renumber integer sets $\mathbf{T}_{\mathbf{C}^\perp}$ and $\mathbf{T}_{\mathbf{C}}$ so that $\mathbf{T}_{\mathbf{C}^\perp} = \{0, 1, \ldots, k-1\}$ and $\mathbf{T}_{\mathbf{C}} = \{k, k+1, \ldots, n-1\}$. Let $\mathbf{t}_{\mathbf{C}^\perp} \subset \mathbf{T}_{\mathbf{C}^\perp}$ and $\mathbf{t}_{\mathbf{C}} \subset \mathbf{T}_{\mathbf{C}}$, where $h = |\mathbf{t}_{\mathbf{C}^\perp}|$ and $h^\perp = |\mathbf{t}_{\mathbf{C}}|$. Let $\mathbf{x}_{\mathbf{t}^\perp} = \{x_i | i \in \mathbf{t}_{\mathbf{C}^\perp}\}$, $\mathbf{x}_{\mathbf{t}} = \{x_i | i \in \mathbf{t}_{\mathbf{C}}\}$, and $\mathbf{x}_* = \mathbf{x}_{\mathbf{t}^\perp} \cup \mathbf{x}_{\mathbf{t}}$. Define $\mathbf{M}$ to be a $k \times k^\perp$ binary matrix where $M_{i,j-k} = 1$ iff $x_i x_j \in p(\mathbf{x})$, and $M_{i,j-k} = 0$ otherwise. Thus $p(\mathbf{x}) = \sum_{i \in \mathbf{T}_{\mathbf{C}^\perp}} x_i (\sum_{j \in \mathbf{T}_{\mathbf{C}}} M_{i,j-k} x_j)$. Let $\mathbf{M}_{\mathbf{t}}$ be a submatrix of $\mathbf{M}$, which comprises only the rows and columns of $\mathbf{M}$ specified by $\mathbf{t}_{\mathbf{C}^\perp}$ and $\mathbf{t}_{\mathbf{C}}$. Let χ_t be the rank of $\mathbf{M}_{\mathbf{t}}$.

Theorem 8 *[18] Let s' be the result of $\prod_{i \in \mathbf{t}_{\mathbf{C}^\perp} \cup \mathbf{t}_{\mathbf{C}}} H(i)$ on $s \in \ell_p$. Then,*

$$PA(s') = 2^{h+h^\perp - 2\chi_t}$$

Corollary 1 *As* $0 \leq \chi_t \leq min(h, h^\perp)$, *it follows that, for* $s \in \ell_p$, *PA*$(s') \geq 2^{|h - h^\perp|}$

In general, PAR$_l$ must consider PA(s) under all LUTs. PA(s) for $s \in \ell_p$ is easily computed. Let the 'HI multispectra' be the union of the power spectra of s under the action of $\prod_{i \in T} H(i)$, for all possible subsets, T, of $\{0, 1, \ldots, n-1\}$.

Theorem 9 *[18] PAR$_l$ of* $s \in \ell_p$ *is found in the HI multispectra of* s.

Theorem 9 means that, for $s \in \ell_p$, we only need compute the 2^n HI transforms to compute PAR$_l$. If PA(s) is optimally low over the HI multispectra, then $s' = m(x)$ is an optimal binary linear code when $T = T_C$ or $T = T_{C^\perp}$.

Definition 15 *The Weight Hierarchy of a linear code* **C**, *is a series of parameters,* d_j, $0 \leq j \leq k$, *representing the smallest blocklength of a linear sub-code of* **C** *of dimension* j, *where* $d_k = n$, $d_1 = d$, *and* $d_0 = 0$.

Theorem 10 *[18] Let* s_c *be the indicator of an* $[n, k, d]$ *binary linear code,* **C**. *Let* $\mathbf{Q} \subset \{0, 1, \ldots, n-1\}$. *Let,*

$$m_\mathbf{Q} = \frac{|\mathbf{Q}| + \log_2(\mu) - n + k}{2}, \quad where \ \mu = PA(s'_c) \quad (7)$$

and $s'_c = \prod_{t \in \mathbf{Q}} H(t)[s_c]$. *Then the Weight Hierarchy of* **C** *is found from the HI multispectra of* s_c, *where* $d_j = min_{\mathbf{Q}|m_Q=j}(|\mathbf{Q}|)$

Quantum measurement projects a system to a subsystem. This allows us to equate a series of quantum measurements with a series of subcodes of C. Let the entanglement order of a system be the size (in qubits) of the largest entangled subsystem of the system. A most-destructive series of j single-qubit measurements over some set of possible measurements on s produces a final state s' such that entanglement order(s) − entanglement order(s') is maximised.

Definition 16 *Stubborness of Entanglement (SE) is a series of parameters,* β_j, $0 \leq j \leq k'$, *representing smallest possible entanglement order,* β_j, *after* $k' - j$ *most-destructive measurements of an n-qubit system, where* $\beta_{k'} = n$, $\beta_0 = 0$.

Theorem 11 *[18] Let* $s \in \ell_p$ *where* s *is LU equivalent to an optimal or near-optimal binary linear code of dimension* $\leq \frac{n}{2}$. *Then Stubborness of Entanglement is equal to the Weight Hierarchy of the code.*

Corollary 2 *Quantum states from* ℓ_p *which have optimum LE and optimum SE are LU-equivalent to binary linear codes with optimum Weight Hierarchy.*

The results of this section suggests the following modification of Conjecture 1.

Conjecture 2 *States with optimal LE can be constructed from (1) if* $L = 2$, *and* $(-1)^{g_j}$ *also has optimal LE,* $\forall j$. *But what* f_{γ_j} *should be chosen?*

5   Discussion and Open Problems

We have highlighted the importance PAR plays (explicitly or implicitly) in current research. We emphasis four areas:

a) Low PAR error-correcting codes for OFDM and CDMA.

b) Highly nonlinear, distinguishable sequence sets for cryptography.

c) Graphical construction primitives for Factor Graphs which represent good error-correcting codes.

d) Classification and quantification of quantum entanglement.

We finish with a list of a few open problems.

- Construction (1) only provides an exact, implementable encoder if the two following sub-problems can be solved:
  - Provide algorithms to generate all permutation polynomials, f_γ, of degree $\mu - 1$. $\mu = 0$ is trivial. Section 2.4 provides an answer for $\mu = 1$. But, for $\mu > 1$ the situation is unclear.
  - Given an algorithm to generate all permutation polynomials, then construction (1) only generates distinct $p(x)$ for $t = 1$. For $t > 1$, the permutation, π, induces extra symmetries which cause many $p(x)$ to be generated more than once. This situation is reflected in (2), which is a strict upper bound for $t > 1$. It remains an open problem to provide an algorithm for $t > 1$ which ensures the generated $p(x)$ are distinct and form the whole code. Such an algorithm would replace of (2) with an exact expression.
- Construct decoders for the above codes.
- It is considered that successful iteration on a Factor Graph requires few short graph cycles. This is ensured if the graph has a large girth. How does one construct Factor Graphs with low PAR_l and large girth?
- Provide a construction for optimally large sets, P, of pure quantum states such that each state satisfies a low upper bound on PAR_l, and where any two members of P are optimally distinguishable. This problem is 'simply' the LUT extension of the problem of low PAR error-correcting codes for OFDM and cryptography.

References

1. Alperin, J.L.,Bell, R.B.: **Groups and Representations,** Graduate Texts in Mathematics, Springer, **162**, pp 39–48, (1995)
2. Brundan, J.: Web Lecture Notes: Math 607, Polynomial representations of GL_n, *http://darkwing.uoregon.edu/~brundan/teaching.html* pp 29–31, Spring (1999)
3. Canteaut, A.,Carlet, C.,Charpin, P.,Fontaine, C.: Propagation Characteristics and Correlation-Immunity of Highly Nonlinear Boolean Functions. EUROCRYPT 2000, Lecture Notes in Comp. Sci., **1807**, 507–522, (2000)
4. Collins, D.,Popescu, S.: A Classical Analogue of Entanglement *http://xxx.soton.ac.uk/ps/quant-ph/0107082* 16 Jul. 2001

5. Davis, J.A.,Jedwab, J.: Peak-to-mean Power Control in OFDM, Golay Complementary Sequences and Reed-Muller Codes. IEEE Trans. Inform. Theory **45**. No 7, 2397–2417, Nov (1999)

6. Eisert, J.,Briegel, H.J.: Quantification of Multi-Particle Entanglement. *http://xxx.soton.ac.uk/ps/quant-ph/0007081 v2* 29 Aug (2000)

7. Forney, G.D.: Codes on Graphs: Normal Realizations. IEEE Trans. Inform. Theory **47**. No 2, 520–548, Feb, (2001)

8. Fuchs, C.A.,van de Graaf, J.: Cryptographic Distinguishability Measures for Quantum-Mechanical States. IEEE Trans. Inform. Theory **45**. No 4, 1216–1227, May (1999)

9. Golay, M.J.E.: Complementary Series. IRE Trans. Inform. Theory, **IT-7**, pp 82–87, Apr (1961)

10. Harrison, M.A.: The Number of Classes of Invertible Boolean Functions. J. ACM, **10**, 25–28, (1963)

11. Jones, A.E.,Wilkinson, T.A.,Barton, S.K.: Block Coding Scheme for Reduction of Peak to Mean Envelope Power Ratio of Multicarrier Transmission Schemes. Elec. Lett. **30**, 2098–2099, (1994)

12. Kschischang, F.R.,Frey, B.J.,Loeliger, H-A.: Factor Graphs and the Sum-Product Algorithm. IEEE Trans. Inform. Theory **47**. No 1, Jan, (2001)

13. Lidl, L.,Niederreiter, H.: **Introduction to Finite Fields and their Applications** Cambridge Univ Press, pp 361–362, (1986)

14. MacWilliams, F.J.,Sloane, N.J.A.: **The Theory of Error-Correcting Codes** Amsterdam: North-Holland. (1977)

15. Parker, M.G.: Quantum Factor Graphs. Annals of Telecom., July-Aug, pp 472–483, (2001), originally 2nd Int. Symp. on Turbo Codes and Related Topics, Brest, France Sept 4–7, (2000), *http://xxx.soton.ac.uk/ps/quant-ph/0010043*, (2000) *http://www.ii.uib.no/∼matthew/mattweb.html*

16. Parker, M.G.,Tellambura, C.: Generalised Rudin-Shapiro Constructions. *WCC2001, Workshop on Coding and Cryptography, Paris(France)*, Jan 8-12, (2001) *http://www.ii.uib.no/∼matthew/mattweb.html*

17. Parker, M.G.,Tellambura, C.: Golay-Davis-Jedwab Complementary Sequences and Rudin-Shapiro Constructions. Submitted to IEEE Trans. Inform. Theory, *http://www.ii.uib.no/∼matthew/mattweb.html* March (2001)

18. Parker, M.G., Rijmen, V.: The Quantum Entanglement of Binary and Bipolar Sequences. Short version accepted for Discrete Mathematics, Long version at *http://xxx.soton.ac.uk/ps/quant-ph/0107106* or *http://www.ii.uib.no/∼matthew/mattweb.html* Jun. (2001)

19. Parker, M.G.,Tellambura, C.: A Construction for Binary Sequence Sets with Low Peak-to-Average Power Ratio. *Submitted to Int. Symp. Inform. Theory, Laussane, Switzerland, (2002), http://www.ii.uib.no/∼matthew/mattweb.html* October (2001)

20. Inequivalent Invertible Boolean Functions for $t = 3$, *http://www.ii.uib.no/∼matthew/mattweb.html*, (2001)

21. Paterson, K.G.: Generalized Reed-Muller Codes and Power Control in OFDM Modulation. IEEE Trans. Inform. Theory, **46**, No 1, pp. 104-120, Jan. (2000)

22. Paterson, K.G.,Tarokh V.: On the Existence and Construction of Good Codes with Low Peak-to-Average Power Ratios. IEEE Trans. Inform. Theory **46**. No 6, 1974–1987, Sept (2000)

23. Paterson, K.G., Sequences for OFDM and Multi-Code CDMA: Two Problems in Algebriac Coding Theory. Hewlett-Packard Technical Report, HPL-2001-146, (2001)

24. Popescu, S.,Rohrlich, D.: On the Measure of Entanglement for Pure States. Phys. Rev. A **56**. R3319, (1997)
25. Rudin, W.,: Some Theorems on Fourier Coefficients. Proc. Amer. Math. Soc., No 10, pp. 855–859, (1959)
26. Shapiro, H.S.: Extremal Problems for Polynomials. M.S. Thesis, M.I.T., (1951)
27. Shepherd, S.J.,Orriss, J.,Barton, S.K.: Asymptotic Limits in Peak Envelope Power Reduction by Redundant Coding in QPSK Multi-Carrier Modulation. IEEE Trans. Comm., **46**, No 1, 5–10, Jan (1998)
28. Sloane, N.J.A.: The On-Line Encyclopedia of Integer Sequences. $(1, 2, 154, \ldots)$, *http://www.research.att.com/∼njas/sequences/index.html*

Attacking the Affine Parts of SFLASH

Willi Geiselmann, Rainer Steinwandt, and Thomas Beth

IAKS/E.I.S.S., Fakultät für Informatik, Universität Karlsruhe,
Am Fasanengarten 5, 76131 Karlsruhe, Germany

Abstract. The signature scheme SFLASH has been accepted as candidate in the NESSIE (New European Scheme for Signatures, Integrity, and Encryption) project. We show that recovering the two secret affine mappings $\mathbb{F}_2^{37} \longrightarrow \mathbb{F}_2^{37}$ in SFLASH can easily be reduced to the task of revealing two linear mappings $\mathbb{F}_2^{37} \longrightarrow \mathbb{F}_2^{37}$. In particular, the 74 bits representing these affine parts do by no means contribute a factor of 2^{74} to the effort required for mounting an attack against the system. This raises some doubts about the design of this NESSIE candidate.

1 Introduction

In [3,4] the asymmetric signature scheme SFLASH has been proposed. It is intended for the use on low-cost smartcards and has been accepted as candidate within the NESSIE (New European Schemes for Signatures, Integrity, and Encryption) project. The secret key in SFLASH consists of a secret 80-bit string Δ, and two affine mappings s, t. However, owing to the verification procedure in this signature scheme knowing Δ is not vital for producing valid signatures, and this contribution shows that the affine parts of the secret affine mappings s and t are vulnerable to a very simple and efficient linear algebra-based attack. The attack makes crucial use of the fact that the secret affine mappings contain only $\{0, 1\}$-entries, and has some conceptual similarity with the successful attacks against ENROOT and SPIFI described in [1].

We show that the public key of SFLASH alone (no signatures are required) is sufficient to reduce the number of candidates for the affine part of the secret key from 2^{74} to typically 2^{11}. When the affine parts of the mappings s and t are known, breaking SFLASH amounts to revealing two linear mappings. Hence, although our attack does not break the system in total, we think it raises some doubts about the design of this NESSIE candidate.

2 Description of SFLASH

We restrict our description to those aspects of SFLASH which are relevant for explaining our attack; for a more detailed description we refer to the SFLASH specification [3].

B. Honary (Ed.): Cryptography and Coding 2001, LNCS 2260, pp. 355–359, 2001.
© Springer-Verlag Berlin Heidelberg 2001

2.1   Parameters of the Algorithm

In SFLASH three finite fields along with corresponding bijections are used:

- $K := \mathbb{F}_2[X]/(X^7 + X + 1)$ along with the bijection

$$
\begin{aligned}
\pi: \quad \{0,1\}^7 &\longrightarrow K \\
(b_0, \ldots, b_6) &\longmapsto \sum_{i=0}^{6} b_i X^i \quad (\bmod\ X^7 + X + 1)
\end{aligned}
$$

- $K' := \pi(\{0,1\} \times \{0\}^6)$ (which is isomorphic to $\mathbb{F}_2$)
- $L := K[X]/(X^{37} + X^{12} + X^{10} + X^2 + 1)$ along with the bijection

$$
\begin{aligned}
\varphi: \quad K^{37} &\longrightarrow L \\
(b_0, \ldots, b_{36}) &\longmapsto \sum_{i=0}^{36} b_i X^i \quad (\bmod\ X^{37} + X^{12} + X^{10} + X^2 + 1)
\end{aligned}
$$

Secret key. The secret key is comprised of three parts:

- $\Delta \in \{0,1\}^{80}$: a secret 80-bit string
- $s = (S_L, S_C)$: an affine bijection $K^{37} \longrightarrow K^{37}$ given by a 37×37 matrix $S_L \in K'^{37 \times 37}$ and a column vector $S_C \in K'^{37}$
- $t = (T_L, T_C)$: an affine bijection $K^{37} \longrightarrow K^{37}$ given by a 37×37 matrix $T_L \in K'^{37 \times 37}$ and a column vector $T_C \in K'^{37}$

For deriving the corresponding public key also the function

$$
\begin{aligned}
F: L &\longrightarrow L \\
\alpha &\longmapsto \alpha^{128^{11}+1}
\end{aligned}
$$

is needed. Moreover, for a bitstring $\lambda = (\lambda_0, \ldots, \lambda_m)$ and integers $0 \leq r \leq s \leq m$ we write $[\lambda]_{r \to s}$ for the bitstring $(\lambda_r, \lambda_{r+1}, \ldots, \lambda_{s-1}, \lambda_s)$. Finally, the concatenation of two tuples $\lambda = (\lambda_0, \ldots, \lambda_m)$, $\mu = (\mu_0, \ldots, \mu_n)$ will be denoted by $\lambda || \mu := (\lambda_0, \ldots, \lambda_m, \mu_0, \ldots, \mu_n)$.

Public key. The public key is the function $G: K^{37} \longrightarrow K^{26}$ defined by

$$
G(X) = [t(\varphi^{-1}(F(\varphi(s(X)))))]_{0 \to 181}.
$$

By construction $(Y_0, \ldots, Y_{25}) = G(X_0, \ldots, X_{36})$ can be expressed in the form

$$
Y_0 = P_0(X_0, \ldots, X_{36})
$$

$$
\vdots
$$

$$
Y_{25} = P_{25}(X_0, \ldots, X_{36})
$$

where each P_i is a polynomial of total degree ≤ 2 with coefficients in K'.

For describing our attack, knowledge about the original signing procedure (which makes use of the secret key) is not necessary. So we omit its description here, and proceed with the description of the (public) verification procedure.

2.2   Verifying a Signature

A signature S of a bitstring M can be represented as a 259-bit string. For verifying the validity, the following steps are to be performed:

1. Compute the 182-bit string

$$V := [\text{SHA-1}(M)]_{0\to159} || [\text{SHA-1}(\text{SHA-1}(M))]_{0\to21}.$$

2. Compute

$$Y := (\pi([V]_{0\to6}), \pi([V]_{7\to13}), \dots, \pi([V]_{175\to181})) \in K^{26}$$
$$Y' := G(\pi([S]_{0\to6}), \pi([S]_{7\to13}), \dots, \pi([S]_{252\to258})) \in K^{26}$$

3. If $Y = Y'$ then the signature is accepted, otherwise it is rejected.

3   Attacking the Affine Parts

First of all it is worth remarking that the public key in SFLASH does not depend on the secret bitstring Δ. Consequently, the above verification procedure does not ensure that the correct value of Δ has been used for computing S. In fact, for forging a signature it is completely sufficient to find affine bijections $s', t' : K^{37} \longrightarrow K^{37}$ such that the public key $G(X)$ satisfies

$$G(X) = [t'(\varphi^{-1}(F(\varphi(s'(X)))))]_{0\to181}.$$

Then a valid signature S' for a bitstring M can be computed as follows:

- compute $Y \in K^{26}$ as above;
- append 11 random elements of K to Y, yielding $\overline{Y}$;
- compute $S' := [s'^{-1}(\varphi^{-1}(F^{-1}(\varphi(t'^{-1}(\overline{Y})))))]_{0\to258}.$

The last 11 components of the affine mapping t' are not used for checking the signature, thus the part of t' corresponding to these 11 components can be chosen arbitrarily such that the resulting t' is invertible.

The aim of the procedure described below is to diminish the number of candidates for the affine parts of s', t' from $2^{37} \cdot 2^{37}$ to (typically) 2^{11} elements. In particular the affine parts of the original secret bijections s and t must be contained in this set of candidates. The attack is based on

Observation 1 *Let* $v := S_L^{-1} \cdot S_C \in K'^{37}$ *and* $\alpha \in K$ *with* $\alpha^7 + \alpha + 1 = 0$.

Then $s((\alpha + 1) \cdot v) = \alpha \cdot S_C \in K^{37}$ *is a vector with coefficients* α *and 0 only, and—owing to the definition of* F—*the vector* $F(\varphi(s((\alpha + 1) \cdot v)))$ *has coefficients* α^2 *and 0 only.*

As the multiplication $T_L \cdot \varphi^{-1}(F(\varphi(s((\alpha + 1) \cdot v))))$ *does not affect this property, the vector* T_C *can be read off directly from*

$$t(\varphi^{-1}(F(\varphi(s((\alpha + 1) \cdot v))))) = T_L \cdot \varphi^{-1}(F(\varphi(s((\alpha + 1) \cdot v)))) + T_C$$

(via the substitution $\alpha^2 \leftarrow 0$*).*

We want to use this observation to derive candidates for T_C and $S_L^{-1} \cdot S_C$ from the given public key $G(X)$. Knowing the correct value of the vectors T_C and $S_L^{-1} \cdot S_C$ we can express the secret parameters s and t in the form

$$s(b_0, \dots, b_{36}) = S_L \cdot ((b_0, \dots, b_{36})^{\mathrm{T}} + \overbrace{S_L^{-1} \cdot S_C}^{\text{known}})$$
$$t(b_0, \dots, b_{36}) = T_L \cdot (b_0, \dots, b_{36})^{\mathrm{T}} + \underbrace{T_C}_{\text{known}} .$$

In other words, breaking SFLASH reduces to finding the linear mappings specified by S_L and T_L.

All the vectors $(\alpha + 1) \cdot \overline{v}$ with $\overline{v} \in K'^{37}$ whose image under the public polynomials contains no α are candidates for $S_L^{-1} \cdot S_C$. For testing this property the quadratic and constant terms of the public key are of no interest $((\alpha + 1) \cdot (\alpha + 1)$ and $0, 1$ do not add up to α). Thus the only relevant parts of the public key are the linear terms. In other words, we are interested in finding all those elements from $\{0, \alpha + 1\}^{37}$ that yield the zero vector when being substituted in the linear part of the public key. Of course, for symmetry reasons, this is equivalent to finding all elements from $\{0, \alpha\}^{37}$ or K'^{37} that yield zero upon substitution in the linear part of the public key. Hence, all we have to do is setting the linear parts of the public key simultaneously to 0; each solution $\overline{v} \in K'^{37}$ of the corresponding homogeneous system of linear equations over K' is a candidate for $S_L^{-1} \cdot S_C$.

If the linear parts of the public key are linearly independent we obtain in this way $2^{37-26} = 2^{11}$ candidates. Of course, if the equations are linearly dependent the number of solutions increases; in several hundred experiments we did with the computer algebra system MAGMA [2] this never happened.

Evaluating the public key at $(\alpha+1) \cdot \overline{v}$ for a candidate $\overline{v}$ for $S_L^{-1} \cdot S_C$ results in a vector whose only non-zero coefficients are α^2 and 1. Moreover, the term 1 occurs in this vector iff there is a 1 in T_C at the appropriate coordinate. In this way we obtain the first 26 entries of the vector T_C corresponding to our candidate $\overline{v}$. As the remaining 11 entries of T_C have no influence on the validity of a signature anyway (and hence can be chosen arbitrarily), this simple procedure reduces the number of candidates for the affine parts in SFLASH from originally 2^{74} to typically 2^{11} possibilities.

4   Conclusion

The above discussion shows that the affine parts of the secret key of SFLASH are vulnerable to a very simple linear algebra-based attack. Namely, instead of considering 2^{74} possible affine parts, an attacker can usually restrict to only 2^{11} candidates. For narrowing the key space in this way knowledge of the public key is sufficient; no signatures have to be known. Although this attack does not break SFLASH in total, we think it raises some questions about the design of this NESSIE candidate.

References

1. F. BAO, R. H. DENG, W. GEISELMANN, C. SCHNORR, R. STEINWANDT, AND H. WU, *Cryptanalysis of Two Sparse Polynomial Based Public Key Cryptosystems*, in Proceedings of PKC 2001, K. Kim, ed., Lecture Notes in Computer Science, Springer, 2001.
2. W. BOSMA, J. CANNON, AND C. PLAYOUST, *The Magma Algebra System I: The User Language*, Journal of Symbolic Computation, 24 (1997), pp. 235–265.
3. J. PATARIN, N. COURTOIS, AND L. GOUBIN, *SFLASH, a fast asymmetric signature scheme for low-cost smartcards. Primitive specification and supporting documentation.* Presented at First Open NESSIE Workshop., November 2000. At the time of writing available electronically at the URL `https://www.cosic.esat.kuleuven.ac.be/nessie/workshop/submissions/ sflash.zip`.
4. ———, *FLASH, a Fast Multivariate Signature Algorithm*, in Progress in Cryptology — CT-RSA 2001, D. Naccache, ed., vol. 2020 of Lecture Notes in Computer Science, Berlin; Heidelberg, 2001, Springer, pp. 298–307.

An Identity Based Encryption Scheme Based on Quadratic Residues

Clifford Cocks

Communications-Electronics Security Group, PO Box 144, Cheltenham GL52 5UE

Abstract. We present a novel public key cryptosystem in which the public key of a subscriber can be chosen to be a publicly known value, such as his identity. We discuss the security of the proposed scheme, and show that this is related to the difficulty of solving the quadratic residuosity problem

1 Introduction

In an offline public key system, in order to send encrypted data it is necessary to know the public key of the recipient. This usually necessitates the holding of directories of public keys. In an identity based system a user's public key is a function of his identity (for example his email address), thus avoiding the need for a separate public key directory. The possibility of such a system was first mentioned by Shamir[4], but it has proved difficult to find implementations that are both practical and secure, although recently an implementation based on elliptic curves has been proposed[3]. This paper describes an identity based cryptosystem which uses quadratic residues modulo a large composite integer.

2 Overview of Functionality

The system has an authority which generates a universally available public modulus M. This modulus is a product of two primes P and Q - held privately by the authority, where P and Q are both congruent to 3 mod 4.

Also, the system will make use of a universally available secure hash function.

Then, if user Alice wishes to register in order to be able to receive encrypted data she presents her identity (e.g. e-mail addresss) to the authority. In return she will be given a private key with properties described below.

A user Bob who wishes to send encrypted data to Alice will be able to do this knowing only Alice's public identity and the universal system parameters. There is no need for a public key directory.

3 Description of the System

When Alice presents her identity to the authority, the hash function is applied to the string representing her identity to produce a value a modulo M such that

B. Honary (Ed.): Cryptography and Coding 2001, LNCS 2260, pp. 360–363, 2001.

the Jacobi symbol $(\frac{a}{M})$ is $+1$. This will be a public process that anyone holding the universal parameters and knowing Alice's identity can replicate. Typically this will involve multiple applications of the hash function in a structured way to produce a set of candidate values for a, stopping when $(\frac{a}{M}) = +1$. Note that the Jacobi symbol can be calculated without knowledge of the factorisation of M. See for example [2].

Thus as $(\frac{a}{M}) = +1$, $(\frac{a}{P}) = (\frac{a}{Q})$, and so either a is a square modulo both P and Q, and hence is a square modulo M, or else $-a$ is a square modulo P, Q and hence M. The latter case arises because by construction P and Q are both congruent to 3 mod 4, and so $(\frac{-1}{P}) = (\frac{-1}{Q}) = -1$. Thus either a or $-a$ will be quadratic residues modulo P and Q. Only the authority can calculate the square root modulo M, and he presents such a root to Alice. Let us call this value r. One way for the authority to determine a root is to calculate

$$r = a^{\frac{M+5-(P+Q)}{8}} \bmod M$$

Such an r will satisfy either $r^2 \equiv a \bmod M$ or $r^2 \equiv -a \bmod M$ depending upon which of a or $-a$ is a square modulo M.

In what follows, I will assume without loss of generality that $r^2 \equiv a \bmod M$. Users wishing to send encrypted data to Alice who do not know whether she receives a root of a or a root of $-a$ will need to double up the amount of keying data they send as described later.

If Bob wants to send an encrypted message to Alice, he first generates a transport key and uses it to encrypt the data using symmetric encryption. He sends to Alice each bit of the transport key in turn as follows:

Let x be a single bit of the transport key, coded as $+1$ or -1.

Then Bob chooses a value t at random modulo M, such that the Jacobi symbol $(\frac{t}{M})$ equals x.

Then he sends $s = (t + a/t) \bmod M$ to Alice.

Alice recovers the bit x as follows:
as $s + 2r = t(1 + r/t) * (1 + r/t) \bmod M$
it follows that the Jacobi symbol $(\frac{s+2r}{M}) = (\frac{t}{M}) = x$.

But Alice knows the value of r so she can calculate the Jacobi symbol $(\frac{s+2r}{M})$, and hence recover x.

If Bob does not know which of a or $-a$ is the square for which Alice holds the root, he will have to replicate the above, using different randomly chosen t values to send the same x bits as before, and transmitting $s = (t - a/t) \bmod M$ to Alice at each step. This doubles the amount of keying data that Bob sends. It would be useful to find a way to avoid having to send this extra information, but at present this is an unsolved problem.

4   Practical Aspects

Computationally, the system is not too expensive. If the transport key is L bits long, then Bob's work is dominated by the need to compute L Jacobi symbols and L divisions mod M. Alice's work mainly consists of computing L Jacobi symbols. For typical parameter values (e.g. $L = 128$ and M of size 1024 bits) and depending upon the implementation this is likely to be no more work than is needed for a single exponentiation modulo M.

The main issue regarding practicality is the bandwidth requirement, as each bit of the transport key requires a number of size up to M to be sent. For a 128 bit transport key, and using a 1024 bit modulus M, Bob will need to send 16K bytes of keying material. If Bob does not know whether Alice has received the square root of a or of $-a$ then he will have to double this. Nevertheless, for offline use such as email this will often be an acceptable overhead.

5   Security Analysis

Clearly, one way to break the system is to factorise M. The fact that this is a weak link means that shared knowledge methods of generating M (see [1] for example) and the use of multiple authorities will be desirable. With shared generation of M it is also feasible to generate the exponent $\frac{M+5-(P+Q)}{8}$ used to compute square roots in a shared fashion, so that no master secret ever needs to exist in a single location.

We study the security of the system on the assumption that M has not been factorised. We consider first a passive attack against the generation of each bit of transport key and show that a weakness would lead to a solution of the quadratic residuosity problem.

Suppose that there is a procedure that recovers x from s without knowing either r or the factors of M. We also assume a constraint on the hash function, that the recovery procedure takes as input the hashed identity a, and can not make separate use of the input to the hash. This excludes obviously weak hash functions, such as one whose final step consists of squaring modulo M. Under this hypothesis the breaking process computes a mapping

$$F(M, a, s) \rightarrow x = (\frac{t}{M})$$

valid whenever $s = (t + a/t) \bmod M$ for some t.

Then consider what the value of F could be if evaluated for an a where the Jacobi symbol $(\frac{a}{M})$ is $+1$, but a is not a square. In this case the Jacobi symbols $(\frac{a}{P})$ and $(\frac{a}{Q})$ will both be -1.

Now, if t was the value used to calculate s, there will be three other values $t1, t2, t3$ giving the same value of s. These are given by:

$$
\begin{array}{ll}
t1 \equiv t \bmod P & t1 \equiv a/t \bmod Q \\
t2 \equiv a/t \bmod P & t2 \equiv t \bmod Q \\
t3 \equiv a/t \bmod P & t3 \equiv a/t \bmod Q
\end{array}
$$

But as $(\frac{a}{P}) = (\frac{a}{Q}) = -1$, then $(\frac{t1}{M}) = (\frac{t2}{M}) = -(\frac{t}{M}) = -(\frac{t3}{M})$.

So, there is no unique $(\frac{t}{M})$ to recover, and so F cannot return $(\frac{t}{M})$ correctly more than half the time whenever a is not a square. Hence we would have a procedure that can distinguish the two cases of $(\frac{a}{M}) = +1$; that is determine whether a is a square or a non square without factoring M. This is the quadratic residuosity problem which is currently unsolved, and is a problem on which a number of other public key systems are based.

Of course, an attacker will in practice be presented with a set of many such terms $(t + a/t) \bmod M$ and possibly also $(t - a/t) \bmod M$ for different values of t. It is desirable that the values of t used are independent and randomly distributed over the set of values consistent with the desired key value. For if successive values of t are related in a systematic way this opens up the possibility of an attack against the set of transmitted values.

The scheme as described so far is vulnerable to an adaptive chosen ciphertext attack. Because the transport key is established one bit at a time, an attacker could take a target transmission and modify the component corresponding to just one bit of transport key at a time, changing it to produce a transport key value known to the attacker. By observing the decrypt to see whether this changes the transport key the attacker could recover the transport key a bit at a time.

I sketch here an outline of how one might block such attacks. The approach is to add redundancy to the transport key establishment data so that only a small proportion of randomly chosen messages will decrypt in a valid way, and arrange that if the recipient is presented with an invalid message then the only output will be the information that the message is invalid. This should be done in a way that prevents an attacker devising challenges which may be of use in an attack. For the system described here we propose sending, suitably encrypted, data that will allow the t values to be reconstructed and then checked by the recipient, along with a cryptographic hash of those t values. This string would be produced separately for the two cases of square root (a and $-a$ respectively) that may be held by the recipient.

References

1. C Cocks *Split Generation of RSA Parameters with Multiple Participants* Proceedings of 6th IMA conference on Cryptography and Coding, Springer LNCS 1355.
2. H Cohen *A Course in Computational Algebraic Number Theory* Springer-Verlag
graduate texts in mathematics 138, 1993
3. D Boneh, M Franklin *Identity-Based Encryption from the Weil Pairing* Advances
in Cryptology - Crypto2001, Springer LNCS 2139
4. A. Shamir *Identity Based Cryptosystems and Signature Schemes* Advances in Cryptology - Proceedings of Crypto '84.

Another Way of Doing RSA Cryptography in Hardware

Lejla Batina and Geeke Muurling

Pijnenburg SECUREALINK B. V.,
Boxtelseweg 26, 5261 NE,
Vught, The Netherlands
{l.batina, g.muurling}@securealink.com

Abstract. In this paper we describe an efficient and secure hardware implementation of the RSA cryptosystem. Modular exponentiation is based on Montgomery's method without any modular reduction achieving the optimal bound. The presented systolic array architecture is scalable in several parameters which makes it possible to implement Compaq's $MultiPrime^{TM}$ in a very efficient way. According to a developed performance model the influence of different parameters is investigated. This platform is optimised for Multiprime as an example for the RSA cryptosystem. In this work we give details about this scheme, which uses three or more factors of the composite N. Security of this scheme, related to this architecture is also presented.

Keywords: Montgomery multiplication, modular exponentiation, MultiPrime, Chinese Remainder Theorem, systolic array, performance model, scalability

1 Introduction

The basic operation in RSA cryptosystems is modular exponentiation, which is based on a repeated modular multiplication. So implementing this operation in an efficient way makes numerous application possible.

In 1985 Peter Montgomery introduced a new method for modular multiplication. This operation is widely used in most cryptographic protocols (public and private key cryptography). The approach of Montgomery avoids the time consuming trial division that is a common bottleneck of other algorithms. His method proved to be very efficient and is the basis of many implementations of modular multiplication, both in software and hardware. In this paper we are interested in a hardware implementation.

Our contribution is in combining a systolic array architecture with Montgomery based RSA implementation. This design has evolved to a secure and efficient cryptographic device used in different applications.

Various definitions are possible, when introducing scalability. It is usually referred to as the ability to process a variety of number lengths at the same time. Although this platform consists of an array of fixed length it is scalable according to the previous definition.

B. Honary (Ed.): Cryptography and Coding 2001, LNCS 2260, pp. 364–373, 2001.

The performance model shows two-fold behaviour. It can be shown that for specific modulus lengths the performance is quadratic in n. For larger modulus size the performance is cubic. The fixed word size causes jumps in the performance curve. More PEs do not always result in a better performance. Smaller parameters make the design more scalable in the sense that it has a better overall performance. A larger number of LNAUs generally improves the performance but does not change the latency. In this work we describe an efficient implementation of modular exponentiation in a systolic array architecture, as part of the RSA cryptography. We use the methods of Montgomery, proved to be very secure in hardware. Namely, we achieved the optimal bound which, with some savings in hardware, omits completely all reduction steps that are presumed to be vulnerable to side-channel attacks. The presented implementation has a world class performance in the sense of speed and power consumption.

The remainder of this paper is organized as follows. Section 2 gives a survey of previous work on systolic arrays and Montgomery based operations in hardware. In Section 3, we give a short introduction to Multiprime and the Chinese Remainder Theorem. In Section 4 we introduce the architecture of the targeted platform and its specific properties, one of which is the scalability of the device. Section 5 describes the performance with an example of 1024 bit RSA and Multiprime. In Section 6 the security of the proposed architecture is discussed. This includes the latest developments on the optimisation of the Montgomery Algorithm. Section 7 concludes the paper.

2   Previous work

Some of the most relevant previous contributions on modular multiplication and systolic arrays are reviewed in this section. The earliest work on hardware implementation of the Modular Multiplication Method (MMM) of P.L.Montgomery is [5]. The authors have shown efficient use of hardware. The work of Iwamura et al. ([8]) followed, as the first one at our knowledge presenting a systolic array performing the modular exponentiation operation using the Montgomery modular multiplication. This work is relevant for our (as for many others) architecture, which however went further in efficiency. The usual bottleneck for hardware implementations of Montgomery's algorithm, is the fact that the number of output bits may exceed the number of input bits. Iwamura et al. derived the following bound for the Montgomery parameter: $R > 2^{n+2}$ where $R = 2^r$. For this value of R the examination of the size of the output each time the Montgomery method is executed, may be omitted. Here, n denotes the number of bits of N. It can be shown how to improve on this with the condition $R > 4N$ [4], which is according to work presented in [16] assumed to be the best possible bound in practice. In Section 6 this will be explained in detail. A scalable architecture is introduced in [3]. There is no limitation on the maximum number of bits manipulated by the multiplier, and the selection of the word size can be freely chosen according to desirable criteria. Our architecture is also considered scalable according to this definition which has various possibilities, such as doing RSA (including

Multiprime) and Elliptic Curves Cryptography on the same chip. However, we are focused more on the broader meaning of scalability, which will be explained further in the following section.

In [15], which is further improved in [16], the author showed that the Montgomery exponentiation method requires no final subtraction, which is very important for fast implementation. Another benefit is that conditional statements, which may be subject to side-channel attacks such as timing attack, power analysis attack etc. may be omitted. Some other results considering a constant time implementations which is presumed first step towards secure hardware solutions are proposed in [7]. However, the result of [16] presents the best possible bound. This bound is also practical and implemented within our architecture.

3   Multiprime

In April 2000 Compaq and RSA Security Inc. announced a new patented technology $MultiPrime^{TM}$ ([2]), as a generalization of the RSA protocol [13]. Instead of a modulus of $N = p * q$, as in traditional RSA system, N is a product of three or more (distinct) prime numbers. The idea was that increasing the number of factors and using the CRT with parallel exponentiators should increase performance. In general, dependence of the performance of modular exponentiation and the length of modulus is not linear. It is approximately cubic or quadratic depending on the implementation, which will be discussed afterwards. The suitable performance model is designed which is represented with the performance curve. This curve is changing from quadratic to cubic behaviour in relation with the length of modulus for every platform.

However, a high level of security has to be preserved. The length of N is not the only relevant factor that provides it, because having smaller factors makes some methods of factoring more efficient. This observation limits the number of prime factors of N. In [2] the tradeoff between efficiency and security results in up to 4 prime factors of N for the modulus lengths of 2048 bits.

4   Scalable Systolic Array

In this section we describe the architecture of the PCC-ISES ([1]). The PCC-ISES is an integrated circuit with an architecture that is very suitable for modular multiplication. For this modular multiplication Montgomery's method is chosen and the notation is as follows:

$$Mont(X, Y) = XYR^{-1} \mod N$$

4.1   Systolic array

The design contains two identical Large Number Arithmetic Units (LNAU), each designed as a systolic array. This array is one-dimensional and consists of a fixed number of Processing Elements (PEs). In the remainder of this paper P denotes the number of PEs. This architecture can be scaled to every desired configuration. A PE consists of some adders and multipliers that can process α bits of X and β bits of Y (α and β are not necessarily of the same length) in one clock cycle. So, in one clock cycle several additions and multiplications can be performed in all PEs which differs from relevant work of other authors. More precisely, a chip based on our architecture performs one loop of the multiplication algorithm in one cycle. (Other authors usually consider one addition or multiplication in one cycle.) In this way the two LNAUs of the PCC-ISES provide extremely fast RSA protocols. Specific commands are defined for modular multiplication and exponentiation, which can be used by the ARM7 processor to access the accelerator. Actually, this hardware accelerator can perform two modular multiplication operations at the same time, which provides the possibility of implementing algorithms in parallel.

4.2 Scalability

Scalabiltiy may be defined as the property that a variety of number lengths can be processed on the same platform. Although the PCC-ISES consists of an array of fixed length it is scalable according to the previous definition. If the operands are "too large" to fit in the available number of PEs the intermediate result of the last PE is fed into the first PE. These intermediate results are temporarily stored in a FIFO memory, if necessary.

One advantage of this scalability is that different modulus lengths can be processed on the same hardware device without having to pad to a larger modulus. This means that Multiprime as well as standard RSA can be used on the same system.

In this paper we only discuss modular exponentiaton. However the PCC-ISES supports more cryptographic functionality. For the sake of completeness a short description of this platform is given.

PCC-ISES has the following characteristics: Embedded Cryptographic Accelerators with 2 LNAUs capable of performing up to 2048-bit modular arithmetic, Embedded microprocessor ARM7, 128 KB embedded RAM, and other features required for various cryptographic applications. For more details the reader is referred to [1].

5   Performance

To examine the behaviour of all PCC-ISES scenarios a performance model has been developed. In the first part of this section this model is presented. In the second part the model is used to display some results for different keylengths and parameter settings.

5.1   Performance model

Depending on the targeted application parameters of this architecture can be chosen. There is always a trade-off between the size of the IC and the performance. To understand the influences of the different parameters a performance model is made. The modular exponentiation is implemented as a repeated Montgomery multiplication in the sense of some of the square-and-multiply methods. Re-coding of the exponent is one possibility to reduce the number of multiplications but this number stays of order of the exponent length [13]. The overall performance is also determined by the performance of a single multiplication. The inputs of the multiplication are divided into words, not necessarily of the same length. The words of X are distributed over the PEs. When there are not enough PEs more rounds are needed. So the number of rounds is $\lceil \frac{n'}{P} \rceil$, where n' is the number of words X of length α bits. In one round each word of Y has to pass all PEs. Each PE calculation takes one clock-cycle. So, in P clock-cycles a word of Y passes all PEs in the array. When the number of words of Y is larger than the number of PEs the FIFO memory is used to store the intermediate results of the last PE. Now each round costs $max(P, n'')$ cycles, where n'' is the number of words of Y of length β bits. The performance (number of clock-cycles for one exponentiation) is now modeled as:

$$\lceil \frac{n'}{P} \rceil \cdot max(P, n'') \cdot c_1 \cdot n$$
$$n' = \lceil \frac{n}{\alpha} \rceil, \; n'' = \lceil \frac{n}{\beta} \rceil \tag{1}$$

Here n is the length of the modulus in bits and $c_1 \cdot n$ is the number of multiplications needed for the exponentiation. The value for c_1 is typically 1.5 for the well known left to right square and multiply exponentiation method. When using exponent re-coding the value of c_1 is usually 1.2 for an exponent with about half of the bits being zero. It is easy to conclude from (1) that for modulus lengths with $n'' < P$ the performance is quadratic in n. The performance is cubic in the modulus length for larger modulus sizes. The fixed word size of α and β bits (the ceiling function in the model) causes jumps in the performance curve (see Figure 3).

The model shows that more PEs is not always the best option. The minimal number of PEs requiring only one round for a certain modulus length seems to be the most efficient solution. On the other hand, the number of PEs is restricted because of maximal implementation complexity. The performance can be further improved by implementing more than one LNAU. The LNAUs perform exponentiations simultaneously and separately, so the performance is proportional in the number of LNAUs.

5.2   Performance figures

When designing a new platform, many factors should be taken into account. First of all, the targeted modulus lengths should be addressed. Different modulus lengths behave differently regarding all parameters involved. The parameters should be chosen in such a way that they are in overall performance the most beneficial for all modulus lengths.

Let us consider an RSA encryption protocol with 1024 bits for 2 or 3 prime factors. In the first case p and q are 512 bits long and for the Multiprime case p, q and r are each 342 bits long. The influence of the number of PEs on the performance in both of these cases is displayed in Figure 1.

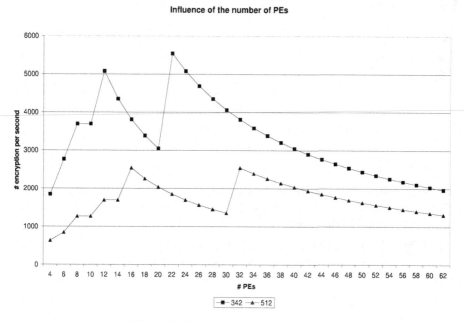

Fig. 1. Influence of the number of PEs

In this graph the peaks represent the extreme values of the ceiling function for the number of rounds. For example the maximal value between 5000 and 6000 encryptions per second stands for the beginning of the interval in which, for the given modulus size, only one round is needed. The second peak in the range of 5000-6000 marks the beginning of the interval for which 2 rounds are required. When one is considering the most efficient (standard) RSA implementation the best choice for the number of PEs would be 32 or 16 as the performance curve has a maximum for these two values. In the case of Multiprime the preferred choice is 22 or 11 PEs. Having in mind doing both types of RSA cryptography on the same platform both curves should be taken into account (Figure 2). In this case the optimal performance is achieved for any of the following numbers: 32 and 16 PEs.

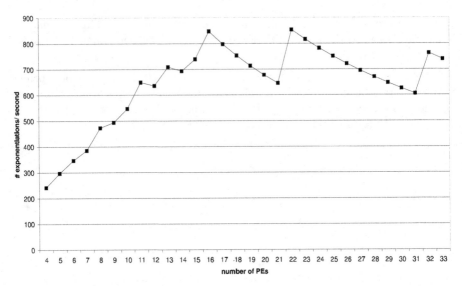

Fig. 2. Sum of the performance-curves for 2 and 3 primes

Figure 3 shows the performance curve for a modular exponentiation on a LNAU with 23 PEs and $\alpha = \beta = 16$. The effect of the ceiling functions can be observed as jumps in the graph.

Similar behavior of the performance figures can be found in [3].

Table 1 presents the timings of the core exponentiations for various modulus lengths.

	mod. exp.	1024 bit N	2048 bit N
342 bits	1400	467	-
512 bits	731	365	18
683 bits	448	-	149
1024 bits	167	167	84
2048 bits	22	-	22

Table 1. Timings on the PCC-ISES with 2 LNAUs and clock-frequency of 50 Mhz.

In the case of 1024 bit RSA, Multiprime will be faster with a factor of approximately 1.27, since the expected timing for three 342 bit exponentiations is 467 per second and for two 512 bit exponentiations is 365 per second. These

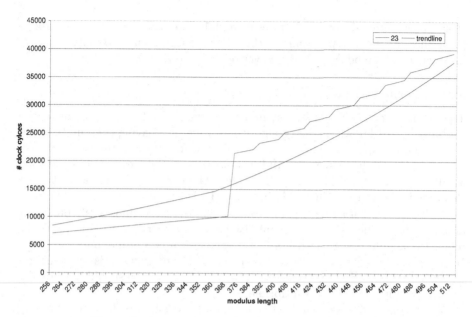

Fig. 3. Performance curve for 23 PEs and its trendline.

timings are not including overhead caused by the re-combination algorithm of CRT.

6   Remarks on bound and security

6.1   Bound for the Montgomery parameter R

In [16] the need of avoiding reduction after each multiplication is addressed. In practice this means that the output of the multiplication can be directly used as an input of the next Montgomery multiplication. We want to find a bound on R such that with $X, Y < 2N$ the output of the Montgomery multiplication $T < 2N$. Write $R \geq kN$, then:

$$T = \frac{XY + mN}{R} = \frac{XY}{R} + \frac{m}{R}N < \frac{4}{k}N + N \qquad (2)$$

where, $m = (XY \mod R)N' \mod R$.

Hence, $T < 2N$ for $k \geq 4$, implying: $4N \leq R$. To guarantee the existence of the modular inverse of R, R and N should be relatively prime. This excludes $4N = R$.

The result of a Montgomery multiplication $XYR^{-1} \mod N < 2N$ when $X, Y < 2N$ and $R > 4N$.

This is the same result as disclosed in [16]. The final round in the modular exponentiation is the conversion to the integer domain, i.e. calculating the

Montgomery function of the last result and 1. The same arguments as above prove that this final step remains within the following bound: $Mont(T, 1) \leq N$. In practice, $A^B \mod N = N$ will never occur since $A \neq 0$.

6.2 Security issues

The security of the RSA crytosystem depends on the difficulty of factoring the modulus N. All known factoring algorithms are either having running time dependent on the size of the factors or only on the size of the modulus N. Hence, when looking at the security of Multiprime one should be especially concerned about the first type of factoring algorithms. The currently fastest algorithm of this type is the Elliptic Curve Method (ECM). It's asymptotic running time is: $exp(O((ln(p))^{\frac{1}{2}} \cdot (lnln(p))^{\frac{1}{2}}))$ [14]. For example, the ECM finds a 167-bit factor of a 768-bit number with the probability 0.63 after spending 6200 Mips-Years, under the assumption that such a factor exists for a two prime modulus ([12]). However, if there are more than two primes, this probability is even higher. Therefore, suggested lower bound for RSA keysizes as in [12] should be carefully reconsidered in the case of Multiprime.

When considering side channel attacks, most attention is directed to timing ([6]) and power analysis attacks ([11]). Ever since P.Kocher introduced a new type of attack ([10]), the so-called Differential Power Analysis attack (DPA for short), reasonable amount of research has been done on this subject. The same author published in 1996 ([9]) one of the relevant papers on time-difference based attacks. Namely, computations performed in non-constant time (usually because of performance optimisations) may leak secret key information. This observation is the basis for timing attacks. On the other hand, power analysis based attacks use the fact that the power consumed at any particular time during a cryptographic operation is related to the function being performed and (possibly sensitive) data being processed.

The benefit of using the method of Montgomery for multiplication is evident for all types of algorithms. The reason for that is the following: a modular reduction step may also be vulnerable and in the case of MMM at most one modular reduction is introduced for every multiplication or per step of exponentiation. In our implementation even these reductions are excluded. By use of an optimal upper bound the number of iterations required in the MMM can be reduced ([17]). More precisely, some savings in hardware that have been included in our architecture avoid the conditional statements while performing the exponentiation. In that way, our implementation of modular exponentiation operates in constant time, which is presumed to significantly reduce the potential risk of the timing attack. However, while using CRT the modular multiplication in general may need a final reduction.

7 Conclusions

We have described one application of RSA cryptosystem in hardware by the use of a systolic array architecture. The used method for modular multiplication is the one of Montgomery, which is proved to be more secure in hardware when considering timing attacks. We showed the possible gain of Multiprime comparing to standard RSA which is related to the platform.

References

1. PCC-ISES datasheet, www.secure-a-link.com /pdfs/isespdf/ isesdata.pdf.
2. www.compaq.com.
3. A.F.Tenca and C.K. Koç. A scalable architecture for Montgomery multiplication. *Lecture Notes in Computer Science, Springer-Verlag*, (1717):94–108, 1999. Cryptographic Hardware and Embedded Systems-CHES 1999.
4. L. Batina and G. Muurling. Montgomery in practice: How to do it more efficiently in hardware. *RSA 2002 - San Jose, to appear*, 2002.
5. S.E. Eldridge and C.D. Walter. Hardware implementation of Montgomery's modular multiplication algorithm. *IEEE Transactions on Computers*, (42):693–9, 93.
6. G. Hachez, F. Koeune, and J.-J. Quisquater. Timing attack: what can be achieved by a powerful adversary? *Proceedings of the 20th symposium on Information Theory in the Benelux*, pages 63–70, May 1999.
7. G. Hachez and J.-J. Quisquater. Montgomery exponentiation with no final subtractions: Improved results. *Lecture Notes in Computer Science, Springer-Verlag*, (1965):293–301, 2000. Cryptographic Hardware and Embedded Systems-CHES 2000.
8. K. Iwamura, T. Matsumoto, and H. Imai. Montgomery modular multiplication method and systolic arrays suitable for modular exponentiation. *Electronics and Communications in Japan*, 77(3):40–50, 1994.
9. P. Kocher. Timing attacks on implementations of Diffie-Hellman, RSA, DSS and other systems. *Lecture Notes in Computer Science, Springer-Verlag*, pages 104–113, 1996. Advances in Cryptology-CRYPTO 96.
10. P. Kocher, J. Jaffe, and B. Jun. Introduction to differential power analysis and related attacks. www.cryptography.com/dpa/technical, 1998.
11. P. Kocher, J. Jaffe, and B. Jun. Differential power analysis. *Lecture Notes in Computer Science, Springer-Verlag*, pages 388–397, 1999. Advances in Cryptology-CRYPTO 99.
12. A. Lenstra and E. Verheul. Selecting cryptographic key sizes. *In Hideki Imai and Yuiliang Zheng, editors, Third International Workshop on Practice and Theory in Public Key Cryptography - PKC 2000*, LNCS 1751.
13. A. Menezes, P. van Oorschot, and S. Vanstone. *Handbook of Applied Cryptography*. CRC Press, 1997.
14. A. Shamir. Rsa for paranoids. *RSA Laboratories' CryptoBytes*, 1(3):1–4, 1995.
15. C.D. Walter. Montgomery exponentiation needs no final subtraction. *Electronic letters*, 35(21):1831–1832, October 1999.
16. C.D. Walter. Safely reducing the number of iterations in Montgomery modular multiplication. 2001. preprint.
17. C.D. Walter and S. Thompson. Distinguishing exponent digits by observing modular subtractions. *Lecture Notes in Computer Science, Springer-Verlag*, (2020):192–207, 2001. Topics in Cryptology - CT-RSA 2001.

Distinguishing TEA from a Random Permutation: Reduced Round Versions of TEA Do Not Have the SAC or Do Not Generate Random Numbers

Julio César Hernández[1], José María Sierra[1], Arturo Ribagorda[1], Benjamín Ramos[1], and J.C. Mex-Perera[2]

[1] Carlos III University, Computer Security Group, 28911 Leganés, Madrid, Spain
{jcesar, sierra, arturo, benja1}@inf.uc3m.es
[2] Bradford University, Cryptography & Computer Communications Security Group, Bradford, UK
J.C.Mex-Perera@brad.ac.uk

Abstract. In this paper the authors present a statistical test for testing the strict avalanche criterion (SAC), a property that cryptographic primitives such as block ciphers and hash functions must have. Random permutations should also behave as good random number generators when, given any initial input, its output is considered part of a pseudorandom stream and then used as an input block to produce more output bits. Using these two ideal properties, we construct a test framework for cyptographic primitives that is shown at work on the block cipher TEA. In this way, we are able to distinguish reduced round versions of it from a random permutation.

1 Introduction

Many cryptographic primitives (a cryptographic hash function, a block cipher with any fixed key, etc...) need to behave as random permutations to be considered usable. In fact, any method of distinguishing them from a random permutation is considered a weakness and the result of a successful cryptanalysis, even though the results may not be directly used in the recovery of plaintext or key bits. In this paper, we will present a method to show how reduced round versions of the block cipher TEA can be distinguished from a random permutation.

1.1 Testing for the Strict Avalanche Criterion

Any random permutation f must have the SAC property, which we can verify with a statistical test. If f has the SAC, then it should pass the test that this pseudocode describes:

B. Honary (Ed.): Cryptography and Coding 2001, LNCS 2260, pp. 374–377, 2001.

```
While (#Texts<NUMTEXT)
{ Randomly generate an input vector V;
  Randomly choose a position p in V;
  Flip the content of the position p in V to get V';
  Calculate  h  the  hamming  distance  between  f(V)  and
f(V') and increase the number observed_values[h];
};
```
Compute the Chi-Square statistic t over the results stored in observed_values[h].

Perform a hypothesis contrast to test if the observed distribution (the t value) is consistent with the expected probability distribution, that should be a Binomial(n, 0.5) being n=length of the output of f.

1.2 Random Permutation Generate Random Numbers

Additionally, if f is a random permutation, independently on the initial vector I_0=IV the recurrence relation:

$$I_0 = IV$$
$$I_{n+1} = f(I_n)$$

must produce a sequence (I_1, I_2, I_3,.....) of essentially random numbers.

1.3 Testing SAC with Autofeeding

This sequence of random numbers can be seen as a stream of random bits, that could be used, for example, to generate V and p in the SAC test described above, a technique we call autofeeding. In this way, we will essentially double-check the strength of the cryptographic primitive. This can be used to mount a framework to test functions.

In this way, any result statistically distinguishable from the expected will prove that the f function has not SAC or is not a good random number generator. In any case, it will be a method of effectively distinguishing f from a random permutation and implies f should be considered not adequate for cryptography.

2   Results

We have tested reduced round versions of the block cipher TEA with the procedure described above. For this particular case, the output of TEA has length 64 bits, so the hamming distance between V and V' is a random variable that should take any value k between 0 and 64 with probability

$$\frac{64!}{k!(64-k)!2^{64}}$$

that is, the distribution of the hamming distance between V and V' should be a binomial with n=64 and p=0.5

To test if this is consistent with the observed results, we performed 10 tests and calculated its average, as seen in the tables below:

Table 1. TEA with 1 round (average of 10 tests) with autofeeding

# Tests	Average
2^6	29.93
2^7	106.41
2^8	240.65

Table 2. TEA with 2 rounds (average of 10 tests) with autofeeding

# Tests	Average
2^6	29.75
2^7	104.19
2^8	231.61

Table 3. TEA with 3 rounds (average of 10 tests) with autofeeding

# Tests	Average
2^6	16.91
2^7	55.66
2^8	163.55
2^9	498.02

Table 4. TEA with 4 rounds (average of 10 tests) with autofeeding

# Tests	Average
2^6	3.20
2^7	10.20
2^8	22.79
2^9	32.01
2^{10}	72.62
2^{11}	118.33
2^{12}	248.20

Table 5. TEA with 5 rounds (average of 10 tests) with autofeeding

# Tests	Average
2^6	5.32
2^7	12.06
2^8	17.20
2^9	19.33
2^{10}	19.24
2^{11}	23.58
2^{12}	21.17
2^{13}	22.90
2^{14}	23.99
2^{15}	26.65
2^{16}	26.50
2^{17}	27.89
2^{18}	30.22
2^{19}	31.15
2^{20}	38.75
2^{21}	42.42
2^{22}	53.15
2^{23}	66.75
2^{24}	92.50
2^{25}	132.56

All the tables conclude approximately when the average of 10 different tests is higher than 93.216 (in light grey), which is the chi-square statistic for an alpha of 99% with 64 degrees of freedom.

3   Conclusions

As seen in the tables above, we can distinguish the block cipher TEA from a random permutation for up to 5 rounds. As expected, increasing the number of rounds also increases the number of encrypted texts needed to observe a significant deviation from the ideal behaviour.

Currently we are working in 6 rounds TEA and we feel quite sure we will be also able to distinguish it from a random permutation. This is probably the limit of this approach on the TEA block cipher. For 7 rounds or more, we believe this approach is not capable of distinguishing TEA from a random permutation, at least in this form.

Anyway, we feel our proposal is useful and will probably produce other interesting results when used to test other cryptographic primitives.

A New Search Pattern in Multiple Residue Method (MRM) and Its Importance in the Cryptanalysis of the RSA

Seyed J. Tabatabaian, Sam Ikeshiro, Murat Gumussoy, and Mungal S. Dhanda

Panasonic, Matsushita Mobile Communications Development of Europe Ltd.,
2 Gables Way, Colthrop, Thatcham, Berk., RG19 4ZD, UK

seyed.tabatabaian@mci.co.uk sam.ikeshiro@mci.co.uk.
murat.gumussoy.@mci.co.uk, mungal.dhanda.@mci.co.uk,
http://www.panasonicmobile.com

Abstract. This paper presents a cryptanalysis attack on the RSA cryptosystem. The method, Multiple Residue Method (MRM), makes use of an algorithm which determines the value of $\phi(n)$ and hence, for a given modulus n where $n = p \times q$, the prime factors can be uncovered. This algorithm calculates and stores all possible residues of p, q and $(p+q)$ in different moduli. It then applies the Chinese Remainder Theorem (CRT) to different combinations of residues until the correct value is calculated, [6]. Further properties in relation to this structure show that improvements in the search process, within the residue of all parameters involved, can be effectively achieved. Besides, it has been established that the security of the RSA is no greater than the difficulty of factoring the modulus n into a product of two secret primes p and q. However, the MRM approaches the factorisation problem from a different angle. This method is aimed at finding towards the $\phi(n)$ in $O(2^{-j} \times n)$, where j is the number of prime moduli. It may also be directed towards the computation of the sum $(p+q)$ and, in the realistic case for the RSA, reduces to $O(2^{-j} \times \sqrt{n})$.

1 Introduction

The concept of public-key cryptography, introduced in 1976 by Diffie and Hellman, provides a proper solution to the problem of key distribution. Their scheme makes use of the apparent difficulty of computing logarithms over finite Galios finite field GF(p) where p is a prime [1]. Since then, many implementations of public-key cryptography have been proposed. One example is the RSA scheme. In the RSA public key cryptosystem, [2], two large primes p and q are chosen randomly to give n where $n = p \times q$. An encryption key, e, is the inverse of the decryption key, d, mod $\phi(n)$. The decryp-

B. Honary (Ed.): Cryptography and Coding 2001, LNCS 2260, pp. 378-386, 2001.

tion key, e, is selected from the interval [2, $\phi(n)$-1], where $\phi(n)$ is the Euler totient function of n and is given by

$$\phi(n) = (p-1)(q-1) \tag{1}$$

e and $\phi(n)$ must be relatively prime. Also e and n are made public but the decryption key, d, is private and is chosen such that

$$e \times d \equiv 1 (\mathrm{mod}\, \phi(n)) \tag{2}$$

In practice p and q are chosen to be $O(\sqrt{n})$ and therefore *general* factorisation methods tend to be more successful than *specific* ones. These two terms, *general* and *specific*, will be explained in the next section.

It is generally assumed that the system's security relies on the difficulty of factoring n. Nevertheless, given the factorisation problem and the length of n, of the order of 200 digits, it is equally hard to compute the value of $\phi(n)$ given n, [2]. Although some of the attacks on the RSA consider finding the value of d by studying different ciphertexts, some others, on the other hand, are aimed at evaluating the two prime factors, p and q. One such attack can be used to factor the modulus when the prime factors of either $p-1$ or $q-1$ are all small [3]. Similarly, another attack can also factor the modulus when the prime factors of either $p+1$ or $q+1$ are small but using a different method [4].

In this paper a new search pattern geared towards finding $(p+q)$ and/or $\phi(n)$ for a given modulus n is presented. Each time a combination of the residues of $(p+q)$ and/or $\phi(n)$ are selected (from different prime fields), have CRT applied using precomputed tables, and are then checked against the square root function in order to find the correct value of $(p+q)$ and/or $\phi(n)$. It has been investigated that there are different search patterns through all possible residues that minimise the search effort: one of these search pattern is discussed in the following sections.

Since the RSA system was introduced there have been many attacks aimed at different weaknesses in the system. To nullify those attacks some constrains have been put on the system selection of parameters of the system. Nevertheless, a general method based on the multiple residues of all parameters involved in the RSA has been introduced [6]. This method discloses the secret parameters of the RSA system regardless of how carefully they may have been chosen. The method is based on the fact that the factorisation problem is effectively decomposed into an arbitrary number of shorter finite field factorisations of the different residues of n. It is apparent that the properties of both $\phi(n)$ and $(p+q)$ may be exploited in terms of the reduced number of possible combinations of residues. The computation of this algorithm can be readily mapped onto a parallel processing architecture and tables of precomputed residues will avoid performing any real-time factorisation operation.

The proof of this structure is based on a Multiple Residue Method (MRM) which relies upon the application of various number theoretic concepts and properties of the prime numbers, moduli, congruences, and the Chinese Remainder Theorem (CRT) [6]. Before discussing the search pattern through the MRM it is essential (necessary) to introduce this method.

2 The Multiple Residue Method

An odd positive integer factoring algorithm can be loosely classified as being either *specific* or *general*, [1]. A specific method uses some special property of the factors being searched for and first discovers those which possess that property. For instance, the divide and factor method, [1], and Pollard ρ-method, [2], will usually find factors which are small before those that are large. The Pollard $p-1$ method, [2], will converge towards a factor p for which the prime factors of $p-1$ are small. On the other hand, in a *general* method, Multiple Residue Method (MRM), the probable computation time to find the factors is independent of their magnitude. For example, in a perfect *general* method as much time will be spent in finding the small factors as in finding any of the other factors.

The basis of the Multiple Residue Method (MRM) is to write the residue of an odd composite number, $n = p \times q$, in terms of the residues of p and q, in different prime fields

$$\langle n \rangle_{m_1} = \langle p_{11}.q_{11} \rangle_{m_1} \vee ... \vee \langle p_{1i_1}.q_{1i_1} \rangle_{m_1}$$

$$\langle n \rangle_{m_2} = \langle p_{21}.q_{21} \rangle_{m_2} \vee ... \vee \langle p_{2i_2}.q_{2i_2} \rangle_{m_2}$$

$$\vdots$$

$$\langle n \rangle_{m_k} = \langle p_{k1}.q_{k1} \rangle_{m_k} \vee ... \vee \langle p_{ki_k}.q_{ki_k} \rangle_{m_k}$$

$$\vdots$$

$$\langle n \rangle_{m_j} = \langle p_{j1}.q_{j1} \rangle_{m_j} \vee ... \vee \langle p_{ji_j}.q_{ji_j} \rangle_{m_j} \tag{3}$$

where $\langle n \rangle_{m_k}$ denotes the residue of n modulo m_k, and m_j is the jth prime, starting from 2, 3, 5, ... Also i_k is the number of residue pairs corresponding to the kth prime. Equation 3 may then be used to calculate the values of residue pairs for different moduli, examples of which are shown in Tables 1, 2, 3 and 4.

Table 1. Combinations of the residue pairs in different prime moduli, $m_j = 2$

$\langle n \rangle_2$	$\langle p, q \rangle_2$
1	(1, 1)

Table 2. Combinations of the residue pairs in different prime moduli, $m_j = 3$

$\langle n \rangle_3$	$\langle p, q \rangle_3$	
1	(1, 1)	(2, 2)
2	(1, 2)	-

Table 3. Combinations of the residue pairs in different prime moduli, $m_j = 5$

$\langle n \rangle_5$	$\langle p, q \rangle_5$		
1	(1, 1)	(2, 3)	(4, 4)
2	(1, 2)	(3, 4)	-
3	(1, 3)	(2, 4)	-
4	(1, 4)	(2, 2)	(3, 3)

Table 4. Combinations of the residue pairs in different prime moduli, $m_j = 7$

$\langle n \rangle_7$	$\langle p, q \rangle_7$			
1	(1, 1)	(2, 4)	(3, 5)	(6, 6)
2	(1, 2)	(3, 3)	(4, 4)	(5, 6)
3	(1, 3)	(2, 5)	(4, 6)	-
4	(1, 4)	(2, 2)	(3, 6)	(5, 5)
5	(1, 5)	(2, 6)	(3, 4)	-
6	(1, 6)	(2, 3)	(4, 5)	-

Also from Equation 3, a similar structure is derived for $\phi(n)$ and $(p+q)$, the result of which is stored in a number j of residue tables such as that shown in Table 5 for a modulus of 7.

Table 5. Residues of $(p+q)$ in mod 7

$\langle n \rangle_7$			$\langle p+q \rangle_7$		
1	2	6	1	5	
2	3	6	1	4	
3	4	0	3	-	
4	5	4	2	3	
5	6	1	0	-	
6	0	5	2	-	

To reach the actual value of $(p+q)$ using these tables, all potential values for $(p+q)$ are first evaluated by applying the CRT to appropriate combinations of these values from the tables. Each combination consists of a set of elements, each of which is drawn out of a separate residue table.

The values of $(p+q)$ should be determined by checking the solution of the following equation for a perfect square

$$p-q=\sqrt{(p+q)^2-4n} \qquad (4)$$

This task can be effectively carried out by using the square-rooting algorithm introduced in [8].

The same technique can be applied to possible residue of $p-1$ and $q-1$, for a given modulus n, to obtain possible values of the Euler totient function, $\overline{\phi(n)}$, in order to determine the correct $\phi(n)$ and thus break the RSA cryptosystem. Similarly the value of $\overline{\phi(n)}$ can be checked using the following equation

$$p-q=\sqrt{(n-1)-\overline{\phi(n)}^2-4\overline{\phi(n)}} \qquad (5)$$

However, the laborious part of this method is finding the right combination of residues. In the following section improvements are made to reduce the computational effort required to determine $(p+q)$ or $\phi(n)$, particularly the search pattern through the residues.

To determine the computational effort required in finding the actual value of $(p+q)$, we need to estimate the number of possible combinations, C. By referring to the precomputed Tables 1, 2, 3, and 4 it can be seen that the number of possible residues in each row of this table is

$$(m_k-1)/2 \le i_k \le (m_k+1)/2 \qquad (6)$$

Hence,

$$C = (\tfrac{m_1 \pm 1}{2})(\tfrac{m_2 \pm 1}{2})...(\tfrac{m_j \pm 1}{2}) \tag{7}$$

where $\pm$ indicates the maximum and minimum number of elements in different prime moduli. However, the results of factoring several large random numbers show that in practice C is given by

$$C \leq 2^{-j} \times \prod_{k=1}^{j} (m_k - 1) \tag{8}$$

Or

$$C \leq 2^{-j} \times m \times S \tag{9}$$

In order to calculate the value of C, the value of S is computed first by using Merten's theorem [7]. This states that

$$S = \prod_{2 \leq m_k \leq m_j} (1 - \frac{1}{m_k}) \approx \frac{e^{-\gamma}}{\ln m_j} \approx \frac{0.5}{\ln m_j} \tag{10}$$

where γ is Euler's constant given by

$$\gamma = \lim_{j \to \infty} (1 + \frac{1}{2} + ... + \frac{1}{m_j} - \ln m_j) \tag{11}$$

Therefore, the maximum number of combinations required to evaluate $(p + q)$ is

$$C \leq 2^{-j} \times \frac{m}{\ln m_j} \tag{12}$$

In order to carry out the square rooting operation given in Equation 4 the overall modulus given by

$$m \leq \prod_{k=1}^{j} m_k \tag{13}$$

needs to exceed $(p - q)$ which in the case of the RSA is of $O(\sqrt{n})$. Hence

$$C \leq O(2^{-j} \times \frac{\sqrt{n}}{\ln m_j}) \tag{14}$$

Table 6 shows the max number of possible combinations for different orders of n.

Table 6. Order of operations for a given n

O(n) in digits	Maximum O(operations)
10	0.3×10^3
20	1.8×10^5
50	4.7×10^{15}
100	1.0×10^{34}
200	1.6×10^{72}

3   Search Pattern

A comparison study has been made between the MRM algorithm and Pollard's $p-1$ method [3]. In this comparison, composites are selected such that some of them favour Pollard's $p-1$ method and some of them do not. The desirable composites for Pollard's $p-1$ algorithm are those for which p has small factors. The undesirable composites are those for which $p-1$ is equal to $2 \times p'$, where p' is a large prime, and also those for which $p-1$ is equal to the product of some small primes raised to large powers, i.e. $p-1=2^\alpha \times 3^\beta \times 5^\gamma \times \cdots$, where α and β are very large and γ is large.

However, all the arithmetic operations involved in MRM algorithm are performed on numbers of size n. Since the labour of multiplying or dividing large numbers normally increases with the square of the length of the numbers involved, this means that one cycle of the MRM algorithm is roughly about 5 times as fast as one cycle of Pollard's algorithm.

The algorithm requires very little space to store the precomputed tables of possible residues in different moduli and these tables could be easily mapped onto a parallel processing architecture. Another important feature of the MRM algorithm is that the only possible outcomes are either the number is composite and its factors are determined or the number is prime. This is unlike some other general factorisation methods where infinite loops may occur revealing nothing about the number.

Furthermore, one may question the possibility of reducing the number of combinations. If so, improvements may be possible in order to speed up this method.

It was shown that the MRM algorithm is comparable to Pollard's $p-1$ method for case when composite number does not favour Pollard's method. The MRM algorithm can go through the residues of $p-1$ and $q-1$ in different prime fields in a pseudo-random manner. This is based on an empirical observation, which shows that it is unlikely for $\phi(n)$ to have the same residue in four or more consecutive prime moduli. It is also shown that $\phi(n)$ can not have factors over the entire prime moduli, m_j, or more than two or three consecutive primes in the prime moduli. However, the search for the value of $\phi(n)$ using the MRM can be carried out in an organised fashion according to various strategies, the purpose of each strategy being to find the right residues of $p-1$ and $q-1$. For instance, the search for the right residue of $p-1$ and $q-1$ could start from the left side of the final table, shown in Fig. 1, (in which all possible residues of $p-1$ and $q-1$ are stored in different prime field knowing n, circles indicate residues). In the final table of the MRM algorithm a different route can be selected for the search through values of $\overline{\phi(n)}$ because all the routes are independent of each other. For instance, if the search starts from the routes in which $p-1$ and $q-1$, have small factors then this method becomes similar to Pollard's $p-1$ method.

Furthermore, a high level of efficiency can be achieved by parallel implementation of MRM, due to the fact that each combination of residues (each route) can be computed and tested entirely independently of each other. Therefore, MRM algorithm can be computed according to the degree of parallelism used. It means that the algorithm can be calculated in semi parallel fashion in which the final part of the algorithm that contains independent instructions are grouped in k groups, where k is the number of processors.

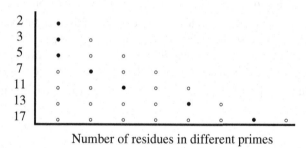

Number of residues in different primes

Fig. 1. Illustrating possible residues of $(p+q)$ in different prime fields.

Moreover, the search can start from the combination of elements most likely to be the correct $\phi(n)$. This method of searching involves going through the combinations having the largest probabilities of being $\phi(n)$ down to the combinations with least probability. For example, see Figure 1 which illustrates the method of searching the various combinations of residues in different prime fields. Dark dots indicate a selection of highest probability of $(p+q)$. From Figure 1 the lowest probability of $(p+q)$ can be determined by selecting the residues combination all from column one.

The algorithm requires very little space to store the precomputed tables of possible residues and these tables could be readily mapped onto a parallel processing architecture. It would thus constitute a general factorisation method since success in factoring a given number, $n = p \times q$, would then be independent of the size of the factors.

4 Conclusion

In this paper suggestions have been made in order to improve algorithm based on the multiple residue method. The MRM algorithm has been introduced as a general approach to break the RSA system and, more generally, in the factorisation of a modulus n which is the product of two primes p and q. This method decomposes the factorisation problem into arbitrary number of n, where $n = p \times q$ and makes use of the CRT to evaluate the Euler totient function or the sum of the two primes.

It was discussed that the MRM algorithm is comparable to Pollard's $p-1$ method for the case when the composite number does not favour Pollard's method, i.e. $p-1$ has a large prime, where $n = p \times q$ and $p-1$ is a product of $2 \times p'$. This is because the MRM algorithm goes through the residues of $p-1$ and $q-1$ in different prime fields in a pseudo random manner according to various strategies, the purpose of each strategy being to find the right residues of $p-1$ and $q-1$.

It was also shown that because of the inherently parallel nature of the MRM algorithm, it is easily possible to group the residues by allocating the task of each group to a processor. Each processor then evaluates $\phi(n)$ from the combination of residues according to its own defined search strategy. Consequently, not only is the order of the algorithm reduced but also the efficiency of search is increased.

Experimental results show that upto 95% success rate is achievable using the MRM of determining $\phi(n)$. This is deemed possible when analysing a pre-selected range of residues through the parallelism technique across small to large prime number (j).

References

1. Diffie ,W., Hellman, M.E.: Privacy and Authentication: An Introduction to Cryptography. Proc. of the IEEE, Vol. 67, No. 3, (1979) 397-427
2. Rivest, R.L., Shamir, A., Adleman, L.: Method for Obtaining Digital Signatures and Public-key Cryptosystems. Com. of the ACM, Vol. 21, No. 2, (1978) 120-126
3. Pollard, J.M.: Theorems on Factoring and Primarily Testing. Proc. Cambridge Philos. Soc., Vol. 76, (1974) 521-528
4. Williams, H.C.: A $p+1$ Method of Factoring. Math. of Comp., Vol. 39, No. 159, (1982) 225-234
5. Wiener, M.J.: Cryptanalysis of Short RSA Secret Exponents. IEEE Trans. on Information Theory, Vol. 36, No. 3, (1990) 553-558
6. Tabatabaian, S.J., Hinton, O.R., Gorgui-Naguib, R.N.: The Use of a Novel Multiple Squared Residue Method in the Cryptanalysis of the RSA. Proc. of the Int. Symp. on Information Theory and Its Applications (ISITA'90) in Coop. with IEEE Information Theory Society, (1990) 975-978
7. Knuth, D.E.: The Art of Computer Programming-Vol. II: Seminumerical Algorithms. Second Edition, Addison-Wesley Publications Co. (1981)
8. Hashemian, R.: Square Rooting Algorithms for Integer and floating-point Number. IEEE Trans., Comput. Vol. 39, NO. 8 (1990) 1025-1029

A New Undeniable Signature Scheme Using Smart Cards

Lee Jongkook[1], Ryu Shiryong, Kim Jeungseop, and Yoo Keeyoung

Computer Engineering Department, E9-508, Kyungpook National University, 1370,
SanGyuk-Dong, Puk-gu, Tae-gu, Korea
kookiss@dreamwiz.com, dryice@purple.knu.ac.kr,
dambi@dittotech.co.kr, yook@knu.ac.kr

Abstract. This paper proposes a new undeniable signature scheme which is based on Chaum-van Antwerpen undeniable signature scheme. In this paper, we extend Chaum-van Antwerpen undeniable signature scheme to take into account smart cards, which are famous for their tamper-resistant feature. And, we also add additional process of authenticating signer to verification and disavowal protocol of Chaum-van Antwerpen undeniable signature scheme. By these modifications, attempts to repudiate or deny a valid signature can be prevented or detected with higher efficiency. With authentication and smart cards, our scheme can be used for settling up disputes over forgeries of signatures, signature repudiations or fraudulent claimants.

1 Introduction

A digital signature of a message is a number, which is dependent on some secret information known only to the signer, like signer's secret key, and additionally on the content of the message to be signed[1]. Digital signatures have to be verifiable; so if a dispute arises as to whether a party signed a document which is caused by either a lying signer trying to repudiate his valid signature, or a fraudulent claimant, an impartial third party should be able to resolve the matter equitably, without requiring access to the signer's secret information or private key.

Generally, digital signature schemes consist of signing algorithm and verifying algorithm. That is, in common digital signature schemes, a signer participates only in making his signature, and the verification is able to be performed without the signer's cooperation or notification. Accordingly, care has to be taken to prevent a signed digital message from being reused, or from being forged by analyzing signing algorithm. And, signers' repudiating their valid signatures and their fraudulent claimants must be prevented, too.

---

[1] Sponsored and supported by Mobile Network Security Technology Research Center, Kyungpook National University, Korea.

B. Honary (Ed.): Cryptography and Coding 2001, LNCS 2260, pp. 387-394, 2001.
© Springer-Verlag Berlin Heidelberg 2001

Attacks based on analyzing algorithm or weak point of protocol, can be blocked by using more powerful and secure crypto algorithm and protocol. However, digital signature schemes can be attacked by signers' intentional denial or repudiation of their valid signatures, because digital signatures can be copied or verified without signers' approval or notification. To provide functionality beyond authentication and non-repudiation, they combine a basic digital signature scheme with a specific protocol, in most instances.

To prevent signers' repudiation or denial, Chaum and van Antwerpen introduced undeniable signature schem, which consists of signing algorithm, verification protocol and disavowal protocol[1][2][3]. In this scheme, no signature can be verified without the signer's cooperation and notification. Accordingly, signers can't repudiate or deny their valid signatures, because they participate in verification of their signatures. However, this scheme is still based on computation, so an invalid signature is accepted as a valid signature, or denial of valid signature is computationally possible by, with a very small probability.

Accordingly, we propose a new undeniable signature scheme based on Chaum-van Antwerpen undeniable signature scheme, to lower or remove the probability of accepting a wrong signature as valid. In our scheme, signing process is unchanged, however, verification protocol and disavowal protocol are reinforced to make our scheme more reliable, by adding authentication using smart cards. Of course, our scheme is still based on computation, accepting a wrong signature as valid or denial of valid signature is still possible, with still less probability.

The rest of this paper is organized as follows. Notations and assumptions related to our scheme are shown in section 2, and we present our new scheme in section 3. Then, we describe and analyze our new scheme in section 4. And, the conclusion is presented in section 5.

2 Notations and Assumptions

In this paper, S represents a signer, V is a verifier and ADV is an adversary in our undeniable signature scheme. All signatures have to be verified by V. Let $p = (2q + 1)$ be a prime number such that q is a prime number. The discrete logarithm problem in Z_p is assumed to be computationally infeasible[4]. α which belongs to Z_p^*, is an element of order q. Let $1 \leq a \leq (q\text{-}1)$ and $\beta = \alpha^a \bmod p$, where a is S' secret value. ID means identification information, which can consist of like S' secret information and unique information of S' smart card. S_k is a secret key maintained by the V's system. And, we assume that a pseudo-random number generator, or PRNG shortly, exists and is available in V's system. $A{\rightarrow}B{:}M$ means A sends a message M to B, and all data transmission is done in cipher. Smart cards are tamper-resistant, so no one can get the contents of smart cards by improper ways. If smart cards are removed while verification protocol or disavowal protocol is proceeding, those protocols must be stopped.

3   A New Undeniable Signature Scheme Using Smart Cards

In our scheme, p, α and β, or public elements, are publicly known to S and V who participate in our scheme. To use smart cards in our scheme, and to allow S to use his smart card, we store those public elements on S' smart card. Accordingly, if S who wants to make his signature, has to insert his smart card into the card reader, because we store all public elements which are used to make his signature, on his smart card.

Our scheme consists of four steps: registration, signing, verification and disavowal step. In undeniable signature scheme, verification protocol and disavowal protocol are important, because V has to prevent S from their denial of his valid signatures, and V from accepting forgeries as valid signatures, by using those two protocols. The following is a detail description of all steps in our scheme, from registration step to disavowal step.

Registration step: To register, S has to submit his ID to V's system. Then V's system calculates the registration information REG for S as Fig. 1:

$$REG = (ID)^{S_k} \bmod p$$

Fig. 1. Description of registration algorithm

After calculating REG, V sends registration information for S, REG to S by secure manner. Each REG is stored on V's system and each S' smart card, and used to authenticate each S. By registration, V can determine S is valid user or signer of V's system, or not. This registration step can be done when S' smart card is issued by V, or registration V's system to S' existent smart card, for confidentiality. Accordingly, ADV can't get S' registration value, REG from S' smart card, because ADV can't access the content of S' smart card. This step differentiates our scheme from Chaum-van Antwerpen undeniable signature scheme.

Signing step: Signing algorithm is so simple and remains in unchanged from Chaum-van Antwerpen undeniable signature scheme. However, it is important that a signature must be generated inside of S' smart card. And, all public elements needed to our scheme are stored on S' smart card, except S' secret information, like S' secret key or signing key, which must be input manually when signature is made. Signing is performed as Fig. 2:

1. S calculates $y = x^a \bmod p$, where x is a message to be signed.

2. $S \rightarrow V$: (x, y)

Fig. 2. Description of signing algorithm

Verification step: This step is similar to the verification protocol of Chaum-van Antwerpen undeniable signature scheme. That is, verification is done with S' cooperation.

However, in our scheme, S has to be authenticated by V. This authentication differentiates our scheme from Chaum-van Antwerpen undeniable signature scheme. Fig. 3 is detailed description of verification protocol:

1. $V \rightarrow S$: r_v, where $r_v \in G$, is random number, timer or nonce, which is chosen by V,

2. $S \rightarrow V$: (z, t), where $z = (r_v)^a \bmod p$, $t = f(r_v, REG) \bmod (p-1)$,

3. V tests if received t is a valid value, or not. If t is a valid value, V continues verification.

4. $V \rightarrow S$: $c = zy^l \beta^m \bmod p$, where l and m which are random numbers in Z_q^*, and are selected by V,

5. $S \rightarrow V$: $d = c^k \bmod p$, where $k = a^{-1} \bmod q$,

6. V accepts y as a valid signature, if and only if $d \equiv r_v x^l \alpha^m \pmod{p}$,

Fig. 3. Description of verification protocol

In Fig. 3, f is an one-way function which is known to S and V. For authentication, we add procedure 1 to 3, to verification protocol of Chaum-van Antwerpen undeniable signature scheme. We use two values, t for validation of S, and z for checking S' cooperation over his signature.

Disavowal step: This step is essential to undeniable signature scheme, because V can settle S' denial of his valid signature by this step, when S repudiates his signature. We also add authentic feature to this step, and full description of this step is shown in Fig. 4:

1. $V \rightarrow S$: r_v, where $r_v \in G$ is random number, timer or nonce, which is chosen by V,

2. $S \rightarrow V$: z, where $z = (r_v)^a \bmod p$, $t = f(r_v, REG) \bmod (p-1)$,

3. V tests if received t is a valid value, or not. If t is a valid value, V continues.

4. $V \rightarrow S$: $c = zy^{e_1} \beta^e \bmod p$, where e_1 and e_2 which are random numbers in Z_q^*, and are selected by V,

5. $S \rightarrow V$: $d = c^k \bmod p$, where $k = a^{-1} \bmod q$,

6. V verifies that d is not congruent to $r_v x^{e_1} \alpha^{e_2} \pmod{p}$,

7. $V \rightarrow S$: $C = zy^{f_1}\beta^{f_2} \bmod p$, where f_1 and f_2 which are random numbers in Z_q^*, and are selected by V,

8. $S \rightarrow V$: $D = C^k \bmod p$, where $k = a^{-1} \bmod q$,

9. V verifies that D is not congruent to $r_v x^{f_1} \alpha^{f_2} \bmod p$,

10. V concludes that y is a forgery if and only if

$$\left(\frac{d\alpha^{-e_2}}{r_v}\right)^{f_1} \equiv \left(\frac{D\alpha^{-f_2}}{r_v}\right)^{e_1} \bmod p.$$

Fig. 4. Description of disavowal protocol

In Fig. 4, f is an one-way function which is known to S and V, like in verification protocol. In Fig. 4, if one of the tests at stage 6 or 9 fails, V can accept y as S' valid signature. And V can regard S as a liar, or believe that S attempt to deny his valid signature, too. Moreover, if the test at stage 10 fails, then S must have used two different value of a at stage 6 and stage 9, or not followed above process properly.

4   Security Analysis

Our scheme must be able to prevent forgeries of S' signature, and, deal with and resolve S' attempt of denial of his valid signature. Central to the above problems are related to compromise or disclosure of S' secret information, i.e. secret key or signing key, a. That is, if S' secret information is compromised, our scheme can't be valid or secure any more. However, our scheme is based on the discrete logarithm problem, or computing discrete logarithms over finite fields is very difficult and complex. So, it is very difficult for ADV to compute a, or S' secret information, from the equation $y = x^a \bmod p$[4].

However, no matter what ADV can't make valid S' signature, our scheme is vulnerable to accepting a wrong signature as a valid one, in itself. That is, V may accept y' as a valid signature for message x with very small probability, where $y' \neq x^a \bmod p$. Same problem is also in Chaum-van Antwerpen undeniable signature scheme, and the probability is $1/q$[2].

In our scheme, signing algorithm is not different from signing algorithm of Chaum-van Antwerpen undeniable signature scheme. Accordingly, our scheme might still regard an invalid signature as a valid one. S can deny or repudiate his signature, because an invalid signature might be accepted as valid, and a signature which is verified successfully, might be a forgery. A detailed description about the probability of accepting a fraudulent signature as valid, is given in Fig. 5.

First, each possible challenge c in verification and disavowal step, corresponds to exactly q ordered pairs (r_v, e_1, e_2). This is because y and β are both elements of the multiplicative group G of prime order q.

Second, S receives the challenge c, he has no way of knowing which of the q possible ordered pairs (r_v, e_1, e_2), which is used to construct challenge c.

Third, let y is not congruent to x^a (mod p). Then any possible response $d \in G$ that S might make is consistent with exactly one of the q possible ordered pairs (r_v, e_1, e_2).

Since α generates G, any element of G as a power of α, where the exponent is defined uniquely modulo q. So, write $c = \alpha^i, d = \alpha^j, x = \alpha^k, y = \alpha^l, r_v = \alpha^s$, where $i, j, k, l, s \in Z_q$ and all arithmetic is modulo p. Consider the following two congruences:

$$c \equiv r_v^{a} y^{e_1} \beta^{e_2} \text{ (mod } p)$$

$$d \equiv r_v x^{e_1} \alpha^{e_2} \text{ (mod } p)$$

This system is equivalent to the following system:
$i \equiv as + le_1 + ae_2$ (mod q)
$j \equiv s + ke_1 + e_2$ (mod q)

This system can be represented as below:
$$\begin{pmatrix} i \\ j \end{pmatrix} = \begin{pmatrix} l & a \\ k & 1 \end{pmatrix} \begin{pmatrix} e_1 \\ (s+e_2) \end{pmatrix} \text{(mod } q)$$

Hence, the coefficient matrix of above system of congruences modulo q has nonzero determinant, and thus there is a unique solution to the system. That is, every $d \in G$ is the correct response for exactly one of the q possible ordered pairs $(e_1, (s + e_2))$. Accordingly, the probability that S gives V a response d that will be verified exactly $1/q$.

Fig. 5. The probability of accepting a fraudulent signature as valid

As described in Fig. 5, the probability of accepting a fraudulent signature as valid is not improved, on the assumption that $(s + e_2)$ is regarded as one variable. However, due to added authentication, the value of s is needed prior to verification protocol or disavowal protocol, and getting directly that s value is infeasible because it depends on solving discrete logarithm problem. Of course, ADV might get the value of $(s + e_2)$, however the complexity of deciding each value is dependent on the size of q. In Fig. 5, the sum of s and e_2 is a constant which is less than q, and there can be so many pairs of such s and e_2. Accordingly, we can say that our scheme can be more secure and reliable than Chaum-van Antwerpen undeniable signature scheme not computationally but logically. In other words, the difference which make our scheme more secure and reliable, depends on the pair of three challenges, s, e_1 and e_2. These challenges are used only once, and changed whenever verification protocol or disavowal protocol is performed, accordingly we can say that our scheme can be more reliable than Chaum-van Antwerpen undeniable signature scheme, without loss of generality.

In our scheme, two responses z and t are used for authenticating S. Because t is made from f and r_v and REG, calculating a right t is impossible without knowing of f and REG. If ADV succeeds in getting t used earlier and knows f, he can hardly get REG due to irreversibility of one-way function f. And, z is used to assure S knows a, his signing key or private key, which is used to sign. ADV can't get S' signing key from z, because of the computational complexity of discrete logarithm problem.

Moreover, S' smart card is tamper-resistant, ADV can't get REG or other information from S' smart card, in improper way. In other words, if S can pass verification protocol, we can say that S must use his valid smart card, know his secret information or signing key and be a registered user of V's system. Accordingly, S who passes verification protocol or disavowal protocol, can hardly deny or repudiate his valid signature.

5   Conclusion

This paper has presented a new undeniable signature scheme which is based on Chaum-van Antwerpen undeniable signature scheme, using smart cards. On the whole, our scheme relies on the difficulty of computing discrete logarithm problems over finite fields, three random challenges, and tamper-resistance of smart cards.

There are two types of data which play a sensitive and crucial role, in our scheme. The one is the signer's signing key which is used whenever his signature is made. If this signer's signing key is compromised, our scheme can't be any more secure or reliable. However, adversaries can hardly get signer's signing key from signer's valid signature, because calculating signer's signing key depends on computational infeasibility of discrete logarithm problem. Accordingly, as far as solving the discrete logarithm problem is infeasible, adversaries can't get signer's signing key and our scheme can be secure.

The other is three random numbers which are used to make challenges. In Chaum-van Antwerpen undeniable signature scheme, two random numbers are used to lead the signer to cooperation on verification of his signature. However, with a very small probability, a wrong signature might be accepted as a valid one. So, we used one more random number to lower the probability of mistakes. This additional random number is used to authenticate the signer of signature, check if the signer is a legitimate user of verifier's system or not. Moreover, if those random numbers are generated using time-stamp, then our scheme can be more effective and withstand replaying attacks.

By added registration feature and random number for challenge, authentication is possible in our scheme. Due to authentication and adopting one more challenge, we can say that the probability of mistake that accepting a wrong signature as a valid one, is lower than Chaum-van Antwerpen undeniable signature scheme, without loss of generality. Moreover, because our scheme introduced registration step, our scheme can be used in user authentication scheme with strict logging feature wherever transaction is occurred, and so on. If signer's private key or signing key is also stored on his smart card, and using biometrics like fingerprint to access that key, then the probability of repudiation over his valid signature can be lowered or removed.

Acknowledgment. Authors would like to thank the program committee and referees of IMA 2001 Cryptography and Coding Conference, for their valuable comments and regards.

References

1. Alfred J. Menezes: Handbook of Applied Cryptography, CRC Press(1977)
2. Douglas R. Stinson: Cryptography Theory and Practice, CRC Press(1995)
3. Bruce Schneier: Applied Cryptography. 2nd edn. John Wiley & Sons, Inc(1996)
4. Leveque, W.: Elementary Theory of Numbers, Dover(1990)

Non-binary Block Inseparable Errors Control Codes

Alexandr Y. Lev, Yuliy A. Lev, and Vyacheslav N. Okhrymenko

Scientific and Engineering Cable Lines Centre
P.O. Box 56, Kiev-110, 03110, Ukraine
lev_katok@profit.net.ua

Abstract. The results of development and research of non-binary block inseparable codes such as mBnq (m > n, q > 2), combining error detection and correction with increase of an information rate are given. The code words are represented by points (vectors) in n-dimensional discrete space with the Minkowski metric (modular metric), appropriate to unsymmetrical multilevel channels. The estimation of the lower and upper bounds of correcting mBnq codes are given. The algorithm of quasi-compact packing of the code words with a given code distance in n-dimensional q-ary space with the Minkowski metric is offered. The block synchronization method on unallowed combinations of q-ary digits appearing on borders of the code words is considered. This method does not reduce an information rate. The estimation of probability of an error in Gaussian channels at usage of 5-ary and 7-ary correcting codes is given. The connection between Hadamard codes and mBnq codes is fixed.

1 Fundamental Concepts and Definitions

Let us use the following designations and definitions:
$N = q^n$ – number of possible n-digit words of the q-ary alphabet;
$M = 2^m$ – number of m-bit binary words – power of mBnq code.
F_b – clock rate of a binary signal;
F_q – clock rate of a q-ary signal;
$\rho = m/n = F_b/F_q$ – coefficient of variation of a specific information rate; at $\rho > 1$ code is named frequency-compact;
t – multiplicity of errors – number of errors in the code word;
h – order of an error – number of discrete levels, on which varies a single code pulse;

$$r = \frac{\log_2 q}{\rho} - 1$$

– information redundancy of mBnq code.

If in the initial binary sequence the digits 0 and 1 are uncorrelated and are equiprobable, then ρ – information quantity, contained in one q-ary digit.

The words selected from a set N for mBnq mapping are named as allowed; the remaining words are named as unallowed. Then

$$\mu = \frac{(N - M)}{M} = \frac{q^n}{2^m} - 1 \tag{1}$$

– number of unallowed words of a mBnq-code come on one allowed word.

B. Honary (Ed.): Cryptography and Coding 2001, LNCS 2260, pp. 395–401, 2001.

Let us name value μ as redundancy of mBnq conversion. Let us enter the concept of a prime mBnq code.

Definition 1. Prime is named m_pBn_pq code, for which at given information redundancy r the length of the code words (blocks) is minimum.

For example, the code 4B3T is prime for q = 3, ρ = 1,33 and r = 0,19; the code 2B1QI – prime for q = 5, ρ = 2 and r = 0,16. The prime codes are, obviously, have minimum redundancy of conversion μ.

It is follows from (1), that the value m of codes cm_pBcn_pq is increased with enlarging c, where c – natural number, i.e. the number of unallowed code words come on one allowed word is increased. Therefore it is true the following.

Statement 1. If there is a prime code with $\rho > 1$, r > 0, then at c major enough there is a frequency-compact correcting code cm_pBcn_pq.

2 Metrics of Code Space

At choice of code space metric of mBnq codes it is necessary to take into account, that for q-ary signals the transmission channel is unsymmetrical, as the error probability of the high orders p_{er} (h > 1) is much less, than error probability of the first order p_{er} (h = 1). To this condition there corresponds the Minkowski metric (modular metric).

At h = 1 distances in metrics of Minkowski and Hamming are coincide.

3 Estimation of a Length of mBnq Codes. Some mBnq Codes

Estimation of a length n of mBnq code, which at given values q and ρ ensures correction, in Minkowski metric, of all errors with multiplicity up to t inclusively, order h=1, looks like

$$2^{n(loq_2q-\rho)} \geq \sum_{i=0}^{2t-1} C^i_{\left\lfloor nloq_2q \right\rfloor -1}, \tag{2}$$

where $\lfloor x \rfloor$ – whole part of number x.

At t=1

$$2^{n(loq_2q-\rho)} \geq \left\lfloor nloq_2q \right\rfloor. \tag{3}$$

Under the formulas (2) and (3) the tables are calculated, in which the parameters of some mBnq codes are listed.

Table 1. Ternary codes (q = 3, t = 1)

Prime code	ρ	r	n_B	Code
1B1T	1,10	0,68	6	6B6T
5B4T	1,25	0,27	16	20B16T
4B3T	1,33	0,19	21	28B21T
3B2T	1,50	0,05	84	126B84T

Table 2. Quintary codes (q = 5)

Prime code	ρ	r	t = 1		t = 2	
			n_B	code	n_B	code
4B3QI	1,33	0,74	3	4B3QI	–	–
3B2QI	1,50	0,55	4	6B4QI	16	24B16QI
7B4QI	1,75	0,33	8	14B8QI	28	49B28QI
2B1QI	2,00	0,16	17	34B17QI	59	118B59QI

Table 3. Septimary codes (q = 7)

Prime code	ρ	r	t = 1		t = 2	
			n	code	n	code
2B1(q = 7)	2,0	0,40	5	10B5(q = 7)	10	20B10(q = 7)
5B2(q = 7)	2,5	0,24	20	50B20(q = 7)		

Table 4. Hexadecimal codes (q = 16)

Prime code	ρ	r	t = 1		t = 2	
			n	code	n	code
5B2(q = 16)	2,5	0,60	2	3B2(q = 16)		
3B1(q = 16)	3,0	0,33	4	12B4(q = 16)	14	42B14(q = 16)
7B2(q = 16)	3,5	0,14	12	42B12(q = 16)		

Note, that (2) gives an estimation of an upper bound of a length of the code words of mBnq codes. At a density packing of the code words in code space mBnq codes at given q and ρ can appear more shortly, than calculated on the formula (2).

As an example we shall give one of possible matrixes of the code words of a code 4B3QI (q = 5).

$$4B3QI = \begin{bmatrix} 040 \\ 220 \\ 021 \\ 241 \\ 421 \\ 201 \\ 132 \\ 312 \\ 043 \\ 223 \\ 443 \\ 403 \\ 024 \\ 244 \\ 424 \\ 204 \end{bmatrix} \tag{4}$$

4 Codes Generated by Generalized Hadamard Matrixes

The class of non-binary correcting codes constructed on the basis of the mathematical apparatus of Generalized Hadamard Matrixes (GHM) is known. Though for these codes ρ < 1, in some cases they can be useful as they are characterized by high correcting ability (big code distance).

The following theorem defines existence and algorithm of construction of non-binary inseparable correcting codes, generated by GHM.

Theorem 1. If there is a Generalized Hadamard Matrix H(q, n), where $q = p^s$, and p – prime number, and s – natural number, then exists q-ary code containing $M = qn$ words with lengths n and with Hamming code distance

$$d_x = \frac{q-1}{q}n.$$

Let us designate these codes as Had(q, n).

Statement 2. There are Hadamard Matrixes $H(p^s, p^s)$ and codes, generated by them, $Had(p^s, p^s)$.

Hadamard Codes are optimum in the sense, that at given q, n and at

$$d_x = \frac{q-1}{q}n,$$

the power of a code M is maximum. Besides they have series of useful properties – they are balance, the lines and columns of the code words matrix are orthogonal. Unlike nonlinear mBnq codes, codes Had(q, n) are linear above a field $GF(p^s)$.

Theorem 2. In the Minkowski metric a code distance of codes $Had(p^s, p^s)$

$$d_m = \frac{p^{2s}-1}{4} \quad \text{at } p \neq 2,$$

$$d_m = 2^{2(s-1)} - 2^{(s-1)} + 2 \quad \text{at } p = 2.$$

As an example we shall give the code words matrix of a code Had(5, 5):

$$Had\,(5,5) = \begin{bmatrix} 01234 \\ 02413 \\ 03142 \\ 04321 \\ 12340 \\ 13024 \\ 14203 \\ 10432 \\ 23401 \\ 24130 \\ 20314 \\ 21043 \\ 34012 \\ 30241 \\ 31420 \\ 32104 \\ 40123 \\ 41302 \\ 42031 \\ 43210 \end{bmatrix} \tag{5}$$

Here $d_x = 4$, $d_m = 6$.

Between Hadamard codes and mBnq codes exist mutual dependence, which is, that code matrixes of mBnq codes can be received as a result of conversion of code matrixes Had(q, n). For example, crossing out in a matrix (5) four lines we shall receive a code 4B5QI with $\rho = 0,8$ and $d_m = 6$; crossing out, except that, the first column, we shall receive a code 4B4QI with $\rho = 1$ and $d_m = 4$.

5 Block Synchronizations of Codes mBnq and Had(q,n)

At decoding mBnq codes and Hadamard codes Had(q, n) the block synchronization is necessary, i.e. in the decoder the borders of the code words should be known. The insertion of special clock signals will reduce efficiency of codes sharply. Therefore for block synchronization of codes mBnq and Had(q, n) as clock signals the combinations of digits will be used which appears only at the turns of the near by code words.

For the explanation of this method of block synchronization we shall consider code matrixes (4) and (5).

The analysis of a code matrix (4) of code 4B3QI display, that the combinations of digits 00, 11, 33, 01, 03, 13, 41, 43 meet only at the turns of the code words. Therefore they can be used as clock signals of block synchronization of a code 4B3QI.

In a code matrix (5) combinations of digits 00, 11, 22, 33, 44 meet only at the turns of the code words. These combinations are clock signals of block synchronization of a code Had(5, 5).

6 Recursion Algorithm of Quasi-Compact Packing

The packing of points in spaces with the Minkowski metric is considered.

Let us enter the following designations:

- $\Pi(n, q)$ – n-dimensional q-ary space (n-dimensional cube);
- d_m – minimum distance between points densely packed in $\Pi(n, q)$;
- $N = qn$ – power of a set of all points in $\Pi(n, q)$ (at $d_m = 1$);
- $\Gamma(l, q)$ – l-dimensional hyperplane in $\Pi(n, q)$ (l-dimensional cube);
- $L_i(l, q, d_m)$ – set of points densely packed in $\Gamma_i(l, q)$ with distance dm;
- $d(L_i, L_k)$ – distance between point sets L_i and L_k.

Recursion algorithm of quasi-compact packing of points in $\Pi(n,q)$ with distance d_m represents step by step procedure.

First step – to construct sets $L_1(l_m, q, d_m)$ and $L_2(l_m, q, d_m)$ at $d(L_1, L_2) = \lfloor d_m/2 \rfloor$, where l_m – minimum dimension of a subspace $\Gamma(l_m, q)$, in which two points with distance d_m can be placed at any rate. A set $Q_1 = L_1(l_m, q, d_m) \cup L_2(l_m, q, d_m)$ we shall call as base structure of the first order.

Second step – to construct sets $L_1(l_{m+1}, q, d_m)$ and $L_2(l_{m+1}, q, d_m)$, using base structure Q_1. To construct base structure of the second order $Q_2 = L_1(l_{m+1}, q, d_m) \cup L_2(l_{m+1}, q, d_m)$.

Third step – using Q_2 to construct sets $L_1(l_{m+2}, q, d_m)$ and $L_2(l_{m+2}, q, d_m)$ and base structure $Q_3 = L_1(l_{m+2}, q, d_m) \cup L_2(l_{m+2}, q, d_m)$.

Etc.; the construction is ended on a step k, at which $l_{m+k-1} = n$.

7 Estimation of Probability of Errors Not Corrected by mBnq Codes

Generally, when the mBnq code allows to correct all errors with multiplicity up to $t = k$ and order up to $h = s$ inclusively, after decoding there are remain errors with multiplicity $k+1 \leq t \leq h$, order $s+1 \leq h \leq q-1$, and full probability of errors $p_{er.\ dec.}$ is determined under the formula

$$P_{er.dec.} = \sum_{i=k+1}^{n} \sum_{j=s+1}^{q-1} p(t=i, h=j) \ , \tag{6}$$

where $p(t = i, h = j)$ – probability of an error with multiplicity $t = i$, order $h = j$ in digital stream at an input of the decoder.

Let us assume, that there is the additive noise in a transmission channel, and signal-to-noise ratio, such, what probability of an error on a digit $p_{er.dig.} \leq 10^{-6}$, $h = 1$. Then expediently to apply a mBnq code error-correcting of multiplicity up to $t = 2$, order $h = 1$. In this case formula (6) becomes simpler and for $t = 1$ we obtain an estimation

$$p_{er.dec.} = p \ (t=1, \ h=1),$$

and for $t = 2$

$$p_{er.dec.} = p \ (t=2, \ h=1).$$

At a Gaussian noise the distribution law of error probabilities in the code word is binomial. Therefore

$$P_{er.dec.} = C_n^{k+1} \cdot p_{er.dig.}^{k+1} \cdot (1 - P_{er.dig.})^{n-(k+1)} \ ,$$

where $k = 1$ at $t = 1$ and $k = 2$ at $t = 2$.

Probability of an error on a digit at $h = 1$

$$P_{er.dig.} = \frac{q-1}{q} \cdot \left[1 - \Phi \left(\frac{1}{q-1} \cdot \frac{U_{dig.}}{\sigma} \right) \right] \ ,$$

where $U_{dig.}$ – amplitude of a maximum code pulse;

σ – meansquare voltage of a Gaussian noise;

$$\Phi(x) = \frac{2}{\sqrt{2\pi}} \cdot \int_0^x e^{-\frac{z^2}{2}} \, \partial z \ .$$

In the table 5 the results of accounts of energy win ΔA, supplied by a code 86B43QI ($r = 2$, $t = 2$) and code 24B12QI ($r = 2$, $t = 1$) as contrasted to by code 2B1QI are given.

Table 5. Energy win of quintary codes

$P_{er.}$	ΔA, dB	
	86B43QI	24B12QI
10^{-6}	3,0	2,0
10^{-7}	3,5	2,4
10^{-8}	3,7	2,5
10^{-9}	3,9	2,6
10^{-10}	4,0	2,7

Cryptanalysis of Nonlinear Filter Generators with $\{0,1\}$-Metric Viterbi Decoding

Sabine Leveiller[1], Joseph Boutros[2], Philippe Guillot[1], and Gilles Zémor[2]

[1] Thales Communication, 66, rue du Fossé Blanc, 92231 Gennevilliers, France
leveille@enst.fr
philippe.guillot@fr.thalesgroup.com
[2] ENST Paris,
46, rue Barrault 75013 Paris, France
boutros@enst.fr
zemor@enst.fr

Abstract. This paper presents a new deterministic attack against stream ciphers based on a nonlinear filter key-stream generator. By "deterministic" we mean that it avoids replacing the non-linear Boolean function by a probabilistic channel. The algorithm we present is based on a trellis, and essentially amounts to a Viterbi algorithm with a $\{0,1\}$-metric. The trellis is derived from the Boolean function and the received key-stream. The efficiency of the algorithm is comparable to Golic et al.'s recent "generalized inversion attack" but uses an altogether different approach : it brings in a novel cryptanalytic tool by calling upon trellis decoding.

Keywords: Boolean functions, stream ciphers, filter generator, Viterbi algorithm, Fourier transform.

1 Introduction

We consider the binary additive stream cipher as depicted in figure 1. Several outputs of a binary linear feedback shift register (LFSR) generator [2] are tapped so as to provide the input of a Boolean function [1][3] that produces a key-stream sequence. The secret-key of the system is defined by the LFSR generator's initial state.

We focus on the left part of figure 1, namely the generation of the key stream : the LFSR produces a PN (pseudo-noise) sequence $(a_i)_{i\in\mathbb{N}}$, and the i'th symbol z_i of the key-stream is obtained from (a_i) by applying f as follows :

$$z_i = f(a_i, a_{i+\lambda_0}, \cdots, a_{i+\lambda_{n-2}})$$

The goal of the cryptanalyst is to recover the initial state $(a_0, ...a_{K-1})$ of the LFSR, given the first N bits of the received key-stream sequence (z_i). It is commonly assumed that the Boolean function f, the feedback polynomial $g(x)$ of the LFSR, and the connection spacings $(\lambda_0, ...\lambda_{n-2})$ between the PN sequence and the nonlinear filter function are known.

B. Honary (Ed.): Cryptography and Coding 2001, LNCS 2260, pp. 402–414, 2001.
© Springer-Verlag Berlin Heidelberg 2001

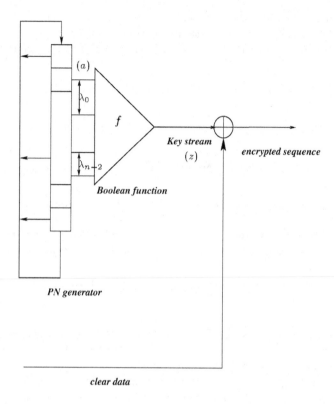

Fig. 1. The cryptographic system.

A number of cryptanalytic attacks on filter generators (and more general combination generators) consist of substituting a binary symmetric channel (BSC) [4][5] for the Boolean function as in figure 2, and the attack is converted into a decoding problem. Most of the literature concerning this topic is based on iterative a posteriori (APP) decoding [6][7][8] also called probabilistic decoding [9] applied to the BSC model in order to decrypt the system. This method benefits from the linear parity-check equations linking input bits a_i and takes advantage of the input/output cross-correlation to succeed in finding the original key (see [10], [11], [12], [13], [14]).

However, a Boolean function is far from being equivalent to a binary symmetric channel. It is true that a balanced function f yields a symmetric behaviour of the transitions from a_i to z_i, i.e. transition probabilities do not depend on particular values, namely $P(z_i = 0|a_i = 0) = P(z_i = 1|a_i = 1)$ and $P(z_i = 1|a_i = 0) = P(z_i = 0|a_i = 1)$, and this would tend to justify the BSC model : but the BSC is a Discrete Memoryless Channel [4][5], while the encryption system in figure 1 has intrinsic memory induced by the connections between the LFSR and the Boolean function f.

This memory degrades the true performance of iterative decoding. Moreover, the knowledge of the Boolean function structure should give us much more information on the system than just a transition probability that characterizes a BSC model.

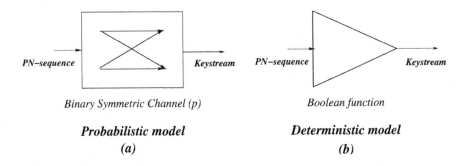

Fig. 2. Information theoretical model and real model.

Therefore, as in [15], [16], [17], we choose to keep the initial deterministic model to take advantage of the exact Boolean function characteristics. In this paper, we present a hard decision algorithm that recovers the initial state of the LFSR. Our strategy builds upon previous work by Anderson [15] in the sense that we strive to recover K "independent" bits of the PN sequence (a_i) which is sufficient to reconstruct the initial state of the LFSR through linear algebra. The way we accomplish this differs however : summarizing, we keep track, as time t varies, of the set of possible states of a sliding window of the PN sequence $(a_t, a_{t+1}, \ldots, a_{t+m})$. The values of this window make up the states of a trellis diagram. After a surprisingly small number t of iterations a sufficient number of individual bits a_i of the PN sequence are determined and the algorithm terminates.

The performance of this algorithm proves to be very good with a rather low complexity, namely $\mathcal{O}(K)2^{\sum \lambda_i}$. This complexity is directly related to the number of inputs to the function f and their spacings (λ_i), but hardly depends on the feedback polynomial $g(x)$. In practice, when f is a resilient function with $n = 8$ input bits, $\lambda_i = 1$, and $g(X)$ is a feedback polynomial of degree 100, only about 200 bits of the key-stream (z_i) are needed to recover the initial state of the LFSR.

The paper is organized as follows. In section 2, we briefly present two known deterministic attacks (Anderson [15] and Golic et al. [17]) based on an extended table and a tree search respectively. Then, we describe in section 3 our Viterbi-like attack based on the Boolean function trellis representation. A generalization of our attack algorithm taking into account all linear forms is given in section 4 where we also propose two improved forward-backward versions. Finally, some numerical results are presented in section 5 before drawing some conclusions.

Notation. The following notation will be used in the sequel :

- The vector space $\{0,1\}^n$ of binary n-tuples is denoted V_n.
- K is the degree of the LFSR connection polynomial g and N is the length of the key-stream sequence.
- the original PN-sequence is denoted $a = (a_i)_{i=0}^{N+K-1}$,
- the key-stream sequence, also called received sequence in the channel decoding terminology, is denoted $z = (z_i)_{i=0}^{N}$,
- f denotes the n-input filtering Boolean function,
- the spacings between the inputs of the Boolean function as illustrated in figure 1 are denoted $(\lambda_0, ... \lambda_{n-2})$,
- N_{dec} is the number of decoded bits, that is, the number of bits of the sequence $\{a_i\}$ recovered by our attack algorithm. These bits are correctly determined with probability one (zero error probability),
- for any set E, the cardinality of E is denoted $card(E)$,
- the memory of f depends on all input bits spread in time between x_1 and x_n. This memory is defined by the integer $\Lambda = 1 + \frac{\sum_{i=0}^{n-2} \lambda_i}{gcd(\lambda_0, \lambda_1, ..., \lambda_{n-2})}$. Thanks to decimation techniques the real memory of the Boolean function $1 + \sum_{i=0}^{n-2} \lambda_i$ can be reduced to Λ. The notation $f(x_1, x_2, ..., x_n)$ is sometimes replaced by $f(x_1, x_2, ..., x_\Lambda)$ which means that we artificially add "fake" or "degenerated" inputs to the function f so as to recover a situation where $\lambda_i = 1$ for every i.

2   Previous Attacks Proposed by Anderson and Golic et al.

We mention now two algorithms that are somehow close to our work : the first one uses correlations between the output and the input of an "augmented" function. The second one is based on the construction of trees, and looks for one possible initialization of the LFSR.

2.1   Block Correlation Attack

Anderson's attack is block-wise oriented [15]. The key idea is to look for particular output patterns and link them to the Boolean function input.

Let us describe this procedure in more detail : consider the filter function f and take the simple case where $\lambda_i = 1$ for all $i = 0 ... n - 2$, so the memory is $\Lambda = n$. Anderson defines an "augmented function" [15],

$$\mathcal{F} : \begin{array}{l} V_{2n-1} \longrightarrow V_n \\ (x_1, ..x_{2n-1}) \longmapsto \Big(f(x_1, ..x_n), f(x_2, ..x_{n+1}), ..f(x_n, ..x_{2n-1}) \Big) \end{array}$$

The dependence between an output sequence and its corresponding input is well represented by $\mathcal{F}$: an output bit z_n depends on n input bits, $a_1, ..., a_n$, and each of these a_i will influence the value of n output z_i. Then, $2n - 1$ inputs will finally influence a total of n output bits, and that's what $\mathcal{F}$ expresses. In the general case where the spacings are not equal to 1, the PN sequence influences a block of Λ bits and a total of 2^Λ output patterns are scanned during the attack (except for certain Boolean function, as mentioned in [16]).

Anderson's algorithm is based on the construction of the augmented table: for each output among the 2^n possible outputs of $\mathcal{F}$, one stores the corresponding inputs and checks whether there is a constant bit over these inputs. The complexity of this construction is therefore 2^{2n-1} when $\lambda_i = 1$.

If all input vectors associated to a given output have a 0 (resp. 1) in the ith position, then the ith bit of the input vector is bound to be a 0 (resp. 1) each time this output is observed in the key-stream.

Decoding enough independent bits of the input sequence enables us to invert the system and recover the initial state of the PN generator by linear inversion. When the function is built in such way that no particular output vector satisfies the above property (these functions exist), it will be impossible to get any 1-probability information on the Boolean function input, and one must either use a further augmented function or follow up the attack by probabilistic decoding.

2.2   Generalized Inversion Attack

The Generalized Inversion Attack, described in [17], is based on a finite tree search. According to the previous notation, for a given initial state, a new tree of height $K - \Lambda + 1$ is constructed. A tree node represents $\Lambda - 1$ bits, the initial state takes $2^{\Lambda-1}$ different values.

The key idea of a forward attack is to expand the tree from time t to time $t + 1$ by looking for the solution of $z_t = f(x_1, x_2, \ldots, x_\Lambda)$. In the latter equation, z_t is known and the state $x_2, \ldots, x_\Lambda$ is specified by the starting node. Such an equation may have no solution, a unique solution or two solutions for the indeterminate x_1. Thus, depending on the number of solutions (0, 1 or 2), we can draw 0, 1 or 2 edges out of the starting node. The backward attack is similar and takes into account the solution relative to x_Λ. The authors [17] suggested to choose exclusively between the forward and the backward attack according to the level of correlation relating z_t to x_1 and x_Λ respectively.

The total number of trees to be examined is $2^{\Lambda-1}$ in the worst case. Once the algorithm succeeds in building a complete tree with a root of $\Lambda-1$ bits and depth $K - (\Lambda - 1)$, the key-stream is recomputed and compared to the observation. If they are the same, the assumed initial memory state is accepted and the attack terminates. Using branching processes theory, Golic et al. show that the typical number of surviving nodes at level $K - (\Lambda - 1)$ is not exponential, but linear in K, which gives a typical complexity $\mathcal{O}(K2^\Lambda)$ for the attack.

3 Trellis Representation and New Simple Deterministic Attack

Our algorithm is associated to a trellis which is built in the following manner: a state in the trellis stands for a Λ-vector input of the Boolean function f. The trellis graph contains 2^Λ states on each section. One state s is connected to two other states located in the next trellis section: the first one corresponds to s shifted once with a 0 as a new bit, and the second one corresponds to s shifted once with a new bit set to 1. For example, the successors of the 5-bit state 01101 would be 00110 and 10110. The 1-bit mapping on the trellis transitions is equal to f applied to the successors of the current state. For the sake of simplicity, we assume that there is no difference of index between a and z, i.e., the received bit z_t corresponds to the last input a_t.

The basic version of our deterministic attack algorithm is described hereafter. Without loss of generality, the Boolean function is taken to be balanced.

1. $t = 0$:
   Initialize $N_{dec} = 0$.
   Exactly half of the trellis states correspond to the received bit z_0 at $t = 0$. Discard all invalid states and store the survivors in a table.
2. $t > 0$:
   Suppose that all surviving states at time $t - 1$ are known.
   At time t, according to the received bit z_t, the surviving states at $t - 1$, and the mapping on the transitions, store only the new valid states matching z_t with their incident branch mapping.
   Check the last bit (most right) of all survivors.
   - If it is constant, equal to b, then
     - $a_t = b$
     - Increment the number of decoded bits, N_{dec}.
     - If $N_{dec} < K$, increment t and return to step 2.
     - If $N_{dec} > K$: if the set of decoded bits contains K independent bits, go to step 3. If not, return to step 2.
   - If the last bit of the survivors is not constant, increment t and go to step 2.
3. Invert the linear system to recover the initial state of the LFSR. The algorithm terminates.

The worse-case complexity of the above algorithm is in $\mathcal{O}(2^\Lambda)$: at each instant, one has to examine at worst $2^{\Lambda-1}$ trellis survivors and each state has 2 out-coming transitions. This algorithm kernel is similar to a Viterbi algorithm, [18], [19], [20], with a hard decision $\{0,1\}$-branch metric : the major difference appears in the add-compare-select unit and the storage unit, we do not process cumulative metrics and we do not store the trellis paths associated to surviving states.

To illustrate the trellis steps, let us take a simple example. We consider the 3-input function defined by the truth-table:

(x_1, x_2, x_3)	Output
000	0
001	0
010	1
011	1
100	1
101	0
110	1
111	0

This function is balanced. Its trellis graph contains $2^3 = 8$ states as illustrated in Figure 3. Let us suppose that the received key-stream sequence is 0010111, and that the corresponding input sequence was generated by a PN-sequence whose polynomial degree is K, where K is arbitrary.

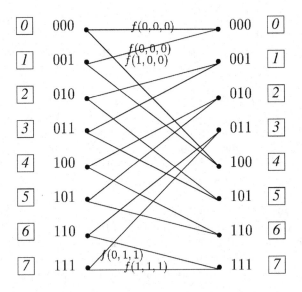

Fig. 3. Trellis of the Boolean function

- t=0 : the surviving states are : 0, 1, 5, 7 and $N_{dec} = 0$.
- t=1 : the surviving states are : 0, 7.
- t=2 : the surviving states are : 3, 4.
- t=3: the surviving states are : 1, 5. These states are all ending with a 1-bit, so $a_3 = 1$ and $N_{dec} = 1$.
- t=4 : the surviving states are : 2, 4, 6. These states are all ending with a 0-bit, so $a_4 = 0$ and $N_{dec} = 2$.
- t=5 : the surviving states are : 2, 3, 6.

- t=6 : the surviving states are : 3. Therefore $a_6 = 1$ and in this particular case, we also have $a_5 = 1$, and $N_{dec} = 4$.

We keep applying this procedure until we determine K independent bits to recover the initial LFSR state.

Comment. Our Viterbi-like attack is related to Anderson's and Golic et al.'s work in the sense that they do not call upon any BSC or otherwise probabilistic modelling. We use the knowledge of the Boolean function, the received key-stream sequence, and the memory of the system to get probability-1 information on the input bits a_i. In this respect our algorithm relates to Anderson's approach, though we use a "convolutional" procedure, as opposed to a "block" strategy: surprisingly, this enables us to significantly shorten the length of the needed observation, as our numerical results will show, while taking the square root of the memory requirements.

4   Improved and Generalized Deterministic Attacks

4.1   A First Simple Improvement : Forward-Backward Version

An improvement of the generalized algorithm is to apply it twice on the same length of observation : once in a forward direction, once in a backward direction. The backward algorithm is a new decoding, with a new trellis (a backward trellis). Therefore, one can expect to get more relationships with the same amount of observation.

This new attack is not a usual "forward and backward" attack: the "forward backward" algorithm developed by [21] uses the information coming from both the forward and the backward trellis run-through to give a probability on one bit in the trellis, while our forward and backward attacks are completely independent. As regards Golic's forward or backward attack, they are only related to the properties of the Boolean function, and they are not done both in the same decoding step.

In our case, the backward attack can be initialized with the survivors remaining at the end of the forward attack, but we gain very little ; moreover, we observed that in most cases, we don't gain much in doing several forward-backward attacks, because most of the time, the sets of survivors at the second iteration almost coincide with the ones found at the first iteration. Therefore, the attack in the second iteration soon becomes strictly identical to the first one, and we don't decode anything more.

4.2   Generalization

Instead of checking only the last bit of the surviving states, one can look for linear combinations between the bits of the survivors. Indeed, each constant linear combination between the bits of the surviving states lead to a linear relation satisfied by terms of the sequence $\{a_i\}$ and thus improves the number of decoded bits. Without loss of generality, we assume in the sequel that the spacings are regular, that is $\Lambda = n$, which simplifies the notation. For our purposes, we define the Boolean function Φ_t (Φ for short when t does not need to be specified) by

$$
\Phi_t : \quad
\begin{aligned}
V_n &\longrightarrow \{0,1\} \\
x &\longmapsto
\begin{cases}
1 \text{ if } x \text{ survives at time } t \\
0 \text{ otherwise}
\end{cases}
\end{aligned}
$$

We will call upon Fourier transform techniques and to this end need some more notation :

The scalar product between two binary vectors x and y is defined on $V_n \times V_n$ by :

$$
(x, y) \longmapsto x.y = \sum_{i=1}^{n} x_i y_i \quad \mod 2
$$

The Walsh transform $\widehat{f}$ [3] of any real-valued function $f : V_n \longrightarrow \mathbb{R}$ is defined by :

$$
\begin{aligned}
\widehat{f} : V_n &\longrightarrow \mathbb{R} \\
u &\longmapsto \sum_{x \in V_n} f(x)(-1)^{u.x}
\end{aligned}
$$

The Hamming weight of a vector equals the number of its nonzero components:

$$
\begin{aligned}
w : V_n &\longrightarrow \mathbb{N} \\
u &\longmapsto w(u) = \sum_{i=1}^{n} u_i
\end{aligned}
$$

A balanced Boolean function is said to be m-resilient, if for any nonzero vector $u \in V_n$ such that $w(u) \leq m$, we have $\widehat{f}(u) = 0$

A Boolean function f is said to be p-degenerated if there exist a Boolean function $g : \{0,1\}^p \longrightarrow \{0,1\}$ and a (p, n)-matrix A, such that : $f = g \circ A$.

The following lemma characterizes linear combinations of bits that are constant on the support $Supp(\Phi) = \{x \ / \ \Phi(x) \neq 0\}$, of Φ.

Lemma 1. If $|\widehat{\Phi}(u)| = \widehat{\Phi}(0)$, then

$$
\begin{aligned}
\lambda_u : V_n &\longrightarrow \{0,1\} \\
x &\longmapsto u.x \ (\mathrm{mod}\ 2)
\end{aligned}
$$

is constant over $Supp(\Phi)$

Proof. By definition, $\widehat{\Phi}(u) = \sum_{x \in V_n} \Phi(x)(-1)^{u.x}$, and $\widehat{\Phi}(0) = \sum_{x \in V_n} \Phi(x)$. Let

$$
\begin{aligned}
A_0 &= \{x \in V_n \ / \ \lambda_u(x) = 0\} \\
A_1 &= \{x \in V_n \ / \ \lambda_u(x) = 1\}.
\end{aligned}
$$

We have therefore

$$\left|\widehat{\Phi}(u)\right| = \left|\sum_{A_0} \Phi(x) - \sum_{A_1} \Phi(x)\right|$$
$$= |card\,(A_0 \cap Supp(\Phi)) - card\,(A_1 \cap Supp(\Phi))|$$
$$\leq card\,(Supp(\Phi))$$

and the equality holds if and only if A_0 or A_1 equals the empty set.   ■

Then, the second step of the algorithm is now :

At time t, according to the received bit z_t, the surviving states at $t - 1$, and the mapping on the transitions, store only the new valid states matching z_t with their incident branch mapping. Evaluate Φ_t as defined above, and its Walsh transform.

- If there is a value of u such that $|\widehat{\Phi_t}(u)| = \widehat{\Phi_t}(0)$ then store u as a new relationship between the bits of the survivors.
    - If the number of linear relationships is inferior to K, increment t and return to step 2.
    - Else, if one can find K independent relationships go to step 3. Else, go back to step 2.
- If not increment t and go back to step 2.

Although this generalization increases the complexity of the algorithm because it requires the computation of $\widehat{\Phi_t}$, which has a complexity of $\mathcal{O}(n2^n)$, it allows us to decrypt the system with a smaller amount of observation. In the basic version, we only checked the last bit of the survivors, because the states at time t are shifted copies of the states obtained before, with a new bit. For the same purpose, in this new version one should only pay attention to u of the form $u = (* * * * 1)$, to avoid counting several times the same relationships at different instants. The basic version appears as a particular case of the generalized algorithm, in which one evaluates only $\widehat{\Phi_t}(0...01)$.

4.3   Second Improvement

This modification takes into account the previous remark : in section 4.1, the forward and the backward attacks are almost independent (almost because the backward starting point we choose can depend on the forward attack).

As in [21], let us combine both attacks to get more information at an instant t : we denote by $Supp_{fwd}\left(\widehat{\Phi_t}\right)$, resp. $Supp_{bwd}\left(\widehat{\Phi_t}\right)$ the support of $\widehat{\Phi_t}$ at time t in the forward, resp. the backward description of the trellis. One runs through the trellis in a forward and in a backward way and memorizes $Supp_{fwd}\left(\widehat{\Phi_t}\right)$, and $Supp_{bwd}\left(\widehat{\Phi_t}\right)$, for all t.

Then, we evaluate $\widehat{\Phi_t}$ on $Supp_{fwd}\left(\widehat{\Phi_t}\right) \cap Supp_{bwd}\left(\widehat{\Phi_t}\right)$.

The intersection of the support is included in each support ; therefore we can expect to get less survivors, and consequently more linear relationships.

5   Numerical Results

We have implemented our algorithms with a filtered LFSR defined by the polynomial $g(x) = 1 + X^2 + X^7 + X^8 + X^{100}$, $K = 100$, and some initializations of the register ; the choice of the initialization hardly affects the performance of the algorithm which mostly relates to the boolean function f. Note that the weight of the polynomial doesn't have any influence either.

The following tables contain the minimum required amount of observation that enable one to get the initialization of the LFSR back. We present some typical results we obtained with resilient functions. These functions have good cryptographic properties and are widely used.

Our first example is with the 2-resilient 5-input function, used by Anderson in his article,

$$f(x) = x_1 + x_2 + (x_1 + x_3)(x_2 + x_4 + x_5) + (x_1 + x_4)(x_2 + x_3)x_5$$

and $\forall i, \; \lambda_i = 1$.

The last two columns give the number of bits required to recover unambiguously the initial state of the LFSR.

	Forward	Forward-Backward
basic algorithm	465	201
generalized algorithm	288	185
2nd improved generalized algorithm	–	185

Our second example is with the 2-resilient 8-input function :

$$f(x) = x_1 x_4 + x_5 + x_6 + x_7 + x_1(x_2 + x_7) + x_2(x_6 + x_7) + x_3(x_6 + x_8)$$
$$+ x_1 x_2(x_4 + x_6 + x_8) + x_1 x_3(x_2 + x_6) + x_1 x_2 x_3(x_4 + x_5 + x_8)$$

and $\forall i, \; \lambda_i = 1$.

	Forward	Forward-Backward
basic algorithm	1044	337
generalized algorithm	899	268
2nd improved generalized algorithm	–	141

Finally, we used different λ_i's, with the same 5-input 2-resilient function as in our first example. The first column gives the value of $(\lambda_0, \lambda_1, \lambda_2, \lambda_3)$. The results are given when applying the generalized algorithm, that takes into account all possible linear forms applied to survivors.

$(\lambda_0, \lambda_1, \lambda_2, \lambda_3)$	Forward	Forward-Backward	2nd Improved Forward-Backward
(2, 3, 1, 2)	299	221	185
(3, 1, 2, 2)	369	270	184
(1, 3, 1, 3)	215	182	157

6   Conclusions

We derive a new deterministic algorithm that enables one to recover the LFSR initial state ; our algorithm is based on the construction of a trellis graph, which is built according to the Boolean function, the connections between the LFSR and the function, and the received key-stream. We gave a generalization of the basic algorithm that slightly increases the computational complexity, but shortens the amount of observed bits of the key-stream sequence. The two modifications of the algorithm rely on a "double" use of the received key-stream, first in a forward way, next in a backward way. They can be done either on the basic version or on the generalized version, and they prove to be efficient in both cases.

The results turn out to be very good, indeed, we generally need very few key-stream bits (of the order of K) and therefore little time to recover the LFSR initial state. The typical complexity of the attack seems therefore to be $\mathcal{O}(K2^\Lambda)$, i.e. very much comparable to the typical complexity of the generalized inversion attack of Golic et al. Compared to the latter, our algorithm has the disadvantage of needing a slightly longer key-stream : however, that it determines individual bits of the PN sequence (a_i) irrespective of their linear structure is structurally simpler, and arguably may be seen as an advantage, since other stream ciphers might try feeding a more complicated type of sequence (a_i) to the boolean function f. Furthermore, we hope that our approach will lead to using other, more involved trellises in cryptanalysis.

References

[1] R.A. Rueppel, *Analysis and Design of Stream Ciphers*. Berlin: Springer-Verlag, 1986.

[2] S.W. Golomb: Shift register sequences, *Holden-Day*, San Francisco, 1967.

[3] W. Meier and O. Staffelbach: "Nonlinearity criteria for cryptographic functions," *Advances in Cryptology-EUROCRYPT'89*, nb 434, Lectures Notes in Computer Science, pp. 549-562, Springer-Verlag, 1990.

[4] R.G. Gallager: *Information theory and reliable communication*. Wiley, New York, 1968.

[5] T.M. Cover, J.A. Thomas: *Elements of information theory*. Wiley series in Telecommunications, 1991.

[6] C. Berrou, A. Glavieux, P. Thitimajshima : "Near Shannon limit error-correcting coding and decoding : turbo-codes," *Proceedings of ICC'93*, Geneva, pp. 1064-1070, May 1993.

[7] N. Wiberg, H.A. Loeliger and R. Kötter: "Codes and iterative decoding on general graphs", *European Trans. on Telecom.*, Vol. 6, Sept/Oct 1995.

[8] J. Hagenauer, E. Offer, L. Papke : "Iterative decoding of binary block and convolutional codes," *IEEE Trans. on Inf. Theory*, vol. 42, no. 2, pp. 429-445, March 1996.

[9] R.G. Gallager *Low Density Parity check codes*. MIT Press, Cambridge, MA, 1963.

[10] W. Meier and O. Staffelbach: " Fast correlation attack on certain stream ciphers," *Journal of Cryptology*, p. 159-176, 1989.

[11] T. Johansson and F. Jönsson: "Improved fast correlation attack on stream ciphers via convolutional codes," *Advances in Cryptology - EUROCRYPT'99*, nb 1592 in Lecture Notes in Cumputer Science, p. 347-362. Springer Verlag, 1999.

[12] T. Johansson and F. Jönsson: "Fast correlation attacks based on turbo code techniques," *Advances in Cryptology - CRYPTO'99*, nb 1666 in Lecture Notes in Cumputer Science, p. 181-197. Springer Verlag, 1999.

[13] M.J. Mihaljević, M. Fossorier, H. Imai: "A Family of Iterative Decoding Techniques for Certain Crypto Applications", submitted to *IEEE Transactions Information Theory*, Dec. 1999 XXXX:

[14] A. Canteaut, M. Trabbia: "Improved fast correlation attacks using parity-check equations of weight 4 and 5," *Advances in Cryptology - EUROCRYPT 2000*, Lecture Notes in Computer Science, Springer Verlag, 2000.

[15] R.J. Anderson,"Searching for the optimum correlation attack", Fast Software Encryption -Leuven'94, *Lectures Notes in Computer Science*, vol. 1008, B. Preneel ed., Springer-Verlag, pp. 137-143, 1995.

[16] J.Dj. Golic, "On the security of Nonlinear Filter Generators," *Proc. Fast Software Encryption-Cambridge'96* D.Gollmann, ed., pp.173-188, 1996.

[17] J.Dj. Golic, A. Clark, E. Dawson, "Generalized Inversion Attack on Nonlinear Filter Generators," *IEEE Transactions on computers*, vol.49, NO. 10, October 2000.

[18] A.J. Viterbi: "Error bounds for convolutional codes and an asymptotically optimum decoding algorithm," *IEEE Trans. Inform. Theory*, vol. 13, pp. 260-269, 1967.

[19] G.D. Forney: "The Viterbi algorithm," *IEEE Proceedings*, vol. 61, pp. 268-278, 1973.

[20] R. Johannesson, K. Sh. Zigangirov: *Fundamentals of convolutional coding*, IEEE Press, February 1999.

[21] L. Bahl, J. Cocke, F. Jelinek, and J. Raviv:"Optimal Decoding of Linear Block Codes for Minimizing Symbol Error Rate,", *IEEE Transactions on Information Theory, pp. 284-287, march 1974.*

Author Index

Ahrens, Andreas   9
Al-Dabbagh, Ahmed   176
Al Jabri, A.   1

Batina, Lejla   364
Benachour, P.   166
Beth, Thomas   355
Borselius, Niklas   239
Boutros, Joseph   402
Brincat, Karl   63

Cai, Yuanlong   205
Cheon, Jung Hee   114
Cocks, Clifford   360
Coulton, P.   158
Crouch, P.A.   329

Daemen, Joan   222
Darnell, Michael   176, 294
Davenport, J.H.   329
Dhanda, Mungal S.   378

Farrell, P.G.   158, 166
Filiol, Eric   85
Fontaine, Caroline   85
Fujiwara, Toru   27

Galán-Simón, F. Javier   128
Geiselmann, Willi   355
Giuliani, Kenneth J.   317
Gong, Guang   317
Guillot, Philippe   402
Gumussoy, Murat   378

Hernández, Julio César   374
Hirst, Simon   38
Honary, Bahram   38, 166, 294

Ikeshiro, Sam   378

Jeungseop, Kim   387
Jongkook, Lee   387
Joye, Marc   99, 114

Kabashima, Yoshiyuki   148, 307
Keeyoung, Yoo   387

Kim, Seungjoo   114
Koeune, François   245
Komninos, N.   294

Lange, Christoph   9
Lev, Alexandr Y.   395
Lev, Yuliy A.   395
Leveiller, Sabine   402
Li, Qi   205
Li, Shujun   205
Li, Wenmin   205
Lim, Seongan   114

MacKay, David J.C.   138
Mambo, Masahiro   114
Martínez-Moro, Edgar   128
Masol, Viktoriya   301
Mex-Perera, J.C.   374
Mitchell, Chris J.   63, 239, 268
Mou, Xuanqin   205
van Mourik, Jort   148
Murayama, Tatsuto   307
Muurling, Geeke   364

Nikov, Ventzislav   191
Nikova, Svetla   191

Okhrymenko, Vyacheslav N.   395

Paire, J.T.   158
Parker, Matthew G.   339
Prissette, Cyril   277

Quisquater, Jean-Jacques   99, 245

Ramos, Benjamín   374
Rantos, Konstantinos   268
Ribagorda, Arturo   374
Rijmen, Vincent   222

Saad, David   148, 307
Schindler, Werner   245
Shiryong, Ryu   387
Siap, Irfan   20
Sierra, José María   374
Smart, N.P.   73
Steinwandt, Rainer   355

Tabatabaian, Seyed J.   378
Tena-Ayuso, Juan G.   128

Vicente, Renato   307

von Willich, Manfred   44
Wilson, Aaron   239

Won, Dongho   114

Yoshida, Maki   27

Zémor, Gilles   402
Zheng, Yuliang   114